Climate Change Management

Series Editor

Walter Leal Filho, International Climate Change Information and Research Programme, Hamburg University of Applied Sciences, Hamburg, Germany

The aim of this book series is to provide an authoritative source of information on climate change management, with an emphasis on projects, case studies and practical initiatives – all of which may help to address a problem with a global scope, but the impacts of which are mostly local. As the world actively seeks ways to cope with the effects of climate change and global warming, such as floods, droughts, rising sea levels and landscape changes, there is a vital need for reliable information and data to support the efforts pursued by local governments, NGOs and other organizations to address the problems associated with climate change. This series welcomes monographs and contributed volumes written for an academic and professional audience, as well as peer-reviewed conference proceedings. Relevant topics include but are not limited to water conservation, disaster prevention and management, and agriculture, as well as regional studies and documentation of trends. Thanks to its interdisciplinary focus, the series aims to concretely contribute to a better understanding of the state-of-the-art of climate change adaptation, and of the tools with which it can be implemented on the ground.

Notes on the quality assurance and peer review of this publication

Prior to publication, the quality of the works published in this series is double blind reviewed by external referees appointed by the editor. The referees are not aware of the author's name when performing the review; the referees' names are not disclosed.

More information about this series at https://link.springer.com/bookseries/8740

Walter Leal Filho · Evangelos Manolas
Editors

Climate Change in the Mediterranean and Middle Eastern Region

Springer

Editors
Walter Leal Filho
FTZ-ALS
HAW Hamburg
Hamburg, Germany

Evangelos Manolas
Department of Forestry and Management
of the Environment and Natural Resources
Democritus University of Thrace
Orestiada, Greece

ISSN 1610-2002 ISSN 1610-2010 (electronic)
Climate Change Management
ISBN 978-3-030-78565-9 ISBN 978-3-030-78566-6 (eBook)
https://doi.org/10.1007/978-3-030-78566-6

This Springer imprint is published by the registered company Springer Nature Switzerland AG
The registered company address is: Gewerbestrasse 11, 6330 Cham, Switzerland

Preface

Climate change is having a much greater impact in the Mediterranean than the global average. In the Paris Climate Agreement, the UN member states pledged to stop global warming at well below two degrees, if possible at 1.5 degrees. This mark, which is expected elsewhere only for 2030 to 2050, has already been reached in the region. The situation could worsen in the coming years if the global community does not limit its emissions.

The above state of affairs illustrates the need for a better and more holistic understanding of how climate change affects countries in the Mediterranean region on the one hand, but also on the many problems it faces on the other, which prevent adaptation efforts. There is also a perceived need to showcase successful examples of how to duly address and manage the many social, economic and political problems posed by climate change in the region, in order to replicate and even upscale the successful approaches used.

It is against this background that the book *Climate Change in the Mediterranean and Middle Eastern Region* has been produced. It contains papers prepared by scholars, practitioners and members of governmental agencies, undertaking research and/or executing climate change projects, and working across the region. It serves the purpose of showcasing some of the works in respect of applied research, field projects and best practice to foster climate change adaptation across the region.

This book is structured in two main parts.

Climate Change Models and Impacts
Climate Change Adaptation and Resilience Initiatives

We would like to thank all the authors and reviewers for making available their experience in their chapters and the willingness to share their ideas. Much can be gained by offering a platform for the debate on climate change in the Mediterranean, in a pragmatic way. By providing their inputs, the authors have made a positive contribution towards a debate which needs to be continued, and reach a depth far beyond what conferences, workshops or seminars may be able to offer.

Hamburg, Germany
Orestiada, Greece
Spring 2022

Walter Leal Filho
Evangelos Manolas

Contents

Climate Change Models and Impacts

Assessment of Future Climate Trend Based on Multi-RCMs Models and Its Impact on Groundwater Recharge of the Mediterranean Coastal Aquifer of Ghis-Nekkor (Morocco) 3
Hanane El Asri, Abdelkader Larabi, and Mohamed Faouzi

Species Distribution Based-Modelling Under Climate Change: The Case of Two Native Wild *Olea europaea* Subspecies in Morocco, *O. e.* subsp. *europaea* var. *sylvestris* and *O. e.* subsp. *maroccana* 21
Jalal Kassout, Jean-Frédéric Terral, Abdeltif El Ouahrani, Mhammad Houssni, Sarah Ivorra, Khalil Kadaoui, Mohamed El Mahroussi, Laure Paradis, and Mohammed Ater

The Frequency of Rare Cyclones in the Eastern Mediterranean and Northeastern Africa as a Sign of Climate Change Using Satellite Imagery, Climate Data Models and GIS-Based Analysis 45
Khaled Mohamed Madkour

Climate Change Impacts on Hydrology in the Mediterranean Part of Slovenia 85
Mauro Hrvatin and Matija Zorn

Climate Change Impacts and the Role of Forestry: Insights from the Mediterranean Region 119
Veronika Andrea

Energy Impacts on Climate Change: Issues, Challenges and Solutions with Clean Conversion Technology 133
Uttara Das, Champa Nandi, Somudeep Bhattacharjee, and Sarbani Mandal

Montenegro and World: Climate Change and Biodiversity Conservation 151
Mira Šorović

Portrayals of Climate Change and Drought in the Politically Oriented Turkish Press: Socialist, Islamist, and Nationalist Accounts of Extreme Weather in 2007 and 2014 169
Mehmet Ali Üzelgün

Climate Changes and Insolation in the Mediterranean Basin: The Case of Montenegro 185
Dragan Burić and Miroslav Doderović

The Conservation Challenge of Traditional Agroecosystems in Morocco: The Case Study of Six Oases Agroecosystems 201
Mhammad Houssni, Jalal Kassout, Abdeltif El Ouahrani, Mohammed El Mahroussi, Vladimiro Boselli, Khalil Kadaoui, Abdelouahab Sahli, Mohamed Kadiri, and Mohammed Ater

Climate Change Adaptation and Resilience Initiatives

Enhancing the Resilience of Oasis Agrosystems to Climate Change in Morocco 227
Jannate Chehbouni, Latifa Elhadioui, Nour-Eddine Benaoda Tlemçani, Latifa Daadaoui, Salma Daoud, and Cherif Harrouni

Climate Change and Agricultural Production in Algeria 249
Mohammed Bouznit, Mohamed Elaguab, Mohammed Mostefa Selt, Mohammed Himrane, and Rachida Aïssaoui

Ready for Climate Change? An Assessment of Measures Adopted by 45 Mediterranean Coastal Cities to Face Climate Change 269
Xira Ruiz-Campillo, Olga Gil, and Cristina García Fernández

Climate Change or Climate Crisis? Investigating the Views of Forestry Students on the Causes, Consequences and Tools for the Mitigation of Climate Change 293
Evangelia Karasmanaki, Georgios Tsantopoulos, and Evangelos Manolas

Development of a Water Security Index Incorporating Future Challenges 313
Deniz Marangoz and Irem Daloglu

An Approach to Adapting Urban Drainage Design to Climate Change: Case of Northern Morocco 331
Mohammed Moujahid, Laila Stour, and Ali Agoumi

Mainstreaming Climate Change Adaptation into Water Development Plan Case of Morocco 345
Laila Oualkacha, Laila Stour, and Ali Agoumi

Consumption Patterns and Public Attitudes Toward Organic Foods: The Role of Climate Change Awareness 365
Elena Raptou and Evangelos Manolas

Climate Change in Lebanon and the Impact to Water Resources 395
Jalal Halwani and Bouchra Halwani

Enhancing Urban Microclimates Towards Climate-Resilient Cities: The Potential of Courtyards 413
Carmen Galán-Marín, Carlos Rivera-Gómez, Jesus Lizana, Jorge Roa-Fernández, Eduardo Diz-Mellado, and Victoria López-Cabeza

Sustainable Land Management for Rural Adaptation in the Mediterranean and Middle Eastern Watersheds 433
Itxaso Ruiz and María José Sanz

Climate Change Impacts on North Africa: Public Health Perspectives 457
Rehab A. Rayan, Mohamed Kamal, Christos Tsagkaris, and Loyle Campbell

Exploring the Beliefs, Concerns and Understandings About Climate Change of Greek University Students from the Social Sciences and Humanities 473
Maria Daskolia

Hydrological Impacts of Projected Climate Change on Northern Tunisian Headwater Catchments—An Ensemble Approach Addressing Uncertainties 499
Hamouda Dakhlaoui, Kirsti Hakala, and Jan Seibert

Addressing Climate Change in the Mediterranean Basin: Some Thoughts for the Future 521
Walter Leal Filho and Evangelos Manolas

Climate Change Models and Impacts

Assessment of Future Climate Trend Based on Multi-RCMs Models and Its Impact on Groundwater Recharge of the Mediterranean Coastal Aquifer of Ghis-Nekkor (Morocco)

Hanane El Asri, Abdelkader Larabi, and Mohamed Faouzi

Abstract The Ghis-Nekkor aquifer is an area highly vulnerable to climate change. Multiple RCMs from CORDEX and RICCAR projects were used in this study to simulate air surface temperature and precipitation changes over the Ghis-Nekkor area. The main objective of this study is to assess the impact of climate change on future groundwater recharge in the Ghis-Nekkor aquifer, based on a deep analysis of available projections from RCMs developed in CORDEX and RICCAR projects. We have focused on the future assessment for the period (2021–2050) and (2051–2080) under the RCP 4.5 and RCP 8.5 scenarios. Additionally, groundwater recharge has been also quantified for the period 1964–2014 based on the observed data. The results of projections from the CORDEX and RICCAR projects produce similar spatial patterns but their characterization of temperature and precipitation are different. It is assumed that it will be on average from multi-RCMs a reduction in precipitation of 17% and an increase in temperature of 0.53 °C by 2050, while by 2080 the average increase of temperature is about 0.9 °C accompanied by a reduction of 22.2% in precipitation. These predicted parameters were subsequently exploited to study future climate change by using the De Martonne aridity index (AI) and the Standardized Precipitation Index (SPI). The latter predicts an increasingly dry climate for the Ghis-Nekkor area which will be more prone to both moderate and severe droughts, but less prone to extreme drought by 2070. Moreover, the results show also a change of climate regime becoming arid at the level of the plain (AI = 9.92 mm/°C) by 2080. The analysis of the impact of these potential future changes on future groundwater recharge aquifer shows that overall predictions estimate a reduction in mean annual recharge of 18% under the RCP 4.5 scenario and 22% under the RCP 8.5 scenario for the period (2021–2080). Furthermore, the groundwater recharge shows a decreasing

H. El Asri · A. Larabi (✉) · M. Faouzi
Regional Water Centre of Maghreb, Laboratory of Analysis and Environmental Modeling and Natural Resources (LAMERN), Mohammadia School of Engineers, Mohammed V University in Rabat, Rabat, Morocco

H. El Asri
e-mail: hanane-elasri@hotmail.fr

M. Faouzi
e-mail: mohamed.faouzi@emi.um5.ac.ma

W. Leal Filho and E. Manolas (eds.), *Climate Change in the Mediterranean and Middle Eastern Region*, Climate Change Management,
https://doi.org/10.1007/978-3-030-78566-6_1

trend as a response to changes in precipitation. Lower values of future recharge are expected for the period (2051–2080) and would decrease more under the RCP 8.5 scenario and reach 26%.

List of Acronyms

RCM	Regional Climate Model
CORDEX	Coordinated Regional Climate Downscaling Experiment
RICCAR	Regional Initiative for the Assessment of Climate Change Impacts on Water Resources and Socio-Economic Vulnerability in the Arab Region
RCA	Rossby Centre Regional Atmospheric Model
HIRHAM	High Resolution Atmospheric Model
RCP	Representative Concentration Pathways
IPCC	Intergovernmental Panel on Climate Change
GCM	Global Climate Model
SPI	Standardized Precipitation Index
AI	Aridity Index
CMIP5	The 5th Coupled Model Inter-comparing Project

1 Introduction

Over the last century, the average temperature of our planet has risen by +1 °C. The (IPCC 2013) predicts an increase in the average temperature of the earth from about 1 °C (RCP 4.5 scenario) to 4.1 °C (RCP 8.5 scenario) by 2100, accompanied by a rise in sea level from 26 to 82 cm. Also, the Mediterranean region, particularly its southern and eastern shores are expected to be affected by climate change more than many other regions in the world during the twenty-first century. Climate specialists anticipate an increase in temperature more than the global average of 3–5 °C by 2100 accompanied by a significant deficit in rainfall of about 35% on the south shore and a rise in sea level which, according to some studies, could be about 35 cm by the end of the century. Furthermore, climate projections suggest that drought is likely to increase and might become more intense in the Mediterranean region which is one of the primary climate change hotspots (Giorgi 2006). As a Mediterranean country, Morocco is exposed to the risk of renewed droughts. This was proved by historical documents relevant to the last centuries (Naciri 1985), which show that most of Morocco has been suffering from precipitation deficit and extended dry periods leading to several economic and environmental impacts as it was the case during the (1944–1945), (1982–1983), (1994–1995), (1998–2000) and (2006–2007) years (Bouaicha and Benabdelfadel 2010). Due to its influence by the Mediterranean front on the north, it is expected that the drought frequency will change in Ghis-Nekkor because of the increase in temperature and significant deficit in rainfall. From this perspective, it is crucial to assess the climate change impact on the Ghis-Nekkor area. To assess the impact of climate change at a local level, the downscaling methods are needed. The Moroccan climate is a relatively contrasting climate influenced by the Atlantic Ocean to the west, the Mediterranean Sea to the north, and the Sahara to the south. The spatial resolution of GCMs in the magnitude of 100–200 km is insufficient to simulate the heterogeneity of the Moroccan climate, nor does it allow the correct reproduction of extreme climate events. Therefore, two approaches are generally used to overcome this problem and are based on statistical and dynamic methods. The statistical methods are based on the stationarity assumption, which is not always verified in the case of Morocco. Hence, this problem can be avoided by using a dynamic approach, i.e. the use of Regional Climate Models (RCMs) (Woillez 2019). In this context, RCMs are practical in the sense that they allow the simulation of future climate with a spatial resolution which is, generally, less than 50 km. Therefore, for our case study, the choice was to work with models from CORDEX (Coordinate Regional Downscaling Experiment) and RICCAR (Regional Initiative for the Assessment of Climate Change Impacts on Water Resources and Socio-Economic Vulnerability in the Arab Region) to assess climate change from 2020 to 2080. The CORDEX's goal is to coordinate an international regional climate project using the downscaling technique to provide a set of climate simulations. These include different global climate model (GCM) simulations, different concentration scenarios greenhouse gas emissions, natural climate variability, and different downscaling methods. The Regional Initiative RICCAR aims to The Regional Initiative RICCAR aims to provide a common platform for addressing and responding to climate change impacts on freshwater resources in the Arab region by serving as the basis for informed dialogue, priority setting and policy formulation on climate change adaptation at the regional level (Chouchani 2013). The CORDEX and RICCAR downscaling activities are based on the latest climate scenarios and predictions generated in the 5th Coupled Model Inter-comparing Project (CMIP5) required in the Fifth Assessment Report (AR5) and follow the Representative Concentration Pathways (RCPs). In this paper, RCP 4.5 and RCP 8.5 are used and for which the choice is justified by the fact that the Ghis-Nekkor is a vulnerable Mediterranean area to climate change, since it is a semi-arid area and involved by the RICCAR initiative. Similar studies were conducted before in Morocco, such as (Cherkaoui 2011; Goury and Chelhaoui 2013). However, most of these studies dealt with climate change, involve only the impact on precipitations. In addition to the future evolution of precipitations and average temperatures carried out by this study, the work focused also on the possible future climate change on aridity and drought conditions through two indices, namely the Standardized Precipitation Index (SPI) introduced by McKee et al. (1993) and the Aridity Index (AI) proposed by De Martonne (1926). The best climate models were selected (El Asri et al. 2019) based on Taylor's Diagram (Taylor 2001) method to study future SPI and AI evolutions. The changes in temperatures and precipitations would have also a direct impact on groundwater recharge of the aquifer. Additionally, the outputs of climate modeling were used as input for computing the future groundwater recharge variation versus time (2020–2080) for the groundwater aquifer of the study area. Indeed, these results are significant to model the impact of climate change on different components of the water balance of the study area, from a qualitative and quantitative point of view, including the extension of the seawater intrusion from the Mediterranean Sea into the aquifer, the salt concentration distribution in the coastal part and the future water balance versus time.

2 Study Area

The study area covers the Ghis-Nekkor plain on the Mediterranean coast localized in the north of Morocco, at 12 km South-East of Al Hoceima city (Fig. 1). It is influenced by the Mediterranean front from the north which is considered as a climate change hotspot. The plain extends over an area of 100 km^2 with an average slope of 1% from South to North.

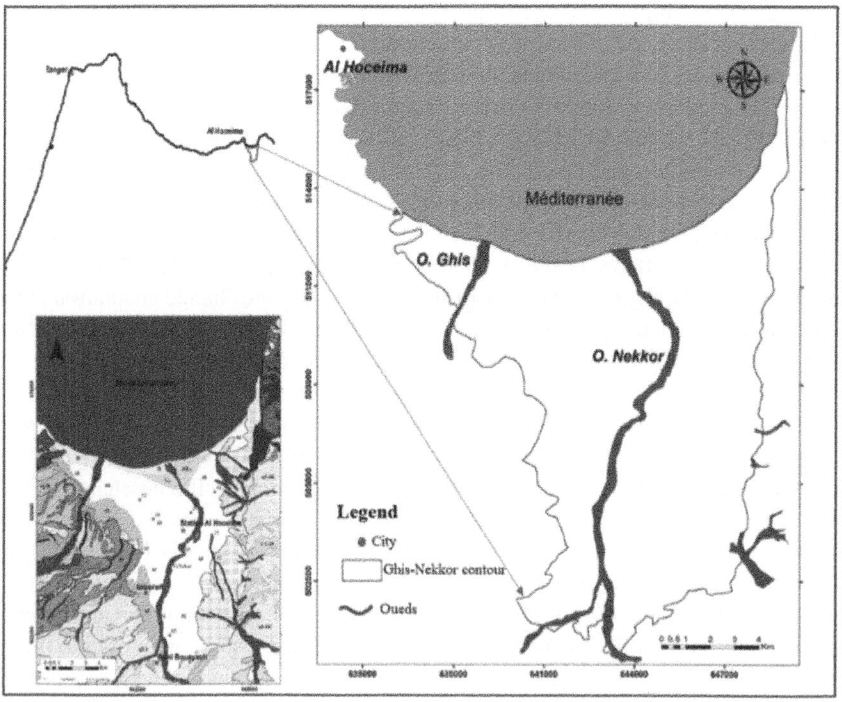

Fig. 1 Geographic location of the study area and geological formations limiting the Ghis-Nekkor area. *Source* El ASRI et al. (2020)

2.1 Characterization of Ghis-Nekkor Climate

The climate of Ghis-Nekkor is semi-arid and it is characterized by a hot and dry summer which is followed by a rainy season and conditioned by two factors: the maritime influence and the existence of the Rif mountain range. The general trend of annual rainfall in the Ghis-Nekkor for the period (1972–2014) is shown in Fig. 2. The average annual rainfall is rather low and oscillates around 300 mm and is characterized by a spatio-temporal disparity. Moreover, the 1980s appeared to be particularly marked by droughts. This is reflected by the deficit in the rainfall experienced during the 1980s years. Besides, the average monthly temperature (Fig. 3) is quite stable during the year as the annual average equals 18 °C. In contrast, the climate of the Ghis-Nekkor is characterized by the succession of two periods: a dry and hot from April to October and a wet from November to March (Jounaid 2001; Iouzzi 2008).

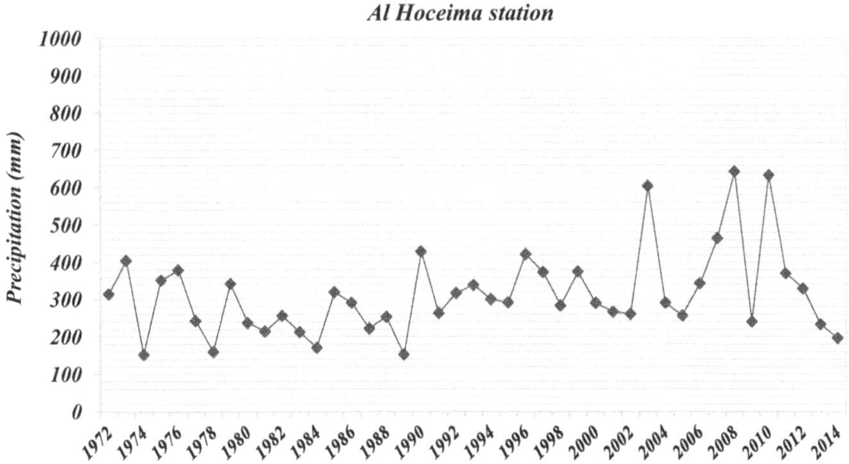

Fig. 2 Evolution of annual precipitation at Al Hoceima station for the period (1982–2014)

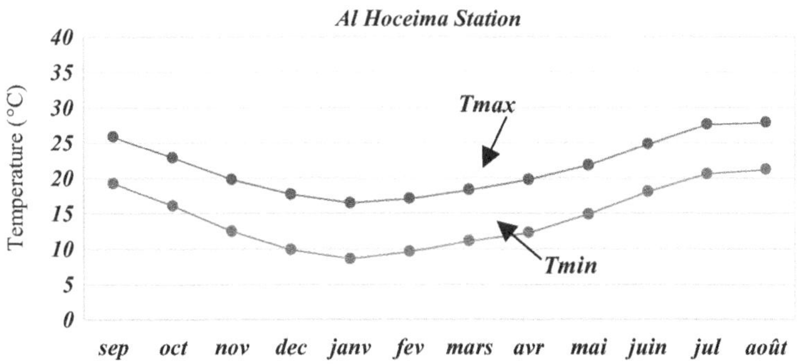

Fig. 3 Evolution of monthly temperature at Al Hoceima station for the period (1982–2012)

2.2 Past Climate Trend

As we mentioned above, the annual precipitation of the 1980s period was the smallest for the observed period (1972–2014) with only 232 mm/yr. The analysis of drought trend is carried out by the SPI and AI indices. The SPI is calculated by Eq. (1):

$$SPI = \frac{(P_i - P_m)}{\sigma} \tag{1}$$

P_i Total precipitation for a period i (mm);
P_m mean of recorded precipitation for the specific period (mm);
σ Standard deviation of recorded precipitation for the specific period (mm).

According to (McKee et al. 1993) a drought event occurs when the SPI is continuously negative and reaches an intensity $\geq$1. The event ends when the SPI becomes positive. Table 1 shows the SPI thresholds that are defined by McKee et al. (1993). De Martonne (1926), defined the aridity index as the ratio of precipitation to the mean temperature (Eq. 2). It can be used to classify the climates of various regions because the ratio of precipitation to temperature provides a method for determining an area's climate regime. In fact, De Martonne has classified the climate according to the AI-values defined in intervals, where the intervals [0–5], [5–10], and [10–20] determine respectively, hyper-arid, arid and semi-arid climates. The calculation of the AI using Eq. 2 for the reference period (1979–2008) shows the existence of a semi-arid climate at Al Hoceima station with a value of 11.75 mm/°C.

$$AI = \frac{P}{(T + 10)}(\text{mm}/^{\circ}\text{C}) \tag{2}$$

P Cumulative annual precipitation (mm);
T Average annual temperature (°C).

The SPI time series at 12-month time steps is given in Fig. 4. Analysis of 12-month SPI time series shows that droughts were quite frequent during the (1980–1984), (1986–1989), (1994–1995). However, several severe dry periods were identified, when only the annual minimum spatially averaged SPI value during 1974–1975 (SPI = −1.86), 1978–1979 (SPI = −1.75) and 1987–1988 (SPI = −1.87) was considered.

Table 1 SPI drought severity classes for wet and dry periods

2.0 +	Extremely wet
1.5 to 1.99	Very wet
1.0 to 1.49	Moderately wet
−0.99 to 0.99	Near normal
−1 to −1.49	Moderately dry
−1.5 to −1.99	Severely dry
−2 and less	Extremely dry

Source McKee et al. (1993)

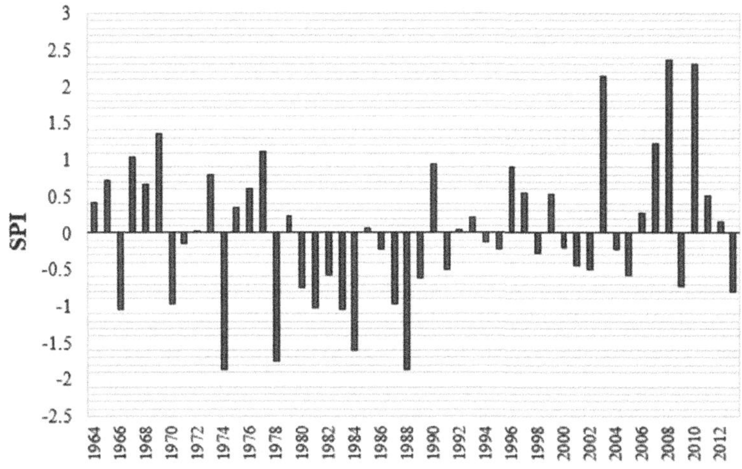

Fig. 4 Time series of the annual mean precipitation at the Ghis-Nekkor plain (1964–2013)

In fact, with only a rate of 10% for moderate wet, the 1980s are the exception, as it is shown in Fig. 4. The spatial distribution of the SPI index for the 1980s, which represents the driest period, is also illustrated in Fig. 5 in terms of maps showing

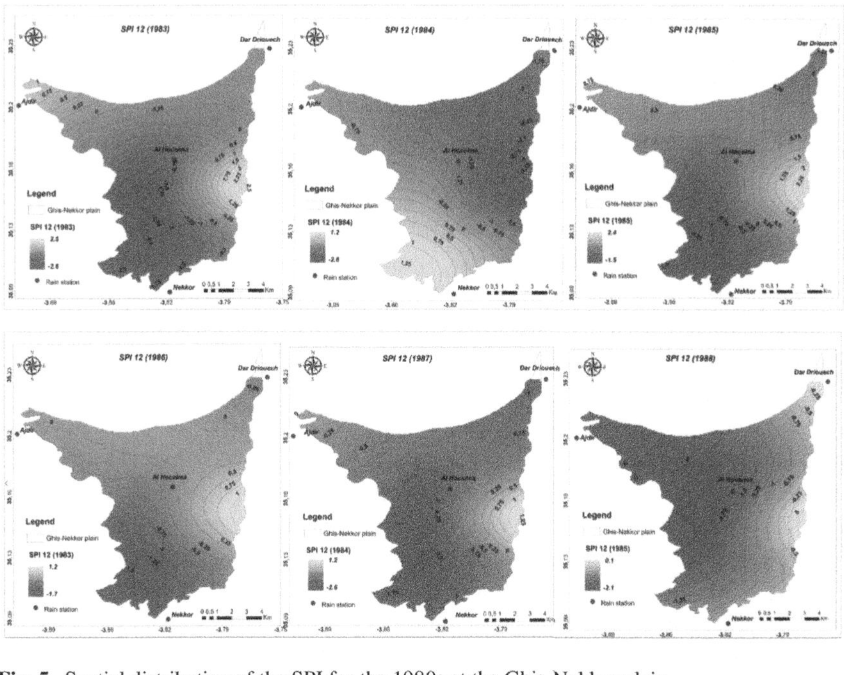

Fig. 5 Spatial distribution of the SPI for the 1980s at the Ghis-Nekkor plain

severe droughts covering the whole study area.

3 Data and Methodology

3.1 Climate Data

The climate variables used for this study are mean annual precipitations and temperatures. The climate data included the historical and future climate projections of these variables from different RCMs (RCA4, HIRHAM5, RACMO22T). The required climate data used in this study are considered for the two RCPs emission scenarios (RCP 4.5 and RCP 8.5). Table 2 describes the available RCMs simulation for RICCAR, while Table 3 shows the different RCMs with driving models used for the study area.

3.2 Limitations of the Study

Although the outputs from RCMs are required to assess future climate change, it is necessary to improve the accuracy of the simulation by the use of the bias-correction method. Accordingly, we need first to find the neighboring points from the grid-box to the observation stations. However, The resolution of the available RCMs (50 km) is particularly restrictive. As only one grid point from available RCMs is closed to the observation station, we provided outputs based on information from a single neighboring grid-box.

3.3 Choice of Scenarios

In view of the 5th IPCC assessment report, the experts defined four representative pathways for the evolution of greenhouse gas (GHG) concentrations. They are called RCPs (Representative Concentration Pathways). The RCPs are consistent with a wide range of possible changes in future anthropogenic GHG emissions, and aim to represent their atmospheric concentrations (Collins et al. 2014). On this basis, climatologists describe the climatic conditions and the impacts of climate change associated with each of these four pathways explained with details in Table 4. The IPCC RCPs retained for this study are: RCP 4.5 (moderate scenario) and RCP 8.5 (extreme scenario). The choice of these two scenarios is justified by the fact that the Ghis-Nekkor plain is a vulnerable Mediterranean area to climate change, located in the south coast and involved by the RICCAR Initiative project.

Table 2 MENA/Arab domain RCM simulations for RICCAR ensembles

RCM	Driving GCM/Reanalysis	Evaluation 1979–2010	Historical 1986–2005	RCP2.6 2006–2100	RCP4.5 2006–2100	RCP8.5 2006–2100	Resolution (km)
RCA4	ERA-INTERIM	x					50
RCA4	EC-EARTH		x	X	x	x	50
RCA4	EC-EARTH		x			x	25
RCA4	CNRM-CM5		x		x	x	50
RCA4	GDFL-ESM2M		x		x	x	50
RCA4	GDFL-ESM2M		x			x	25
HIRAM	GDFL-ESM2M		x				25
REMO	MPI-ESM-LR		x				50

Source RICCAR main report (2017)

Table 3 Available RCMS and driving models used for the study area

RCM	Driving model	Model_id	Experiment	Actual period	Future period
RCA4	EC-EARTH	SMHI_RCA4	Historical; RCP 4.5; RCP 8.5	1950–2010	2010–2100
RACMO22T	EC-EARTH	KNMI_RACMO22T	Historical; RCP 4.5; RCP 8.5	1950–2010	2010–2100
HIRHAM5	CNRM-CM5	DMI_HIRHAM5	Historical; RCP 4.5; RCP 8.5	1980–2010	2010–2100
RCA4	CNRM-CM5	SMHI_RCA4	Historical; RCP 4.5; RCP 8.5	1986–2005	2006–2100
RCA4	GDFL-ESM2M	SMHI_RCA4	Historical; RCP 4.5; RCP 8.5	1986–2005	2006–2100

Table 4 The new IPCC scenarios expressed in terms of radiative forcing

Nom	Radiative forcing	Concentration (ppm)	Pathway
RCP 8.5	>8.5 W/m^2 in 2100	1370	Rising
RCP 6.0	~6 W/m^2 post 2100	850	Rising
RCP 4.5	~4.5 W/m^2 post 2100	650	Stabilization without overshoot
RCP 2.6	3 W/m^2 before 2100 declining to 2.6 W/m^2 by 2100	490	Peak and decline

Source Summary of the 5th IPCC report (2013)

3.4 Methodology

This research was based on the following methodology summarized in the diagram depicted in Fig. 6, where the selected RCMs for the study area is used to extract the future evolution of the studied climatic parameters under two scenarios (RCP 4.5) and (RCP 8.5). From these climatic variables, we can calculate the climatic indices SPI and AI that characterize the future climate trend of the Ghis-Nekkor area. Besides, the future groundwater recharge was predetermined and its future behavior was also analyzed. These steps are described in details in the following sections.

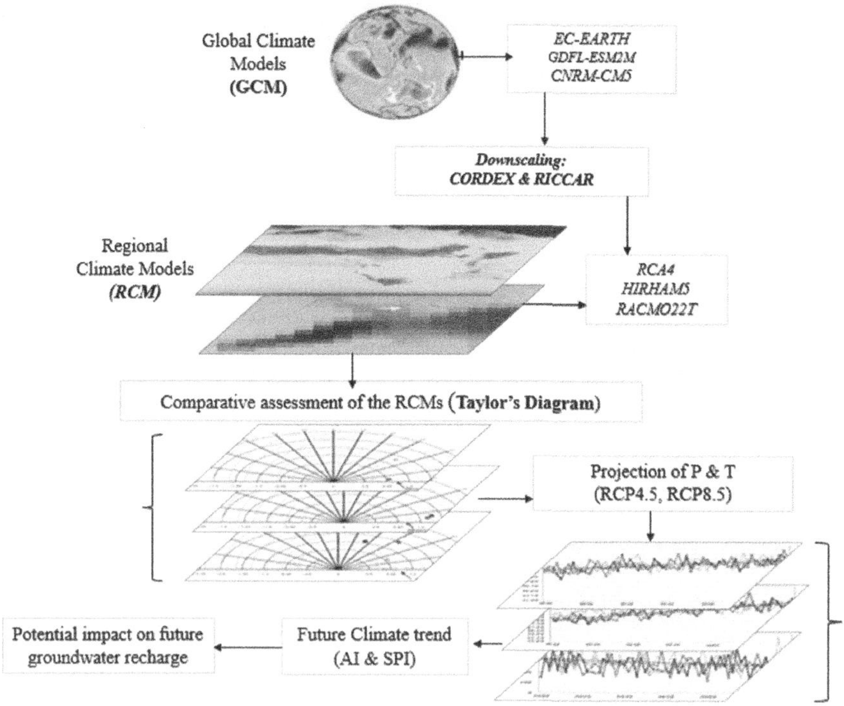

Fig. 6 Methodology of the work

4 Results and Discussions

4.1 Projections of Precipitation and Temperature

A study of the future evolution of temperature and precipitation through RCMs from the CORDEX project was carried out (El Asri et al. 2019). The analysis of available projections from CORDEX RCMs (RACMO22T, HIRHAM5) and RICCAR (RCA4) produce similar spatial patterns (Figs. 7 and 8) but their characterization for temperatures and precipitations are different. Climate change would affect upward the temperature as demonstrated in Fig. 7a, b. It is then expected from the RACMO22T model an increase in the mean annual temperature of 0.5 °C for the period (2021–2050) and an increase of about 0.75 °C for the period (2051–2080) under the RCP 4.5 scenario. While the average increase for the RCA4 model is about 0.57 °C for the period (2021–2050) and 1.05 °C for the period (2051–2080) (Fig. 7a). For the RCP 8.5 scenario, Fig. 7b indicates an increase in the mean annual temperature from RACMO22T model of 0.65 °C for the period (2021–2050) and an increase of about 1 °C for the period (2051–2080). The results from the RCA4 model show an

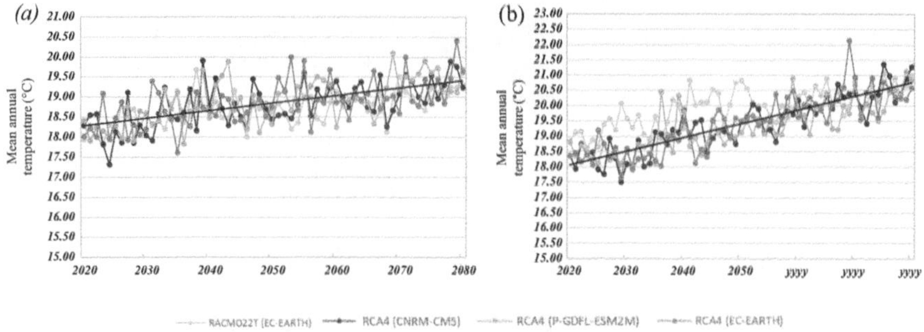

Fig. 7 Future evolution of mean annual temperature from CORDEX RCMs (RACMO22T, HIRHAM5) and RICCAR (RCA4) following the RCP4.5 (**a**) and RCP8.5 (**b**) scenarios

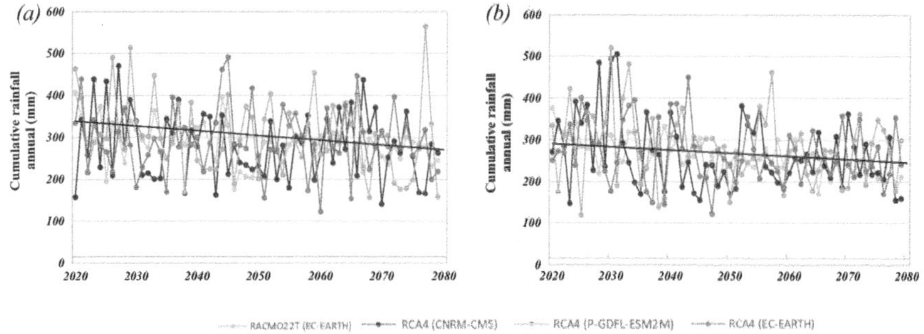

Fig. 8 Future evolution of cumulative annual rainfall from CORDEX RCMs (RACMO22T, HIRHAM5) and RICCAR (RCA4) following the RCP4.5 (**a**) and RCP8.5 (**b**) scenarios

average increase in the mean annual temperature of 0.73 °C for the period (2021–2050) and 1.7 °C for the period (2051–2080) (Fig. 7b). The warming trend seems clear on average annual temperature but also a decrease in cumulative annual rainfall is expected by all the selected RCMs models as illustrated in Fig. 8. Indeed, the results from Fig. 8 show that by 2080, both RCP 4.5 and RCP 8.5 scenarios predict a decrease in precipitation in the Ghis-Nekkor area as is illustrated by the trend lines plotted in Fig. 8a, b.

4.2 Computing AI and SPI Indices

Accordingly, the situation of rising temperatures and reducing rainfall would affect the drought duration and climate aridity. Indeed, the future estimation of AI-values (De Martonne 1926) and the SPI-values (McKee et al. 1993) predict the appearance of arid climate and intensification of drought episodes at the level of the study area by the year 2080 (El Asri et al. 2019).

Table 5 Aridity index for different periods using RCMs (RACMO22T and HIRHAM5) under the RCP 4.5 scenario

AI classification	*Semi-arid*		*Arid*
AI value (Scenario RCP 4.5)	Reference (1982-2013) ***11.75 mm/°C***	Near future (2021-2050) ***11.44 mm/°C***	Future (2051 – 2080) ***9.92 mm/°C***

Aridity Index [10 - 20]
Aridity Index [5 – 10]

Source El Asri et al. (2020)

The AI index is generated for two future periods (2021–2050) and (2051–2080) while the SPI index is calculated for the future period of (2021–2070). The future evolution for both indices is established following the RCP 4.5 climate change scenario. The AI was calculated for the future horizons (2021–2050) and (2051–2080), using temperature and precipitation produced from RACMO22T and HIRHAM5 respectively. The results given in Table 5 predict a value of 11.44 mm/°C and 9.92 mm/°C for respectively (2021–2050) and (2051–2080) periods. According to the De Martonne classification, the value of 11.44 mm/°C would be for the near future (2050) in the interval [10–20] which characterizes a semi-arid climate, while the value of 9.92 mm/°C would be in the interval [5–10] by the horizon (2051–2080), which defines an arid-climate. Finally, the obtained results for the two future periods show the appearance of the arid climate at the level of the study area for the period (2051–2080) according to the RCP4.5 (El Asri et al. 2019).

A reduction in precipitation accompanied by an increase in temperature due to climate change will also affect the future drought trend. To assess the impact of climate change on drought conditions in the Ghis-Nekkor plain, we calculate the 12-month SPI time series through the projection of precipitation. Projected changes in precipitation and temperature simulated by multi-RCMs models (HIRHAM5, RACMO22T, and RCA4) outputs show a downward trend for precipitation compared to the past period. The results show that the average annual rainfall in the plain might reduce up to 17% and 22.2% respectively by 2050 and 2080 under the RCP 4.5 scenario as illustrated by Fig. 8a. While, for RCP 8.5, Fig. 8b shows a pronounced downward trend for precipitation (up to 24%) for the whole period (2021–2080). The obtained results indicate that the drought severity and duration (SPI < 0) will increase under the RCP4.5 scenario for the period from (2040 to 2070) as it is shown in Fig. 9. The Ghis-Nekkor area will experience dry periods from (2044 to 2049), from (2049 to 2056) and from (2066 to 2070), and their intensity varies from moderate to severe. The period (2040–2070) shows a high frequency of drought compared to (2021–2040), with respectively 23% of moderate dry and 19% of severe dry. Otherwise, the positive values of the SPI will be more frequent with a diminution of 4% for the moderate dry. The RCP 4.5 climate change scenario indicates that the drought conditions of the period (2021–2070) will be more severe with a higher frequency.

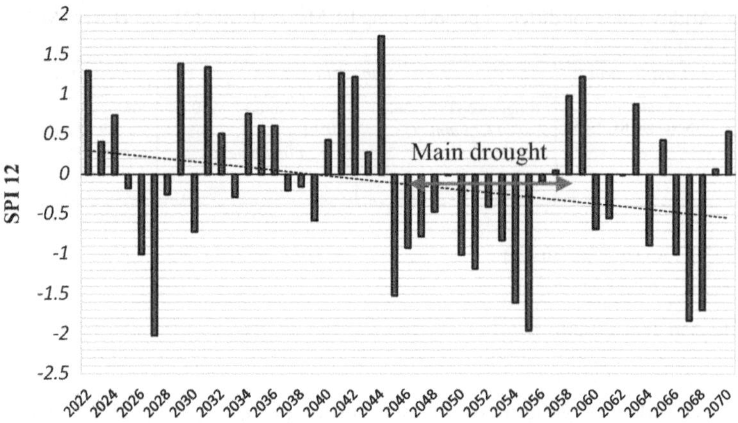

Fig. 9 SPI time series at 12 months provided by the projected precipitation from HIRHAM5 model under the RCP 4.5 scenarios for the future period (2021–2070). The straight line is the SPI annual mean trend for this period

The rate predicted is about 12%; it doubles the rate for the reference period (1965–2014). Despite the high frequency of the dry period expected by 2070, their intensity would remain moderate to severe without reaching an extreme degree (El ASRI et al. 2019).

4.3 Impact of Climate Change on the Aquifer Groundwater Recharge

In this paper, the study focuses also on estimating the groundwater recharge volume from future precipitation, as it is an important component in the water balance. The future aquifer recharge is estimated through rainfall projections from HIRHAM5 (CORDEX) and RCA4 (RICCAR) developed for both (RCP 4.5 and RCP 8.5) scenarios for (2021–2080). Figures 10 and 11 show the future estimated aquifer recharge variability under respectively RCP 4.5 and RCP 8.5 scenarios for the period (2021–2080). The results of future recharge groundwater from RCMs from CORDEX and RICCAR produce similar spatial patterns. Indeed, the expected groundwater recharge by infiltration from precipitation is estimated to 3 Mm^3 per year (Mm^3/yr) and 2.7 Mm^3/yr under the RCP 4.5 scenario for respectively (2021–2050) and (2051–2080) for the HIRHAM5 RCM. For RCA4 from the RICCAR project, the expected groundwater recharge by infiltration from precipitation is estimated to 3.01 Mm^3/yr and 2.82 Mm^3/yr under the RCP 4.5 scenario for respectively (2021–2050) and (2051–2080).

It is important to note that the groundwater recharge reduction will also occur during the same period where the climate arid regime changes also. If we consider the

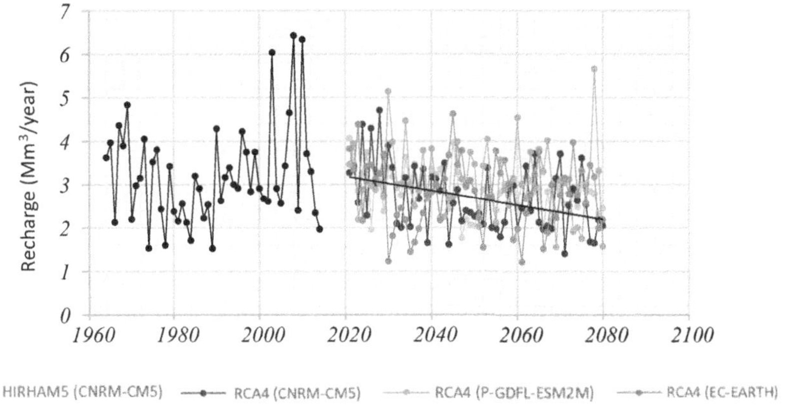

Fig. 10 Estimated aquifer recharge (Mm^3) for RCP 4.5 scenarios in the Ghis-Nekkor area for (2021–2080). The straight line is the mean trends in recharge versus time

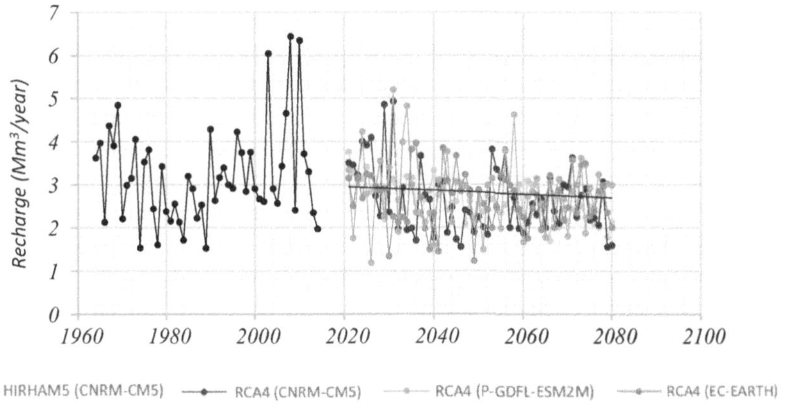

Fig. 11 Estimated aquifer recharge (Mm^3) for RCP8.5 scenarios in the Ghis-Nekkor area for (2021–2080). The straight line is the mean trends in recharge versus time

whole period (2021–2080), the mean annual groundwater recharge will be approximately 2.9 Mm^3/yr and 2.63 Mm^3/yr under respectively the RCP 4.5 and RCP 8.5 scenarios for the HIRHAM5 model. The two plots in Figs. 10 and 11 follow almost the same trend for future simulated recharge, but the results under RCP 8.5 in Fig. 11 are characterized by low recharge values for the period (2065–2080) compared to simulation results under the RCP 4.5 scenario in Fig. 10. The results from Fig. 11 show also that the future aquifer recharge would be more affected for both periods (2049–2056) and (2066–2080) and confirm that the groundwater recharge will be decreasing more over time under RCP 8.5 than RCP 4.5.

5 Conclusion

In this study, the projections of future climate change were evaluated by the RCMs models from CORDEX and RICCAR Projects. The results of projections from the CORDEX and RICCAR projects produce similar spatial patterns. The outputs of the selected models show an increase in the average temperature of 0.71 °C and a decrease in precipitation of 19.6% for the period (2021–2080). The situation of increasing temperature and decreasing precipitation anticipate more droughts and aridity at the Ghis-Nekkor plain as it has been shown through the values of the SPI and AI indices. The analysis of drought intensity showed that the study area is more prone to both moderate and severe droughts but less prone to extreme drought. Also, the predicted change of the aridity index, from 11.44 mm/°C by 2050 to 9.92 mm/°C by 2080, imposes an arid climate in the study area. The decrease in future groundwater recharge in the Ghis-Nekkor aquifer can be consequence of a projected decrease in precipitation. Future groundwater recharge will be more affected by climate change and it is expected to be less than what has been experienced in the past periods. The present results will be used later to model the impact of climate change on the future evolution of the Ghis-Nekkor water table from a quantitative and qualitative point of view including the extension of the seawater intrusion from the Mediterranean sea into the aquifer and the future water balance. Additionally, the limitation of the spatial resolution will be overcome with the use of a 12 km resolution from RICCAR outputs. The improvement of the spatial resolution is expected to be completed in a few months which will improve the updated results relevant to the study area. Based on the new results to be obtained from the groundwater model (under view), solutions under planning scenarios will be proposed for adaptive strategies and best schemes for groundwater resources management in the study area, to deal with climate change vulnerability, such as artificial aquifer recharge from potential unconventional water resources available in the study area.

Acknowledgements This research is a part of the PhD of the first author under the supervision of the second author and was funded by the MEDAQCLIM project in the framework of the ERANET-MED programme (FP7). Special Thanks to UN-ESCWA at Beirut who has provided RICCAR Data.

References

Bouaicha R, Benabdelfadel A (2010) Variabilité et gestion des eaux de surface au Maroc. Sécheresse en ligne 21(1e):325–326. https://doi.org/10.1684/sec.2010.0276

Cherkaoui A (2011) Désagrégation des données via l'outil SDSM pour la projection des changements climatiques futurs dans la haute Moulouya, Maroc. Projet de fin d'études, Université Hassan II Mohammedia, Casablanca, Maroc

Chouchani C (2013) Strengthening national capacities of conflict-affected countries to manage water scarcity and drought. Expert group and inception meeting of UNDA project

Collins M et al (2014) The new concentration driven RCP scenarios, and their extension: chapter 12: long-term climate change: projections, commitments and irreversibility, archived in: IPCC AR5 WG1 2013, pp 1045–1047

De Martonne E (1926) Aréisme et indice d'aridité. Comptes Rendus Académie des Sciences 181:1395–1398

El Asri H, Larabi A, Faouzi M (2019) Climate change projections in the Ghis-Nekkor region of Morocco and potential impact on groundwater recharge. Theor Appl Climatol 135. https://doi.org/10.1007/s00704-19-02834-8

El Asri H, Larabi A, Faouzi M (2020) Configuration hydrogéologique et impact potential des changements climatiques sur la recharge future et l'état piézométrique de l'aquifère de Ghis-Nekkor (Maroc). Techniques et Sciences Méthodes. https://doi.org/10.1007/s00704-19-02834-8

Giorgi F (2006) Climate change hotspots. Geophys Res Lett 33:L08707. https://doi.org/10.1029/2006gl025734

Goury M, Chelhaoui Y (2013) Evaluation comparative des modèles météorologiques entre une région de plaine et une région de montagne et évaluation des scénarios des changements climatiques à l'échelle locale en perspective d'adaptation des projets en développement. Université Mohammed V, Rabat, Maroc

Iouzzi N (2008) Hydrogéologie et modélisation de la nappe de Ghis-Nekkor (Provinces d'El Hoceima-Nador), Mémoire DESS. Université Mohammed V de Rabat, Ecole Mohammadia d'Ingénieurs, Rabat, p 133

IPCC (2013) The physical science basis. Contribution of working group I to the fifth Assessment report of the intergovernmental panel on climate change. In: Stocker TF, Qin D, Plattner G-K, Tignor M, Allen SK, Boschung J, Nauels A, Xia Y, Bex V, Midgley PM (eds) Cambridge University Press, Cambridge, 1585 pp

Jounaid H (2001) Hydrogéologie de la nappe de Ghis-Nekkor, Thèse de Doctorat. Université Mohammed V de Rabat, Ecole Mohammadia d'Ingénieurs, Rabat, p 145

McKee TB, Doesken NJ, Kleist J (1993) The relationship of drought frequency and duration of timescales. In: Proceedings of the 8th conference on applied climatology, Anaheim, California, 17–22 Jan 1993, pp 179–184

Naciri M (1985) Calamites naturelles et fatalités historique. Science et changements planétaires/Sécheresse 1(1):1–16

RICCAR (2017) (Regional Initiative for the assessment of climate change impact on water resources and socio-economic vulnerability in the arab region). arab climate change assessment report, United Nations Economic and Social Commission for Western Asia (ESCWA) – Main Report. Beirut, E/ESCWA/SDPD/2017/RICCAR/Report. Available at: https://www.unescwa.org/publications/riccararab-climate-change-assessment-report

Taylor KE (2001) Summarizing multiple aspects of model performance in a single diagram. J Geophys Res 106:7183–7192

Woillez MN (2019) Revue de littérature sur le changement climatique au Maroc: observations, projections et impacts. Papiers de recherche AFD, 2019–2018, Juillet 2019

Collins M et al (2013) The new concentration driven RCP scenarios, and their extensions, chapter 12, long-term climate change: projections, commitments and irreversibility, archived in IPCC AR5 WG1 2013, pp 1045–1047
De Martonne E (1926) Aréisme et indice d'aridité. Comptes Rendus Académie des Sciences 181:1395–1398
El Asri H, Larabi A, Faouzi M (2019) Climate change projections in the Ghris Nekkor region of Morocco and potential impact on groundwater recharge. Theor Appl Climatol 135. https://doi.org/10.1007/s00704-019-02831-4
El Asri H, Larabi A, Faouzi M (2019) Cartographie des sécheresses et analyse du potentiel des ressources en eau du bassin Ghris par les indices SPI et l'approche de Thiessen. In: Methods and techniques…
Giorgi F (2006) Climate change hot-spots. Geophys Res Lett 33:L08707. https://doi.org/10.1029/2006GL025734

Species Distribution Based-Modelling Under Climate Change: The Case of Two Native Wild *Olea europaea* Subspecies in Morocco, *O. e.* subsp. *europaea* var. *sylvestris* and *O. e.* subsp. *maroccana*

Jalal Kassout, Jean-Frédéric Terral, Abdeltif El Ouahrani, Mhammad Houssni, Sarah Ivorra, Khalil Kadaoui, Mohamed El Mahroussi, Laure Paradis, and Mohammed Ater

Abstract Climate change is expected to greatly alter and modify the ecological conditions of plant growth and distribution, particularly in the Mediterranean Basin, considered as one of the most vulnerable zone to global warming in the world. In this chapter, we look at the biogeography of the olive tree, an emblematic species of the Mediterranean Basin, represented in Morocco by two wild subspecies: *Olea europaea* subsp. *europaea* var. *sylvestris*, the ancestor of all the olive varieties and widely distributed in the country, and *Olea e.* subsp. *maroccana*, endemic in a restricted southwestern area. We hypothesis, within the context of future warming, an increase of *O. e.* subsp. *e.* var. *sylvestris* distribution area is expected, while for *O. e.* subsp. *maroccana*, an alteration of its distribution is predicted, increasing seriously the risk of extinction. In order to assess the current and future potential geographic distribution of the two wild olive species in Morocco, *a* species distribution based-modelling was performed to understand the relationships between species distributions and climatic factors, on the basis of field data and 19 climatic variables. Two representative concentration pathways, RCP4.5 and RCP8.5, were used to forecast the future distribution of the two wild olive subspecies in 2050 and 2070. To avoid multicollinearity, the highly correlated climatic variables ($r > 0.9$, Pearson correlation coefficient) were deleted from the independent variables list. The Jackknife test was carried out to evaluate the relevance of the climatic variables for predictive modeling. The maximum entropy model for the current distribution of both species provides a satisfactory result, with a high value of the Area Under Curve equal

J. Kassout (✉) · A. El Ouahrani · M. Houssni · K. Kadaoui · M. El Mahroussi · M. Ater
Department of Biology, Agrobiodiversity Team, Faculty of Sciences Tétouan, Abdelmalek Essaadi University, Tétouan, Morocco

J.-F. Terral · S. Ivorra · L. Paradis
Institut des Sciences de L'Evolution, Equipe Dynamique de La Biodiversité, Anthropo-écologie, Université de Montpellier, Montpellier, France

J. Kassout · J.-F. Terral · S. Ivorra · L. Paradis · M. Ater
Associated International Laboratory (IRP), EVOLEA, INEE-CNRS, CNRST, Montpellier, France

W. Leal Filho and E. Manolas (eds.), *Climate Change in the Mediterranean and Middle Eastern Region*, Climate Change Management,
https://doi.org/10.1007/978-3-030-78566-6_2

to 0.980 (±0.001) for *Olea europaea* subsp. *europaea* var. *sylvestris* and equal to 0.997 (±0.001) for *Olea europaea* subsp. *maroccana*. Jackknife test indicates that precipitation and temperature variables play a significance role in wild olive species biogeographical dynamics in Morocco. The study results confirm our hypothesis of an expansion of *O. e.* subsp. *e.* var. *sylvestris* suitable area and the threatened aspect of *Olea e.* subsp. *maroccana* under climate change scenarios. The approach used in this study is promising to predict the potential distribution of wild olive species, and can be an effective tool to support conservation and restoration programs.

1 Introduction

The ongoing global changes poses considerable challenges for natural ecosystems. The rapid rate of changing climate is likely to affect plant species growth and distribution, and its impacts are expected to reduce their long-term persistence and resilience (Parmesan and Yohe 2003; Thuiller et al. 2005; Allen et al. 2010; Vellend et al. 2017). For instance, climate change impacts include changes in plant species distribution (Lenoir et al. 2010; Doblas-Miranda et al. 2017), widespread tree mortality (Allen et al. 2010) and loss of key habitat of many species (Pereira et al. 2010; Dullinger et al. 2012). In the Mediterranean region, considered as a valuable biodiversity hotspot, the climate scenarios predict an increase in temperature and a decrease in precipitation (Gibelin and Deque 2003; Driouech et al. 2010; Polade et al. 2014; IPCC 2014; Lionello and Scarascia 2018). Such changing contexts could affect plant communities with a prominent threat to growing conditions of trees and forest ecosystems (Bréda et al. 2006; Choat et al. 2012; Matesanz and Valladares 2014; Lloret et al. 2016). The complexity of climate change impacts on naturel ecosystems could disrupt current and future geographical distribution pattern of plants (Lenoir et al. 2008) which could greatly affect terrestrial energy and water fluxes, and therefore CO_2 emissions (van Bodegom et al. 2014; Forzieri et al. 2020).

Within the context of climate change, improving our understanding of current and future species habitat distributions appears to be necessary. In fact, considerable progress has been achieved toward improving species distribution models (SDMs), which reflecting spatial and geographical representation of species presence probability or abundance (Elith and Leathwick 2009; Booth 2017). The habitat suitability models are mainly based on quantifying the species-environment relationship and are widely used as prominent tools to understand past and future climate change impacts on biodiversity (Guisan and Zimmermann 2000; Guisan and Tuiller 2005; Guisan et al. 2017). SDMs approaches can provide information on the potential distribution of species under specific environmental conditions and within specific study area (Guisan et al. 2017). In fact, SDMs provide a comprehensive basis for species conservation and naturel resource management (Elith and Burgman 2002; Qin et al. 2017), thus, suitable for potential land use of some endangered species under current and future climate (Franklin 2009; Sinclair et al. 2010). Currently, a considerable number of SDMs method were available and widely used such as,

BIOCLIM (bioclimatic modeling), Domain (domain environmental envelope), GAM (generalized additive models), MARS (multivariate adaptive regression splines) and Maxent (Maximum entropy) (Elith et al. 2006; Guisan et al. 2017; Pecchi et al. 2019). Among these models, Maxent (Phillips et al. 2006) provides a reliable habitat suitability results when modelling widely distributed or more rare species (Elith et al. 2006; Harte 2011; Moukrim et al. 2019) and enabled with high capacity to deal with complex interactions between response and predictor variables (Elith et al. 2006, 2011). In fact, the maximum entropy model has been widely used to predict macroecological patterns (Harte et al. 2008; Harte 2011), species abundance distributions (White et al. 2012), trait-based community assembly (Sonnier et al. 2010; Shipley et al. 2011) and species ecological niche at multiple scales (Elith et al. 2010; Peterson et al. 2011; Guisan et al. 2017). Hence, the main advantages of Maxent include (i) requires presence-only data to generate good performance outputs, (ii) can directly produce habitat suitability maps and (iii) can evaluate the importance of environmental variables using the jackknife test (Phillips et al. 2004, 2006, 2017b; Elith et al. 2006).

In this work, Maxent model will be applied to olive tree, which is the most iconic and emblematic species of the Mediterranean basin, with high valuable economic importance and cultural significance (Zohary et al. 2012; Kaniewski et al. 2012). The genus *Olea* (Oleaceae) contains approximately 40 species and sub-species (Green and Wickens 1989). Actually, six subspecies of wild olive, also known as 'olive complex' are recognized (Médail et al. 2001; Green 2002). In Morocco, two wild olive subspecies are recognized, *Olea europaea* subsp. *europaea* var. *sylvestris* (known as oleaster) and *Olea europaea* subsp. *maroccana* (known as Moroccan olive) (Médail et al. 2001; Green 2002). *O. e.* subsp. *e.* var. *sylvestris* is widely distributed in the Mediterranean region, considered a distinctive element of its flora, which have been persisted in the region since the late Tertiary (Palamarev 1989; Carrión et al. 2010). Its potential distribution area is considered as indicators of the Mediterranean climate (Rubio de Casas et al. 2002; Blondel et al. 2010; Besnard et al. 2013; Gianguzzi and Bazan 2019). Oleaster, highly genetically diversified forms of olive, is considered as the wild progenitor of the Mediterranean cultivated olive (Besnard et al. 2013, 2017). In Morocco, oleaster (*O. e.* susbp. *e.* var. *sylvestris*) is broadly distributed under different bioclimates, altitudes and soils types, from humid and subhumid in northern areas to semi-arid areas in the south where its populations become scarce (Benabid 1984; Kassout et al. 2019). Oleaster is representing an iconic tree of the Moroccan agroecosystems for different purposes such as food (oil and edible fruit), shade for subsistence crops, fodder for cattle and religious observances (Hmimsa and Ater 2008; Aumeeruddy-Thomas et al. 2014; Ater et al. 2016). Regarding the historical, ecological and the genetic importance of oleaster, there is no doubt that this iconic tree is a valuable natural resource with a great genetic diversity for potential breeding programs (Belaj et al. 2011; Besnard et al. 2013, 2017; Kassout et al. 2019). Whereas, *O. e.* subsp. *maroccana* is an endemic subspecies considered as relictual to the southwest of the country and mainly linked to the argan (*Argania spinosa,* Sapotaceae) and ârâr trees (*Tetraclinis articulata*, Cupressaceae), and coexist in sympatry with the oleaster (Médail et al. 2001; Kassout 2019). The Moroccan olive (*O. e.* subsp.

maroccana) is geographically isolated within a restricted distribution area under semi-arid climate (Médail et al. 2001; Garcia-Verdugo et al. 2009; Kassout 2019). Despite their interest, wild olive subspecies populations are under several threats from climatic and anthropogenic pressures, especially in the light of rapid global warming. Therefore, it is crucial to understand possible response of wild olive to climate change and model the area of its potential distribution under current and future climate change scenarios.

In the present study, we used Maxent to predict the distribution of these emblematic and iconic subspecies and their response to climate change. Specifically, we aimed to use species distribution models (SDMs) to answer two main questions: (1) what are the key climatic factors that affect current *Olea europaea* subspecies distribution and suitable area in Morocco? and (2) how the future climate scenarios would affect their habitat suitability? Finally, our study will provide a theoretical basis for management and conservation decisions and possible plantation efforts.

2 Methodology

2.1 Study Area

This study was carried out in Morocco, situated in the extreme west part of the Mediterranean region within the latitude and longitude of 32° 00 N, 05° 00 W, considered as a biodiversity hotspot (Médail and Quézel 1997; Myers et al. 2000). The study area covers the distribution area of two wild olive subspecies in Morocco: *Olea e.* subsp *e.* var. *sylvestris* and *Olea e.* subsp. *maroccana.* The field work follows a latitudinal transect of 1000 km crossing three climatic zones, ranging from humid/subhumid in the north to semi-arid conditions in the south (UNEP 1997; Kassout et al. 2019; Kassout 2019). Thus, the field survey covers a wide range of habitats and vegetation types including forests, matorrals and shrub-lands (Barbéro et al. 1981; Benabid 1985; Benabid and Fenane 1994; Médail et al. 2001; Taleb and Fennane 2019).

2.2 Occurrence Records

Occurrence data of *Olea europaea* subspecies (*Olea e.* subsp *e.* var. *sylvestris* and *Olea e.* subsp. *maroccana*) were collected during the field surveys between 2012 and 2019. Each observation was recorded with GPS (Global positioning system) coordinates (latitude and longitude). Records from published literature were also collected (Barbéro et al. 1981; Benabid 1984; Benabid and Fennane 1994; Médail et al. 2001; Ali et al. 2017; Ater et al. 2016; Kassout et al. 2016, 2019; Kassout 2019). The records of *Olea* e. subsp. e. var. *sylvestris* from the open access GBIF digital database (https://www.gbif.org/) were considered as well. Finally, we got 447 occurrence points for *Olea e.* subsp *e.* var. *sylvestris* and 90 occurrence points for *Olea e.* subsp. *maroccana.*

2.3 Current and Future Climatic Data

Nineteen bioclimatic variables were extracted from the Worldclim database version 2.0 (Table 1, Fick and Hijmans 2017) with a spatial resolution of approximately 1 km^2 (30 arc seconds). The bioclimatic variables consisted of monthly averages of temperature and precipitation covering the period between 1970 and 2000. Looking to their relevant influence on plant growth, productivity and physiological processes (Moles et al. 2014; Li et al. 2016; Madani et al. 2018), the extracted 19 bioclimatic variables are largely used in species distribution models (Hijmans et al. 2005; Elith et al. 2011; Guisan et al. 2013). The bioclimatic variables were tested for multi-collinearity using a Pearson's correlation test (Graham 2003). Multicollinearity of variables can results in over-fitting of species distribution models and lead to miss interpretation of variables contribution (Schröder 2008; Elith et al. 2010). The highly correlated variables with $r > 0.9$ were eliminated from data before species distribution modeling (Dormann et al. 2013). Pearson's correlation coefficient was calculated with the open R software v 3.5.1 (R Core Team 2018).

Future bioclimatic variables projections by the year 2050 (average for 2041–2060) and 2070 (average for 2061–2080) were used to predict future distribution of *Olea europaea* subspecies in Morocco. In this study, two representative concentration pathways (RCPs) trajectories were used corresponding to an increase by 4.5 and 8.5 W/m^2 by 2100. From the Global Circulation Models (CGMs), and for all projections, we used the Community Climate System Model 4.0 (CCSM4) from the CMIP5 (Coupled Model Inter-comparison Project Phase 5) model developed by the Intergovernmental Panel on Climate Change (IPCC) at approximately 1 km^2 resolution (Gent et al. 2011; IPCC 2014). The CCSM4 has been reported to better predict precipitation and temperature variables (Gent et al. 2011).

2.4 Habitat Suitability Modelling

The Maxent software version 3.4.1 (Phillips et al. 2017a) was used in this study to develop a dynamic species niche models for *O. e.* subsp. *e.* var *sylvestris* and *O. e.* subsp *maroccana,* using species occurrences and climatic variables. This modelling approach is based on the Maximum entropy algorithm allowing to estimate habitat suitability and ecological niche for both studied wild olive subspecies. In our models, 10 bootstrap runs with a 25% random test percentage were used, which is appropriate for estimating species suitable presence (Phillips et al. 2017b). This procedure allows to use 75% of presence data to construct and calibrate the models and 25% to test and

Table 1 Bioclimatic variables used as Maxent environmental factors to model current and future habitat suitability for *Olea e.* subsp. *e.* var. *sylvestris* and *Olea e.* subsp. *maroccana*

Bioclimatic variable	Description	Unit	% contribution*	
			Olea e. subsp. *e.* var. *sylvestris*	*Olea e.* subsp. *maroccana*
Bio1	Annual mean temperature	°C	3.06	–
Bio2	Mean diurnal range (mean of monthly max. and min. temp.)	°C	2.54	14.75
Bio3	Isothermality ((Bio2/Bio7) × 100)	Unitless	3.75	20.64
Bio4	Temperature seasonality (standard deviation × 100)	Unitless	1.32	–
Bio5	Maximum temperature of warmest month	°C	0.88	0.22
Bio6	Minimum temperature of coldest month	°C	0.64	–
Bio7	Temperature annual range (Bio5–Bio6)	°C	–	–
Bio8	Mean temperature of wettest quarter	°C	1.49	–
Bio9	Mean temperature of driest quarter	°C	0.85	–
Bio10	Mean temperature of warmest quarter	°C	–	–
Bio11	Mean temperature of coldest quarter	°C	–	–
Bio12	Annual precipitation	mm	76.73	36.76
Bio13	Precipitation of wettest period	mm	–	–
Bio14	Precipitation of driest period	mm	0.84	0.46
Bio15	Precipitation seasonality (CV)	Unitless	5.61	0.22
Bio16	Precipitation of wettest quarter	mm	–	–
Bio17	Precipitation of driest quarter	mm	–	–
Bio18	Precipitation of warmest quarter	mm	2.23	–

(continued)

Table 1 (continued)

Bioclimatic variable	Description	Unit	% contribution*	
			Olea e. subsp. *e.* var. *sylvestris*	*Olea e.* subsp. *maroccana*
Bio19	Precipitation of coldest quarter	mm	–	–

*Percent contribution of bioclimatic variables used after multicollinearity test results

evaluate the accuracy and the predictive ability of the models (Phillips and Dudik 2008). For each subspecies, the average of all runs were used as final models.

The Maxent model was performed with an auto-feature option (linear, quadratic, product, threshold and hinge methods) and the cloglog output format (ranging from 0 to 1), a maximum number of iterations of 5000 with a convergence threshold of 10^{-5}, a maximum number of background points as 10,000, and a regularization parameter value of 1. The use of a cloglog transform provides a stronger theoretical justification than the logistic transform which replaces it by default (Phillips et al. 2017b). For current (1970–2000) and projected (RCP4.5 and RCP8.5) climate change scenarios by 2050 and 2070, the jackknife test was used to determine the climatic variables that influence significantly the potential distribution of the studied subspecies (Elith et al. 2011; Li et al. 2016). The jackknife test run the model with removing each variable one by one using other variables in order to show the individual importance and contribution of each variable. Maxent model was run with 145 occurrence points for *O. e.* subsp. *e.* var *sylvestris* reduced from 447 and 52 points for *O. e.* subsp. *maroccana* reduced from 90 occurrence points by removing duplicate presence within 1 km^2 using the spThin R package (Aiello-Lammens et al. 2015). This procedure allows to reduce spatial autocorrelation caused by sample bias (Boria et al. 2014; Kwon et al. 2016). Binary maps for current and future occurrences were used to calculate the suitability and unsuitability habitat using the maximum training sensitivity plus specificity threshold for *O. e.* subsp. *e.* var *sylvestris,* which is more appropriate for large distributed species (Liu et al. 2013) and the 10-percentil training presence threshold for *O. e.* subsp *maroccana,* which is appropriate for rare and endemic species (Liu et al. 2005; Escalante et al. 2013). The results of the analyses were visualized in QGIS 2.18.20 (QGIS Development Team 2012).

2.5 Model Evaluation

The area under the receiving operator curve (AUC) was used to test the final models performance (Delong et al. 1988; Elith et al. 2006; Phillips and Dudik 2008; Merow et al. 2013). High AUC values indicates that the model can distinguish between presence and background points, and therefore indicates a better performance of the model (Phillips et al. 2004; Elith et al. 2006). AUC values less than 0.5 indicates a model with random prediction, values between 0.5 and 0.7 are considered low accuracy, values between 0.7 and 0.9 are adequate and values higher than 0.9 indicates excellent accuracy of the predicted model (Pearce and Ferrier 2000; Elith et al. 2006; Yang et al. 2013; Kurpis et al. 2019).

3 Results

3.1 Climatic Variables Used for Maxent Models

After multicollinearity test ($r < 0.9$, Pearson correlation coefficient), 12 variables for *Olea e.* subsp *e.* var. *sylvestris* and 6 for *Olea e.* subsp. *maroccana* were used as input climatic variables in Maxent models (Table 1). Among the selected climatic variables, six were common: mean diurnal range (Bio2), isothermality (Bio3), maximum temperature of warmest month (Bio5), annual precipitation (Bio12), precipitation of driest period (Bio14) and precipitation seasonality (Bio15). Therefore, mean annual temperature (Bio1), temperature seasonality (Bio4), minimum temperature of the coldest month (Bio6), mean temperature of the wettest quarter (Bio8), mean temperature of the driest quarter (Bio9) and precipitation of the warmest quarter (Bio18) were used for modelling *Olea e.* subsp. *e.* var. *sylvestris* (Table 1).

3.2 Models Performance

The Maxent models for *Olea europaea* subspecies distribution in Morocco provided an excellent performance with an AUC (values of the Area Under Curve) training and testing values greater than 0.9, for the current as for the future climatic scenarios (Table 2). The average AUC of the replicates runs of current climatic conditions was 0.980 ± 0.001 for *Olea e.* subsp. *e.* var. *sylvestris* with a maximum training sensitivity plus specificity threshold of 0.100. Thus, for *Olea e.* subsp. *maroccana* AUC mean value was 0.997 ± 001 with a 10-percentil training presence threshold of 0.445. The obtained AUC values reflect a reliable prediction for the studied subspecies habitat suitability. The future suitability distribution of *Olea europaea* subspecies predicted by the models shows high AUC values for both tested RCPs by 2050 and 2070 representing a good success of the models (Table 2). These results show that the climatic variables used for modelling wild olive subspecies in Morocco led to an excellent prediction.

Table 2 Area under the receiving operator curve (AUC) values of Maxent models performance for *Olea e.* subsp. *e.* var. *sylvestris* and *Olea e.* subsp. *maroccana* under current and future climate change RCPs scenarios by 2050 and 2070

		Olea e. subsp. *e.* var. *sylvestris*		*Olea e.* subsp. *maroccana*	
		AUC^{*}_{mean} training	AUC^{*}_{mean} test	AUC^{*}_{mean} training	AUC^{*}_{mean} test
Current (1970–2000)		0.980 ± 0.001	0.973 ± 0.004	0.997 ± 001	0.996 ± 0.001
Projected	2050—RCP4.5	0.950 ± 0.004	0.932 ± 0.015	0.993 ± 0.001	0.991 ± 0.003
	2050—RCP8.5	0.951 ± 0.006	0.935 ± 0.012	0.994 ± 0.001	0.993 ± 0.002
	2070—RCP4.5	0.952 ± 0.003	0.936 ± 0.013	0.992 ± 0.001	0.992 ± 0.003
	2070—RCP8.5	0.948 ± 0.008	0.935 ± 0.012	0.992 ± 0.002	0.989 ± 0.004

*Mean values of 10 bootstrap runs
AUC mean values of training and test data are given after partitioning subspecies presence-only data into training (75%) and test (25%) with standard deviation

3.3 Jackknife Test and Environmental Variable Importance

The results of the jackknife test indicate that among the 12 climatic variables used as input to model current habitat suitability of *Olea e.* subsp. *e.* var. *sylvestris*, annual precipitation (Bio12) was the most influential and contributed with 76.7% to the distribution model (Table 1; Fig. 1a). Precipitation seasonality (Bio15), isothermality (Bio3) and mean annual temperature (Bio1) contributed with 5.6%, 3.7% and 3%, respectively. Thus, jackknife analysis shows a considerable change in *Olea e.* subsp. *e.* var. *sylvestris* model when annual mean temperature (Bio1) and mean temperature of wettest quarter (Bio8) were used in isolation (Fig. 1a). The maximum temperature of the warmest month (Bio5) had also a substantial gain when used separately (Fig. 1a). Mean diurnal range (Bio2) and precipitation of the warmest quarter (Bio18) had a substantial influence on the model with respectively 2.5% and 2.2% of contribution, meanwhile, mean temperature of wettest quarter (Bio8) and temperature seasonality (Bio4) contributed with 1.4% and 1.3%, respectively. However, the lowest contribution was recorded for the maximum temperature of the warmest month (Bio5, 0.8% of contribution), mean temperature of the driest quarter (Bio9, 0.8%), precipitation of the driest period (Bio14, 0.8%) and the minimum temperature of the coldest month (Bio6, 0.6%) as shown in Table 1. Nevertheless, these contributions are heuristically defined and depend on the particular path that Maxent algorithm uses to generate optimal solution (Yang et al. 2013).

Regarding *Olea e.* subsp. *maroccana*, annual precipitations (Bio12), maximum temperature of the warmest month (Bio5) and isothermality (Bio3) contributed with 36.7%, 27.1% and 20.6% respectively to the suitability distribution model (Table 1; Fig. 1b). The jackknife test confirmed these results, and showed the importance of the maximum temperature of the warmest month (Bio5), annual precipitations (Bio12) and mean diurnal range (Bio2) when used separately from other climatic variables (Fig. 1b). Mean diurnal range (Bio2) also had a substantial influence on the habitat

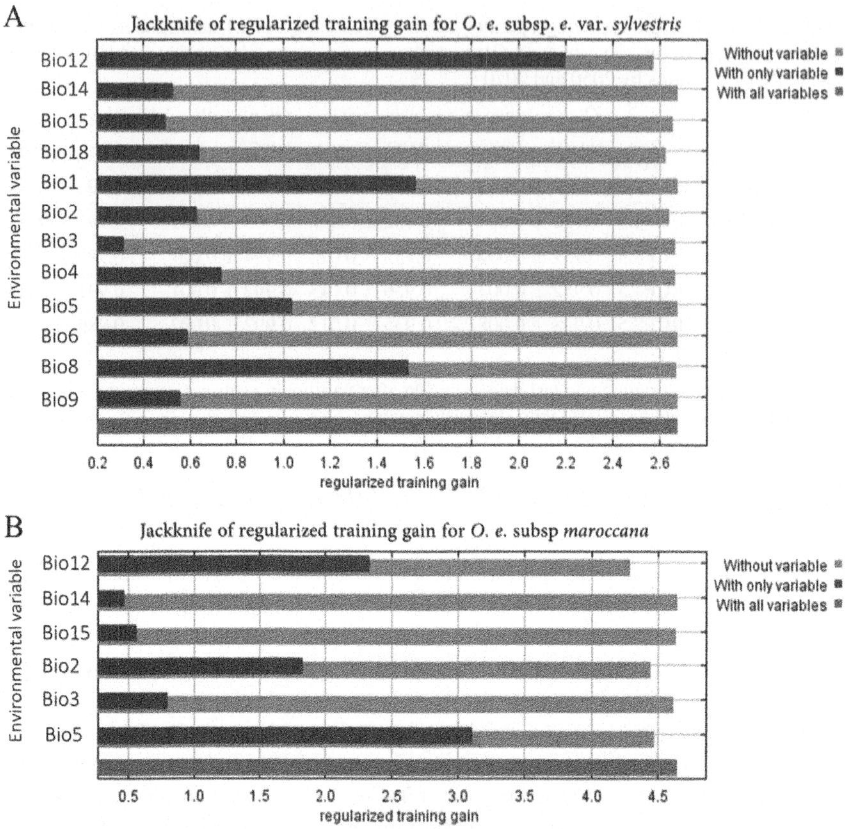

Fig. 1 The Jackknife test for evaluating the relative importance of environmental variables for *Olea e.* subsp. *e.* var. *sylvestris* (**a**) and *Olea e.* subsp. *maroccana* (**b**)

model, which was 14.7%. Precipitations of the driest period (Bio14) and precipitation seasonality (Bio15) contributed only with 0.4% and 0.2%, respectively.

3.4 Predicted Habitat Suitability for the Olea europaea *Subspecies in Morocco*

The ecological niche model shows that suitable habitat produced by the Maxent models of current climate conditions was 144 889 km^2 for *Olea e.* subsp. *e.* var. *sylvestris* and 6 623 km^2 for *Olea e.* subsp. *maroccana* (Table 3; Fig. 2).

Habitat suitability modelling of *Olea europaea* subspecies in our study area shows significant changes under global climate models (GCMs) of CCSM4 considering the

Table 3 Suitable area in km^2 and percentage of current occurrences within suitable area of *Olea e.* subsp. *e.* var. *sylvestris* and *Olea e.* subsp. *maroccana* under current and future climate change RCPs scenarios by 2050 and 2070

Scenario		*Olea e.* subsp. *e.* var. *sylvestris*		*Olea e.* subsp. *maroccana*	
		Area (Km^2)	Percentage	Area (Km^2)	Percentage
Current					
1970–2000		144,889	21.23	6623	0.97
Future					
2050	RCP4.5	148,341	21.74	3678	0.54
	RCP8.5	152,815	22.39	4072	0.60
2070	RCP4.5	146,047	21.40	3860	0.57
	RCP8.5	166,029	24.33	5213	0.76

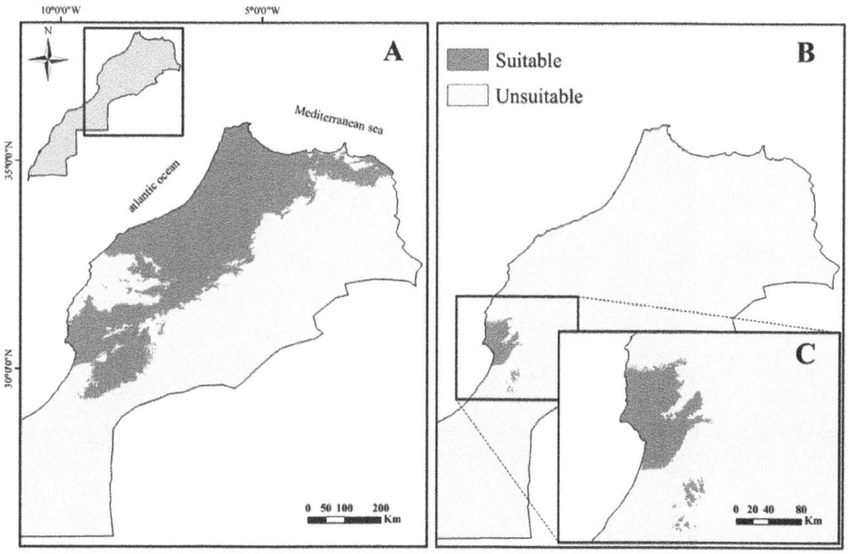

Fig. 2 Habitat suitability for *Olea e.* subsp. *e.* var. *sylvestris* (**a**) and *Olea e.* subsp. *maroccana* (**b**, **c**) in Morocco based on current bioclimatic variables

two pathways RCP4.5 and RCP8.5 by 2050 and 2070, compared to the current distribution (Table 3; Figs. 3a and 4). Overall, the variation between current and future models of *Olea e.* subsp. *e.* var. *sylvestris* distribution shows an increase in suitable area under all RCPs by 2050 and 2070 (Figs. 3a and 4). However, RCP8.5 models shows higher suitable area by 2050 and 2070, respectively with 152 815 km^2 (22.39%) and 166 029 km^2 (24.33%), compared to RCP4.5 with 148 341 km^2 (21.74%) and 146 047 km^2 (21.40%) (Table 3). Meanwhile, the future suitable area of *Olea e.* subsp. *maroccana* shows a decrease under all RCPs by 2050 and 2070 (Figs. 3b and 5). However, RCP4.5 models shows high decrease rate compared to

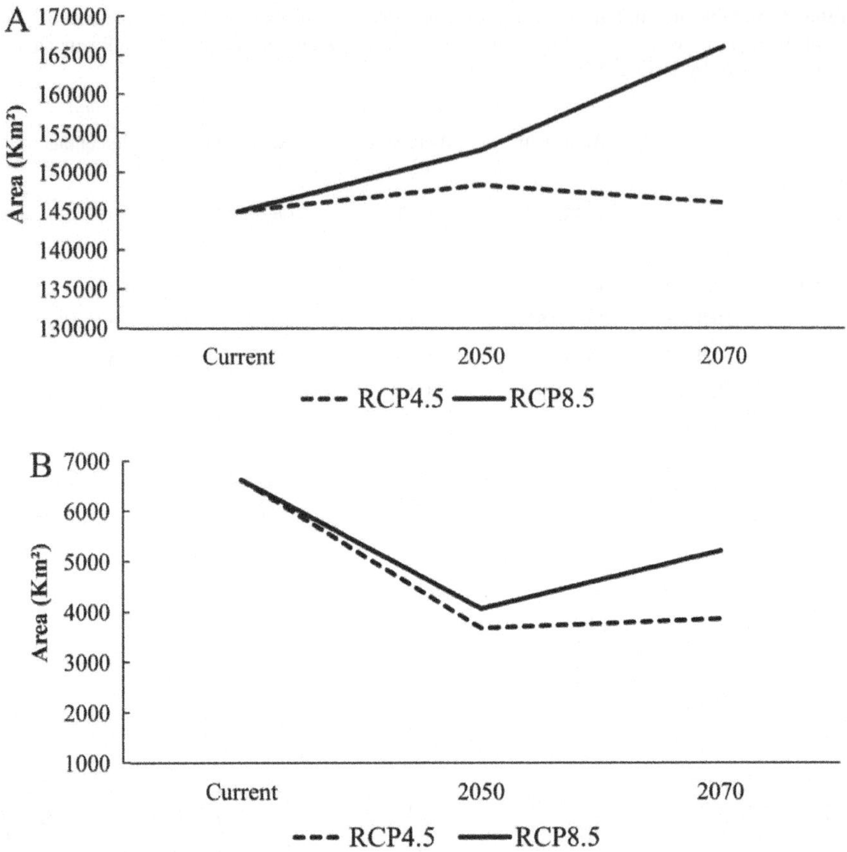

Fig. 3 Comparison of suitability area for *Olea e.* subsp. *e.* var. *sylvestris* (**a**) and *Olea e.* subsp. *maroccana* (**b**) for both RCP scenarios. Suitable area in km^2

RCP8.4 by 2050 and 2070 for this endemic subspecies (Table 3; Figs. 3b and 5). Regarding jackknife analysis, the annual precipitations (Bio12) represent the most bioclimatic variable contributing to future models of *Olea e.* subsp *e.* var. *sylvestris* by 2050 and 2070 under both RCPs models (Fig. 6). For *Olea e.* subsp. *maroccana*, the two most contributing variables are the maximum temperature of the warmest month (Bio5) and the annual precipitations (Bio12). The isothermality (Bio3) and the mean diurnal range (Bio2) have a substantial contribution to RCPs models (Fig. 7).

4 Discussion

In the last decades, many studies appointed an unprecedented rate of biodiversity loss in many ecosystems around the globe, including the Mediterranean region, as

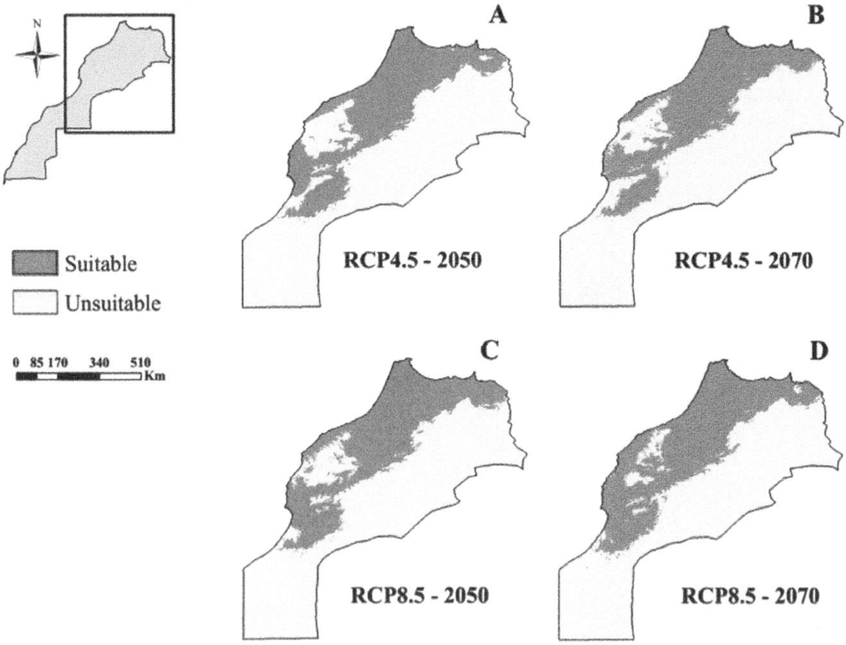

Fig. 4 Projected suitable distribution range for *Olea e.* subsp. *e.* var. *sylvestris* in RCP4.5 by 2050 (**a**) and 2070 (**b**) and RCP8.5 by 2050 (**c**) and 2070 (**d**)

a result of climate change, vegetation and land degradation, factors from several complex human-induced changes (Sala et al 2000; Thomas et al. 2004; Chapin et al. 2011; Dawson et al. 2011). In fact, the substantial influence of climate change on the Mediterranean ecosystems, considered as one of the largest biodiversity hotspot, may reduce the geographical distribution range of many plant species, negatively affects their long-term resilience, and thus become endangered or even extinct. Consequently, ecologists are continuously looking to understand and identify suitable areas under which plant species may flourish, as a way to understand their dynamic and biogeographical history/distribution in relation to environmental factors (Dawson et al. 2011; Brotons 2014). In this study, we highlighted the current and future habitat suitability of the emblematic *Olea e.* subsp *e.* var. *sylvestris* and the endemic *Olea e.* subsp. *maroccana* in relation to climatic factors. Thus, we explored the impacts of climate change on the geographical range of these two highly important wild olive subspecies.

The results obtained in our study identify the current as well as the future *Olea europaea* subspecies habitat suitability distribution in Morocco within 1 km^2 spatial resolution. The average AUC values of 10 bootstrap runs retrieved from the used Maxent modelling procedure, for current and future predictions of both subspecies, were greater than 0.9 indicating the high accuracy and reliability of the final models (Anderson et al. 2003; Elith et al. 2006; Kurpis et al. 2019). Hence, as in this study, the

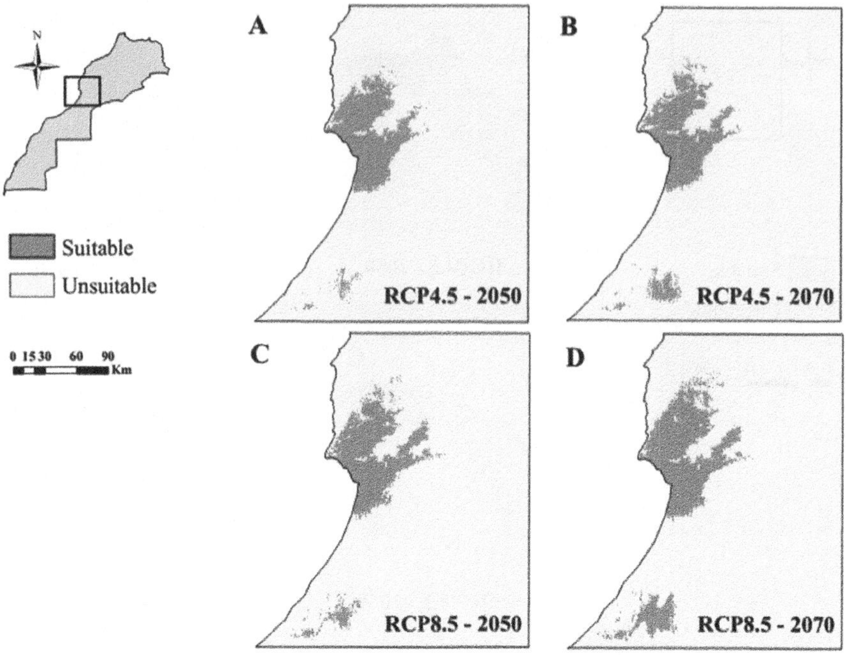

Fig. 5 Projected suitable distribution range for *Olea e.* subsp. *maroccana* in RCP4.5 by 2050 (**a**) and 2070 (**b**) and RCP8.5 by 2050 (**c**) and 2070 (**d**)

maximum entropy approach have shown high accuracy to predict plant species potential habitat (Casalegno et al. 2010; Yang et al. 2013; Guisan et al. 2017; Kurpis et al. 2019). Moreover, habitat suitability results for both wild olive subspecies fits with our field observations and the known distribution reported in literature (Emberger 1971; Barbéro et al. 1981; Benabid 1984; Benabid and Fennane 1994; Médail et al. 2001; Ali et al. 2017; Ater et al. 2016; Kassout et al. 2016, 2019; Kassout 2019). This suggest that the modelling results of climatic suitable area were highly consistent with the potentially occupied area of wild olive subspecies in Morocco. The accuracy of the obtained models is consistent with previous finding suggesting that climatic factors play key role in the establishment of ecological niche, and thus the distribution of wild olive subspecies (Ionesco and Sauvage 1969; Terral et al. 2004; Moles et al. 2014; Kassout et al. 2019). In fact, wild olive subspecies (*Olea europaea* L.) were found under different climates ranged from arid to humid within various naturel conditions in Africa, Asia and Europe (Green 2002; Besnard et al. 2017). This capacity provide wild olive trees with an important ability to gain more suitable habitat even under challenging environmental conditions, which may explained by their high genetic diversity and ecological plasticity (Terral et al. 2004; Besnard et al. 2017; Kassa et al. 2019; Kassout et al. 2019).

Our models of the current potential distribution of *Olea europaea* subspecies, based only on presence data, shows that annual precipitations (Bio12) has the largest

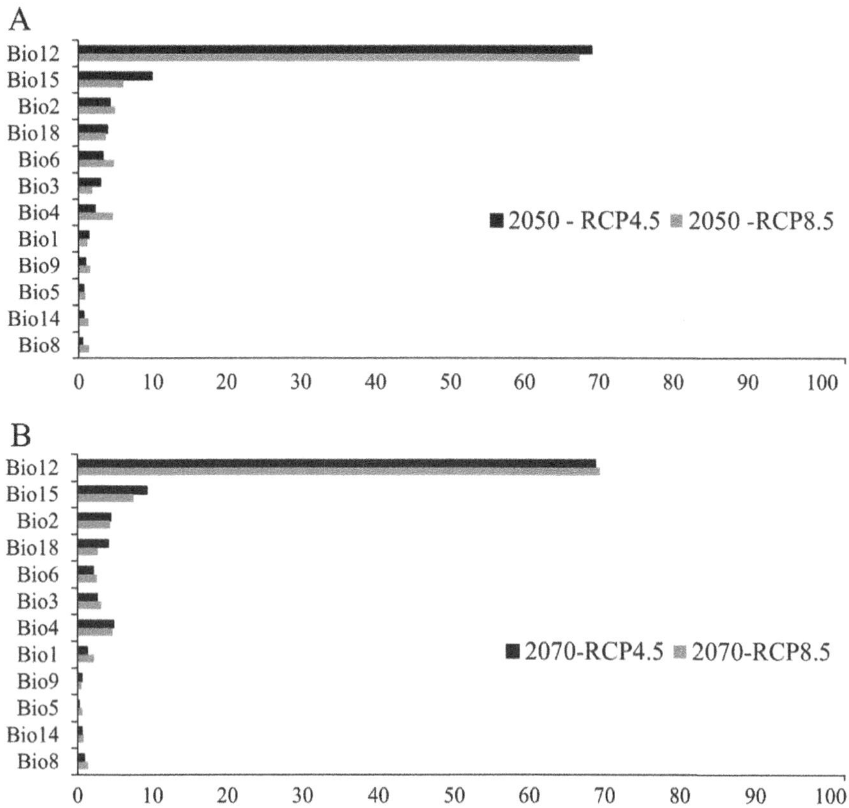

Fig. 6 Bioclimatic variables contribution to projected models for *Olea e.* subsp. *e.* var. *sylvestris* by 2050 (**a**) and 2070 (**b**) under and RCP4.5 and RCP8.5

impact on both subspecies suitable area with a contribution of 76.73% and 36.76% for *Olea e.* subsp *e.* var. *sylvestris* and *Olea e.* subsp. *maroccana*, respectively. While, suitable habitat models of the endemic *Olea e.* subsp. *maroccana* shows that the maximum temperature of the warmest month (Bio5) and isothermality (Bio3) has significant importance in the definition of current suitable area, which was only 6623 km^2 (0.97%). These results are not surprising known that precipitation and temperature are the most common factors shaping plant communities distribution and diversity (Moles et al. 2014) with direct implication in the ecological and evolutionary trajectory of plant species (Terral et al. 2004; Donoghue 2008; Frank 2011). The weak influence of temperature variables on the distribution range of *Olea e.* subsp *e.* var. *sylvestris* may result from the important adaptability of this subspecies to a broad range of temperatures, which explain the obtained suitable area of 144,889 km^2. While, the high contribution of temperature variables (Bio5 and Bio3), in addition to precipitation variables, in the distribution model of *Olea e.* subsp. *maroccana* can be explained by the continentality influence within its restricted area. Recently,

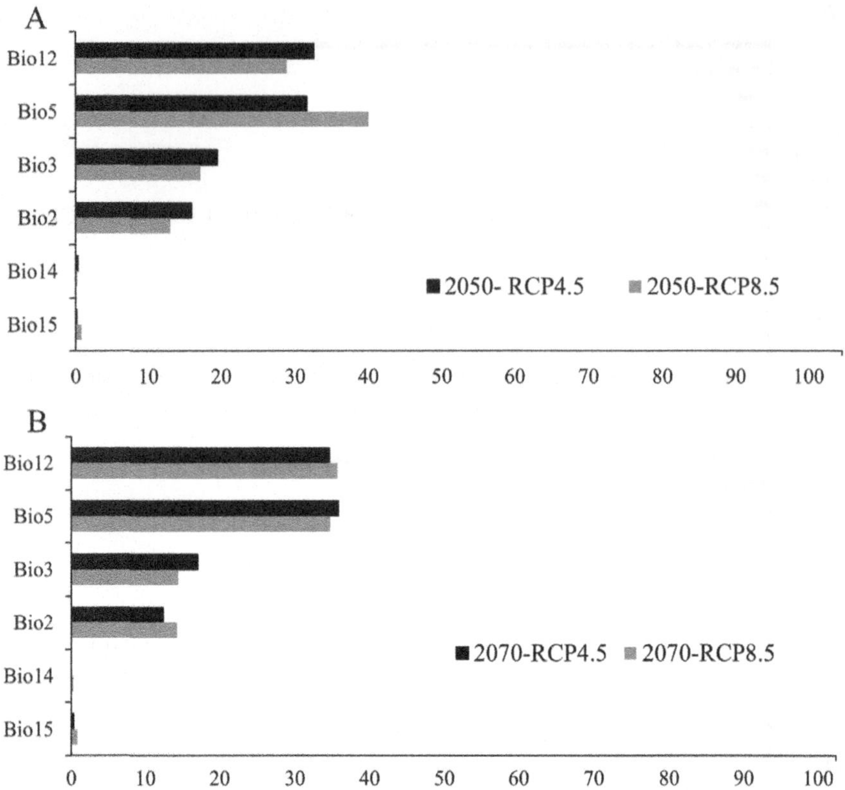

Fig. 7 Bioclimatic variables contribution to projected models for *Olea e.* subsp. *maroccana* by 2050 (**a**) and 2070 (**b**) under and RCP4.5 and RCP8.5

Kassout et al. (2019) demonstrated that annual precipitations have significant impacts on *Olea e.* subsp. *e.* var. *sylvestris* competitive ability and consequently an important capacity to gain more suitable area. Thus, Terral et al. (2004) and Kassout (2019) demonstrated that *Olea europaea* subspecies wood anatomical traits are enabled with an important ecological plasticity, which provide wild olive trees with a significant adaptation capability to stressful conditions especially to water availability within their distribution range in Morocco.

The obtained models under predicted scenarios of climate change by 2050 and 2070 shows different pattern of habitat suitability for each wild olive subspecies in Morocco. Our results showed that *Olea e.* subsp. *e.* var. *sylvestris* will expand under both RCPs scenarios by 2050 and 2070 compared to current distribution. For example, under RCP8.5, current potential suitable area will increase from 144 889 km^2 to 152,815 and 166,029 km^2 by 2050 and 2070, respectively. The predicted suitable area of *Olea e.* subsp. *e.* var. *sylvestris* under current conditions cover more that 21% of the studied area, which lies under different climate and soil types. However, most of the suitable area are transformed to cultivation lands, which limit the dispersal ability of this subspecies. Meanwhile, the future predicted suitable area represents an important proportion of land occupied by multiple vegetation types and series, which indicates that *Olea e.* subsp. *e.* var. *sylvestris* can be the most spreading plant species without considering anthropogenic factors. These results were consistent with previous finding stated that climate change scenarios might improve habitat suitability of plant species in some part of the world (Donoghue and Edwards 2014; Petitpierre et al. 2015; Liu et al. 2016). Thus, Mousavi et al. (2017) demonstrated that precipitation is the most important climatic factor determining the distribution of Olive around the Mediterranean region. At the contrary, *Olea e.* subsp. *maroccana* shows a decline in habitat suitability under climate change scenarios. In fact, suitable area of this endemic subspecies shows a decrease from 6,623 km^2 to 3678 km^2 and 3860 km^2 by 2050 and 2070, respectively, under RCP4.5. Therefore, the Maxent analysis showed that precipitations (Bio12) coupled with temperature factors (Bio5, Bio3 and Bio2) limit the future potential distribution of *Olea e.* subsp. *maroccana*. These results are consistent with some findings showed that suitable area of restricted and endangered plant species might decrease at different scales under climate change scenarios (Yi et al. 2016; Qin et al. 2017; Meng et al. 2018; López-Tirado et al. 2020; Ye et al. 2020). In fact, climate change is expected to reduce the potential distributions of many plant species because of reduced capacity to adapt under new challenging conditions, such as extreme aridity (Pereira et al. 2010; Dawson et al. 2011; Doblas-Miranda et al. 2017). The loss of future habitat suitability of *Olea e.* subsp. *maroccana* might be greatly influenced by increasing temperature and decreasing precipitations within its restricted distribution area (Driouech et al. 2010; IPCC 2014). Furthermore, extreme water scarcity and prolonged drought periods are considered as the most limiting environmental factors for plants species, such as for *Olea e.* subsp. *maroccana,* which can badly affect their potential to adapt to such conditions (Gates 1980; Michaletz et al. 2015; Kassout et al. 2019). However, the response of wild olive subspecies to climate change, as many plant species, will depend on a variety of factors such as the intensity of human activities, the phenotypic plasticity, genetic diversity and competition ability (Phillips et al. 2006; Parmesan and Hanley 2015; Fisher et al. 2018).

5 Conclusion

Our study provides a foundation for additional research using *Olea europaea* subspecies as a study model to elucidate how environmental factors are shaping the geographical distributions of plant species. The obtained models can improve our understanding of wild olive biogeography and history according to climatic conditions. In the light of the ongoing global change, the important potential suitable area of *Olea e.* subsp. *e.* var. *sylvestris* under current and future scenarios highlights the possibility of using this subspecies as forest tree in plantation efforts, aiming to improve forest system resilience. Considering the relatively restricted geographic range of *Olea e.* subsp. *maroccana*, our study underline a high risk of extinction facing this endemic subspecies, and point out the necessity to set an urgent conservation strategies to allow for preservation of its genetic and morphological diversity.

Funding This research was supported by the EcoGenOlea international program (2015–2017), the Associated International Laboratory (IRP) EVOLEA (INEE-CNRS France/CNRST Morocco), Toubkal grants (15/04, n° 32525WH) and the Moroccan Excellency research Grants (024UAE2014).

References

Aiello-Lammens ME, Boria RA, Radosavljevic A, Vilela B, Anderson RP (2015) spThin: an R package for spatial thinning of species occurrence records for use in ecological niche models. Ecography 38:541–545. https://doi.org/10.1111/ecog.01132

Ali N, Tavoillot J, Besnard G et al (2017) How anthropogenic changes may affect soil-borne parasite diversity? Plant-parasitic nematode communities associated with olive trees in Morocco as a case study. BMC Ecol 17:4. https://doi.org/10.1186/s12898-016-0113-9

Allen CD, Macalady AK, Chenchouni H, Bachelet D, McDowell N, Vennetier M, Cobb N (2010) A global overview of drought and heat-induced tree mortality reveals emerging climate change risks for forests. For Ecol Manage 259(4):660–684. https://doi.org/10.1016/J.FORECO.2009.09.001

Anderson RP, Lew D, Peterson AT (2003) Evaluating predictive models of species distributions: criteria for selecting optimal models. Ecol Model 162:211–232

Ater M, Barbara H, Kassout J (2016) Importance des variétés locales, de l'oléastre et des pratiques traditionnelles de l'oléiculture dans la région de Chefchaouen (Nord du Maroc). In: Ater M, Essalouh L, Ilbert H, Moukhli A, Khadari B (eds) L'Oléiculture au Maroc de la préhistoire à nos jours: pratiques, diversité, adaptation, usages, commerce et politiques. CIHEAM—Options Méditerranéennes A, Montpellier, vol 118, pp 109–121

Aumeeruddy-Thomas Y, Hmimsa Y, Ater M, Khadari B (2014) Beyond the divide betweeen wild and domesticated: spaciality, domesticity and practices pertaining to fig (Ficus carica L.) and olive (Olea europaea L.) agroecosystems among Jbala communities in Northern Morocco. In: Chevalier A, Marinova E, Peña-Chocarro L (eds) Plants and people. Choices and diversity through time. Oxbow Books, Oxford, pp 191–197

Barbéro M, Quézel P, Rivas-Martínez S (1981) Contribution à l'étudedes groupements forestiers et préforestiers du Maroc. Phytocoenologia 9:311–412

Belaj A, León L, Satovic Z, De La Rosa R (2011) Variability of wild olives (Olea europaea subsp. europaea var. sylvestris) analyzed by agromorphological traits and SSR markers. Sci Hortic 129:561–569

Benabid A (1984) Etude phytoécologique des peuplements forestiers et préforestiers du Rif centro-occidental (Maroc). Institut scientifique Université Mohammed V, Rabat

Benabid A (1985) Les écosystèmes forestiers préforestiers et présteppiques du Maroc: diversité, répartition biogéographique et problèmes posés par leur aménagement. Foret Mediterranéenne VII:53–64

Benabid A, Fennane M (1994) Connaissances sur la végétation du Maroc: Phytogéographie, phytosociologie et séries de végétation. Lazaroa 14:21

Besnard G, Khadari B, Navascués M et al (2013) The complex history of the olive tree: from Late Quaternary diversification of Mediterranean lineages to primary domestication in the northern Levant. Proc R Soc B 280:20122833

Besnard G, Terral J-F, Cornille A (2017) On the origins and domestication of the olive: a review and perspectives. Ann Bot 121:385–403. https://doi.org/10.1093/aob/mcx145

Blondel J, Aronson J, Bodiou JY, Boeu G (2010) The Mediterranean region: biological diversity through time and space. Oxford University Press, Oxford, p 376

Booth TH (2017) Assessing species climatic requirements beyond the realized niche: some lessons mainly from tree species distribution modelling. Clim Change 145:259–271

Boria RA, Olson LE, Goodman SM, Anderson RP (2014) Spatial filtering to reduce sampling bias can improve the performance of ecological niche models. Ecol Model 275:73–77. https://doi.org/10.1016/j.ecolmodel.2013.12.012

Bréda N, Huc R, Granier A, Dreyer E (2006) Temperate forest trees and stands under severe drought: a review of ecophysiological responses, adaptation processes and long-term consequences. Ann For Sci 63(6):625–644

Brotons L (2014) Species distribution models and impact factor growth in environmental journals: methodological fashion or the attraction of global change science. PLoSOne. https://doi.org/10.1371/journal.pone.0111996

Carrión Y, Ntinou M, Badal E (2010) Olea europaea L. in the North Mediterranean Basin during the Pleniglacial and the Early-Middle Holocene. Quat Sci Rev 29:952–968

Casalegno S, Amatulli G, Camia A, Nelson A, Pekkarinen A (2010) Vulnerability ofPinus cembra L. In the Alps and the Carpathian mountains under present and future climates. For Ecol Manage 259:750–761. https://doi.org/10.1016/j.foreco.2009.10.001

Chapin FSI, Matson PA, Vitousek PM (2011) Principles of terrestrial ecosystem ecology, 2nd edn. Springer, New York

Choat B, Jansen S, Brodribb TJ et al (2012) Global convergence in the vulnerability of forests to drought. Nature 491(7426):752–755. https://doi.org/10.1038/nature11688

Dawson TP, Jackson ST, House JI, Prentice IC, Mace GM (2011) Beyond predictions: biodiversity conservation in a changing climate. Science 332:53–58

Delong ER, Delong DM, Clarke-Pearson DL (1988) Comparing the areas under two or more correlated receiver operating characteristic curves: a nonparametric approach. Biometrics 44:837–845

Doblas-Miranda E et al (2017) A review of the combination among global change factors in forests, shrublands and pastures of the Mediterranean Region: beyond drought effects. Glob Planet Change 148:42–54

Donoghue MJ (2008) A phylogenetic perspective on the distribution of plant diversity. Proc Natl Acad Sci USA 105:11549–11555

Donoghue MJ, Edwards EJ (2014) Biome shifts and niche evolution in plants. Annu Rev Ecol Evol Syst 45:547–572

Dormann CF, Elith J, Bacher S, Buchmann C, Carl G, Carré G, Marquéz JRG, Gruber B et al (2013) Collinearity: a review of methods to deal with it and a simulation study evaluating their performance. Ecography (Cop.) 36:027–046. https://doi.org/10.1111/j.1600-0587.2012.07348.x

Driouech F, Déqué M, Sanchez-Gomez E (2010) Weather regimes-Moroccan precipitation link in a regional climate change simulation. Global Planet Change 72:1–10

Dullinger S et al (2012) Extinction debt of high-mountain plants under twenty-frst-century climate change. Nat Clim Chang 2:619–622

Elith J et al (2006) Novel methods improve prediction of species' distributions from occurrence data. Ecography 129–151

Elith J, Burgman M (2002) Predictions and their validation: rare plants in the Central Highlands, Victoria, Australia. In: Predicting species occurrences: issues of accuracy and scale. Island Press, Washington, USA, pp 303–313

Elith J, Leathwick JR (2009) Species distribution models: ecological explanation and prediction across space and time. Annu Rev Ecol Systemat 40:677–697. https://doi.org/10.1146/annurev.ecolsys.110308.120159

Elith J, Kearney M, Phillips S (2010) The art of modelling range-shifting species. Methods Ecol Evol 1:330–342

Elith J, Phillips SJ, Hastie T, Dudík M, Chee YE, Yates CJ (2011) A statistical explanation of MaxEnt for ecologists. Divers Distrib 17:43–57. https://doi.org/10.1111/j.1472-4642.2010.00725.x

Emberger L (1971) Carte phytogéographique du Maroc. Masson et Cie

Escalante T, Rodríguez-Tapia G, Linaje M, Illoldi-Rangel P, González-López R (2013) Identification of areas of endemism from species distribution models: threshold selection and Nearctic mammals. TIP Revista Especializada en Ciencias Químico-Biológicas 16:5–17
Fick S, Hijmans R (2017) WorldClim 2: new 1-km spatial resolution climate surfaces for global land areas. Int J Climatol 37:4302–4315
Fisher RA, Koven CD, Anderegg WRL, Christoffersen BO, Dietze MC, Farrior CE et al (2018) Vegetation demographics in earth system models: a review of progress and priorities. Glob Chang Biol 24:35–54
Forzieri G, Miralles DG, Ciais P et al (2020) Increased control of vegetation on global terrestrial energy fluxes. Nat Clim Chang 10:356–362. https://doi.org/10.1038/s41558-020-0717-0
Frank SA (2011) Natural selection. II. Developmental variability and evolutionary rate. J Evol Biol 24:2310–2320
Franklin J (2009) Mapping species distributions: spatial inference and prediction. Cambridge University Press, Cambridge, UK
Garcia-Verdugo C, Fay MF, Granado-Leal C, De Casas RR, Balaguer L, Besnard G, Vargas P (2009) Genetic diversity and differentiation processes in the ploidy series of Olea europaea L.: a multiscale approach from subspecies to insular populations. Mol Ecol 18(3):454–467
Gates DM (1980) Biophysical ecology. Springer-Verlag, Berlin, p 611
GBIF. Global biodiversity information facility. Available at: http://www.gbif.org/. Accessed 13 May 2020
Gent PR et al (2011) The community climate system model version 4. J Climate 24:4973–4991. https://doi.org/10.1175/2011JCLI4083.1
Gianguzzi L, Bazan G (2019) The Olea europaea L. var. sylvestris(Mill.) Lehr. forests in the Mediterranean area. Plant Sociol 56:3–34. https://doi.org/10.7338/pls2019562/01
Gibelin A-L, Déqué M (2003) Anthropogenic climate change over the Mediterranean region simulated by a global variable resolution model. Clim Dyn 20(4):327–339. https://doi.org/10.1007/s00382-002-0277-1
Graham MH (2003) Confronting multi-collinearity in ecological multiple regression. Ecology 84:2809–2815
Green PS (2002) A revision of Olea (Oleaceae). Kew Bull 57:91–140
Green PS, Wickens GE (1989) The Olea europaea complex. In: The Davis and Hedge Festschrift (ed) Kit an. Edinburgh University press, pp 287–299
Guisan A, Tuiller W (2005) Predicting species distribution: offering more than simple habitat models. Ecol Lett 8:993–1009
Guisan A, Zimmermann NE (2000) Predictive habitat distribution models in ecology. Ecol Modell 135:147–186
Guisan A et al (2013) Predicting species distributions for conservation decisions. Ecol Lett 16:1424–1435
Guisan A, Thuiller W, Zimmermann NE (2017) Habitat suitability and distribution models. With applications in R. Cambridge University Press, Cambridge
Harte J (2011) Maximum entropy and ecology: a theory of abundance, distribution, and energetics. Oxford University Press, London
Harte J et al (2008) Maximum entropy and the state variable approach to macroecology. Ecology 89:2700–2711
Hijmans RJ, Cameron SE, Parra JL, Jones PG, Jarvis A (2005) Very high resolution interpolated climate surfaces for global land areas. Int J Climatol 25:1965–1978
Hmimsa Y, Ater M (2008) Agrodiversity in the traditional agrosystems of the Rif mountains (North of Morocco). Biodiversity 9(1–2):78–81
Ionesco T, Sauvage C (1969) Fichier des espèces climax. Al Awamia, pp 105–124
IPCC (2014) In: Core Writing Team, Pachauri RK, Meyer LA (eds) Climate change 2014: synthesis report. Contribution of working groups I, II and III to the fifth assessment report of the Intergovernmental Panel on Climate Change. IPCC, Geneva, Switzerland, p 151

Kaniewski D, Van Campo E, Boiy T, Terral JF, Khadari B, Besnard G (2012) Primary domestication and early uses of the emblematic olive tree: palaeobotanical, historical and molecular evidences from the Middle East. Biol Rev 87:885–899

Kassa A, Konrad H, Geburek T (2019) Molecular diversity and gene flow within and among different subspecies of the wild olive (Olea europaea L.): a review. Flora 250:18–26

Kassout J (2019) Réponse de traits fonctionnels foliaires et anatomiques chez l'olivier sauvage (Olea europaea L.) dans un environnement hétérogène et changeant: le modèle du Maroc soumis aux changements globaux. Biologie végétale. Université de Montpellier (France), Université Abdelmalek Essaâdi (Maroc) Thèse de doctorat

Kassout J, Barbara H, Ivorra S, Terral J-F, Ater M (2016) Etude préliminaire de la variation de caractères anatomiques du bois d'une forme spontanée et de sept variétés traditionnelles d'olivier (Olea europaea L.) de la région Nord du Maroc (Chefchaouen et Ouazzane). In: Ater M, Essalouh L, Ilbert H, Moukhli A, Khadari B (eds) L'Oléiculture au Maroc de la préhistoire à nos jours: pratiques, diversité, adaptation, usages, commerce et politiques. CIHEAM—Options Méditerranéennes A, Montpellier, vol 118, pp 109–121

Kassout J, Terral J-F, Hodgson J, Ater M (2019) Trait-based plant ecology a flawed tool in climate studies? The leaf traits of wild olive that pattern with climate are not those routinely measured. PLoS ONE 14(7): https://doi.org/10.1371/journal.pone.0219908

Kurpis J, Serrato-Cruz MA, Arroyo TPF (2019) Modeling the effects of climate change on the distribution of Tagetes lucida Cav. (Asteraceae). Glob Ecol Conserv 20:11

Kwon H, Kim B, Jang G (2016) Modelling the spatial distribution of wildlife animals using presence and absence data. Contemp Probl Ecol 9:515–518

Lenoir J, Gégout JC, Marquet PA, De RP, Brisse H (2008) A significant upward shift in plant species optimum elevation during the 20th century. Science 320:1768–1771

Lenoir J, Gégout J-C, Guisan A et al (2010) Going against the flow: potential mechanisms for unexpected downslope range shifts in a warming climate. Ecography 33(2):295–303. https://doi.org/10.1111/j.1600-0587.2010.06279.x

Li GQ, Du S, Wen ZM (2016) Mapping the climatic suitable habitat of oriental arborvitae (Platycladus orientalis) for introduction and cultivation at a global scale. Sci Rep, UK, p 6

Liu C, Berry P, Dawson T, Pearson R (2005) Selecting thresholds of occurrence in the prediction of species distributions. Ecography 28:385–393

Liu C, White M, Newell G (2013) Selecting thresholds for the prediction of species occurrence with presence-only data. J Biogeogr 40:778–789. https://doi.org/10.1111/jbi.12058

Liu N, Feng Y, Guan KY (2016) Potential distribution prediction of Calligonum L. in China. J Arid Land Resour Environ. 30(3):112–120. https://doi.org/10.13448/j.cnki.jalre.2016.089

Lionello P, Scarascia L (2018) The relation between climate change in the Mediterranean region and global warming. Reg Environ Chang 18:1481–1493. https://doi.org/10.1007/s10113-018-1290-1

Lloret F, de la Riva EG, Pérez-Ramos IM, Marañón T, Saura-Mas S, Díaz-Delgado R, Villar R (2016) Climatic events inducing die-off in Mediterranean shrublands: are species' responses related to their functional traits? Oecologia 180(4):961–973. https://doi.org/10.1007/s00442-016-3550-4

López-Tirado J, Vessella F, Stephan J et al (2020) Effect of climate change on potential distribution of Cedrus libani A. Rich in the twenty-first century: an Ecological Niche Modeling assessment. New Forests. https://doi.org/10.1007/s11056-020-09798-y

Madani N, Kimball JS, Ballantyne AP, Affleck DLR, van Bodegom PM, Reich PB et al (2018) Future global productivity will be affected by plant trait response to climate. Sci Rep 8:2870

Matesanz S, Valladares F (2014) Ecological and evolutionary responses of Mediterranean plants to global change. Environ Exp Bot 103:53–67. https://doi.org/10.1016/J.ENVEXPBOT.2013.09.004

Médail F, Quézel P (1997) Hot-spots analysis for conservation of plant biodiversity in the Mediterranean Basin. Ann Mo Bot Gard 84:112–127

Médail F, Quézel P, Besnard G, Khadari B (2001) Systematics, ecology and phylogeographic significance of Olea europaea L. subsp. maroccana (Greuter & Burdet) P. Vargas et al. a relictual olive tree from South West Morocco. Bot J Linn Soc 137:249–266

Meng YH, Xu X, Jing XL, Xu GB (2018) Potential distribution modeling and analysis of Disanthus Maxim. Actae Ecol Sin 39:2816–2825
Merow C, Smith MJ, Silander JA (2013) A practical guide to MaxEnt for modeling species' distributions: what it does, and why inputs and settings matter. Ecography
Michaletz ST, Weiser MD, Zhou J, Kaspari M, Helliker BR, Enquist BJ (2015) Plant thermoregulation: energetics, trait-environment interactions, and carbon economics. Trends Ecol Evol 30:714–724. https://doi.org/10.1016/j.tree.2015.09.006
Moles AT, Perkins SE, Laffan SW et al (2014) Which is a better predictor of plant traits: temperature or precipitation? J Veg Sci 25:1167–1180
Moukrim S, Lahssini S, Rhazi M et al (2019) Climate change impacts on potential distribution of multipurpose agro-forestry species: Argania spinosa (L.) Skeels as case study. Agroforest Syst 93:1209–1219. https://doi.org/10.1007/s10457-018-0232-8
Mousavi S, Mariotti R, Bagnoli F, Costantini L, Cultrera NG, Arzani K, Pandolfi S, Vendramin GG et al (2017) The eastern part of the Fertile Crescent concealed an unexpected route of olive (Olea europaea L.) differentiation. Ann Bot 119:1305–1318
Myers N, Mittermeier RA, Mittermeier CG, da Fonseca GAB, Kent J (2000) Biodiversity hotspots for conservation priorities. Nature 403(6772):853–858. https://doi.org/10.1038/35002501
Palamarev E (1989) Palaeobotanical evidences ofthe Tertiary history and origin of the Mediterranean schlerophyll dendroflora. Plant Syst Evol 162:93107
Parmesan C, Hanley ME (2015) Plants and climate change: complexities and surprises. Ann Bot 116:849–864. https://doi.org/10.1093/aob/mcv169
Parmesan C, Yohe G (2003) A globally coherent fingerprint of climate change impacts across natural systems. Nature 421(6918):37–42. https://doi.org/10.1038/nature01286
Pearce J, Ferrier S (2000) Evaluating the predictive performance of habitat models developed using logistic regression. Ecol Model 133:225–245
Pecchi M, Marchi M, Burton V, Giannetti F, Moriondo M, Bernetti I, Bindi Mb Chirici G (2019) Species distribution modelling to support forest management. A Lit Rev 2(411):1–12
Pereira HM, Leadley PW, Proença V, Alkemade R, Scharlemann JPW et al (2010) Scenarios for global biodiversity in the 21st century. Science 330:1496–1501
Peterson AT et al (2011) Ecological niches and geographic distributions, monographsin population biology No 49. Princeton University Press, Princeton, NJ, USA
Petitpierre B, MacDougall K, Seipel T, Broennimann O, Guisan A, Kueffer C (2015) Will climate change increase the risk of plant invasions into mountains? Ecol Appl 26:530–544
Phillips SJ, Dudik M (2008) Modeling of species distributions with Maxent: new extensions and a comprehensive evaluation. Ecography 31:161–175
Phillips SJ, Dudík M, Schapire RE (2004) A maximum entropy approach to species distribution modeling. In: Proceedings of the 21st international conference on machine learning. Association for Computing Machinery, Banff, Canada, p 83
Phillips SJ, Anderson RP, Schapire RE (2006) Maximum entropy modeling of species geographic distributions. Ecol Model 190:231–259. https://doi.org/10.1016/j.ecolmodel.2005.03.026
Phillips SJ, Anderson RP, Dudík M, Schapire RE, Blair ME (2017a) Opening the black box: an open-source release of Maxent. Ecography 40(7):887–893
Phillips SJ, Dudík M, Schapire RE (2017b) Maxent sofware for modeling species niches and distributions (Version3.4.1). Available at: www.biodiversityinformatics.amnh.org/open_source/maxent/. Accessed: 8 May 2020
Polade SD, Pierce DW, Cayan DR, Gershunov A, Dettinger MD (2014) The key role of dry days in changing regional climate and precipitation regimes. Sci Rep 4(1):4364. https://doi.org/10.1038/srep04364
QGIS Development Team (2012) QGIS geographic information system. Open Source Geospatial Foundation Project. Available at: http://qgis.org
Qin AL, Liu B, Guo QS, Bussmann RW, Ma F, Jian ZJ, Xu GX, Pei SX (2017) Maxent modeling for predicting impacts of climate change on the potential distribution of Thuja sutchuenensis Franch. An extremely endangered conifer from southwestern China. Glob Ecol Conserv 10:139e146

Rubio De Casas R, Balaguer L, Manrique E, Pérez ME, Vargas P (2002) On the historical presence of the wild olives Olea europaea L. var. sylvestris (Miller) Leh. In: the Eurosiberian North of the Iberian Peninsula. An del Jardín Botánico, Madrid

Sala OE, Chapin FSI, Armesto JJ, Berlow E, Bloomfield J et al (2000) Global biodiversity scenarios for the year 2100. Science 287:1770–1774

Schröder B (2008) Challenges of species distribution modeling belowground. J Plant Nutr Soil Sci 171:325–337. https://doi.org/10.1002/jpln.200700027

Shipley B, Laughlin DC, Sonnier G, Otfinowski R (2011) A strong test of a maximum entropy model of trait- based community assembly. Ecology 92:507–517

Sinclair S, White M, Newell G (2010) How useful are species distribution models for managing biodiversity under future climates. Ecol Soc 15:8

Sonnier G, Shipley B, Navas ML (2010) Plant traits, species pools and the prediction of relative abundance in plant communities: a maximum entropy approach. J Vegetable Sci 21:318–331

R Core Team (2018) R: A language and environment for statistical computing. R Foundation for Statistical Computing, Vienna, Austria. Available at: https://www.R-project.org/

Taleb MS, and Fennane M (2019) Vascular plant communities of Morocco, Phytoecology, Ecology and Geography. Springer International Publishing, p 161

Terral J-F, Badal E, Heinz C, Roiron P, Thiébault S, Figueiral I (2004) A hydraulic conductivity model points to post-Neogene survival of the Mediterranean Olive in riparian habitat. Ecology 85(11):3158–3165. https://doi.org/10.1890/03-3081

Thomas CD, Cameron A, Green RE, Bakkenes M, Beaumont LJ, Collingham YC et al (2004) Extinction risk from climate change. Nature 427:145–148

Thuiller W, Lavorel S, Araujo MB (2005) Niche properties and geographical extent as predictors of species sensitivity to climate change. Glob Ecol Biogeogr 14(4):347–357. https://doi.org/10.1111/j.1466-822X.2005.00162.x

UNEP (1997) World Atlas of desertification. Middleton L, Thomas D (eds), 2nd edn. Arnold, London, 182 p

van Bodegom PM, Douma JC, Verheijen LM (2014) A fully traitsbased approach to modeling global vegetation distribution. Proc Natl Acad Sci USA 111:13733–13738. https://doi.org/10.1073/pnas.1304551110

Vellend M, Baeten L, Becker-Scarpitta A, Boucher Lalonde V, McCune J, Messier J et al (2017) Plant biodiversity change across scales during the Anthropocene. Annu Rev Plant Biol 68:563–586

White EP et al (2012) Characterizing species abundance distributions across taxa and ecosystems using a simple maximum entropy model. Ecology 93:1772–1778

Yang XQ, Kushwaha SPS, Saran S, Xu J, Roy PS (2013) Maxent modeling for predicting the potential distribution of medicinal plant, Justicia adhatoda L. in Lesser Himalayan foothills. Ecol Eng. https://doi.org/10.1016/j.ecoleng.2012.12.004

Ye XZ, Zhao GH, Zhang MZ, Cui XY, Fan HH, Liu B (2020) Distribution pattern of endangered plant Semiliquidambar cathayensis (Hamamelidaceae) in response to climate change after the last interglacial period. Forests 11:434. https://doi.org/10.3390/f11040434

Yi YJ, Cheng X, Yang ZF, Zhang SH (2016) Maxent modeling for predicting the potential distribution of endangered medicinal plant (H. riparia Lour) in Yunnan, China. Ecol Eng 260–269

Zohary D, Hopf M, Weiss E (2012) Domestication of plants in the old world: the origin and spread of cultivated plants in Southwest Asia, Europe, and the Mediterranean Basin. Oxford University Press, Oxford

The Frequency of Rare Cyclones in the Eastern Mediterranean and Northeastern Africa as a Sign of Climate Change Using Satellite Imagery, Climate Data Models and GIS-Based Analysis

Khaled Mohamed Madkour

Abstract Eastern Mediterranean and Northeastern Africa (EMNEA) are exposed to atmospheric depressions that affecting their climatic conditions during the period between October and May of each year "wet season", and the passage of cyclones is a rare event. In the wet season 2019/2020, Egypt which located in the EMNEA region, exposed to two cyclones that passed from the south to the north in a rare event as well, the first was in October 2019 and called "Medicane" and the second was in March 2020 and called "Dragon". The study deals with the frequency of the rare cyclones in Egypt during the wet season 2019/2020 as a sign of climate change in the EMNEA region using MODIS/AQUA satellite imagery, climate satellite imagery by the PERSIANN project and CHIRPS model, outputs of the ECMWF climate model as well as outputs of the Global Climate Forecast System (GFS). The study concluded that these cyclones were a rare weather condition represented in a heavy rain volume was about 543 million m^3 and gusty winds 65 km/h for the "Medicane" cyclone and about 950 million m^3 and a cyclonic wind speed of 120 km/h for the "Dragon" cyclone. In addition, these cyclones were associated with a set of factors that coincided in a rare manner and contributed to the engender and development Which can be considered as a Sing of climate change. These factors were represented in the northern oscillation movement of the Sudan depression and the Red Sea Trough (RST) accompanied by a deepening of a cold depression in the upper atmosphere of 500 hpa, and a simultaneous rise in the Sea Surface Temperature (SST) of the eastern Mediterranean region in the Medicane cyclone, and the convergence of the polar and tropical jet stream in the upper atmosphere above the EMNEA region. In addition to the correlation of these factors with the major pressure oscillations in the world and the occurrence of the El-Nino.

K. M. Madkour (✉)
Geography and GIS Department, Faculty of Arts, Ain Shams University, Cairo, Egypt
e-mail: Khaled.madkour@art.asu.edu.eg

W. Leal Filho and E. Manolas (eds.), *Climate Change in the Mediterranean and Middle Eastern Region*, Climate Change Management,
https://doi.org/10.1007/978-3-030-78566-6_3

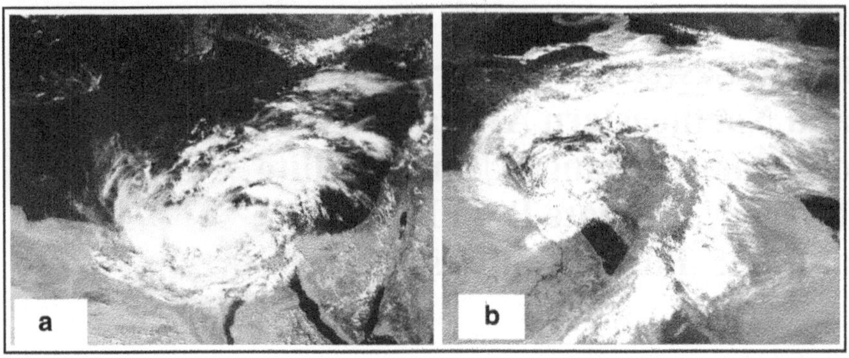

Fig. 1 The rare cyclones over Egypt in the wet season 2019/2020, **a** medicane Oct. 2019, **b** dragon Mar. 2020. *Source* MODIS/AQUA imagery

1 Introduction

Eastern Mediterranean and Northeastern Africa (EMNEA) are exposed to atmospheric depressions that affecting their climatic conditions, mainly it tracked western to eastern routes which brings from northern of Atlantic and north southern routes bring from Polar regions, in addition to atmospheric depressions which generated in Cyprus island and Aegean Sea (Flocas et al. 2010). Egypt located in the EMNEA and its north areas above 30 N and north coast region affected by many atmospheric depressions move from the West to East for the period from October to May every year "wet season"[1] and results in rains with about 120 mm annually and sometimes the hail (FAO 2016). These depressions vary in their sources, strength, movement tracks, and the extent of their impact on the Egyptian climatic conditions. At the same time, cyclones are very rare events in Egypt and the EMNEA region.

Though Egypt was exposed to two cyclones "Medicane[2]" on 22–26 October 2019 and "Dragon" on 11–14 March 2020, which classified at the first degree of Befort scale of Cyclones, it started as strong depressions and developed into a cyclone. Figure 1 clearly shows Medicane and Dragon cyclones were characterized by severe instability in the accompanying weather conditions, in addition to heavy rains, which reached hail in most parts of the EMNEA region. Moreover, these cyclones followed unusual tracks except in very rare cases as seen from Fig. 2.

Generally, Egypt exposed to the atmospheric depressions activity from mid-October every year which affects the weather conditions in Egypt and characterized

[1] The wet season is the period extended from October to May which characterized by atmospheric depression activity and is associated with many climatic conditions including rains (Xoplaki et al. 2004).

[2] Medicane consists of two parts (Mediterranean - Hurricane) and it is a cyclone in the first category according to NASA classification and Greek meteorological authority, and it is one of the rare weather phenomena that occur in the Mediterranean Sea, and it is often referred to as the Mediterranean tropical cyclone or the Mediterranean hurricane.

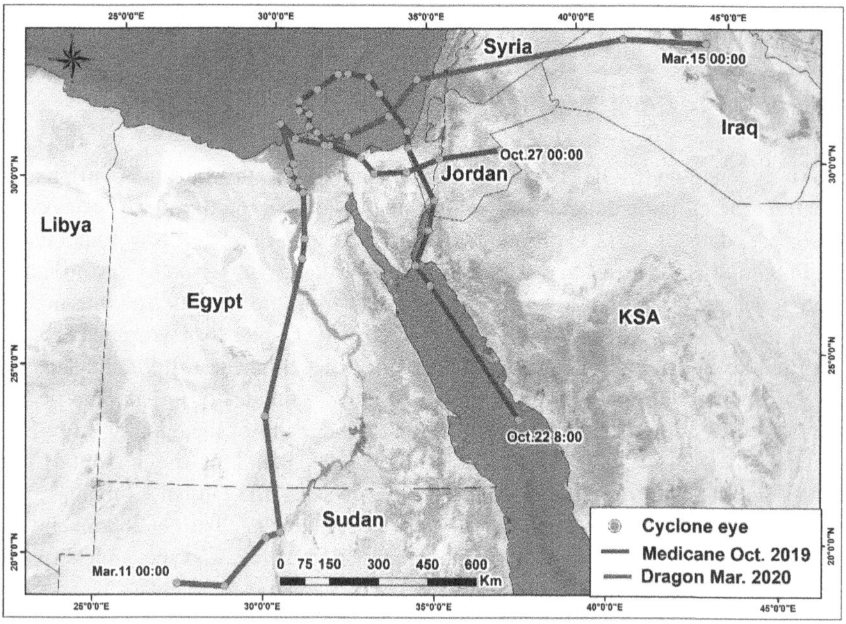

Fig. 2 The rare cyclones track over Egypt in the wet season 2019/2020

with the weakness during that period despite their increased frequency, and those depressions continue in their activity until the end of May where about 90.1% of depressions occurred in this period (Zaghloul 1977). Spring is the highest season in the number of depressions with about 36.4% of the total number of atmospheric depressions per year, followed by winter and autumn with about 35.8 and 22% consequently (Al-Fiqi 1999). Furthermore, the spatial distribution of the sources of depressions are about 53% of the annual number of depressions coming from the northwestern of the Mediterranean, and about 38% coming from the Sahara Desert (El-Asrag and Hassan 1997), and about 9% of it generated in the Central and Eastern Mediterranean which are mostly as secondary depressions that originated in other regions and are renewed in the Eastern Mediterranean.

The main objective of this study is to monitor the extent of the change in the strength, intensity, and tracks of the cyclones during the study period and the impact of climate change on it via the impact of General Air Circulations (GAC) and Pressure Oscillations (PO) on the movement, strength and the frequency of the rare cyclones in Egypt. The study assumed that the current climate change has a direct and indirect impact in creating those exceptional weather conditions associated with cyclones under study and their correlation with the change in the PO. The study aims to determine the characteristics of cyclones during the wet season 2019/2020 "Medicane" and "Dragon" in addition to determining the change in the strength, intensity, and tracks of cyclones under study about the general behavior of the usual depressions during their passage over Egypt in the wet season, and the factors affecting on it.

2 Data and Methods

2.1 Satellite Images

The study relied on (MODIS/AQUA Clouds 5-min L2/swath 1 km and 5 km) images to extract the climatic data of the cyclones understudy at the timing of their occurrence, e.g. data of surface pressure levels (SPL) and Land Surface Temperature (LST), Cloud Base Level and Cloud Top Height data, with a spatial resolution of 1 km × 1 km. In addition to extracting rain data from the climatic satellite images by the PERSIANN project, which was created by the Center for Hydrometeorology and Remote Sensing (CHRS) at the University of California (http://chrsdata.eng.uci.edu/), where the model provides data for daily estimates of the amount of rain in a range extending between 60 degrees north and south of the equator during the period from 1983 to 2020. This model is based on calculating the amount of rain using FridSat-B1 infrared data and modifying it using the Global Climate Model (GCM) data, and the spatial resolution of a single visible is 28 km × 28 km, and the current study relied on data extracted from the PERSIANN-CCS type, and its spatial resolution was 2.7 km × 2.7 km.

The study relied on satellite images from the Climate Hazards InfraRed Precipitation with Station data (CHIRPS) model issued by the University of California's Climate Hazards Center (https://data.chc.ucsb.edu/products/CHIRPS-2.0/) The model provides daily data for the amount of rain for the period from 1981 to 2020 in a range extending between the two latitudes of 50° south and north of the equator, and its spatial resolution is 5.55 km × 5.55 km, this is to extract the daily rain quantities associated with cyclones under study.

Furthermore, the study calculated the daily rain quantities coinciding with the cyclones under study from PERSIANN-CCS and CHIRPS satellite images, using the following equations (Badawy et al. 2016):

$$R_{vol} = \left(R_{cum}/100_{mm/m}\right)i_z$$

Where:

R_{vol} Rain volume in cubic meters (m^3), R_{cum} Rain accumulation per mm, and i_z the size of cell per square meters (m^2).

In addition, calculating the intensity of rain R_{int} for every 6 h in conjunction with the cyclones under study through the following equation (Food Control Section 2003):

$$R_{int} = 7.44 \times P6 \times D^{-0.645}$$

where

R_{int} Rain intensity, $P6$ Rain fall within 6 h, and D the period of storm in minutes.

2.2 *Climatic Models*

The study relied on the climate models data of the European Center for Medium-Range Weather Forecasts (ECMWF 2020), which were used to monitor the stages of development and track the path of the cyclones under study (https://www.ecmwf.int/), as well as the outputs of the Global Forecast System (GFS) to track the paths of the cyclones under study and the associated climatic conditions (https://www.ventusky.com/ and https://www.windy.com/). The study relied on data from the US Ocean Protection Agency NOAA on the main vibrations in the eastern Pacific (El Nino/Lanina) during the period 1950–2020 (NOAA 2020a, b) (https://origin.cpc.ncep.noaa.gov/products/analysis_monitoring/ ensostuff/ONI_v5.php).

2.3 *Rain Data*

The study relied on data of annual rainfall quantities in Egypt for the period 1950–2020, which was obtained through the NASA space agency (NASA 2020) for climate data through the following website: (https://power.larc.nasa.gov/data-access-viewer/), to determine the correlation between MPV and the frequency of cyclones in Egypt.

3 Analysis and Discussion

3.1 *Medicane Oct. 2019*

In September 2019, several meteorological agencies were warned of the possibility of a strong hurricane in the eastern Mediterranean, starting from the south of the island of Cyprus and heading south towards the northern Egyptian coasts, then eastward towards the Palestinian territories (NASA 2020). On 22nd October 2019, Egypt was exposed to a first-degree cyclone that formed over the eastern Mediterranean region, south of Cyprus, and was accompanied by an exceptional tropical storm that NASA called "Medicane" after the Greek meteorological authority. The amount of rain reached 200 mm in some Areas through the cyclone passed, as seen from Fig. 1.

Furthermore, the occurrence of the Medicane is not an exceptional event, as it is a frequent phenomenon and formation in the western and central of the Mediterranean, but it is an exceptional and rare event in the eastern Mediterranean, in addition, its activity reaches the peak between September and January at the most time (Nastos et al. 2015; Cavicchia et al. 2014; Fecht 2017). The Medicane name is rooted in the Hellenic National Meteorological Service (HNMS) as the largest meteorological agency covering the Mediterranean region (OMM 2019), which named Medicane to a Mediterranean storm similar to tropical cyclones (HNMS 2017), This is called deep tropical storms that occur in the Mediterranean and have a clear cyclone eye that lasts for up to 72 h, and wind speeds range from 47 to 180 km/h, and the radius of this cyclone ranges between 70 and 200 km, and its passage takes a period ranging between 12 h and 5 days, with distances ranging from 700 to 3000 km (Nastos et al. 2015; Cavicchia and van Storch 2011).

Moreover, the Medicane has a clear frequency in the regions of the western and central Mediterranean, as the occurrence of the tropical storm Medicane in September 1969 (Winstanley 1970) and on the Medicane in September 1996 in Spain's Balearic Islands region (Nastos et al. 2017) and many other exceptional tropical storms during the dates January 1982, September 2006, November 2014 (Cavicchia and van Storch 2011, 2014; Nastos et al. 2017). In addition to many studies that have indicated the frequency of tropical Medicane storms in the Mediterranean, the most important of which are (Ernst and Matson, 1983; Rasmussen and Zick, 1987; Luque et al., 2007; Moscatello et al., 2008; Miglietta et al., 2013; Conte et al., 2011). As well as, Meteorological records indicate that approximately 100 Medicane tropical storms have occurred frequently since the 1940s, and usually occur during the autumn season when the sea surface temperature rises more than 15 ° C (HNMS, 2017).

Figure 3 illustrates the Medicane occurrence Locations in the Mediterranean during the period 1983–2003, numbering 12 tropical cyclones, most of which reached the first category on the Saffir-Simpson Hurricane Scale (Wieczorek, 2015). All

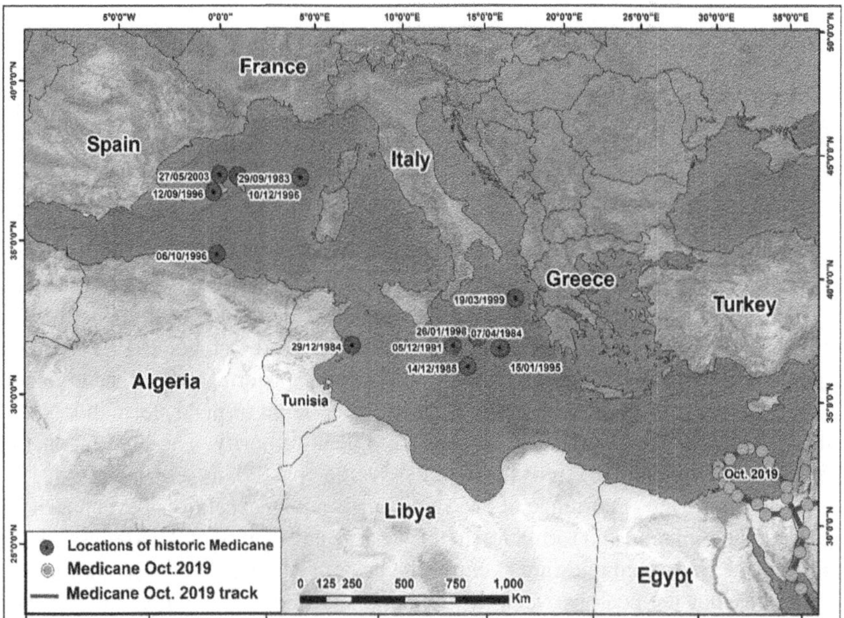

Fig. 3 The geolocations of medicane cyclones in the mediterranean from 1983 to 2003 compared to the medicane of Oct. 2019. *Source* Tous and Romero (2012)

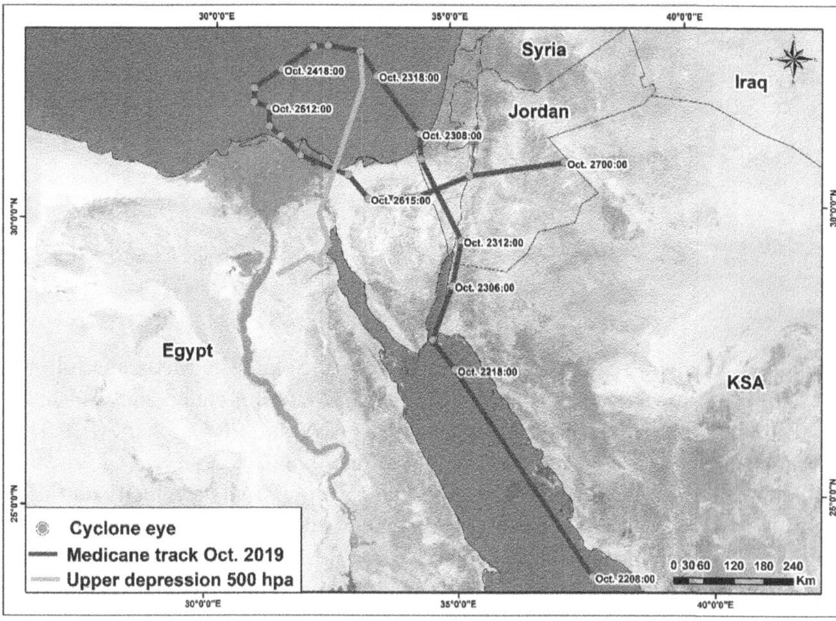

Fig. 4 Medicane cyclone and the upper depression track on Egypt and East mediterranean in Oct. 2019. *Source* GFS & ECMWF models data obtained from (https://www.ventusky.com/, https://www.windy.com/)

Medicane cyclones were in the western and central of the Mediterranean and didn't exceed the eastern region and the longitude of 20E to east. In addition, they didn't exceed the southern region and the latitude 35 N to the south, as seen in Fig. 4. furthermore, all these factors give the Medicane of October 2019 an exceptional case in the eastern Mediterranean.

Accordingly, Fig. 4 illustrates that the cyclone began to form at 8:00 AM on 22nd October 2019 in the Red Sea region at the western of the city of Yanbu Saudi Arabia with the form of an arm or a northern extension of the warm and moist Sudanese depression, which called the Red Sea Trough (RST) (Alpert et al, 2006). Then the cyclone took a northerly course towards the entrance to the Gulf of Aqaba, which the cyclone followed in its path until 8:00 Am on 23rd October 2019 northeast of the city of Aqaba, then the cyclone took a northwestern path until it centered above the eastern Mediterranean water at the north of the city of Rafah at 12:00 PM at the same day. the cyclone continued its movement toward the northwest until it coalesced with a cold depression in the upper layers of the atmosphere in a layer of 500 hpa at 00:00 AM on 24th October 2019, and a clear eye was formed for the Medicane cyclone. Then, both upper depression and surface cyclone began in the simultaneous movement towards the east, then the south and the southeast above the Mediterranean waters off the coast of the Egyptian delta until they penetrated into the Egyptian territories at 02:00 AM on 26th October 2019, passing through the cities

Table 1 The variation in the speed of medicane cyclone october 2019

Day	Distance (km)	Time (h)	Speed (km/h)
22/10/2019	673	24	28.04
23/10/2019	850	24	24.16
24/10/2019	201	24	8.37
25/10/2019	215	24	8.94
26/10/2019	502	24	31.41

Source GFS & ECMWF models data obtained from (https://www.ventusky.com/ and https://www.windy.com/)

of Gamasa, Damietta, South Port Said, and Central Sinai. The Medicane followed an eastern path towards areas south of the Dead Sea until it ended and vanished in an area on the Jordanian-Saudi border at 00:00 AM on 27th October 2019. Thus, the Medicane has traveled 2171 km within a time of 112 h at an average speed of 19.4 km/h, although the cyclone was exposed to a variation in its velocity during the period of its occurrence. Where the speed ranged between 8.4 to 31.4 km/h, as seen in Table 1. this cyclone classified in the category of Medicane cyclones, as the wave height during its occurrence ranged between 4.5 meters in the Abu Qir area and 3.9 meters in Port Said (Salah 2019).

Figure 5 illustrated that wind speed ranged between 35 and 65 km/h from the beginning and the end of the cyclone compared to the period of maturity of the cyclone this places it in the eighth category (storm) on the Beaufort wind scale, hence the atmospheric pressure was at the lowest value at 14:00 PM on 25th October 2019, and reached to 1004 hpa in the cyclone eye. The cloud base level decreased to the lowest about 70 meters above land during the period between 24 to 26 October 2019 compared to more than 7,000 meters during the pre-cyclone period, specifically on October 18, 2019, which indicates the strength of the cyclone, as seen from Fig. 6.

Furthermore, the rain volume was about 543.7 million cubic meters on the Egyptian lands excluding the volume of rain on the Mediterranean water during the movement of the Medicane cyclone, and the daily rain volume was ranged between 40.2 and 280.3 million m^3 for days 24th and 26th October 2019 respectively. The average intensity of rain was 248 mm/h, and the intensity of the rain ranged between 92 and 640 mm/h for days 24th and 26th October 2019 respectively, and the average precipitation was about 412 m^3/km^2. This heavy rain resulted in a great impact on the occurrence of water accumulations on roads in the various governorates of Egypt, in addition to the occurrence of torrents in wadis, as seen from Table 2 and Fig. 7.

3.2 *Dragon Mar. 2020*

Egypt was exposed by a cyclone during the period from 11 and 14 March 2020 and called "Dragon" cyclone in relation to the shape of the clouds accompanying it, the origin of the cyclone is an arm and a part of the Sundanese depression which

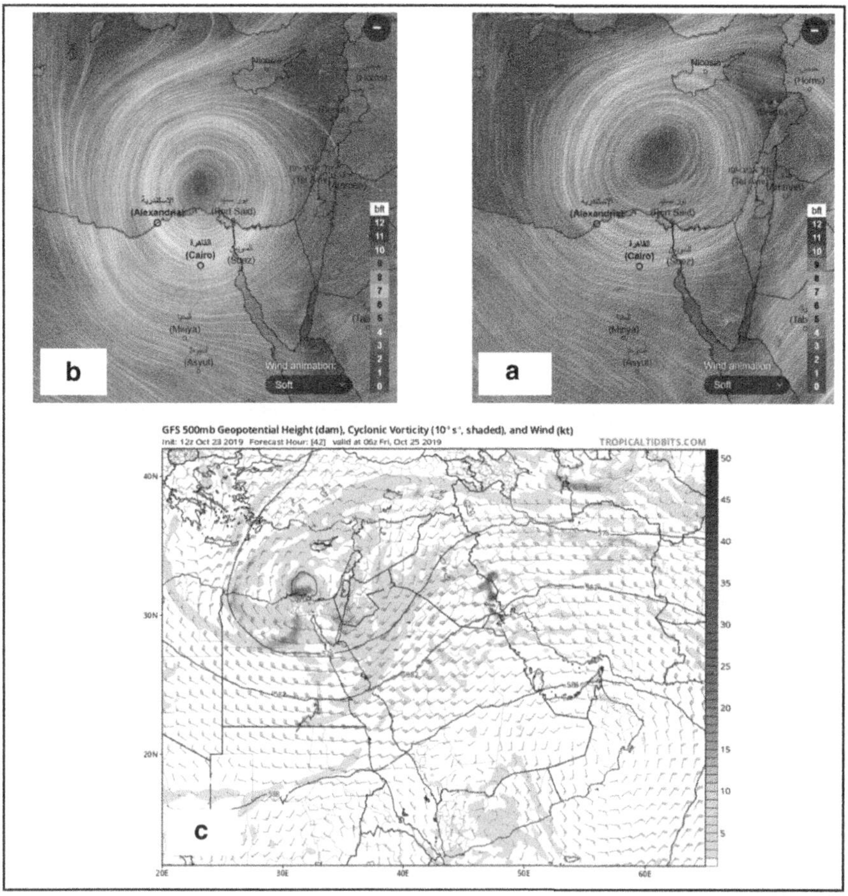

Fig. 5 Characteristics of the Medicane cyclone from 22nd to 26th Oct. 2019, **a** wind speed in Beaufort Scale in 24th Oct, **b** wind speed in beaufort scale in 25th Oct, **c** cyclonic vorticity and wind speed for the upper depression 500 hpa from 23rd to 25th Oct. *Source* GFS & ECMWF models data obtained from (https://www.ventusky.com/, https://www.windy.com/)

characterized by humidity and warm front. Figure 8 illustrates the track of the cyclone which tracked from the south of the Egyptian lands towards the Mediterranean along the western side of the Nile Valley and its deltas in a rare event, and it can be said It has not happened in that path before. The cyclone began its movement from the Darfur region in northwestern Sudan at 00:00 AM on 11th March 2020, and began to move east and then north towards the Egyptian-Sudanese borders until it began to penetrate into Egyptian territory at 00:00 AM on 12th March 2020, and reached to the west of Cairo at 14:00 PM, and at 18:00 PM the cyclone moving from south to north coalesced with a cold depression in the upper atmosphere of 500 hpa resulting in an increase in the strength and depth of the cyclone and formatting the cyclone eye and an increase in wind speeds that reached 120 km/h which recorded at the

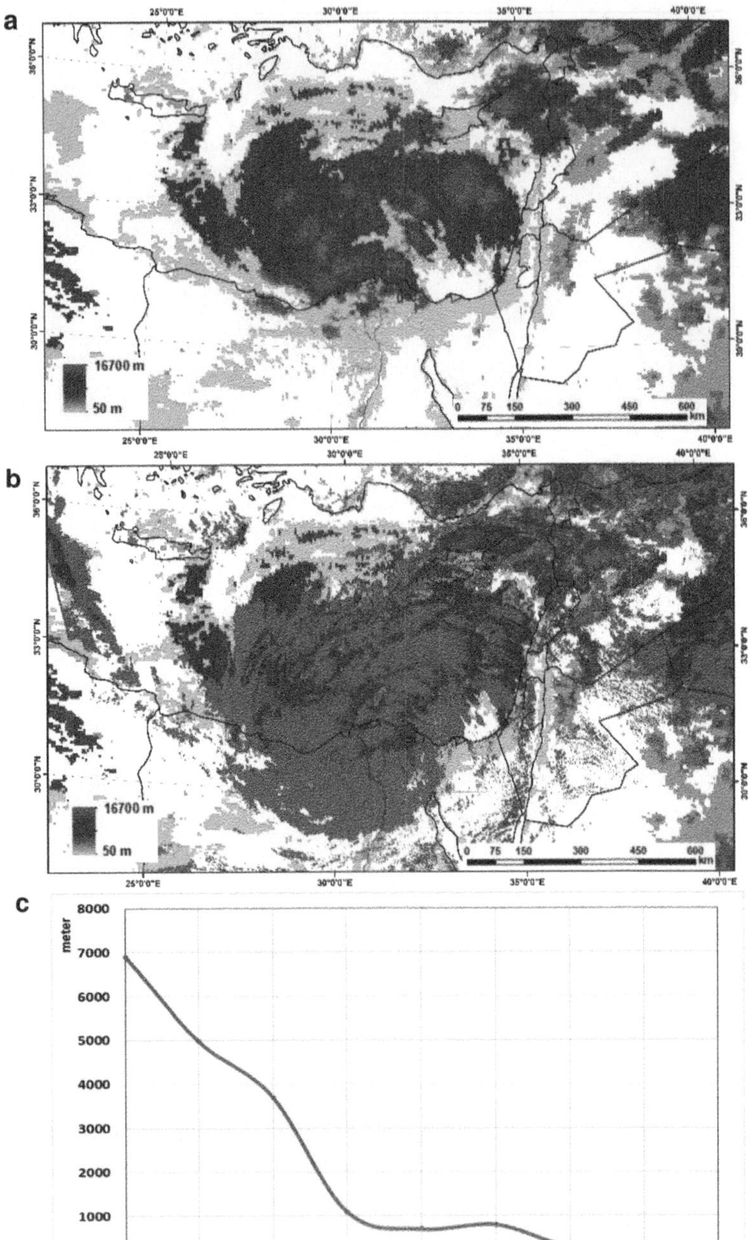

Fig. 6 Cloud base high for the Medicane cyclone from 22nd to 26th Oct. 2019, **a** 24th Oct, **b** 25th Oct, **c** develop the cloud base high from 18th to 25th Oct. *Source* MODIS/AQUA images

Table 2 Rain volume (m^3) and intensity (mm/h) and the Avg. precipitation (m^3/km^2) for the "Medicane" cyclone October 2019

Day	R_vol (m^3(	R_int (mm/h)	Avg. Precipitation (m^3/km^2)
22/10/2019	83,494,912	191	63
23/10/2019	82,398,840	188	62
24/10/2019	40,230,220	92	30
25/10/2019	57,260,948	131	43
26/10/2019	280,324,224	640	212
Total	543,709,144	248	412

Source CHIRPS images obtained from (https://data.chc.ucsb.edu/products/CHIRPS-2.0/)

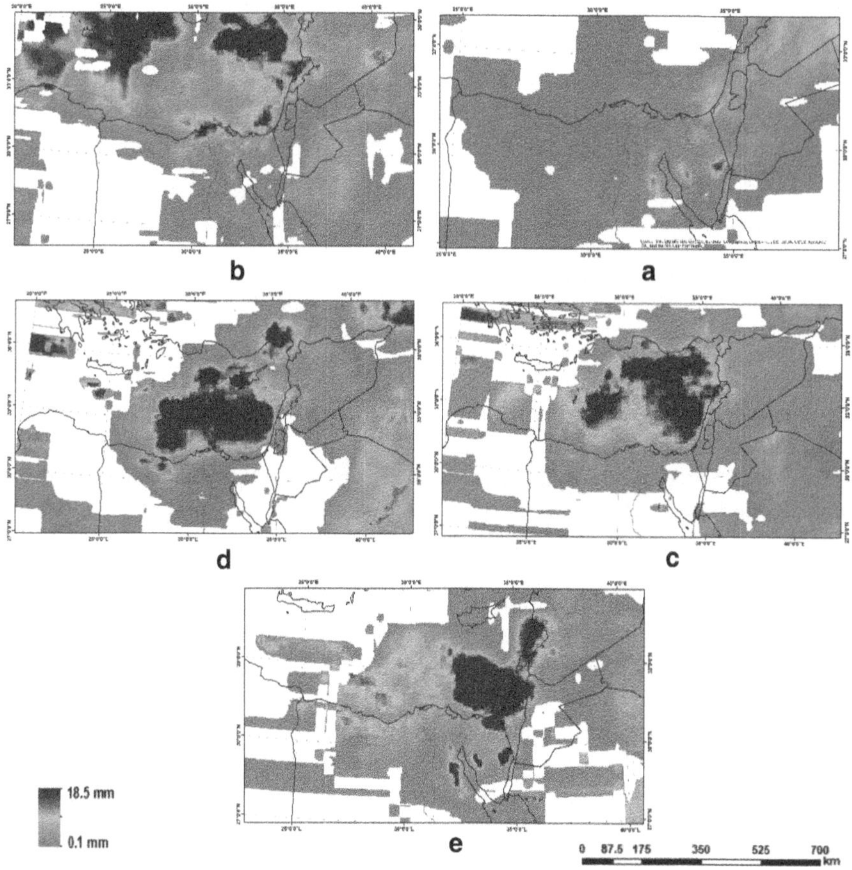

Fig. 7 The rain volume of the Medicane cyclone from 22nd to 26th Oct. 2019, **a** 22nd Oct, **b** 23rd Oct, (C) 24th Oct, (D) 25th Oct, (E) 26th Oct 2019. *Source* CHIRPS images obtained from (https://data.chc.ucsb.edu/products/CHIRPS-2.0/)

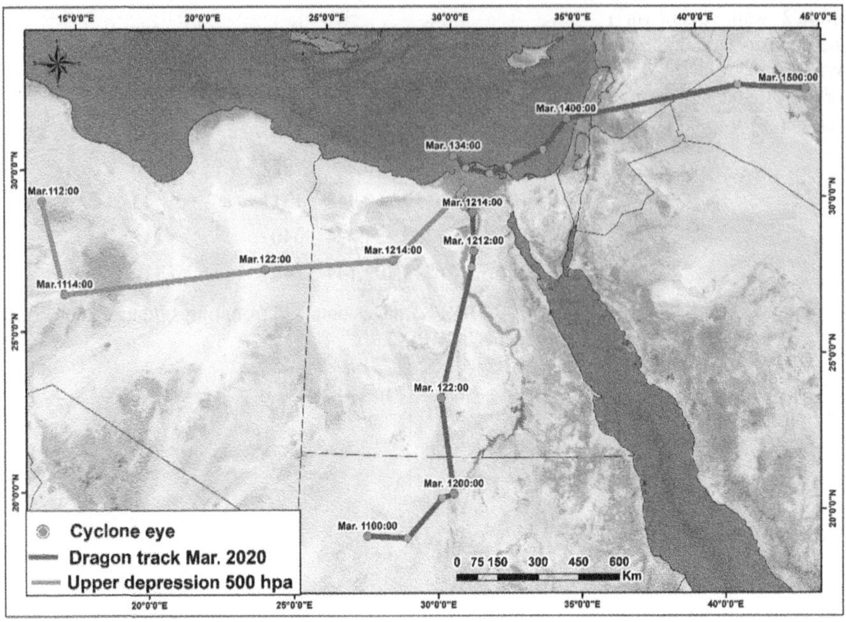

Fig. 8 Dragon cyclone and the upper depression track on Egypt and East Mediterranean in Mar. 2020. *Source* GFS & ECMWF models data obtained from (https://www.ventusky.com/, https://www.windy.com/)

12 categories at the Beaufort wind scale at 23:00 Pm on 12th March. In addition to Land Surface Pressure (LSP) decreased to reach 986 hpa at the cyclone eye, which is the lowest value of atmospheric pressure occurring on Egyptian Land for a long time, Cloud base level decreased to below 50 m during the period between 12 and 13 March 2020, as seen from Fig. 9.

Furthermore, the cyclone has traveled 3083 km within a time of 96 h with an average speed of 32 km/h, although the cyclone has fluctuations in its velocity between 1.95 and 48.4 km/h, as seen from Table 3, and the rain volume was about 1 billion cubic meters of water, excluding the rain volume which fell above the Mediterranean water and outside the Egyptian territories, and the daily rain volume ranged between 648.1 and 36.9 million cubic meters between 12 and 14 March, respectively. In addition, The average intensity of rain was about 543 mm/h and ranged between 1564 and 85 mm/h between 12 and 14 March, respectively and the average precipitation per km^2 was about 720 m^3/km^2 above the Egyptian land, as seen from Table 4 and Fig. 10.

Accordingly, the cyclone has torrents which cut off many roads and destroying hundreds of homes were in torrential torrents areas in Al-Saff area south of Helwan in the south of Cairo. As a result, the Nile River watercolor changed for several days and the silt deposits increased as a result of the runoff of floodwaters in Wadis that end up directly to the Nile. In addition to the effect of the cyclone wind speed on

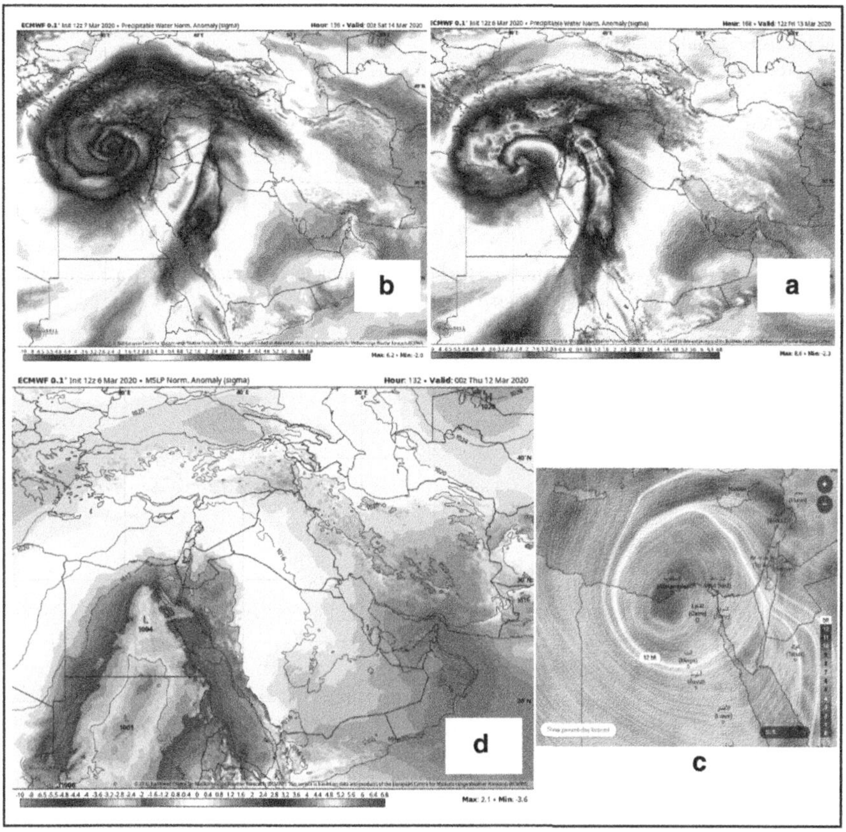

Fig. 9 Characteristics of Dragon cyclone from 12th to 14th Mar. 2020, **a** cyclonic vorticity for 13th Mar, **b**Cyclonic vorticity for 14th Mar, **c** wind speed in beaufort scale in 12th Mar, **d** Sudan depression northern extension for 12th Mar. 2020. *Source* GFS & ECMWF models data obtained from (https://www.ventusky.com/, https://www.windy.com/)

Table 3 The variation in the speed of Dragon cyclone March 2020

Day	Distance (km)	Time (h)	Speed)km/h)
11/3/2020	46.5	24	1.94
12/3/2020	1162	24	48.42
13/3/2020	140	24	5.83
14/3/2020	254	24	10.6

Source GFS & ECMWF models data obtained from (https://www.ventusky.com/ and https://www.windy.com/)

Table 4 Rain volume (m^3) and intensity (mm/h) and the Avg. precipitation (m^3/km^2) for the "Dragon" cyclone March 2020

Day	R_vol (m^3(	R_int (mm/h)	Avg. Precipitation (m^3/km^2)
11/3/2020	42,607,852	97	32
12/3/2020	684,101,504	1564	518
13/3/2020	186,319,264	426	141
14/3/2020	36,937,788	85	28
Total	949,966,408	543	720

Source CHIRPS images obtained from (https://data.chc.ucsb.edu/products/CHIRPS-2.0/)

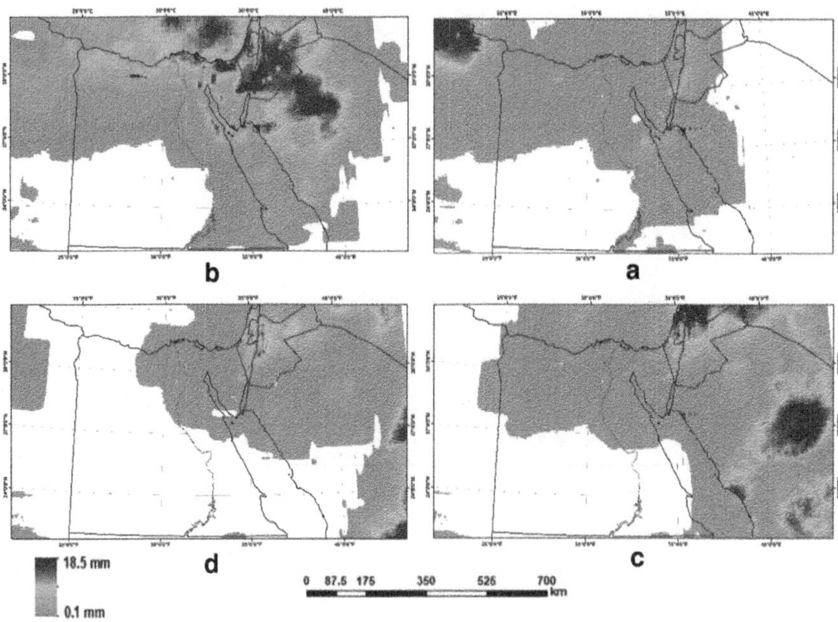

Fig. 10 The rain volume of Dragon cyclone from 11th to 14th Mar. 2020, **a** 11th Mar, **b** 12th Mar, **c** 13th Mar, **d** 14th Mar. 2020. *Source* CHIRPS images obtained from (https://data.chc.ucsb.edu/products/CHIRPS-2.0/)

uprooting many huge trees from their roots in many regions in Egypt, as seen from Fig. 11.

Fig. 11 The impact of Dragon cyclone on Egypt from 11th to 14th Mar. 2020, **a** changing the Nile River waters color, **b** the destruction of houses in the villages of Al-Woudi and Al-Desmi in Al-Saff District, **c** the Al-Orouba tunnel sank on Salah Salem Road in Cairo, and the water rose to more than two meters, **d** the collapses of roads in Al-Obour city in Qalyubia Governorate as a result of the torrents, **e** the uprooting of trees in the governorate Qena as a result of high wind speed

3.3 Changing in Strength, Frequency, and Tracks of Cyclones Under Study About the General Behavior of the Atmospheric Depressions in Egypt

Cyclones in the Mediterranean region are large tropical storms with a diameter of up to 2000 km with surface winds ranging between 4-8 m/s (15-30 km/h) and exceeding 33 m/s (119 km/h) sometimes, LSP reaches 970 hpa compared to the normal atmospheric pressure 1013 hpa, and its chronological life ranges between 3 and 10 days, and it moves from west to east (HMSO 1962). Furthermore, its characterized by shorter life cycles and smaller spatial ranges than tropical and subtropical cyclones that originate in the Atlantic Ocean, with a radius of about 500 km, and an average lifespan of 1 or 2 days (Flocas et al. c). These Mediterranean cyclones arise as a result of the dynamic interaction between warm, moist tropical air masses and cold polar air masses, where the warm air rises in a circular cyclonic perpendicular to the atmosphere and combines with the cooler upper atmosphere air causing a tornado, and this process helps to transfer the excess energy from the lower widths To high-profile presentations (Trigo et al. 2002).

In Egypt, the recurrence of the Medicane on the north coast region from 10 to 14 December 1961, and LSP was about 1001 hpa in northern Alexandria on 12th December 1961. The source of this cyclone was an air depression coming from central Europe, specifically from the northern Italian region and its extension to western Cyprus. In addition to another Medicane from 15 to 19 March 1998, where LSP was about 990 hpa in western Matrouh on 15th March 1998 (Osetinsky and Alpert2006). In addition, the 2nd November 1994 storm is one of the violent depressions in Egypt which has destructive effects as a result of the torrential rains causing the torrents and the high-speed winds. this storm is strongly similar to the Dragon cyclone, where it began as a medium depression coming from southern Europe accompanied by a cold air mass, coincided with the great activity of Sudan depression accompanied by a warm air mass, and resulted in a severe state of instability. And torrential rains quantities that Egypt had not witnessed in 60 years prior to this date. The average of rain was about 30 mm in Cairo as the largest rate of rain since the storm of 29th November 1957 which was about 18 mm and associated with one of the largest flood-related disasters in Egypt, the Dronka incident in Asyut. (Nazmi 2009).

Table 5 and Fig. 12 illustrates the rain volume falling during the November 1994

Table 5 Rain volume (m^3) and intensity (mm/h) and the Avg. precipitation (m^3/km^2) for the deep depression of November 1994

Day	R_vol (m^3)	R_int (mm/h)	Avg. Precipitation (m^3/km^2)
01/11/1994	533.332.992	1215	404
02/11/1994	832.144.576	1898	630
Total	1.365.477.568	1557	1034

Source CHIRPS images obtained from (https://data.chc.ucsb.edu/products/CHIRPS-2.0/)

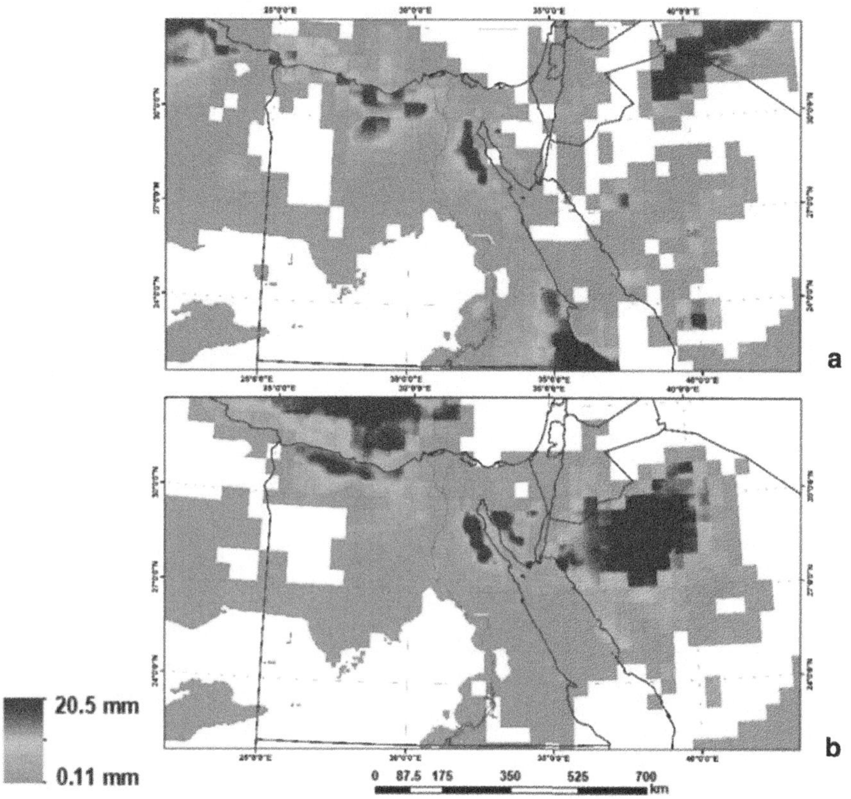

Fig. 12 The rain volume of the deep depression of 1st and 2nd Nov. 1994, **a** 1st Nov, **b** 2nd Nov. 1994. *Source* CHIRPS images obtained from (https://data.chc.ucsb.edu/products/CHIRPS-2.0/)

storm compared to the Medicane 2019 and Dragon 2020 cyclones by analyzing CHIRPS satellite images, it was about 1.4 billion m^3 fells on 2 days 2nd and 3rd November 1994 with rain intensity of about 1557 mm/h, compared to the rain volume for the Medicane 2019 which represented about 39.8% of the rain volume of 1994 storm and Dragon 2020 which represented about 69.6% of the rain volume of the storm.

Figure 13 shows the change in the tracks of cyclones under study from their usual tracks during the same period of occurrence of the year, it is noted that the Medicane 2019 represented an exceptional case in its track that began on the Red Sea, west of the city of Yanbu at the eastern of Saudi Arabia, and its northern direction until It is connected to the cold upper air depression in the eastern Mediterranean and its rotation around its center and its movement from west to east and then return to movement from east to west, with the exception of the usual tracks of the autumn depressions affecting Egypt usually in that period of the year, which moves in tracks from west to the East. At the same time, Dragon cyclone 2020 has the same exception

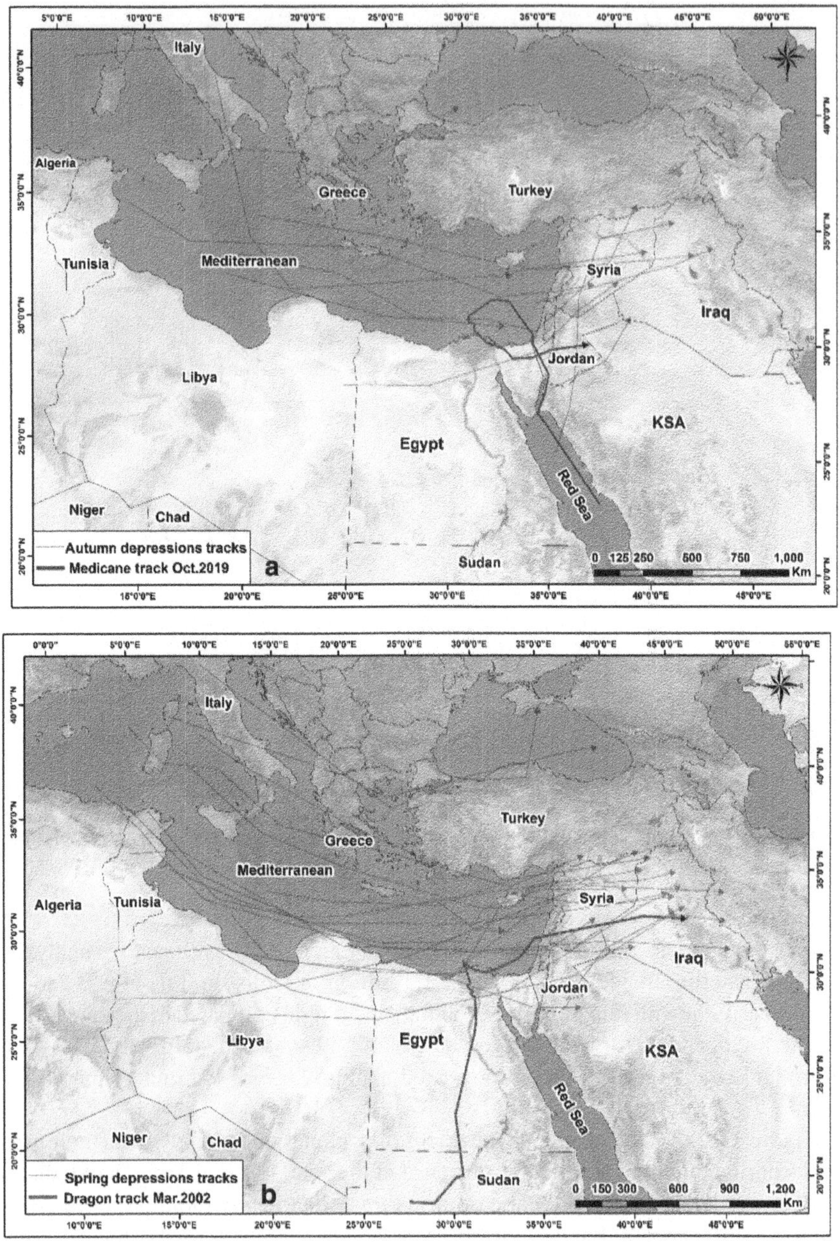

Fig. 13 Changing in the tracks of cyclone under study from depressions usual tracks during the same period in the year, **a** Medicane Oct. 2019, **b** Dragon Mar. 2020. *Source* ECMWF & GFS images obtained from (https://www.ventusky.com/, https://www.windy.com/) and (Sutton. L. J., Curry. P. A., 1931)

in its track compared to tracks of spring depressions affecting Egypt during that period of the year. the Dragon cyclone track began from the Darfur region in the northwest of the Republic of Sudan and moved north until it joined the cold upper depression in the western Egyptian delta, which increased its intensity and strength. Although, it is rare for the center of the depressions and cyclones pass south from the latitude of Cairo during winter and spring, and It is rare for depressions and cyclones to pass through northern Egypt during autumn, and its frequency is about every 10 years for the period 1958–1977 (Al-Feki 1999).

Furthermore, Spring depressions in North Africa has a change in their tracks during the period 1940–1990 in addition to the change in their speed, place, and time of formation, and the increase in frequency during the period 1940–1963 and the downward trend in the frequency of the spring depressions and the associated dust storms during the period 1964-1990, and the increase in the strength and depth of the spring depressions and a decrease in pressure values at their centers from 1002.5 to 999 hpa (El-Asrag et al. 1997). In addition, Spring depressions have a decrease in their number compared to the increase in their strength and intensity and the increase in the rain volume by a greater than the rate of the increase in the number of rainy days, thus increasing the intensity of the rain, which may be due to the shifting of climatic regions towards the north as a result of climate changes (Hassan 2009).

Accordingly, the cyclones that Egypt was exposed to during the wet season 2019/2020 is an exception not only in the track of those cyclones from south to north, driven by the strength of the arm of the Sudan depression or the Red Sea Trough (RST). But the exception extended to include the recurrence of those cyclones twice during the same season, with a time difference was only 4 months between them, instead of a time interval between 3 and 5 years for the occurrence of the cyclonic exceptional cases in the eastern Mediterranean especially in winter and autumn, where the frequency of Medicane in the eastern Mediterranean was at December 1985, October 1990, and December 1993 (Hussein 1994). Moreover, the study deduced that the time interval for the occurrence of these cyclones has been reduced to 4 months, in addition to the occurrence of the Dragon cyclone during the month of March 2020 in a transitional period between the winter and spring seasons.

3.4 Factors Affecting the Occurrence of Rare Cyclones in Egypt

There are many factors that affected the occurrence of cyclones under study e.g. the seasonal oscillation of the Sudan depression, the northern extension of the RST, and the location of the atmospheric depression in the upper atmosphere of 500 hpa in the eastern Mediterranean, in addition to the difference in the sea surface temperatures of the Mediterranean and the Red Sea, the role of the polar and subtropical jet stream, the role of sunspots, and the influence of the El Nino. Furthermore, rains in the Mediterranean region are associated with tropical storms and semi-cyclonic depressions during the winter season, which related to the fluctuations of the air pressure system in the North Atlantic (NAO) when the upper air depressions form and their interaction with the mountainous regions or the atmospheric depressions in the North Mediterranean or wet systems originating from the tropical system in the eastern Mediterranean. In addition, another type of rainstorm in the eastern Mediterranean is associated with tropical plumes, which are a long strip of clouds that extend from the ITCZ to 30-40 N accompanied by a deep upper basin in the subtropical jet stream (Krichak et al. 1997a; b; Krichak and Alpert 1998; Dayan et al. 2001; Kahana et al. 2002; Ziv et al. 2004b). Moreover, any change in the axis of the sub-orbital jet stream from its original position leads to the formation of a large warm and moist air mass that reaches very high levels in the atmosphere over the Middle East (Dayan and Abramski 1983).

3.4.1 The Role of the Red Sea Trough (RST) and the Sudan Depression on the Occurrence of Rare Cyclones in Egypt

The Red Sea Trough (RST) is one of the most important phenomena of tropical weather interactions that affect the southeastern Mediterranean, especially during the autumn and spring seasons, and it is part of the seasonal Sudan depression and its extension, and those northern or southern oscillations of the Red Sea depression arise as a result of the vibrations of the northern and southern Sudanese depression. And that the intensity and duration of rain spells in the eastern Mediterranean are related to the interaction between the upper and lower jet streams in the troposphere, where the jet streams increase the development of moisture complexes and convection and the development of the cyclone shape resulting from disturbances associated with strong wind shear or sudden change in the velocity of the horizontal and vertical winds In addition to the disturbance of the tropopause that allows stratospheric air to penetrate downward toward the troposphere (Alpert et al. 2006). In addition, the RST affected the genesis and evolution of tropical storms in Egypt, Palestine, and Syria as a result of the deepening of Sudan's seasonal depression towards the north and providing it to the eastern Mediterranean region with warm south-easterly air mass during the autumn season, and this fluctuation sometimes in the form of widening or extend towards the north, and this depth reaches its maximum in the eastern Mediterranean and then soon It deviates southward under the influence of cold, westerly or northwestern air masses coming from the upper latitudes which increase the incidence of thunderstorms in the eastern Mediterranean (Al Fandy 1948; Levi 1970).

Moreover, the operations of the RST incursions into the eastern Mediterranean have increased in intensity and have doubled since 1970 from about 50 days/year to about 100 days/year (Alpert et al. 2004a, b). In addition, the RST is the source of the formation and origin of one-fifth of the Mediterranean cyclones with about 20.5% of the total number of them (Romem et al. 2007). Figure 14 shows the movement of the Sudan depression on 11th March 2020 and the strength of the sources of moisture

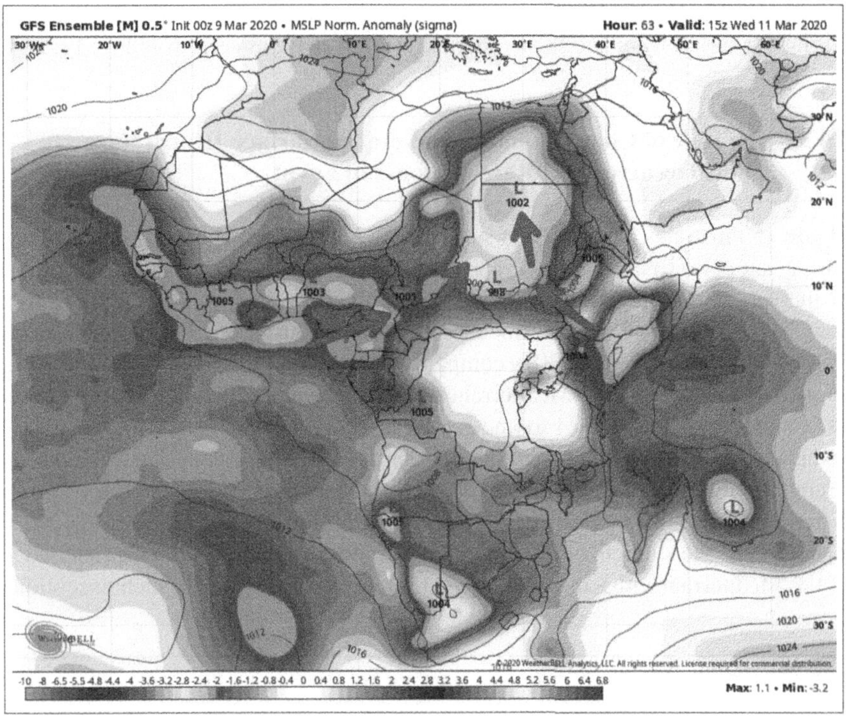

Fig. 14 Sudan depression northern movement in 11th Mar. 2020. *Source* ECMWF & GFS images obtained from (https://www.ventusky.com/, https://www.windy.com/)

that worked to feed and deepen the cyclone during the period from 11 to 14 March 2020.

3.4.2 The Role of the Upper Atmospheric Depression 500 Hpa on the Occurrence of Rare Cyclones in Egypt

The upper atmospheric depression 500 hpa is called the "Mediterranean Oscillation (MO)" or the dipole pressure oscillation which occurs when an extreme High pressure in the western Mediterranean, also coincides with a deep Low depression in its east and vice versa (Feidas et al. 2007). This phenomenon arises as a result of an exceptional rise in temperature and the accompanying extreme low pressure in the eastern Mediterranean, accompanied by a decrease in temperature and an exceptional rise in atmospheric pressure in the western Mediterranean, resulting in northerly and westerly winds over the eastern Mediterranean and is called the negative face of the oscillation (Maheras, Kutiel, 1999). These fluctuations were at their negative face in the Mediterranean during the period of the cyclones under study, and the upper air depression was located 500 hpa above the eastern Mediterranean region, especially during the period from late September to October 2019, as seen from Fig. 17.

3.4.3 The Role of Climate Change and the Sea Surface Temperature on the Occurrence of Rare Cyclones in Egypt

Figure 15 illustrates the change in surface water temperatures of the Mediterranean during the period between 1982 and 2020, where the surface water temperatures of the eastern Mediterranean region have increased by about 2 and 2.5 °C from their natural rates and reached their highest value During the period between 1st October and mid-November 2019 compared to the twentieth century. The sea surface temperature of the eastern Mediterranean recorded values ranging between 26 and 28 °C compared to 22 and 24 for the central region, and about between 17 and 20 °C in the western Mediterranean region, this contributed to the increase in the depth and strength of the Medicane cyclone in the eastern Mediterranean, as the warm surface waters are the most supportive of the unusual Medicane storms (Korosec 2019). Accordingly, the strength of this cyclone illustrates the impact of the thermal budget of the Mediterranean basin and North Africa in the formation of such exceptional

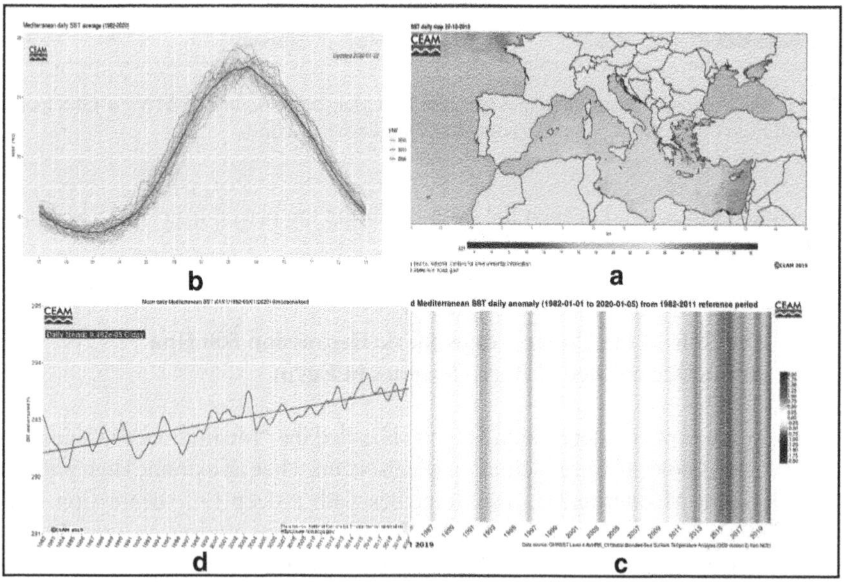

Fig. 15 Changing in the Sea Surface Temperature (SST) of the Mediterranean during the period 1982–2020 **a** SST 22nd Oct. 2019, **b** the monthly change of the SST during the period 1982–2020, **c** The change of the cumulative temperature of the Surface water during the period 1982–2020, **d** the trend of the SST during the period 1982–2020. *Source* (NCEI, 2020) National Centers for Environmental Information (https://www.ncei.noaa.gov/)

phenomena, as these cyclones arise over the warmest waters of the Mediterranean during that period of the year.

Furthermore, the Mediterranean basin is the most vulnerable region to the impacts of climate change and it is one of the most sensitive climates in the world due to its location in a transitional zone between mid-latitudes and subtropics and the most affected by high temperatures due to climate changes, which may reach about 2.2 °C by the year 2040, about 3 degrees Celsius by 2050, and nearly 3.8 °C Celsius by 2100 from pre-industrial levels in the nineteenth century (Kuglitsch et al.2010; Jacob et al. 2014). The current rate of increase in the average ann2002ual temperature in the Mediterranean region reached 1.5 °C in 2019 compared to the rates during the pre-industrial revolution (1880–1899), this increase exceeds the global rates of temperature increase, which amounted to 1.1 °C in the world (MedECC 2019, P5 & IPCC 2013). In addition, the sea surface temperature expected to rise by about 0.4 °C for every 10 years during the period from 1985–2006, and that increase reached 0.3 degrees C for every 10 years in the western basin and 0.5 in the eastern basin of the Mediterranean (Nykjaer 2009). Where the Mediterranean region is one of the most important hot spots that will be exposed to climate change in the future (Giorgi 2006), and increase the occurrence of this regional variation in the frequency of "Medicane" and the occurrence of more extreme events in the eastern Mediterranean region and an increased likelihood of their occurrence during October in general (Romero and Emanuel 2017).

3.4.4 The Role of General Air Circulations and the Main Pressure Oscillations on the Occurrence of Rare Cyclones in Egypt

The eastern Mediterranean region affects by the East Atlantic (EA) oscillation during the winter, while the western region affected by the North Atlantic Oscillation (NAO) (Hurrell, 1996; Alpert et al. 2006; Hatzianastassiou et al. 2009; Trigo et al. 2006). In addition to, there are a close correlation between El-Nino oscillations and the climatic variations of Europe and the Mediterranean, including a decrease in temperatures and an increase in precipitation in the southern and eastern parts of the Mediterranean and increase cyclones and deep depressions on the central and southern parts of the Mediterranean, including the most pronounced of these values were during 1983 and 1992, in addition to decreased cyclones and hurricanes over Central Europe according to the European Grosswetterlagen model and, conversely, during the La-Nina phenomenon (Hess and Brezowsky 1969; Bronnimann 2007; van Loon and Madden, 1981; Gouirand and Moron, 2003; Moron and Plaut, 2003). Furthermore, the effect of El-Nino, which occurs in Southeast Asia and the Western Pacific, on Mediterranean rains which correlated with a non-linear model between El-Nino and the increased of rainfall in the eastern Mediterranean (Sardeshmukh et al., 2000; Pozo-Va´zquez et al., 2005a; Xoplaki 2002; Mariotti et al. 2005; Alpert et al. 2006). And a non-linear correlation between the occurrence of El-Nino and the Extratropical SLP during the winter months from December to March (Wu and Hsieh, 2004a, 2004b). and the strong correlation between the El-Nino and the climate of the northern

Red Sea and the southern Mediterranean over a period of 245 years and that there is an influence of El-Nino on the climate of the Middle East and North Red Sea (Felis et al. 2000; Rimbu et al. 2003).

Moreover, there are a decrease in rain volumes in southwestern Europe and the Black Sea during the cold La-Nina and an increase in rain rates during the warm El-Nino in the same regions (Fraedrich and Muller, 1992). In addition to, the strong positive correlation between increased rainfall during the period from October to March and the El Nino phenomenon during 25 years of study in the area of Israel in the southern Mediterranean, where the amounts of rain were more than their normal rates by about 25%, the maximum of which was in 1992, with a correlation rate of 0.59, while La Nina worked to prevail in drought conditions (Price et al, 1998), and the western Mediterranean receives 10% more rain during the autumn and early winter periods preceding El-Nino (Mariotti et al. 2002; Shaman and Tziperman 2011), while increasing the rainfall during late autumn in the eastern Mediterranean region, especially after the mid-seventies (Price et al. 1998; Alpert et al. 2006), as the climate change in the late seventies, altered the atmospheric circulation and Teleconnection Patterns in the Northern Hemisphere (Trenberth 1990; Trenberth and Hurrell 1994). In addition cyclones and Tropical storms in the Mediterranean region have increased in their intensity during the autumn and spring as a result of flowing humid air from the Indian Ocean and tropical Africa, as an indirect impact of El-Nino in the Mediterranean (Krichak et al, 2013).

As well, El-Nino affects the stratosphere and the tropopause over the equatorial Pacific, where the height of the tropopause is lower in the western region of the Pacific and higher in the eastern equatorial region of the Pacific during the El-Nino period compared to the La-Nina periods, where the cold tropopause shifts to the east due to a shift in convective activity and its influence the stratosphere in the Extratropical range (Hatsushika and Yamazaki 2001). In addition to, the relation between the occurrence of El-Nino and reducing the activity of North Atlantic Oscillations (NAO), where El-Nino effects on shifting of the low-pressure centers to the northeast Atlantic (Bronnimann et al. 2007b).

Egypt is affected by El-Nino by cooling the downward air from the upper layers, which increases the formation of strong High pressure in Northwest Africa and Western Europe and increases the ascendancy of the Azorean High, which impedes the movement of Mediterranean depressions and cyclones from the western to the eastern Mediterranean regions which giving way to the movement of the Sudan depression to the north and northeast with its content of humidity and its warm air, and on the other side, the cold polar air masses and upper depressions are pushing towards eastern Europe and the eastern Mediterranean and moving away from the height pressure centers in western Europe and northwest Africa. As a result, a cyclone convergence occurs between the two fronts and results in an increase in the depth of the depression to develop into a storm or a cyclone at times.

Accordingly, During the wet season 2019/2020, El-Nino Index reached its maximum value with about 1.5, and the associated increase in the amount of rain in the southwestern USA, Peru, and Central Europe regions, in addition to drought in the East of the USA and Australia, which led to the outbreak of devastating forest fires in eastern Australia and the Amazon forest fires in eastern South America in the summer and fall of 2019. This can be explained by the emergence and movement of cyclone in October 2019 and March 2020, as El-Nino modifies the air movement in general and results in changes in normal weather patterns in many regions of the world, the most important of which is the change of the position of the jet stream in the northern hemisphere, The temperature rises above average in North America, Europe, East Asia and the northern hemisphere in general, causing heavy rains than in parts of Europe and the Middle East.

Furthermore, the current study applied a correlation coefficient between historical rain data on Egypt for the period between 1957-2020 and the East Pacific Oscillation Index El-Nino during the same period and there is a positive correlation between the rain volume fell and El-Nino with values ranging between 0.11 and 0.30, for the rain volume in Alexandria for the period 1957-2020, and El-Arish for the period 1987-2020, respectively, and about 0.21 in Cairo for the period 1958-2020, and about 0.17 for the total rain volume in Egypt for the period 1986-2020, as seen from Fig. 16.

Moreover, there is a positive correlation of 0.58 between the winter rain in northern Israel and El-Nino during the last quarter of the twentieth century, and the presence of an increase in the amounts of rain during El-Nino years than the normal rate of rain (Colin et al. 1998), and also a correlation between the increase in autumn rains over the Iberian Peninsula and El-Nino (Rocha 1999), and the positive relationship between rain during the spring and autumn seasons and the El-Nino in the southern,

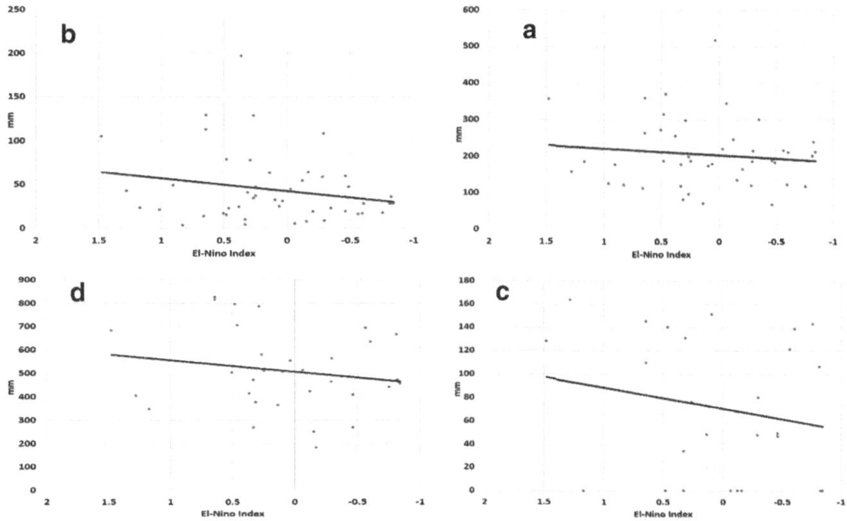

Fig. 16 The correlation between El-Nino index and the rain volume in Egypt for the period 1957–2020 **a** Alexandria, **b** Cairo, **c** AL-Arish, **d** total of Egypt. *Source* Rain data obtained from (NASA 2020 https://power.larc.nasa.gov/data-access-viewer/) and El-Nino Index data obtained from ((NOAA 2020a, b https://origin.cpc.ncep.noaa.gov/products/analysis_monitoring/ensostuff/ONI_v5.php)

central and eastern Mediterranean regions during the period 1951–1998 (Hasanean 2004), and a correlation between annual rain in northern and central Israel and El Nino during the period 1950-2004 (Yosef et al. 2009). In addition to the significant correlation between the occurrence of the El-Nino and the precipitation in winter in Israel, where the amount of rain during it increases in contrast to the winter rains during La-Nina, in which the rain volume is decreases to below their rates(Yakir et al. 1996; Price et al. 1998).

As well, there are about 45% of the annual variation in winter precipitation in Israel and the Eastern Mediterranean results from El-Nino oscillation, which related to the position of the jet stream in the upper layer of the troposphere at a latitude of 30 N above the eastern Mediterranean during the winter, El-Nino effects on shifting the jet stream to the south of its location over the Mediterranean by about 50-100 km and, during the La-Nina the jet stream has shifted to the north of its position. In addition increases the rain volume during El-Nino by about 75% for the period 1970-2000 compared to drought conditions and decreases the rain volume by about 75% during La-Nina, and increases the frequency and intensity of El-Nino events and the rainfall in the eastern Mediterranean since mid-seventies (Alpert et al. 2006). This related to the shifting in the global climate system during the seventies which may have led to a stronger link between Pacific events and midlatitudes over the past period (Trenberth and Hoar 1997; Wuethrich 1995), in addition, El-Nino works to change the movement of the circulation of the atmosphere and increase the strength of the orbital jet currents north and south of the equator (Walker 1923).

The Mediterranean and North Africa region is affected by two jet streams, the first is the polar jet stream, and its path is located in the north of the Mediterranean and shifted southwards above the Mediterranean and North Africa during the northern winter with the flow of polar cold air masses in the upper levels of the atmosphere, and the other is the Subtropical jet stream which shifts northward over the region during the northern summer and is accompanied by a state of stability in the air (Shehadeh 1986; Al-Fiqi 1999).

Moreover, when the polar jet stream converges during its maximum southern displacement accompanied by an Atlantic depression with the subtropical jet stream in its northern displacement in early spring with a surface air depression in the eastern Mediterranean, a severe state of instability arises that leads to a deepening of the depression which develops to forming cyclones and storms. In the case of Medicane October 2019, it is noted that those conditions related to the maximum southern displacement of the polar jet stream occurred during the autumn season of 2019 and its converged with the sub-tropical jet stream at its maximum northern displacement, which represents an exception in the timing of occurrence and convergence, as seen from Fig. 17, and this coincided with the upper depression in the eastern Mediterranean and the RST extending northward and resulted in the formation of Medicane cyclone. In addition, during El-Nino, the subtropical jet stream is at its strongest stage compared to its weakening during the La-Nina periods (Bengtsson et al. 1996; Roeckner et al. 1996; Martineu et al. 1999), this is largely consistent with the case of Medicane cyclone October 2019.

Figure 18 shows the positive side of the NAO and the negative side of the Mediter-

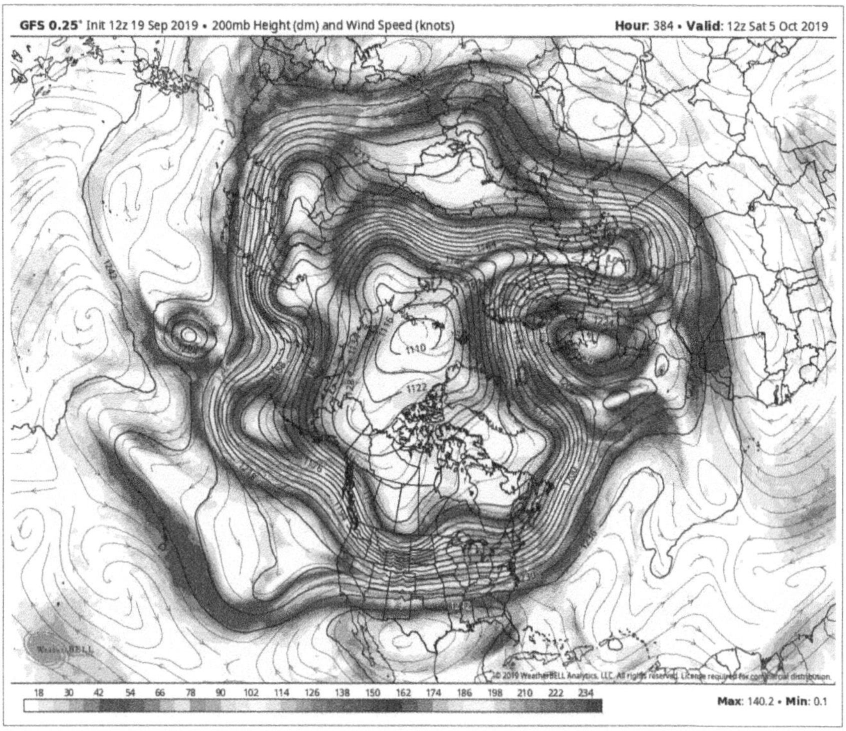

Fig. 17 The confluence of the polar jet stream with the sub-tropical jet stream during the period from 19th September to the end of October 2019. *Source* ECMWF & GFS images obtained from (https://www.ventusky.com/, https://www.windy.com/)

ranean oscillation (MO) at the end of September 2019, where the atmospheric pressure rose significantly in the range of the Azorean high pressure in western Europe and the high-pressure of "North Sea-Caspian Pattern (NCP)" and decrease the atmospheric pressure lower than normal values in the Icelandic low-pressure range in northern Europe, giving way to an incursion of cold polar air masses south towards the eastern Mediterranean and gusts strong westerly winds and strong tropical storms and cyclones with warm, humid air and heavy rain over central, southeastern, and northern Europe, and the eastern Mediterranean regions as a result of these large pressure differences. All these conditions coincided with the negative side of the Arctic Oscillation (AO) where the values of the oscillation decrease and unusually high pressure dominates the polar circle, with unusually low pressure over the mid-latitudes north of the 45 N, as seen in Figs. 18 and 19, this results in weaker than usual westerly winds in the upper atmosphere and called "polar vortex", and the cold polar air zone is shifting towards the south, and the subtropical regions are exposed to weaker trade winds - which increases the strength of the El-Nino -, and cyclones are frequent and increase the rain volume in the eastern of the Mediterranean. In addition to the role of the Mediterranean oscillation (MO) or the dipole

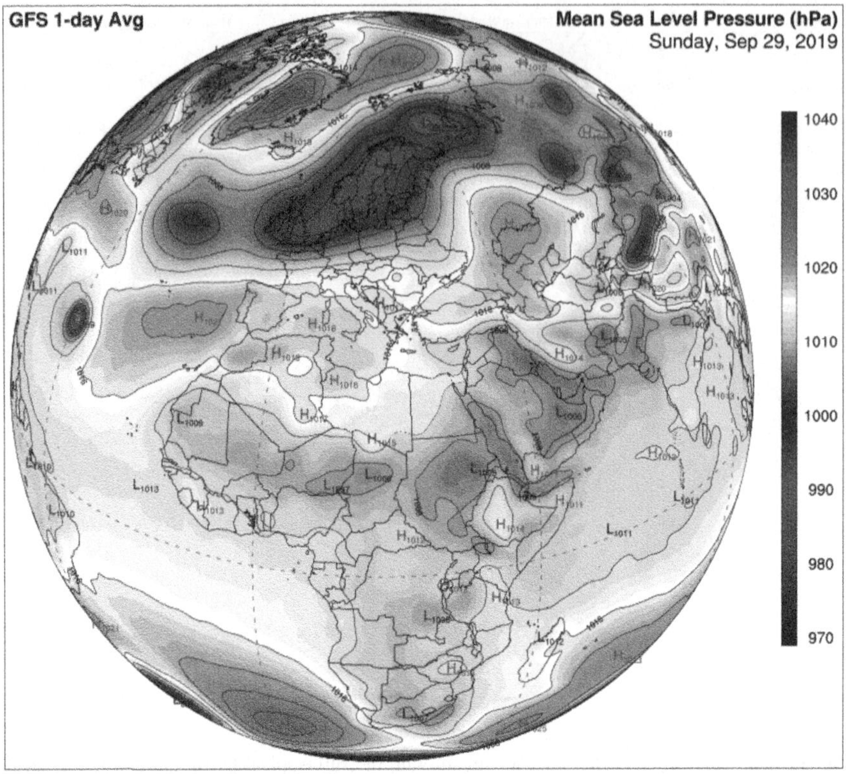

Fig. 18 the Synoptic situation for pressure oscillations during the end of September 2019 that affecting the medicane cyclone in Egypt and the Eastern of the mediterranean. *Source* ECMWF & GFS images obtained from (https://www.ventusky.com/, https://www.windy.com/)

oscillation in its negative side with a deep upper depression of 500 hpa on the eastern Mediterranean region, and its impact on generating tropical storms and cyclones in the eastern Mediterranean.

Furthermore, North Africa and Southern Europe regions are exposed to a great activity of cyclones and heavy rain and the tracks of storms shift towards the Mediterranean when the difference between the Icelandic low-pressure and the Azorean high-pressure reaches large degrees in the values of NAO especially during the period October-April, where these oscillations caused changes in the distribution of temperature, humidity, intensity, number and frequency of storms, and their strength and frequency (Kalyn et al. 2005; David et al. 1998).

In addition, El-Nino increases the frequency of the Stratospheric Sudden Warnings (SSW), which in turn increases the negative AO frequency that often occurs after the SSW events and that more warming in the stratosphere which affected more negative AO events and more extreme weather events over the Eurasia and Mediterranean region (Butler and Polvani 2011).

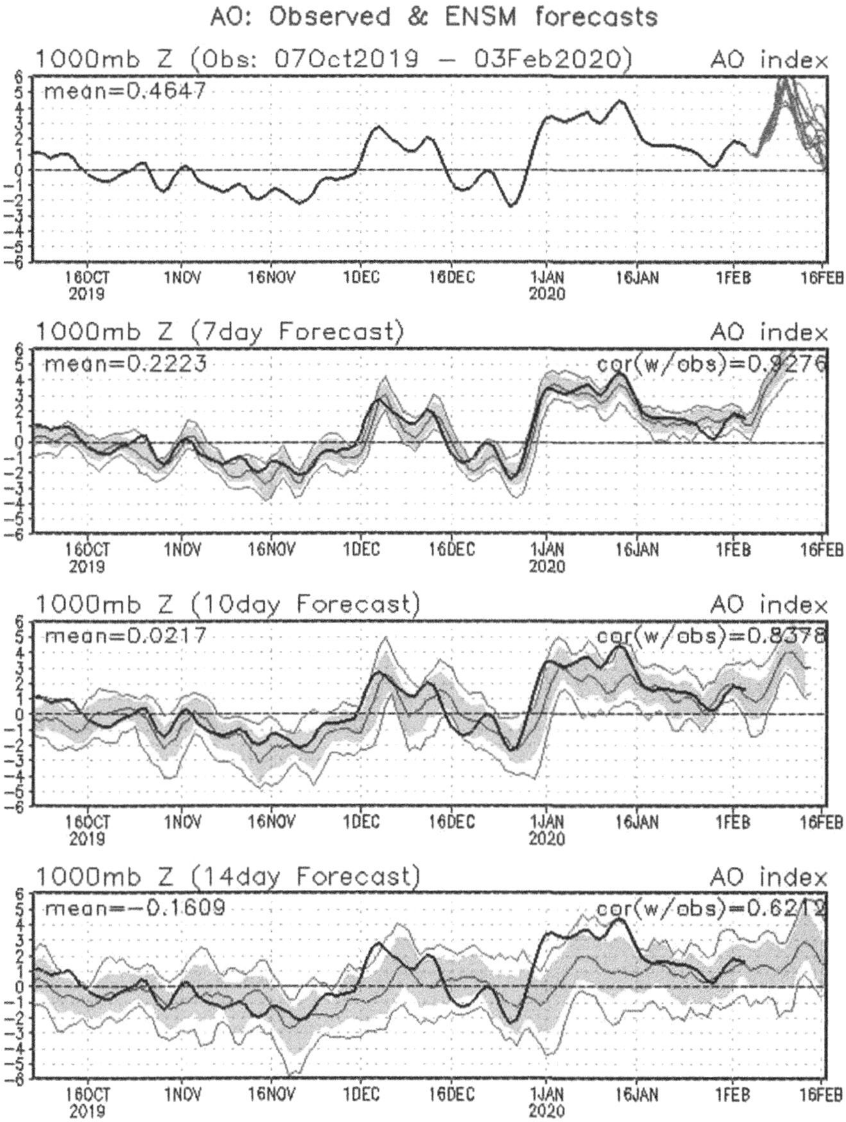

Fig. 19 The Change in the AO Index at its negative side between 16th October and early December 2019. *Source* ECMWF & GFS images obtained from (https://www.ventusky.com/ and https://www.windy.com/)

4 Conclusion

The main objective of this study is to monitor and observe the impact of climate change on the frequency of cyclones in the Eastern Mediterranean and the North-eastern of African in the wet season 2019/2020 in Egypt as a case study, and the study concluded that the cyclones which passed over Egypt during the rainy season 2019/2020 were associated with a set of factors that supported their development in this rare event Which can be considered as a sign of climate change. those factors that rarely coincided as the northern oscillation of the Sudan depression and the RST was accompanied by an upper deep cold depression, coinciding with a rise in the SST of the EMNEA region in Medicane October 2019, and the convergence of the polar and tropical jet current in shows above Egyptian lands. And the correlation of these factors with the occurrence of El-Nino, coinciding with the positive side of NAO, the negative side of AO, the negative side of MO, and the negative side of the Caspian Sea - North Sea Oscillation.

The study concluded that the Dragon cyclone March 2020 was deeper and had a greater impact than the Medicane October 2019 depression, which is related to the increase in the general air circulation strength in the Mediterranean basin and the increase in the maximum thermal gradient in the front between the polar and tropical masses, this was accompanied by the emergence of the upper low-pressure basin which increases the cyclone movement and makes it deeper and longer.

The study deduced that an increase in the depth and strength of the depressions in the Eastern Mediterranean region and Northeast Africa with increasing global warming and climatic changes associated with the change in the intensity and strength of El-Nino and general pressure oscillations, and increase the SST above its averages.

References

Abdel-kader AA (1987) On the genesis and structure of winter deprssions. Bull Fac Arts Cairo Univ xxxxiv and xxxxv 1:111–132

Ahrens CD (2008) Essentials for meteorology an invitation to the atmosphere, 5th edn. Thomson Learning Academic Resource Center, USA, p 198

Al-Fiqi AA (1999) The wind in Egypt, study in climatic geography. Master Thesis. Faculty of Arts. Ain Shams University

Alpert P, Baldi M, Ilani R, Krichak SO, Price C, Rodo X, Saaroni H, ZivB Kishcha P, Barkan J, Mariotti A, Xoplaki E (2006) Relations between climate variability in the Mediterranean region and the tropics: ENSO, South Asian and African monsoons, hurricanes and Saharan dust. In: Lionello P, Malanotte-Rizzoli P, Boscolo R (eds) Mediterranean climate variability, vol 4. Elsevier, Amsterdam, pp 149–177

Alpert P, Ziv B (1989) The Shara cyclone, observations and some Theoretical considerations. J Geophys Res 94:18495–18514

Alpert P, Osetinsky I, Ziv B, Shafir H (2004a) Semi-objective classification for daily synoptic systems: application to the EM climate change. Int J Climatol 24:1001–1011

Alpert P, Osetinsky I, Ziv B, Shafir H (2004b) A new seasons definition based on the classified daily synoptic systems: an example for the EM. Int J Climatol 24:1013–1021

Ambaum MHP, Hoskins BJ, Stephenson DB (2001) Arctic oscillation or north Atlantic oscillation? Am Meteorol Soc J Climate 14:3495–3507. https://doi.org/10.1175/1520-0442(2001)014%3c3495:AOONAO%3e2.0.CO;2
Andronova NG, Schlesinger ME (2000) Causes of global temperature changes during the 19th and 20th centuries. Geophys Res Lett 27:2137–2140
Anna W (2015) Medicanes—the hurricane of the mediterranean?", DWD, wetterdienst.de, 1 September 2015, Accessed 15 December 2019, available at: https://www.wetterdienst.de/Deutschlandwetter/Thema_des_Tages/1880/medicanes-die-hurrikane-desmittelmeeres
Arpe K, Bengtsson L, Golitsyn GS, Mokhov II, Semenov VA, Sporyshev PV (2000) Connection between Caspian Sea level variability and ENSO. Geophys Res Lett 27:2693–2696
Artale V, Calmanti S, Malanotte-Rizzoli P, Pisacane G, Rupolo V, Tsimplis M (2006) Chapter 5 The Atlantic and mediterranean sea as connected systems. Dev Earth Environ Sci 283–323. https://doi.org/10.1016/s1571-9197(06)80008-x
Attaher SM, Medany MA, El-Gindy A (2010) Feasibility of some adaptation measures of on-farm irrigation in Egypt under water scarcity conditions. Options Mediterraneennes 2010(95):307–312
Badawy MM, Abdel Aziz AO, Mamtimin B (2016) Flash floods in the Sahara: a case study for the 22 January 2013 flood in Qena. Egypt Geomatics, Natural Hazards and Risk 7–1:215–236
Baldwin MP, Dunkerton TJ (1999) Propagation of the Arctic Oscillation from the stratosphere to the troposphere. J Geophys Res 104(30):937–30 946, doi:https://doi.org/10.1029/1999JD900445
Ag Barnston, Re Livezey (1987) Classification, Seasonality and Persistence of Low Frequency Atmospheric Circulation Patterns. Mon Weather Rev 115:1083–1126
Bengtsson L, Arpe K, Roeckner E, Schulzweida U (1996) Climate predictability experiments with a general circulation model. Clim Dyn 12:261–278
Bjerknes J (1966) A possible response of the atmospheric Hadley circulation to anomalies of ocean temperature. Tellus 18:820–829
Boscolo R (eds) Mediterranean climate variability. Elsevier
Bro¨nnimann S (2007) Impact of El Nin˜ O–Southern Oscillation on European Climate, Reviews of Geophysics, 45, RG3003/ 2007, the American Geophysical Union
Bronnimann S, Xoplaki E, Casty C, Pauling A, Luterbacher J (2007) ENSO influence on Europe during the last centuries. Clim Dyn 28:181–197
Butler AH, Polvani LM (2011) El Niño, La Niña, and stratospheric sudden warmings: A reevaluation in light of the observational record: El Niño, La Niña, and SSWS. Geophys Res Lett 38:L13807. https://doi.org/10.1029/2011GL048084
Butler AH, Polvani LM, Deser C (2014) Separating the stratospheric and tropospheric pathways of El Niño-Southern Oscillation teleconnections. Environ. Res. Lett. 9: https://doi.org/10.1088/1748-9326/9/2/024014
Cavicchia and von Storch. (2011). "The Simulation of Medicanes in A High-Resolution Regional Climate Model", Springer-Verlag, vol. 39, November 2011, Climate Dynamic, https://doi.org/10.1007/s00382-011-1220-0
Cavicchia L, von Storch H, Gualdi S (2014) A Long-Term Climatology of Medicanes. Climate Dyn 43:1183–1195. https://doi.org/10.1007/s00382-013-1893-7
Christiansen B (2001) Downward propagation of zonal mean zonal wind anomalies from the stratosphere to the troposphere: Model and reanalysis. J Geophys Res 106, 27 307–27 322. https://doi.org/10.1029/2000JD000214
Cohen J, Barlow M (2005) The NAO, the AO, and global warming: how closely related? J Clim 18:4498–4513
Cohen J, Frei A, Rosen RD (2005) The role of boundary conditions in AMIP-2 simulations of the NAO.J. Climate 18:973–981
Colin P, Stone L, Huppert A, Rajagopalan B, Alpert P (1998) A possible Link Between El Nino and Precipitation in Israel, Geophysical ResearchLetters,Vol.25, No.21
Cook BI Anchukaitis KJ, Touchan R, Meko DM, Cook ER (2016) Spatio-temporal drought variability in the Mediterranean over the last 900 years, Journal of Geophysical Research: Atmospheres https://doi.org/10.1002/2015jd023929

Copernicus. (2020). Near surface meteorological variables from 1979 to 2018 derived from bias-corrected reanalysis, European Union, available at: https://cds.climate.copernicus.eu/cdsapp#!/dataset/derived-near-surface-meteorological-variables?tab=form

Cramer W, Guiot J, Fader M, Garrabou J, Gattuso J-P, Iglesias A, Lange MA, Lionello P, Lla-sat MC, Paz S, Peñuelas J, Snoussi M, Toreti A, Tsimplis MN, Xoplaki E. (2018). Climate change and interconnected risks to sustainable development in the Mediterranean. Nature Climate Change 8, 972-980, https://doi.org/10.1038/s41558-018-0299-2. Available at: https://ufmsecretariat.org/wp-content/uploads/2019/10/MedECC-Booklet_EN_WEB.pdf

Dai. P., Benkui. T. (2017). The Nature of the Arctic Oscillation and Diversity of the Extreme Surface Weather Anomalies It Generates. American Meteorological Socity. J. Climate (2017) 30 (14): 5563–5584. https://doi.org/10.1175/JCLI-D-16-0467.1

David W.j. Thomson, John M. Wallace. (1998) The Arctic Oscillation Signature in The Wintertime Geopotential Height and Temperature Fields, Geophysical Research Letters, Vol.25, N.9

Dayan U, Abramski R (1983) Heavy rain in the Middle East related to unusual Jet Stream properties. Bull Am Meteor Soc 64:1138–1140

Dayan U, Ziv B, Margalit A, Morin E, Sharon D (2001) A severe autumn storm over the Middle East: synoptic and mesoscale convection analysis. Theor. Appl. Clim. 69:103–122

Delworth TL, Mann ME (2000) Observed and simulated multidecadal variability in the Northern Hemisphere. Climate Dyn. 16:661–676

Deser C (2000) On the teleconnectivity of the “Arctic Oscillation”. Geo phys. Res. Lett. 27:779–782. https://doi.org/10.1029/1999GL010945

Donald Ahrens C (2008) Essentials for Meteorology an Invitation to The Atmosphere, 5th edn. Thomson learning Academic Resource Center, USA

El Dessouky TM (1980) Forecasting Winter rainfall over Egypt using the (1000–500 mb) thickness and the 500 mb anamolies in Autumn and Summer. Meteorological Department, Cairo

El Guindy MF (1971) Dynamical Interaction between Troposphere and Lower Stratosphere in Spring 1967. Meteorological Department, Cairo

El-Asrag, A. M. (1999). Climate Change over Egypt and Its Relevance to Global Change, The fourth conference meteorology and sustainable development to 21 st century 7-9 march, the Egyptian Meteorology Authority, pp. 85-114

El-Asrag, A. M. and Hassan, A. S. (1997). Climatological Study of Spring Depressions Over North Africa In Last Five Foregoing Decades., Conference of Meteorology and Environmental Cases; March 2–6, Egyptian Meteorological Authority, pp. 113-126

El-Fandy MG (1948) the effect of the Sudan monsoon low on the development of thundery condition in Egypt. J., R. Met. Soc, Vol, Palestine and Syria, Quart, p 74

El-Fandy MG (1949) Oscillation of the Sudan monsoon low. J., R. Met. Soc., Vol, Quart, p 94

Enfield DB, Mestas-Nunez AM, Trimble PJ (2001) The Atlantic multidecadal oscillation and its relation to rainfall and river flows in the continental U.S. Geophys Res Lett 28:2077–2080

European Centre for Medium-Range Weather Forecasts ECMWF. (2020). Available at: https://www.ecmwf.int/

FAO. (2016). Egypt, Geography, Climate and Population. Food and Agriculture Organization. Available at: http://www.fao.org/nr/water/aquastat/countries_regions/egy/print1.stm

Feidas H, Noulopoulou Ch, Makrogiannis T, Bora-Senta E (2007) Trend Analysis of Precipitation Time Series in Greece and Their Relationship with Circulation Using Surface and Satellite Data : 1955–2001. Theor. Appl. Climatol. 87:155–177

Feldstein SB, Dayan U (2008) Circum global teleconnections and wave packets associated with Israeli winter precipitation. Q J R MeteorolSoc 134:455–467

Felis, T., J. Pa¨tzold, Y. Loya, M. Fine, A. H. Nawar, and G. Wefer. (2000). A coral oxygen isotope record from the northern Red Sea documenting NAO, ENSO, and North Pacific teleconnections on Middle East climate variability since the year 1750, Paleoceanography, 15, 679–694

Flocas, H.A., J. Kouroutzoglou, K. Keay and M. Hatzaki1. (2009). Cyclonic Tracks Over the Eastern Mediterranean In the Present Climate, 312–319

Flocas, H. A., Simmonds, I., Kouroutzoglou, J., Keay, K., Hatzaki, M., Bricolas, V., & Asimakopoulos, D. (2010). On Cyclonic Tracks over the Eastern Mediterranean, Journal of Climate, *23*(19), 5243–5257. Available at: https://journals.ametsoc.org/view/journals/clim/23/19/2010jcli3426.1.xml
Flood Control Center (2003) San Diego County Hydrology Manual, 322 p. San Diego County Flood Control Advisory Commission, San Diego
Fraedrich K, Muller K (1992) Climate anomalies in Europe associated with ENSO extremes. Int J Climatol 12:25–31
Funk Chris, Peterson Pete, Landsfeld Martin, Pedreros Diego, Verdin James, Shukla Shraddhanand, Husak Gregory, Rowland James, Harrison Laura, Hoell Andrew, Michaelsen Joel (2015) The climate hazards infrared precipitation with stations—a new environmental record for monitoring extremes. Scientific Data 2(150066):2015. https://doi.org/10.1038/sdata.2015.66
Giorgi F. (2006). Climate change hot spots. Geophys Res Lett
Givati A, Rosenfeld D (2013) The Arctic Oscillation, climate change and the effects on precipitation in Israel. Atmos Res 132–133:114–124
Gordana Jovanovic. (2008). The influence of Arctic and North Atlantic Oscillation on Precipitation Regime in Serbia, IOP Conf. Series,Earth and Environmental Science, 4, P.3.(3)
Gouirand I, Moron V (2003) Variability of the impact of El Nin˜o–Southern Oscillation on sea-level pressure anomalies over the North Atlantic in January to March (1874–1996). Int J Climatol 23:1549–1566
Greatbatch Richard J (2000) The North Atlantic Oscillation, Stochastic Environmental Research and Risk Assessment, Montreal, Version as of May 4. Entretiens Jacques-Cartier, Montreal
H. M. S. O. (1962). Weather in the Mediterranean I: general meteorology. 2nd ed. London, 362 p
Hafez YY, Hasanean HM (2000) The Variability of Wintertime Precipitation in The Northern Coast of Egypt and its Relationship with The North Atlantic Oscillation. Cairo University, Egypt, September, ICEHM
Hasanean H.M. (2004). Precipitation Variability Over the Mediterranean and Its Linkage With El Niño Southern Oscillation (ENSO), Journal of Meteorology,Vol.29,No.289, May/ June
Hassan, W. Abbas. (2009). “Change in some elements of climate in the Nile Delta during the twentieth century, a study using geographic information systems”. a master’s thesis. Faculty of Arts. Ain Shams University
Hatsushika H, Yamazaki K (2001) Interannual variations of temperature and vertical motion at the tropical tropopause associated with ENSO. Geophys Res Lett 28:2891–2894
Hatzianastassiou N, Gkikas A, Mihalopoulos N, Torres O, Katsoulis BD (2009) Natural versus anthropogenic aerosols in the eastern Mediterranean basin derived from multiyear TOMS and MODIS satellite data. J Geophys Res 114:D24202. https://doi.org/10.1029/2009JD011982
Hellenic National Meteorological Service. (2017). “Significant Weather and Climatic Events in Greece during 2017”, HNMS Report, 2017
Hess, P., and H. Brezowsky. (1969). Katalog der Grosswetterlagen Europas. 2. Neu bearbeitete und erga¨nzte Auflage, Ber. Dtsch. Wetterdienstes, 15(113), 56 pp
Houghton JT (1991) (1993). Scientific Assessment of Climate Change: Summary of the IPCC Working Group I Report, Paper. In: Jager J, Ferguson HL (eds) Climate Change: Science. Cambridge University Press, Cambridge, First Published, Impacts and Policy, pp 23–44
Hurrell J (1996) Influence of variations in extratropical wintertime teleconnections on Norhtern Hemisphere temperature. Geo Res Lett 23(6):665–668
Ignatius G. Rigor, John M. Wallace. (2004). Variations in The Age of Arctic Sea-ice and Summer Sea-ice Extent, Geophysical Research Letters, Vol. 31,P.1.(4)
Ineson S, Scaife AA (2009) The role of the stratosphere in the European climate response to El Niño. Nat Geosci 2:32–36. https://doi.org/10.1038/ngeo381
Intergovernmental Panel on Climate Change (IPCC). (2007). Climate Change: The Physical Science Basis, eds. S
Intergovernmental Panel on Climate Change (IPCC). (2013). Annex I: Atlas of Global and Regional Climate Projections [van Oldenborgh GJ et al. (eds.)]. In: Climate Change 2013: The Physical

Science Basis. Contribution of Working Group I to the Fifth Assessment Report of the Intergovernmental Panel on Climate Change [Stocker TF et al. (eds.)]. Cambridge University Press, Cambridge, United Kingdom and New York, NY, USA

Ionita M., Boroneanț C., Chelcea S. (2015). Seasonal modes of dryness and wetness variability over Europe and their connections with large scale atmospheric circulation and global sea surface temperature, Clim. Dyn., DOI 10.1007/s00382-015-2508-2

Jacob D (2014) EURO-CORDEX: new high-resolution climate change projections for European impact research. Reg Environ Change 14(2):563–578

Gordana Jovanovic, Reljin I, Reljin B (2008) The Influence of Arctic and North Atlantic Oscillation on Precipitation Regime in Serbia, IOP Conf. Earth and Environmental Science, Series, p 4

Kahana R, Ziv B, Enzel Y, Dayan U (2002) Synoptic climatology of major floods in the Negev Desert. Israel. Int. J. Clim. 22:822–867

Kodera K, Kuroda Y (2004) Two teleconnection patterns involved in the North Atlantic/Arctic Oscillation. Geophys Res Lett 31:L20201. https://doi.org/10.1029/2004GL020933

Korosec. M. (2019). Rising potential for shallow warm core development in the far eastern Mediterranean, Oct 24-25th, Server Weather Europe, available at: https://www.severe-weather.eu/mcd/very-warm-eastern-mediterranean-sea/

Krichak SO, Alpert P (2005a) Decadal trends in the East Atlantic/WestRussia pattern and the Mediterranean precipitation. Int J Climatol 25:183–192

Krichak SO, Alpert P (2005b) Signatures of the NAO in the atmospheric circulation during wet winter months over the Mediterranean region. Theor Appl Climatol 82(1–2):27–39

Krichak SO, Kishcha P, Alpert P (2002) Decadal trends of main Eurasian oscillations and the Mediterranean precipitation. Theor Appl Climatol 72:29–220

Krichak, S. O., & Alpert, P. (1998). Role of large-scale moist dynamics in November 1-5, 1994 hazardous Mediterranean Weather, 103(D16), 19453-19468

Krichak SO, Alpert P, Krishnamurti TN (1997a) Interaction of topography and tropospheric flow - a possible generator for the Red Sea trough? Meteorol. Atmosph. Phys. 63(3–4):149–158

Krichak SO, Alpert P, Krishnamurti TN (1997b) Red Sea trough/cyclone development - numerical investigation. Meteorol. Atmosph. Phys. 63(3–4):159–170

Krichak SO, Breitgand JS, Gualdi S, Feldstein SB (2013) Teleconnection–extreme precipitation relationships over the Mediterranean region. Theoretical and applied climatology, Springer, https://doi.org/10.1007/s00704-013-1036-4

Kuglitsch FG (2010) Heat wave changes in the eastern Mediterranean since 1960. Geophys Res Lett 37(4):L04802

Kutiel H, Maheras P, Guika S (1996) Circulation Indices Over the Mediterranean and Europe and Their Relationship with Rainfall Conditions Across the Mediterranean. Theor. Appl. Climatol. 54:125–138

Kutiel H, Maheras P, Turkes M, Paz S (2002) North Sea – Caspian Pattern (NCP) – An upper Level Atmospheric Teleconnection Affecting the Eastern Mediterranean – Implications on The Regional Climate. Theor. Appl. Climatol. 72:173–192

Kutiel H, Paz S (1998) Sea Level Pressure Departures in the Mediterranean and their Relationship with Monthly Rainfall Conditions in Israel. Theor. Appl. Climatol. 60:93–109

Lamb Peter J., Peppler Randy A. (1987). North Atlantic Oscillation:Concept and Application, Bulletin of The American Meteorological Society, Vol. 68, No. 10

Lasheen AM (1971) On the significance of the Humidity field in the northern Africa and in adjacent areas. Meteorological Department, Cairo

Levi M (1970) the Synoptics of Hail in Israel. Israel J Earth Sci 19:77–83

Li, L., Bozec, A., Somot, S., Béranger, K., Bouruet-Aubertot, P., Sevault, F., & Crépon, M. (2006). Chapter 7 Regional atmospheric, marine processes and climate modelling. Developments in Earth and Environmental Sciences, 373–397. https://doi.org/10.1016/s1571-9197(06)80010-8

Lionello, P., Bhend, J., Buzzi, A., Della-Marta, P. M., Krichak, S. O., Jansà, A., Trigo, R. (2006). Chapter 6 Cyclones in the Mediterranean region: Climatology and effects on the environment.

Developments in Earth and Environmental Sciences, 325–372. https://doi.org/10.1016/s1571-9197(06)80009-1

Lionello, P., Malanotte-Rizzoli, P., Boscolo, R., Alpert, P., Artale, V., Li, L., Xoplaki, E. (2006). The Mediterranean climate: An overview of the main characteristics and issues. Developments in Earth and Environmental Sciences, 1–26. doi:10.1016/s1571-9197(06)80003-0

López-Moreno J., Vicente-Serrano S.M., Morán-Tejeda E., Lorenzo J., El-Kenawy A., Zabalza, J., Beniston M. (2010). Impact of NAO on Winter Temperature and Precipitation Modes in The Mediterranean Mountains: Implications for Snow Cover, ESF-Medclivar Workshop on " Hydrological, Socioeconomic and Ecological Impacts of The North Atlantic Oscillation in The Mediterranean Region " 24 - 27 May, Zaragoza, Spain

Lopez-Moreno JI, Vicente-Serrano SM (2008) Extreme phases of the wintertime North Atlantic oscillation and drought occurrence over Europe: a multi-temporal-scale approach. J Clim 21:1220–1243

Luterbacher, J., Xoplaki, E., Casty, C., Wanner, H., Pauling, A., Küttel, M., Ladurie, E. L. R. (2006). Chapter 1 Mediterranean climate variability over the last centuries: A review. Developments in Earth and Environmental Sciences, 27–148. https://doi.org/10.1016/s1571-9197(06)80004-2

Maheras P, Kutiel H (1999a) Spatial and temporal variations in the temperature regime in the mediterranean and their relationship with circulation during the Last Century. Int J Climatol 19:745–764

Maheras P, Kutiel H (1999b) Spatial and temporal variations In the temperature regime In tthe mediterranean and their relationship with circulation during the last century. Int J Climatol 19(745–764):763

Manzano A (2019) Analysis of the atmospheric circulation pattern effects over SPEI drought index in Spain, Atmosp Res 24 July, 104630

Marcella Marc P, Eltahir Elfatih AB (2008) The Hydro climatology of Kuwait: explaining the variability of rainfall at seasonal and interannual time scales, J Hydrometeorol 9

Mariotti A, Zeng N., Lau K.M. (2002). Euro-Mediterranean Rainfall and ENSO—a seasonally varying relationship. Geophys Res Lett 29:12. https://doi.org/10.1029/2001gl014248

Mariotti A, Ballabrera-Poy J, Zeng N (2005) Tropical influence on Euro-Asian autumn rainfall variability. Clim Dyn 24:511–521

Martineu C, Caneill JY, Sadourny R (1999) Potential predictability of European winters from the analysis of seasonal simulations with an AGCM. J Clim 12:3033–3061

MedECC (2019) Risks associated to climate and environmental changes in the Mediterranean region, a preliminary assessment by the MedECC Network Science-policy interface, Union of the Mediterranean

Mestas-Nuñez AM, Enfield DB (1999) Rotated global modes of non-ENSO sea surface temperature variability.J. Climate 12:2734–2746

Miksovsky J, Brazdil R, Trnka M, PiSoft P (2019) Long-term variability of drought indices in the Czech Lands and effects of external forcing and large-scale climate variability modes. Climate of the Past 15:827–847

Moron M, Plaut G (2003) The impact of El Nin˜o Southern Oscillation upon weather regimes over Europe and the North Atlantic boreal winter. Int J Climatol 23:363–379

NASA (2020) Power Data Access Viewer, Climate data for Egypt. Available at: https://power.larc.nasa.gov/data-access-viewer/

Nastos PT, Karavana-Papadimou and Matsangouras IT (2015) Tropical-Like Cyclones in the mediterranean: impacts and composite daily means and anomalies of synoptic conditions. In: 14th International conference on environmental science and technology, Rhodes, Greece, 3-5 Sept 2015, CEST2015_00407

Nastos PT, Papadimou KK, Matsangouras IT (2017) Mediterranean tropical-like cyclones: impacts and composite daily mans anomalies of synoptic patterns. Atmos Res 208:156-166. https://doi.org/10.1016/j.atmosres.2017.10.023

Nazmi NM (2009) The flood disaster in Egypt, focusing on a study of one of the stricken villages, "Darnaka" village in Assiut Governorate". J Eng Sect Al-Azhar Univ 4(12). Faculty of Engineering. Al-Azhar University. Cairo

Nieto R, de la Torre L, NoguerolM., Añel JA, Gimeno L (2008) A climatology based on reanalysis of baroclinic developmental regions in the extratropical northern hemisphere. Annals of the New York Academy of Sciences, vol 1146, pp 235-255

NOAA (2020a) ENSO: Recent Evolution, Current Status and Predictions, Climate Prediction Center/ NCEP. Available at: https://www.cpc.ncep.noaa.gov/products/analysis_monitoring/lanina/enso_evolution-status-fcsts-web.pdf

NOAA (2020b) Cold & Warm Episodes by Season, Historical El Nino/ La Nina episodes (1950-resent), available at: https://origin.cpc.ncep.noaa.gov/products/analysis_monitoring/ensostuff/ONI_v5.php

North West Knowledge (2020) Climate Data Download, Egypt. Available at: https://climate.northwestknowledge.net/NWTOOLBOX/formattedDownloads.php?fbclid=IwAR2sLTy82aZRA_3iEHdd5N_r-bE1QJEiEKns3JsJ5Y0t1J19Qu1mOWC-egU

Nykjaer L (2009) Mediterranean Sea surface warming 1985-2006. Climate Res 39:11–17

OMM-JCOMM-GMDSS/World Marine Weather Forecast (2017) Limit of Met areas—2017. WMO OMM, Accessed 15 December 2019. Available at: http://weather.gmdss.org/metareas.html

Osetinsky I, Alpert P (2006) Calendaricities and multimodality in the Eastern Mediterranean cyclonic activity, Nat Hazards Earth Syst Sci 6:587–596. Available at: www.nat-hazards-earth-syst-sci.net/6/587/2006/

Lamb Peter J, Peppler Randy A (1987) North Atlantic Oscillation: concept and application. Bull Am Meteorol Soc 68(10):1218

Pozo-Va´zquez D, Esteban-Parra MJ, Rodrigo FS, CastroDiez Y (2001) The association between ENSO and winter atmospheric circulation and temperature in the North Atlantic region. J Clim 14:3408–3420

Price C, Stone L, Huppert A, Rajagopalan B, Alpert P (1998) A possible link between El Nino and precipitation in Israel. Geophys Res Lett 25:3963–3966

Quadrelli R, Pavan V, Molteni F (2001) Winter-time variability of Mediterranean precipitation and its links with large-scale circulation anomalies. Clim Dyn 17(5–6):457–466

Radinovic D (1987) Mediterranean Cyclones and their influence on the weather and climate. WMO, PSMP, NO, p 24

Reiter E (1963) Jet Stream Meteorology. Harvard Univ. Press, London

Richard JG (2000) The North Atlantic oscillation, stochastic environmental research and risk assessment. Entretiens Jacques-Cartier, Montreal, p 2

Rigor Ignatius G, Wallace John M (2004) Variations in the age of Arctic Sea-Ice and Summer Sea-ice Extent, Geophys Res Lett 31

Rimbu N, Lohmann G, Felis T, Patzold J (2003) Shift in ENSO teleconnections recorded by a northern Red Sea coral. J Clim 16:1414–1422

Rocha A (1999) Low-frequency variability of seasonal rainfall over theIberian peninsula and ENSO. Int J Climatol 19:889–901

Roeckner E, Oberhuber JM, Bacher A, Christoph M, Kirchner I (1996) ENSO variability and atmospheric response in a global coupled atmosphere-ocean GCM. Clim Dyn 12:737–754

Rogers JC, McHugh MJ (2002) On the separability of the North Atlantic oscillation and the Arctic oscillation. Climate Dyn. 19:599–608

Jc Rogers (1990) Patterns of low frequency monthly sea level pressure variability (1899–1996) and associated wave cyclone frequencies. J Clim 3:1364

Rojas M, Li L, Kanakidou M, Hatzianastassiou N, Seze G, Le Treut H (2013) Winter weather regimes over the Mediterranean region: their role for the regional climate and projected changes in the twenty-first century. Clim Dyn 2013(41):551–571. https://doi.org/10.1007/s00382-013-1823-8

Romem M, Ziv B, Saaroni H (2007) Scenarios in the development of Mediterranean cyclones, Adv Geosci 12:59–65. Available at: www.adv-geosci.net/12/59/2007/

Romero R, Emanuel K (2017) Climate change and hurricane like extratropical cyclones: projections for North Atlantic Polar Lows and medicanes based on CMIP5 models. Am Meteorol Soc 30

Salah H (2019) Mediterranean cyclones and offshore structures. Int Res J Eng Technol (IRJET) 06(12)

Sarah F (2017) What we Know about Medicanes Hurricane-Like Storms in the Mediterranean. Phys.org, November 2017, Accessed 14 December 2019, available at: https://phys.org/news/2017-11-edicaneshurricane-like-storms-mediterranean.html

Sardeshmukh PD, Compo GP, Penland C (2000) Changes of probability associated with El Nin˜o. J Clim 13:4268–4286

SCS (US SOIL CONSERVATION SERVICE). (1986). Urban Hydrology for Small Watersheds. Technical Report 55 (Springfield, VA: USDA). Available at: https://www.nrcs.usda.gov/Internet/FSE_DOCUMENTS/stelprdb044171.pdf

Shaman J, Tzipperman E (2011) An Atmospheric teleconnection linking ENSO and Southwestern European precipitation. J Clim 24:124–139

Shehadeh N (1986) The rainy season in the eastern basin of the mediterranean Sea and Arab Asi". Bulletin of the geography department research bulletin. Fac Arts Kuwait Issue 89

Shehaheh N (1973) The spatial Variation of Rainfall Spectra in Eastern the United States, Unpublished Ph.D. Thesis, Indiana University, Bloomington, U.S.A

Stephenson DB, Pavan V, Collins M, Junge MM, Quadrelli R (2006) North Atlantic Oscillation response to transient greenhouse gas forcing and the impact on European winter climate: a CMIP2 multi-model assessment. Climate Dyn 27:401–420

Stoner AK, Hayhoe K, Wuebbles DJ (2009) Assessing general circulation model simulations of atmospheric teleconnection patterns. J Climate 22(16): 4348–4372. https://doi.org/10.1175/2009JCLI2577.1

Sutton LJ, Curry PA (1931) Meteorological Atlas of Egypt. Egypt. Physical Department, Giza, Survey of Egypt

Thompson DWJ, Wallace JM (2000) Annular modes in the extratropical circulation. Part I: Month-to-month variability. Am Meteorol Soc J Climate 13:1000–1016. https://doi.org/10.1175/1520-0442(2000)013%3c1000:AMITEC%3e2.0.CO;2

Thompson DWJ, Wallace JM (1998) The Arctic Oscillation signature in the wintertime geopotential height and temperature fields. Geophys Res Lett 25:1297–1300. https://doi.org/10.1029/98GL00950

Toreti A, Naveau P (2015) On the evaluation of climate model simulated precipitation extremes. Environ Res Lett 10:

Toreti A (2013) Projections of global changes in precipitation extremes from coupled model intercomparison project phase 5 models. Geophys Res Lett 40:4887–4892

Tous, M. and Romero, R. (2012). Meteorological environments associated with medicane development. Int J Climatol 33:1–14. https://doi.org/10.1002/joc.3428

Trenberth KE (1990) Recent observed interdecadal climate changes in the Northern Hemisphere. Bull Amer Meteor Soc 71:988–993. https://doi.org/10.1175/1520-0477(1990)071%3c0988:ROICCI%3e2.0.CO;2

Trenberth KE, Hoar TJ (1997) El Nino and climate chdi UgQ. Geo phys. Res Lett 24:3057–3060

Trenberth KE, Hurrell JW (1994) Decadal atmosphere–ocean variations in the Pacific. Climate Dyn. 9:303–319. https://doi.org/10.1007/BF00204745

Trigo R, Xoplaki E, Zorita E, Luterbacher J, Krichak SO, Alpert P, Jacobeit J, Saenz J, Fernandez J, Gonzalez-Rouco F, Garcia-Herrera R, Rodo X, Brunetti M, Nanni T, Maugeri M, Turkes M, Gimeno L, Ribera P, Brunet M, Trigo IF, Crepon M, Mariotti A (2006a) Relations between variability in the Mediterranean region and mid-latitude variability. In: Lionello P, Malanotte-Rizzoli P, Boscolo R (eds) Mediterranean climate variability, vol 4. Elsevier, Amsterdam, pp 179–226

Trigo IF, Bigg GR, Davies TD (2002) A Climatology of cyclogenesis mechanisms in the Mediterranean. Mon Weather Rev 130:549–569

Trigo R, Xoplaki E, Zorita E, Luterbacher J, Krichak SO, Alpert P, Mariotti A (2006) Chapter 3 Relations between variability in the Mediterranean region and mid-latitude variability. Dev Earth Environ Sci 179–226. https://doi.org/10.1016/s1571-9197(06)80006-6
Tsimplis MN, Zervakis V, Josey SA, Peneva EL, Struglia MV, Stanev EV, Theocharis A, Lionello P, Malanotte-Rizzoli P, Artale V, Oguz T (2006) Chapter 4 Changes in the oceanography of the Mediterranean Sea and their link to climate variability. Dev Earth Environ Sci 227–282. https://doi.org/10.1016/s1571-9197(06)80007-8
Ulbrich U, Christoph M (1999) A shift of the NAO and increasing storm track activity over Europe due to anthropogenic greenhouse gas forcing. Clim Dyn 15:551–559
Ulbrich U, May W, Li L, Lionello P, Pinto JG, Somot S (2006) Chapter 8 The Mediterranean climate change under global warming. Dev Earth Environ Sci 399–415. https://doi.org/10.1016/s1571-9197(06)80011-x
UNEP (1992) Climate change fact sheet. Available at: www.unep.ch
van Loon H, Madden RA (1981) The Southern Oscillation. part I. Global associations with pressure and temperature in northern winter. Mon Weather Rev 109:1150–1162
Vicente-Serrano SM (2016) The Westerly Index as complementary indicator of the North Atlantic oscillation in explaining drought variability across Europe. Clim Dyn 47:845–863. https://doi.org/10.1007/s00382-015-2875-8
Walker GT (1923) Correlation in seasonal variations of weather. VIII. A preliminary study of world-weather. Mem. Indian Meteor Dep 24(Part 4):75–131
Walker GT, Bliss EW (1932) World weather V Mem Roy Meteor Soc 4:53–84
Wallace JM (2000) North Atlantic Oscillation/annular mode: two paradigms—one phenomenon. Q J R Meteorol Soc 126:791–805
Wallace JM, Thompson DWJ (2002) The Pacific center of action of the Northern Hemisphere annular mode: Real or artifact? J. Climate 15:1987–1991. https://doi.org/10.1175/1520-0442(2002)015%3c1987:TPCOAO%3e2.0.CO;2
Wang H, Chen Y, Pan Y, Li W (2015) Spatial and temporal variability of drought in the arid region of China and its relationships to teleconnection indices. J Hydrol 523:283–296
Winstanley. D. (1970). The North African Flood Disaster, Sep. 1969, Royal meteorological society, 1st published Sep. 1970 https://doi.org/10.1002/j.1477-8696.1970.tb04128.x
World Bank. (2014). Turn Down the Heat: Confronting the New Climate Normal. Washington, DC: World Bank. Licence: CC BY-NC-ND 3.0IGO. Available at: http://documents.worldbank.org/curated/en/317301468242098870/Main-report
World Meteorological Organization. (1998). Guide to Wave Analysis and Forecasting. WMO no. 702, Geneva, Switzerland, 2nd Edition
Wu A, Hsieh WW (2004) The nonlinear association between ENSO and the Euro-Atlantic winter sea level pressure. Clim Dyn 23:859–868
Wuethrich, B. (1995). El Nino goes critical. New Scientist, 4, (Feb), 32-35
Xoplaki E. (2002). Climate Variability over The Mediterranean, PhD,Geography Faculty of Science, the University of Bern, Switzerland
Xoplaki E, Luterbacher J, Gonz´alez-Rouco JF (2006) Mediterranean Summer Temperature and Winter Precipitation, Large-Scale Dynamics, Trends, IL Nuovo Cimento, Societ'a Italiana di Fisica, Vol. 29 C, N. 1,Gennaio-Febbraio
Xoplaki E, Gonzalez-Rouco JF, Luterbacher J, Wanner H (2004) Wet season Mediterranean precipitation variability: influence of large-scale dynamics and trends. Clim Dyn 23:63–78. https://doi.org/10.1007/s00382-004-0422-0
Xoplaki E (2002) Climate variability over the Mediterranean, Ph.D. thesis, Univ. of Bern, Bern, Switzerland
Yakir D, Lev-Yadun S, Zangvil A (1996) El Nino and tree growth near Jerusalem over the last 20 years. Glob Change Biol 2:101–105
Yosef Y, Hadas Saaroni, Pinhas Alpert (2009) Trends in Daily Rainfall Intensity Over Israel 1950/1-2003/4. The Open Atmospheric Science Journal 3:196–203

Ziv B (2001) A subtropical rainstorm associated with a tropical plume over Africa and the Middle East. Theor Appl Clim 69(1/2):91–102
Ziv B, Dayan U, Sharon D (2004b) A mid-winter, tropical extreme floodproducing storm in southern Israel: synoptic scale analysis. Meteorol Atmos Phys 10.1007/s00703-003-0054-7
Zohdy H (1971) On the interaction between extratropical and tropical disturbances over Africa as seen from satellite pictures. Meteorological Department, Cairo

Climate Change Impacts on Hydrology in the Mediterranean Part of Slovenia

Mauro Hrvatin and Matija Zorn

Abstract Although Slovenia has only 47 km of coastline in the northern Adriatic, the natural and social influences of the Mediterranean extend much further inland, into the southeastern Alps and northern Dinaric Mountains. Temperature measurements in the Mediterranean part of Slovenia show climate changes similar to those observed in most parts of Europe. The average air temperature is rising, and the increase has been most noticeable in recent decades. Climate change is also reflected in the hydrological characteristics because minimum, mean, and maximum discharges are decreasing at most gauging stations. This chapter examines the trends in average annual temperature, annual precipitation, days with precipitation over 0.1 mm, and days with snow cover, and the trends in minimum annual discharges, mean annual discharges, maximum annual discharges, and absolute maximum annual discharges, using the Mediterranean part of Slovenia as an example. Changes in the monthly discharge regimes are also presented. The period studied is from 1961 to 2018.

1 Introduction

Climate change is one of the underlying causes of hydrological changes because the climate and hydrological systems are interconnected (Kundzewicz 2008). Responses to temperature and precipitation changes thus also impact hydrological characteristics (Hartmann et al. 2013). Changes in precipitation may result in increased or decreased river discharges, and higher temperatures, which can cause snow to turn into rain, may lead to changes in discharge regimes. Hydrological responses to climate change vary by region, but the trends for Europe and the Mediterranean area indicate a likely increase in dryness (Hartmann et al. 2013), which can also result in lower discharges in southern and southeastern Europe (Stagl et al. 2014;

M. Hrvatin · M. Zorn (✉)
ZRC SAZU Anton Melik Geographical Institute, Ljubljana, Slovenia
e-mail: matija.zorn@zrc-sazu.si

M. Hrvatin
e-mail: mauro.hrvatin@zrc-sazu.si

W. Leal Filho and E. Manolas (eds.), *Climate Change in the Mediterranean and Middle Eastern Region*, Climate Change Management,
https://doi.org/10.1007/978-3-030-78566-6_4

Climate 2017). In this region, this is primarily influenced by higher temperatures and decreasing annual precipitation trends (Tošić et al. 2016).

Slovenia is located in south-central Europe, where four major natural landscape regions merge: the Mediterranean, the Alps, the Pannonian Basin, and the Dinaric Mountains (Perko et al. 2020). All these regions are characterized by increasing temperatures. In the Alps, this is reflected in less snow and more rain, and earlier snow melting, which causes changes in the river discharge regimes (Bard et al. 2015; Zorn et al. 2020). Something similar applies to rivers in the Dinaric Mountains (Čanjevac and Orešić 2018). In turn, rivers in the western Pannonian Basin (Čanjevac and Orešić 2018; Hrvatin and Zorn 2020c) and the northeastern Mediterranean area are characterized by decreasing mean annual discharges (Kovačič 2016; Čanjevac and Orešić 2018). In general, decreasing river discharges are typical of the entire Mediterranean area (Ludwig et al. 2009; Masseroni et al. 2020). Despite their natural geographical differences, a decreasing mean annual discharge trend can be observed across all Slovenian regions; this is the result of climate change and an increase in the proportion of forested land (Hrvatin et al. 2020).

This chapter examines the trends in average annual temperature, annual precipitation, days with precipitation over 0.1 mm, and days with snow cover, and the trends in minimum annual discharges, mean annual discharges, maximum annual discharges, and absolute maximum annual discharges, using the Mediterranean part of Slovenia as an example.

2 Methodology

2.1 Study Area

Mediterranean landscapes extend across southwest Slovenia along the Adriatic and in its hinterland, between Italy to the northwest and Croatia to the south. They cover 8.6% of Slovenia. Toward the east, they gradually change into Dinaric landscapes, and to the north, in the Soča Valley, they approach Alpine landscapes (Perko and Ciglič 2020). A purely Mediterranean character can only be ascribed to the belt several kilometers wide along the 47 km Adriatic coast, but the natural and social-geographical effects of the Mediterranean extend much further into the interior (Perko 1998; Staut et al. 2007; Zorn et al. 2013). Therefore, the area presented in this article is somewhat larger (Fig. 1): it covers just under one-fifth of Slovenia and includes the entire basins of the Slovenian (Mediterranean) rivers belonging to the Adriatic watershed. The entire study area is characterized by a sub-Mediterranean precipitation regime (Ogrin 1996).

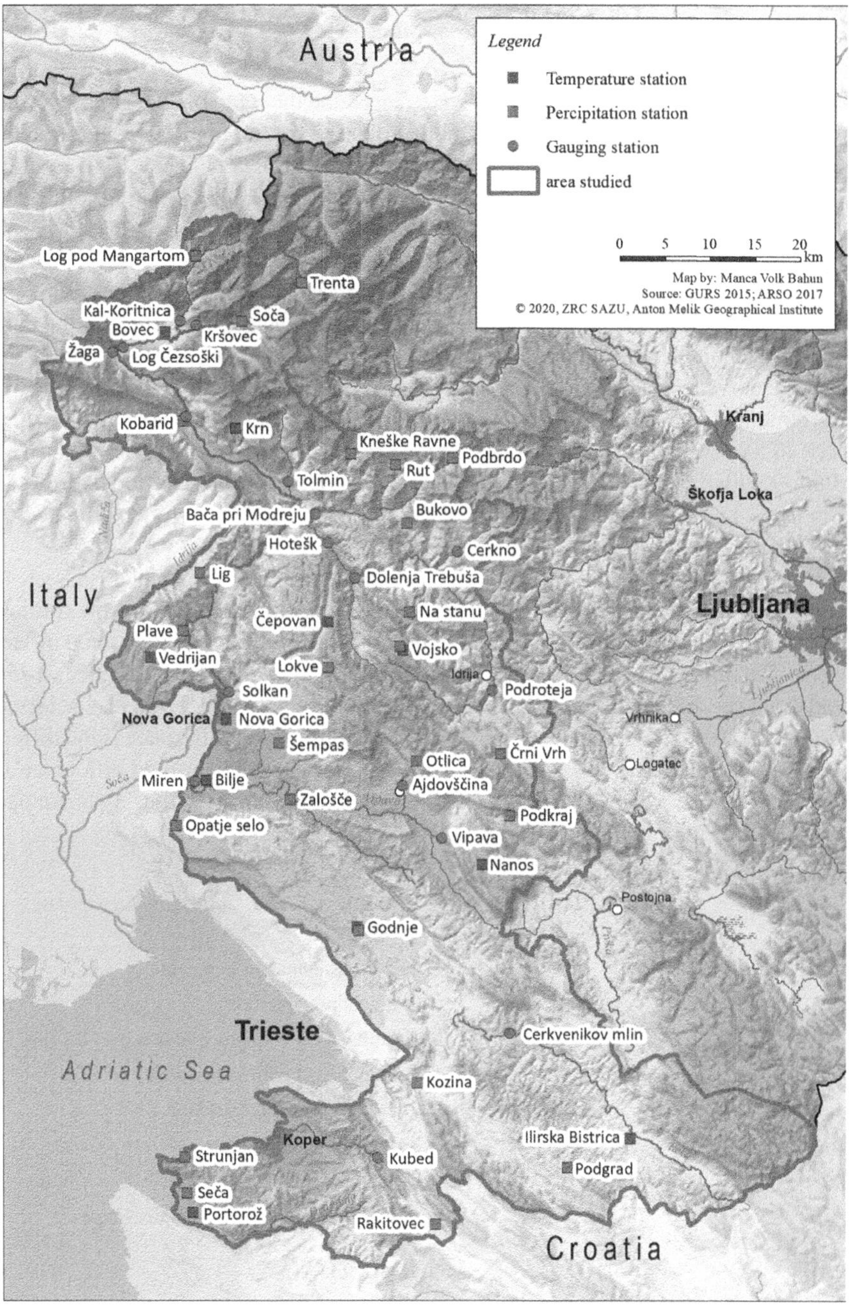

Fig. 1 Locations of temperature, precipitation, and gauging stations included in the analysis. *Source* authors

2.2 Methods

Two non-parametric methods were used to determine the trend of change in selected climate and hydrological variables (Table 1) between 1961 and 2018: the Mann–Kendall test and Theil–Sen estimator (Sen's slope estimator). They were used to analyze data from selected temperature, precipitation, and gauging stations (Tables 2 and 3, Fig. 1). For calculating the Mann–Kendall test and Sen's slope estimator values, the free online software MAKESENS 1.0 (Mann–Kendall test for trend and Sen's slope estimates; Salmi et al. 2002) was used. The "Mann–Kendall test is a nonparametric test used to detect monotonic trends. It is not sensitive to outliers in the data and is based on the test statistics. A positive test statistic implies an increasing trend, and a negative test statistic indicates a decreasing trend. Sen's slope estimator is the most frequent non-parametric test used for detecting linear time trends" (Hrvatin and Zorn 2020c: 163; see also Salmi et al. 2002; Kraner Šumenjak and Šuštar 2011). The Mann–Kendall test is widely used to detect trends in meteorological and hydrologic data (e.g., Yue and Wang 2004; Hamed 2008; Gocic and Trajkovic 2013; da Silva et al. 2015; Tošić et al. 2016; Liu et al. 2020).

In addition to the Mann–Kendall test and Sen's slope values, the tables (in Sect. 4) with climate and hydrological variables also show the confidence levels, the (initial) 1961 trend value, the (final) 2018 trend value, and absolute and relative trend differences. The initial 1961 trend value represents the value of a selected variable in 1961 read from the trend line, and the final 2018 trend value is the value of a selected variable in 2018 read from the trend line. The absolute trend difference is the difference between the final and initial trend values, and the relative trend difference is the difference between the final and initial trend values expressed as a percentage (Hrvatin and Zorn 2020c).

Table 1 Climate and hydrological variables examined

Variable type	Variable
Climate	Average annual air temperature
	Annual precipitation
	Annual days with precipitation over 0.1 mm
	Annual days with snow cover
Hydrological	Minimum annual discharge
	Mean annual discharge
	Maximum annual discharge
	Absolute maximum annual discharges

Table 2 Weather stations with the time series analyzed

Station type	Weather station	Municipality	Elevation, m	Time series	No. of annual measurements
Temperature	Bovec	Bovec	426	1961–2018	58
	Krn	Kobarid	910	1961–2018	58
	Čepovan	Nova Gorica	593	1961–2018	58
	Vojsko	Idrija	1,070	1961–2018	58
	Vedrijan	Brda	258	1961–2018	58
	Nova Gorica	Nova Gorica	112	1961–2018	58
	Bilje	Miren-Kostanjevica	51	1961–2018	58
	Nanos	Vipava	905	1961–2018	58
	Godnje	Sežana	295	1961–2018	58
	Ilirska Bistrica	Ilirska Bistrica	424	1961–2018	57
	Portorož Airport	Piran	2	1961–2018	58
Precipitation	Log pod Mangartom	Bovec	650	1961–2018	55
	Trenta	Bovec	640	1961–2018	56
	Soča	Bovec	485	1961–2018	58
	Kobarid	Kobarid	240	1961–2018	58
	Kneške Ravne	Tolmin	739	1961–2018	54
	Podbrdo	Tolmin	503	1961–2018	56
	Rut	Tolmin	695	1961–2018	58
	Bukovo	Cerkno	715	1961–2018	57
	Lig	Kanal ob Soči	620	1961–2018	55
	Na Stanu	Cerkno	989	1961–2018	53
	Plave	Kanal ob Soči	86	1961–2018	55
	Vojsko	Idrija	1,070	1961–2016	56
	Lokve	Nova Gorica	955	1961–2018	55
	Šempas	Nova Gorica	85	1961–2018	57
	Črni Vrh	Idrija	688	1961–2018	56
	Otlica	Ajdovščina	835	1961–2018	54
	Zalošče	Nova Gorica	67	1961–2018	55
	Podkraj	Ajdovščina	796	1961–2018	57
	Opatje Selo	Miren-Kostanjevica	165	1961–2018	58
	Godnje	Sežana	295	1961–2018	55
	Kozina	Hrpelje-Kozina	500	1961–2017	53

(continued)

Table 2 (continued)

Station type	Weather station	Municipality	Elevation, m	Time series	No. of annual measurements
	Strunjan	Piran	2	1961–2018	57
	Podgrad	Ilirska Bistrica	596	1961–2018	55
	Seča	Piran	2	1961–2018	58
	Rakitovec	Koper	520	1961–2018	57

3 Data

3.1 Climate Data

Climate data were obtained from the Slovenian Environment Agency (Arhiv meteoroloških 2020). The analysis included eleven temperature stations and twenty-five precipitation stations in the Mediterranean part of Slovenia (Table 2, Fig. 1) that had measured data spanning several decades.

3.2 Hydrological Data

Hydrological data were obtained from the Slovenian Environment Agency (Arhiv hidroloških 2020). The analysis included seventeen stations in the Mediterranean part of Slovenia (Table 3, Fig. 1) that had measurement data spanning several decades.

4 Results

4.1 Climate Variables

The following were analyzed in terms of climate variables (Table 1): (1) average annual air temperature trends, (2) annual precipitation trends, (3) trends in annual days with precipitation over 0.1 mm, and (4) trends in annual days with snow cover.

4.1.1 Average Annual Air Temperature

The average annual air temperature trends from 1961 to 2018 were similar at all eleven temperature stations examined, and they indicate a clear increasing trend (Table 4, Fig. 2). There is also an exceptionally high confidence level, which is 99.9% at all temperature stations.

Table 3 Rivers with the time series analyzed

River	Gauging station	Municipality	Elevation, m	Time series	No. of annual measurements	Drainage area, km^2	Mean annual discharge 1961–2018, m^3/s
Koritnica	Kal-Koritnica	Bovec	405	1961–2017	56	86.04	7.18
Soča	Kršovec	Bovec	402	1961–2018	56	157.90	11.52
Soča	Log Čezsoški	Bovec	341	1961–2018	57	324.74	24.47
Učja	Žaga	Bovec	342	1961–2018	58	50.21	3.38
Soča	Kobarid	Kobarid	196	1961–2018	58	437.02	33.75
Tolminka	Tolmin	Tolmin	168	1961–2018	53	73.37	7.90
Bača	Bača pri Modreju	Tolmin	164	1961–2018	58	142.31	6.82
Cerknica	Cerkno	Cerkno	296	1961–2018	54	40.30	1.75
Trebuša	Dolenja Trebuša	Tolmin	186	1961–2018	56	54.70	2.70
Idrijca	Podroteja	Idrija	327	1961–2018	57	112.84	8.99
Idrijca	Hotešk	Tolmin	161	1961–2018	57	442.83	23.76
Soča	Solkan	Nova Gorica	54	1961–2018	57	1,572.80	93.88
Hubelj	Ajdovščina	Ajdovščina	107	1961–2018	56	93.14	2.93
Vipava	Vipava	Vipava	96	1961–2018	58	131.90	6.49
Vipava	Miren	Miren–Kostanjevica	36	1961–2018	57	589.96	17.87
Reka	Cerkvenikov mlin	Divača	342	1961–2018	58	377.89	8.06
Rižana	Kubed	Koper	61	1961–2018	55	204.50	3.94

Table 4 Average annual air temperature trends, 1961–2018. The trend difference as a percentage is calculated based on the absolute (Kelvin) temperature scale

Temperature station	Mann–Kendall test, Z	Confidence level, %	Sen's slope, *Q*	1961 trend value, °C	2018 trend value, °C	1961–2018 trend difference, °C	1961–2018 trend difference, %
Bovec	6.19	99.9	0.037	8.70	10.83	2.13	0.76
Krn	6.20	99.9	0.038	7.01	9.20	2.19	0.78
Čepovan	6.04	99.9	0.037	7.90	10.02	2.12	0.76
Vojsko	5.92	99.9	0.040	5.23	7.48	2.25	0.81
Vedrijan	6.10	99.9	0.036	12.08	14.13	2.05	0.72
Nova Gorica	6.45	99.9	0.038	11.36	13.50	2.14	0.75
Bilje	6.51	99.9	0.038	11.22	13.39	2.17	0.76
Nanos	6.47	99.9	0.042	6.44	8.86	2.42	0.87
Godnje	6.56	99.9	0.042	9.99	12.38	2.39	0.84
Ilirska Bistrica	5.88	99.9	0.029	9.32	10.95	1.63	0.58
Portorož Airport (Fig. 2)	6.29	99.9	0.038	12.05	14.20	2.15	0.75

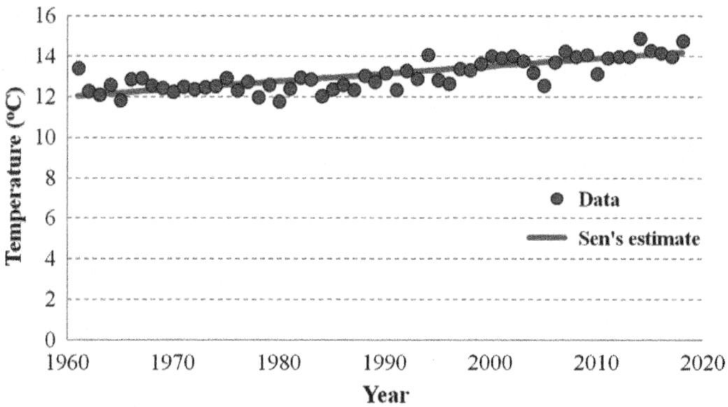

Fig. 2 Average annual temperature trend at the Portorož Airport temperature station, 1961–2018. *Source* authors

During the period studied (i.e., 1961–2018), the temperature at the selected temperature stations rose by 0.029–0.042 °C a year on average, which means that over the past six decades the temperatures at these stations increased by 1.63–2.42 °C. The absolute temperature difference was the smallest at Ilirska Bistrica, where the temperature rose by 1.63 °C, and the greatest at Nanos, where it rose by 2.42 °C.

4.1.2 Annual Precipitation

In contrast to the increasing temperature trends, the annual precipitation trends from 1961 to 2018 indicate a decreasing trend at twenty-three of the twenty-five precipitation stations studied (Table 5, Fig. 3). The confidence level is very modest because at thirteen stations it does not even reach 90%, which is far from statistical significance (95%). All twelve stations with a 95% confidence level show a decreasing annual precipitation trend.

Table 5 Annual precipitation trends, 1961–2018

Precipitation station	Mann–Kendall test, *Z*	Confidence level, %	Sen's slope, *Q*	1961 trend value, mm	2018 trend value, mm	1961–2018 trend difference, mm	1961–2018 trend difference, %
Log pod Mangartom	−0.61	<90.0	−2.043	2,509.82	2,393.34	−116.48	−4.64
Trenta	−1.60	<90.0	−5.226	2,298.73	2,000.82	−297.91	−12.96
Soča	0.89	<90.0	3.870	2,309.42	2,530.04	220.62	9.55
Kobarid	−1.63	<90.0	−5.501	2,777.41	2,463.85	−313.56	−11.29
Kneške Ravne	−2.54	95.0	−8.018	3,164.54	2,707.53	−457.01	− 14.44
Podbrdo	−1.34	<90.0	−5.102	2,340.03	2,049.22	−290.81	−12.43
Rut	0.35	<90.0	1.361	2,413.15	2,490.73	77.58	3.21
Bukovo	−3.14	99.0	−6.746	2,181.87	1,797.38	−384.49	−17.62
Lig	−0.58	<90.0	−1.717	2,349.20	2,251.36	−97.84	−4.16
Na Stanu	−2.62	99.0	−6.536	2,048.34	1,675.79	−372.55	−18.19
Plave	−2.57	95.0	−7.344	2,107.86	1,689.26	−418.60	−19.86
Vojsko (Fig. 3)	−2.13	95.0	−6.287	2,569.49	2,211.11	−358.38	−13.95
Lokve	−1.86	90.0	−6.727	2,562.92	2,179.46	−383.46	−14.96
Šempas	−0.23	<90.0	−0.751	1,500.08	1,457.25	−42.83	−2.85
Črni Vrh	−1.46	<90.0	−4.365	2,641.26	2,392.47	−248.79	−9.42
Otlica	−2.25	95.0	−8.176	2,568.44	2,102.38	−466.06	−18.15
Zalošče	−1.44	99.0	−2.969	1,574.76	1,405.52	−169.24	−10.75
Podkraj	−1.71	90.0	−4.427	2,282.77	2,030.43	−252.34	−11.05
Opatje Selo	−0.32	<90.0	−1.000	1,418.10	1,361.10	−57.00	−4.02
Godnje	−1.54	<90.0	−3.833	1,475.03	1,256.53	−218.50	−14.81
Kozina	−3.15	99.0	−6.322	1,613.70	1,253.36	−360.34	−22.33
Strunjan	−2.06	95.0	−2.706	1,035.90	881.63	−154.27	−14.89
Podgrad	−2.45	95.0	−6.161	1,677.11	1,325.95	−351.16	−20.94
Seča	−2.05	95.0	−3.244	1,054.55	869.65	−184.90	−17.53
Rakitovec	−2.09	95.0	−4.330	1,627.61	1,380.78	−246.83	−15.16

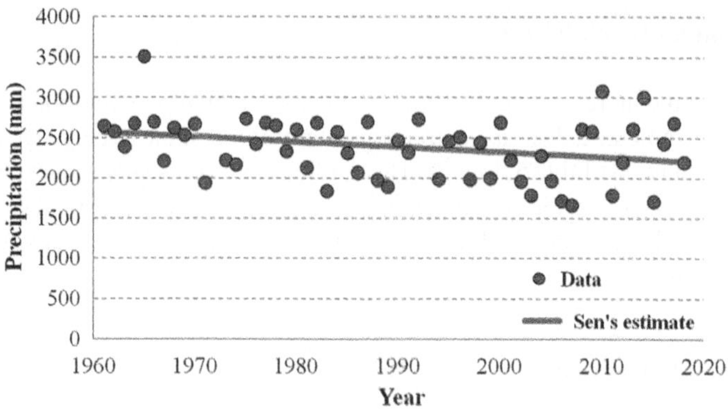

Fig. 3 Annual precipitation trend at the Vojsko precipitation station, 1961–2018. *Source* authors

Most differences in the annual precipitation are moderate. At seven precipitation stations out of twenty-five, these differences do not exceed 10%, and at twenty-three they do not exceed 20%. A decreasing annual precipitation trend is the most distinct at Kozina (−360.3 mm or −22.3%), Podgrad (−351.2 mm or −20.9%), and Plave (−418.6 mm or −19.9%), and an increasing trend is the most evident at Soča (220.6 mm or 9.6%) and Rut (77.6 mm or 3.2%).

4.1.3 Annual Days with Precipitation Over 0.1 mm

From 1961 to 2018, the annual number of days with precipitation over 0.1 mm increased at sixteen precipitation stations, decreased at four, and remained the same at five (Table 6, Fig. 4). The confidence level is very modest because at twenty-one stations it does not even reach 90%. At the remaining four precipitation stations (i.e., at Trenta, Bukovo, Šempas, and Podgrad) the confidence level is at least 99.0%, with a distinctly increasing trend.

Seventeen stations show smaller negative or positive trend deviations, not reaching 10%. The decrease in the annual days with precipitation over 0.1 mm is the most distinct at Seča (−13.7 days or −12.5%), Zalošče (−10.9 days or −9.0%), and Soča (−9.0 days or −6.3%), and the increase is the greatest at Trenta (33.5 days or 25.8%), Bukovo (28.5 days or 23.5%), and Podgrad (23.7 days or 21.3%).

4.1.4 Annual Days with Snow Cover

Between 1961 and 2018, the number of days with snow cover greatly decreased at all precipitation stations where it snows every year (Table 7, Fig. 5). The confidence level at most of these stations is high and statistically significant.

At seven stations (Plave, Šempas, Zalošče, Opatje Selo, Godnje, Strunjan, and Seča), the snow cover is more of an exception and therefore their trends cannot be

Table 6 Trends in annual days with precipitation, 1961–2018

Precipitation station	Mann–Kendall test, Z	Confidence level, %	Sen's slope, Q	1961 trend value, days	2018 trend value, days	1961–2018 trend difference, days	1961–2018 trend difference, %
Log pod Mangartom	1.65	90.0	0.200	130.60	142.00	11.40	8.73
Trenta	4.54	99.9	0.587	129.54	163.01	33.47	25.83
Soča	−1.53	<90.0	−0.158	142.92	133.92	−9.00	−6.30
Kobarid	0.09	<90.0	0.000	140.00	140.00	0.00	0.00
Kneške Ravne	1.49	<90.0	0.167	149.08	158.58	9.50	6.37
Podbrdo	−0.01	<90.0	0.000	148.00	148.00	0.00	0.00
Rut	−0.07	<90.0	0.000	147.00	147.00	0.00	0.00
Bukovo	3.33	99.9	0.500	121.50	150.00	28.50	23.46
Lig	−0.82	<90.0	−0.118	138.06	131.35	−6.71	−4.86
Na Stanu	0.84	<90.0	0.138	131.21	139.05	7.84	5.97
Plave	−0.05	<90.0	0.000	120.00	120.00	0.00	0.00
Vojsko	1.53	<90.0	0.229	150.58	163.62	13.04	8.66
Lokve	1.19	<90.0	0.222	140.67	153.33	12.66	9.00
Šempas (Fig. 4)	2.63	99.0	0.340	112.06	131.47	19.41	17.32
Črni Vrh	0.54	<90.0	0.103	146.59	152.47	5.88	4.01
Otlica	1.80	90.0	0.348	134.65	154.48	19.83	14.73
Zalošče	−1.20	<90.0	−0.190	120.14	109.29	−10.85	−9.03
Podkraj	1.76	90.0	0.286	142.00	158.29	16.29	11.47
Opatje Selo	1.35	<90.0	0.160	107.60	116.72	9.12	8.48
Godnje	0.13	<90.0	0.000	131.00	131.00	0.00	0.00
Kozina	1.52	<90.0	0.247	114.75	128.84	14.09	12.28
Strunjan	1.09	<90.0	0.155	102.45	111.30	8.85	8.64
Podgrad	2.59	99.0	0.417	111.33	135.06	23.73	21.31
Seča	−1.56	<90.0	−0.240	109.76	96.08	−13.68	−12.46
Rakitovec	0.94	<90.0	0.167	106.17	115.67	9.50	8.95

compared with other stations. At thirteen of the remaining eighteen precipitation stations, the number of days with snow cover decreased by 30–60%, or on average by thirty-six days a year. It decreased by less than 30% at Otlica (−12.7%), Podbrdo (−17.2%), and Rakitovec (−22.8%), and by more than 60% at Podgrad (−69.4%) and Bukovo (−63.8%).

The duration of snow cover expressed in the number of days decreased the most at Log pod Mangartom (−58.3 days), Lokve (−57.0 days), and Vojsko (−50.2 days).

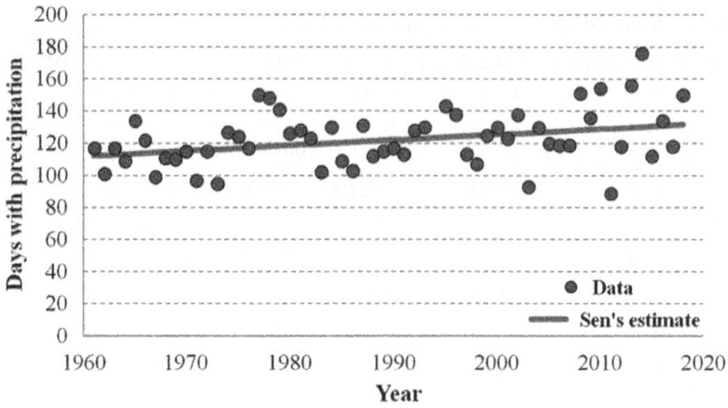

Fig. 4 Trends in annual days with precipitation at the Šempas precipitation station, 1961–2018. *Source* authors

4.2 *Hydrological Variables*

The following were examined in terms of hydrological variables (Table 1): (1) minimum annual discharge trends, (2) mean annual discharge trends, and (3) maximum annual discharge trends.

4.2.1 Minimum Annual Discharge

The minimum annual discharge trends from 1961 to 2018 were decreasing for all the watercourses analyzed except the Reka (Table 8, Fig. 6). The confidence level varies greatly: it does not exceed 90% at eleven gauging stations, it reaches 95% for the Bača, Trebuša, and upper Vipava, and at least 99% for the lower Idrijca, Cerknica, and Rižana.

Except for the lower reaches of the Soča, the minimum annual discharge in the period observed decreased by 0.04–2.44 m^3/s or 40–2,440 l/s. Four gauging stations recorded a relative decrease of up to 10%, seven stations recorded a relative decrease of 10–20%, three stations of 20–30%, and two stations over 30%. The greatest absolute decrease was recorded for the Soča at Solkan, where the discharge decreased by 10.25 m^3/s, and the relative difference was the greatest for the Rižana at Kubed, where the minimum annual discharge decreased by 46.9%. For the Reka, which is the only Slovenian Mediterranean river with a positive trend, the minimum annual discharge increased by a minute 0.02 m^3/s.

4.2.2 Mean Annual Discharge

The mean annual discharge trends from 1961 to 2018 were decreasing for all rivers and at all the gauging stations observed (Table 9, Fig. 7). The confidence level varies

Table 7 Trends in the annual days with snow cover, 1961–2018

Precipitation station	Mann–Kendall test, Z	Confidence level, %	Sen's slope, Q	1961 trend value, days	2018 trend value, days	1961–2018 trend difference, days	1961–2018 trend difference, %
Log pod Mangartom	−3.69	99.9	−1.023	113.02	54.69	−58.33	−51.61
Trenta	−2.96	99.0	−0.800	106.50	60.90	−45.60	−42.82
Soča	−3.05	99.0	−0.813	91.84	45.53	−46.31	−50.43
Kobarid	−1.96	95.0	−0.311	34.47	16.71	−17.76	−51.53
Kneške Ravne	−2.98	99.0	−0.611	64.17	29.33	−34.84	−54.29
Podbrdo	−0.79	<90.0	−0.108	35.81	29.65	−6.16	−17.19
Rut	−2.89	99.0	−0.571	59.43	26.86	−32.57	−54.80
Bukovo	−4.05	99.9	−0.857	76.57	27.72	−48.85	−63.80
Lig	−2.13	95.0	−0.229	25.97	12.94	−13.03	−50.18
Na Stanu	−2.75	99.0	−0.593	105.33	71.56	−33.77	−32.06
Plave	0.19	<90.0	0.000	4.00	4.00	0.00	0.00
Vojsko	−3.90	99.9	−0.880	152.23	102.05	−50.18	−32.96
Lokve (Fig. 5)	−3.39	99.9	−1.000	111.50	54.50	−57.00	−51.12
Šempas	−0.76	<90.0	0.000	1.00	1.00	0.00	0.00
Črni Vrh	−3.69	99.9	−0.840	104.64	56.76	−47.88	−45.76
Otlica	−0.55	<90.0	−0.156	70.28	61.38	−8.90	−12.67
Zalošče	−2.22	95.0	−0.005	1.10	0.81	−0.29	−26.03
Podkraj	−1.76	90.0	−0.383	68.56	46.69	−21.87	−31.90
Opatje Selo	−0.29	<90.0	0.000	1.00	1.00	0.00	0.00
Godnje	−2.92	99.0	−0.171	12.20	2.43	−9.77	−80.08
Kozina	−1.20	<90.0	−0.115	17.11	10.54	−6.57	−38.41
Strunjan	−2.48	95.0	−0.003	0.21	0.06	−0.15	−71.74
Podgrad	−2.58	99.0	−0.342	28.08	8.59	−19.49	−69.41
Seča	−1.95	90.0	0.000	0.10	0.10	0.00	0.00
Rakitovec	−0.86	<90.0	−0.042	10.42	8.04	−2.38	−22.82

from only 90% or lower at nine stations to 95% for the Bača, Idrijca, Hubelj, and upper Vipava, and to at least 99% for the Cerknica, Trebuša, and Rižana.

During the period studied, the mean annual discharge of the Slovenian Mediterranean rivers decreased by 0.35–10.03 m^3/s. Most rivers recorded a relative decrease of 5–25%, with a major decrease only observed for the Cerknica and Rižana. Between 1961 and 2018, the absolute trend difference in the mean annual discharge was the greatest for the Soča at Solkan, where the discharge decreased by 10.03 m^3/s, and

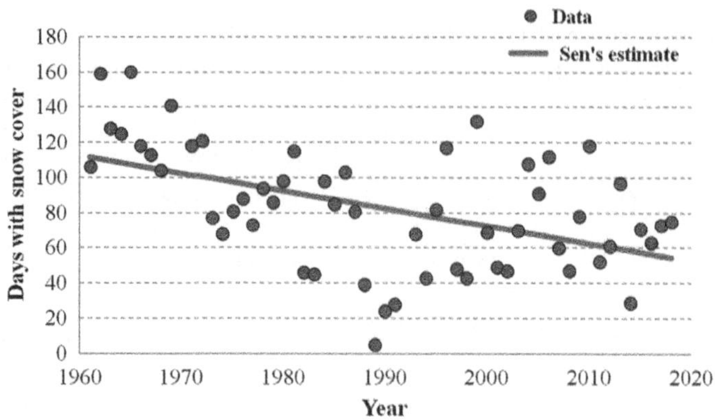

Fig. 5 Trend in annual days with snow cover at the Lokve precipitation station, 1961–2018. *Source* authors

the greatest relative trend difference can be observed for the Rižana at Kubed, where the discharge decreased by 36.5%.

4.2.3 Maximum Annual Discharge

The maximum annual discharge trends from 1961 to 2018 were decreasing for all the watercourses analyzed except for the Soča at Kobarid (Table 10, Fig. 8). The confidence level is low for rivers that have their source in the Julian Alps (e.g., the Koritnica, Soča, and Tolminka), not even reaching 90%. In contrast, the confidence level for the rivers in the direct hinterland of the Adriatic Sea (e.g., the Hubelj, Reka, and Rižana) is at least 95%. During the period observed, the maximum annual discharge largely decreased by 5–30%. A decrease over 40% was recorded for the Cerknica, Trebuša, upper Idrijca, and Rižana. The greatest absolute decrease was recorded for the Idrijca at Podroteja, where the discharge decreased by 25.77 m^3/s, and the relative difference was the greatest for the Trebuša at Dolenja Trebuša, where the discharge decreased by 49.1%. An increase in the maximum annual discharge was only observed for the Soča at Kobarid, where the discharge increased by 5.80 m^3/s or 5.7%.

4.3 Discharge Regimes

Long-term changes in temperature and precipitation not only affect the volume of the minimum, mean, and maximum discharge, but also have a significant impact on changes in the discharge regime (Hrvatin 1998; Frantar 2005; Frantar and Hrvatin 2005; Hrvatin and Zorn 2017b, 2020a, b). Among the climate indicators examined,

Table 8 Minimum annual discharge trends, 1961–2018

River	Gauging station	Mann–Kendall test, Z	Confidence level, %	Sen's slope, Q	1961 trend value, m^3/s	2018 trend value, m^3/s	1961–2018 trend difference, m^3/s	1961–2018 trend difference, %
Koritnica	Kal-Koritnica	−1.05	<90.0	−0.006	4.40	4.07	−0.33	−7.43
Soča	Kršovec	−1.48	<90.0	−0.016	6.11	5.18	−0.93	−15.20
Soča	Log Čezsoški	−1.07	<90.0	−0.022	13.05	11.80	−1.25	−9.54
Učja	Žaga	−1.58	<90.0	−0.004	1.22	1.00	−0.22	−18.10
Soča	Kobarid	−1.37	<90.0	−0.030	17.50	15.81	−1.69	−9.63
Tolminka	Tolmin	−0.18	<90.0	−0.001	3.25	3.21	−0.04	−1.20
Bača	Bača pri Modreju	−2.16	95.0	−0.011	3.46	2.84	−0.62	−17.89
Cerknica	Cerkno	−3.55	99.9	−0.006	0.97	0.62	−0.35	−35.98
Trebuša	Dolenja Trebuša	−2.56	95.0	−0.006	1.29	0.95	−0.34	−26.55
Idrijca	Podroteja	−1.31	<90.0	−0.005	2.54	2.24	−0.30	−11.74
Idrijca	Hotešk	−3.17	99.0	−0.043	9.44	7.00	−2.44	−25.87
Soča	Solkan	−2.49	<90.0	−0.180	41.35	31.10	−10.25	−24.79
Hubelj	Ajdovščina	−1.50	<90.0	−0.003	1.00	0.82	−0.18	−17.64
Vipava (Fig. 6)	Vipava	−2.44	95.0	−0.008	2.27	1.83	−0.44	−19.56
Vipava	Miren	−1.65	90.0	−0.013	5.04	4.30	−0.74	−14.74
Reka	Cerkvenikov mlin	0.11	<90.0	0.000	1.87	1.89	0.02	0.83
Rižana	Kubed	−3.69	99.9	−0.008	1.00	0.53	−0.47	−46.92

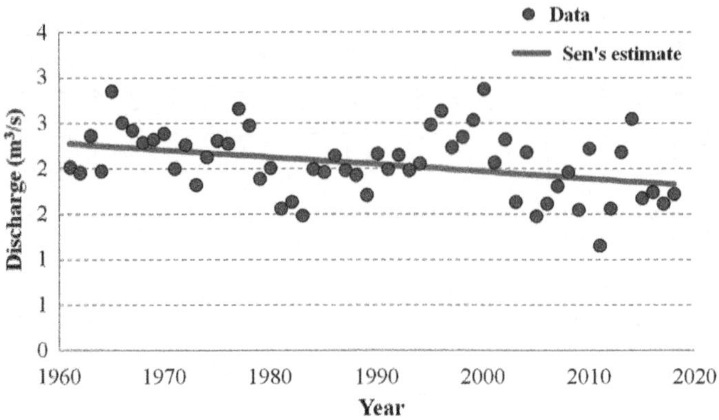

Fig. 6 Minimum annual discharge trend of the Vipava river at the Vipava gauging station, 1961–2018. *Source* authors

changes in temperature seem especially important because they affect evapotranspiration, as do changes in the number of days with snow cover because they strongly affect all discharge regimes with a distinct snow share. In the Mediterranean part of Slovenia, the snow share is more or less important for rivers with headwaters in the Julian Alps and the high Dinaric plateaus, such as the Trnovo Forest Plateau (*Trnovski gozd*) and the Snežnik Plateau.

In classifying the discharge regimes based on the 1961–1990 data set (Hrvatin 1998), rivers in the Mediterranean part of Slovenia were categorized into three groups. The Koritnica, Soča, Učja, and Tolminka, which have their headwaters high up in the Julian Alps, were categorized among rivers with a snow–rain regime. Due to snow melting, their main discharge maximum was in late spring, and their secondary maximum was the result of autumn rain. Due to snow retention, their main discharge minimum was in winter, and due to stronger evapotranspiration their secondary minimum was in the middle of the summer. Snow was retained for four to five months.

The Bača, Idrijca, Cerknica, Trebuša, Vipava, and Hubelj, which have their source in the prealpine hills or in the foothills of high Dinaric plateaus, had a rain–snow regime. Their main discharge maximum was in the fall due to autumn rain, and their secondary maximum was in the spring, when the snow was melting. They had their main discharge minimum in the summer and a secondary one in the winter. Snow was retained for one to three months.

The Reka and Rižana, which are the only two proper Mediterranean rivers among all the ones studied, had a simple rain regime. Their discharge maximum was in the fall, after which the volume of water did not decrease significantly up until April. They reached their discharge minimum in the middle of the summer.

After nearly thirty years, many rivers have experienced a significant change in their discharge regime. The Koritnica, Soča, Učja, and Tolminka now have their main discharge maximums in November and no longer in late spring. The volume of snow melting in the mountains in May is now too small to exceed the rainfall at the end of

Table 9 Mean annual discharge trends, 1961–2018

River	Gauging station	Mann–Kendall test, *Z*	Confidence level, %	Sen's slope, *Q*	1961 trend value, m^3/s	2018 trend value, m^3/s	1961–2018 trend difference, m^3/s	1961–2018 trend difference, %
Koritnica	Kal-Koritnica	– 1.54	<90.0	– 0.018	7.58	6.53	– 1.05	– 13.81
Soča	Kršovec	– 1.81	90.0	– 0.037	12.67	10.57	– 2.10	– 16.56
Soča	Log Čezsoški	– 1.84	90.0	– 0.066	25.82	22.03	– 3.79	– 14.67
Učja	Žaga	– 1.15	<90.0	– 0.007	3.53	3.15	– 0.38	– 10.80
Soča	Kobarid	– 0.68	<90.0	– 0.035	34.54	32.58	– 1.96	– 5.68
Tolminka	Tolmin	– 0.68	<90.0	– 0.006	7.74	7.39	– 0.35	– 4.50
Bača	Bača pri Modreju	– 2.05	95.0	– 0.024	7.61	6.26	– 1.35	– 17.75
Cerknica	Cerkno	– 3.18	99.0	– 0.011	1.99	1.37	– 0.62	– 31.19
Trebuša	Dolenja Trebuša	– 3.50	99.9	– 0.016	3.15	2.24	– 0.91	– 29.00
Idrijca	Podroteja	– 2.07	95.0	– 0.044	9.89	7.36	– 2.53	– 25.54
Idrijca	Hotešk	– 1.99	95.0	– 0.066	26.05	22.31	– 3.74	– 14.36
Soča	Solkan	– 0.94	<90.0	– 0.176	97.64	87.61	– 10.03	– 10.27
Hubelj	Ajdovščina	– 2.32	95.0	– 0.011	3.23	2.58	– 0.65	– 20.19
Vipava	Vipava	– 2.30	95.0	– 0.020	7.12	5.98	– 1.14	– 16.06
Vipava	Miren	– 0.96	<90.0	– 0.033	18.44	16.56	– 1.88	– 10.19
Reka	Cerkvenikov mlin	– 1.61	<90.0	– 0.029	8.99	7.35	– 1.64	– 18.21
Rižana (Fig. 7)	Kubed	– 3.08	99.0	– 0.031	4.80	3.05	– 1.75	– 36.45

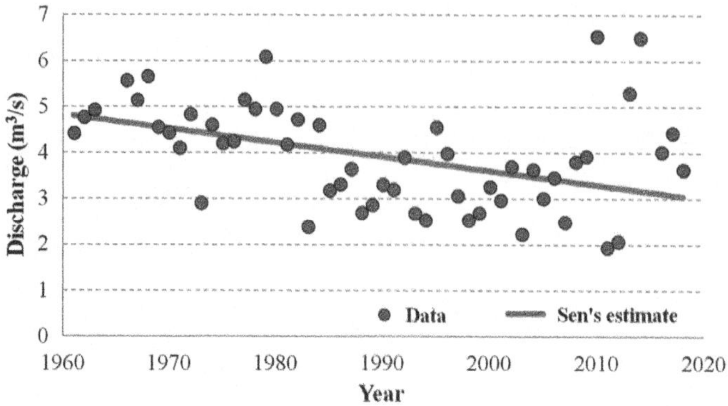

Fig. 7 Mean annual discharge trend of the Rižana river at the Kubed gauging station, 1961–2018. *Source* authors

the fall. Only the winter discharge minimum, which is still more pronounced than the summer minimum, continues to indicate that these rivers used to have a snow–rain regime.

The discharge regimes of the Bača, Idrijca, Cerknica, Trebuša, Vipava, and Hubelj have also changed. The former April high waters have been replaced by a November rain maximum. The main discharge minimums for the rivers with a rain–snow regime have remained in August and they have become even more pronounced; the secondary winter discharge minimums are already above the annual average.

Changes in the discharge regimes of the Reka and Rižana have been relatively small. Their main discharge maximums have remained in November and their main discharge minimums have remained in July. Because of a smaller and shorter effect of snow cover melting on the Snežnik high karst plateau, the secondary discharge maximum for the Reka has moved from April to February. Similar conditions can be observed for the Rižana, which obtains a portion of its water from the Brkini Hills, which used to receive significantly more snow. The secondary maximum for the Rižana has moved from March to February.

A comparison of the discharge regimes based on the 1961–1990 data set with the discharge regimes based on the 1991–2018 data set primarily shows the following differences (Tables 11, Fig. 9):

- For rivers that used to have a pronounced snow–rain regime (i.e., the Koritnica, Soča, Učja, and Tolminka), the main discharge maximum has moved from the spring to the fall, which shows a significantly reduced impact of snow cover and the gradual transition into a rain–snow regime;
- The impact of winter snow retention has been significantly reduced and now only remains significant for the rivers that have their main discharge minimums in winter (i.e., the Koritnica and upper Soča),
- For rivers that used to have a pronounced river–snow regime (i.e., the Bača, Idrijca, Cerknica, Trebuša, Vipava, and Hubelj), the main discharge maximum has also

Table 10 Maximum annual discharge trends, 1961–2018

River	Gauging station	Mann–Kendall test, *Z*	Confidence level, %	Sen's slope, *Q*	1961 trend value, m^3/s	2018 trend value, m^3/s	1961–2018 trend difference, m^3/s	1961–2018 trend difference, %
Koritnica	Kal-Koritnica	−1.21	<90.0	−0.036	15.70	13.66	−2.04	−12.98
Soča	Kršovec	−1.43	<90.0	−0.091	34.36	29.20	−5.16	−15.03
Soča	Log Čezsoški	−1.45	<90.0	−0.258	77.94	63.20	−14.74	−18.91
Učja	Žaga	−0.60	<90.0	−0.016	17.78	16.88	−0.90	−5.04
Soča	Kobarid	0.48	<90.0	0.102	102.16	107.96	5.80	5.68
Tolminka	Tolmin	−1.54	<90.0	−0.066	26.82	23.08	−3.74	−13.94
Bača	Bača pri Modreju	−2.98	99.0	−0.165	29.10	19.71	−9.39	−32.26
Cerknica	Cerkno	−3.36	99.9	−0.057	7.34	4.11	−3.23	−43.97
Trebuša	Dolenja Trebuša	−3.75	99.9	−0.139	16.10	8.20	−7.90	−49.08
Idrijca	Podroteja	−3.39	99.9	−0.441	60.06	34.29	−25.77	−42.91
Idrijca	Hotešk	−1.80	90.0	−0.417	115.38	91.60	−23.78	−20.61
Soča	Solkan	−0.57	<90.0	−0.440	334.92	309.87	−25.05	−7.48
Hubelj	Ajdovščina	−3.28	99.0	−0.078	13.77	9.31	−4.46	−32.40
Vipava	Vipava	−3.02	99.0	−0.123	29.24	22.24	−7.00	−23.93
Vipava	Miren	−1.22	<90.0	−0.173	82.98	73.10	−9.88	−11.91
Reka (Fig. 8)	Cerkvenikov mlin	−2.46	95.0	−0.211	41.41	29.38	−12.03	−29.05
Rižana	Kubed	−3.30	99.9	−0.148	19.57	11.14	−8.43	−43.08

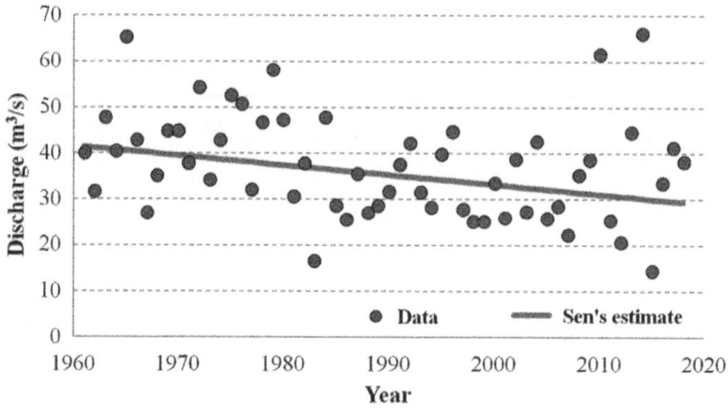

Fig. 8 Maximum annual discharge trend of the Reka river at the Cerkvenikov mlin gauging station, 1961–2018. *Source* authors

shifted from the spring to the fall, which indicates a gradual transition into a rain regime;

- Summer minimums are becoming more pronounced for all rivers;
- The November and December water levels exceed the annual average for many rivers, implying that winter is "running late."

The intensity of changes in the monthly discharge coefficients for individual rivers during the three-decade periods of 1961–1990 and 1991–2018 were also ascertained with the Pearson correlation coefficient (Table 11). The results indicate that the degree of correlation between both sets fluctuates between 0.73 and 0.97. Values ranging between 0.73 and 0.80 were ascertained for the Soča (at Solkan, Kobarid, and Log Čezsoški), Učja, Koritnica, and Cerknica. The basins of these rivers extend high into the mountainous regions, where both the volume and duration of the snow cover has been significantly decreasing over the past decades. Somewhat higher coefficients, ranging between 0.83 and 0.89, were determined for the Soča (at Kršovec), Bača, Hubelj, Tolminka, Vipava (at Vipava), and Idrijca. As a rule, their basins cover lower hill ranges and karst plateaus, where in the past the snow cover was more modest than in the high mountains. The highest Pearson's correlation coefficients, ranging from 0.91 to 0.97, are typical of the Vipava (at Miren), Trebuša, Reka, and Rižana. These are mostly rivers with basins in Mediterranean low hills, where snow cover is rare or even exceptional.

The fact that changes in the discharge regimes of the Mediterranean rivers between the 1961–1990 and 1991–2018 periods were primarily the result of changes in snow cover duration and volume is also evident for the Hubelj (Fig. 9). Its November maximum has already significantly exceeded the spring high waters, which primarily result from melting snow on the Trnovo Forest high karst plateau. Due to high temperatures and strong evapotranspiration, its main summer minimum has become even more pronounced, and, due to a smaller volume of snow retained, its secondary winter minimum is already nearly at the level of the average annual discharge.

Table 11 1961–1990 and 1991–2018 monthly discharge coefficients (blue shading shows the main and secondary minimums, and red shading shows the main and secondary maximums)

River	Gauging station	Data set	Jan	Feb	Mar	Apr	May	Jun	Jul	Aug	Sep	Oct	Nov	Dec	Pearson's *r*
Koritnica	Kal-Koritnica	1961–1990	0.57	0.55	0.62	1.17	1.65	1.50	1.12	0.81	0.95	1.05	1.22	0.79	0.79
		1991–2018	0.72	0.58	0.70	1.02	1.31	1.15	0.99	0.74	1.00	1.36	1.52	0.92	
Soča	Kršovec	1961–1990	0.47	0.42	0.53	1.13	1.94	1.73	1.06	0.75	0.97	1.08	1.24	0.68	0.83
		1991–2018	0.64	0.46	0.64	1.15	1.55	1.18	0.88	0.62	1.02	1.40	1.55	0.90	
Soča	Log Čezsoški	1961–1990	0.53	0.47	0.59	1.15	1.79	1.62	1.06	0.79	0.98	1.07	1.23	0.72	0.80
		1991–2018	0.62	0.42	0.66	1.18	1.48	1.15	0.87	0.64	1.02	1.40	1.62	0.93	
Učja	Žaga	1961–1990	0.71	0.65	0.96	1.54	1.47	1.04	0.64	0.63	0.91	1.07	1.48	0.89	0.73
		1991–2018	0.90	0.67	1.01	1.25	1.03	0.72	0.53	0.47	0.91	1.47	1.87	1.16	
Soča	Kobarid	1961–1990	0.58	0.52	0.66	1.20	1.69	1.48	0.98	0.75	0.99	1.10	1.28	0.79	0.78
		1991–2018	0.70	0.54	0.71	1.14	1.36	1.07	0.83	0.60	1.01	1.41	1.66	0.97	
Tolminka	Tolmin	1961–1990	0.58	0.52	0.66	1.14	1.84	1.60	0.89	0.64	0.90	1.13	1.33	0.77	0.87
		1991–2018	0.69	0.58	0.73	1.19	1.58	1.13	0.73	0.57	1.02	1.26	1.64	0.89	
Bača	Bača pri Modreju	1961–1990	1.02	0.95	1.08	1.40	1.15	0.98	0.62	0.52	0.72	1.01	1.40	1.15	0.85
		1991–2018	1.07	0.96	1.07	1.12	0.93	0.79	0.66	0.49	0.78	1.20	1.61	1.32	
Cerknica	Cerkno	1961–1990	1.02	1.02	1.20	1.40	1.01	0.92	0.67	0.61	0.77	1.07	1.28	1.04	0.80
		1991–2018	1.10	1.10	1.15	1.16	0.83	0.69	0.62	0.45	0.82	1.11	1.50	1.48	
Trebuša	Dolenja Trebuša	1961–1990	1.08	0.97	1.18	1.38	1.02	0.88	0.60	0.55	0.73	1.09	1.35	1.16	0.94
		1991–2018	1.09	1.08	1.22	1.23	0.93	0.73	0.54	0.40	0.77	1.11	1.49	1.40	
Idrijca	Podroteja	1961–1990	1.12	0.98	1.28	1.37	0.81	0.79	0.58	0.54	0.78	1.10	1.50	1.16	0.89
		1991–2018	1.06	1.16	1.33	1.09	0.72	0.56	0.43	0.39	0.90	1.13	1.66	1.57	
Idrijca	Hotešk	1961–1990	1.07	1.02	1.24	1.42	0.92	0.87	0.58	0.50	0.73	1.07	1.41	1.17	0.87
		1991–2018	1.08	1.10	1.22	1.10	0.78	0.61	0.50	0.45	0.84	1.16	1.67	1.49	
Soča	Solkan	1961–1990	0.87	0.80	0.94	1.29	1.32	1.17	0.76	0.62	0.85	1.06	1.36	0.97	0.75
		1991–2018	0.94	0.83	0.95	1.12	1.07	0.84	0.64	0.48	0.90	1.26	1.72	1.26	
Hubelj (Fig. 9)	Ajdovščina	1961–1990	0.91	0.95	1.21	1.62	1.06	0.89	0.50	0.43	0.72	1.18	1.38	1.16	0.86
		1991–2018	1.11	0.99	1.17	1.33	0.89	0.59	0.42	0.32	0.79	1.18	1.66	1.55	
Vipava	Vipava	1961–1990	1.01	1.02	1.27	1.49	0.96	0.91	0.55	0.53	0.71	1.01	1.35	1.20	0.87
		1991–2018	1.09	1.10	1.21	1.18	0.83	0.68	0.46	0.42	0.82	1.17	1.59	1.43	
Vipava	Miren	1961–1990	1.21	1.12	1.20	1.37	0.87	0.85	0.47	0.41	0.66	1.08	1.44	1.32	0.91
		1991–2018	1.19	1.22	1.15	1.09	0.77	0.59	0.36	0.29	0.73	1.09	1.88	1.66	
Reka	Cerkvenikov mlin	1961–1990	1.34	1.33	1.29	1.34	0.82	0.66	0.26	0.27	0.53	1.08	1.64	1.44	0.95
		1991–2018	1.41	1.43	1.19	1.04	0.70	0.43	0.22	0.24	0.54	1.02	1.99	1.79	
Rižana	Kubed	1961–1990	1.31	1.30	1.34	1.33	0.83	0.76	0.27	0.32	0.67	0.91	1.53	1.43	0.97
		1991–2018	1.30	1.53	1.31	1.15	0.82	0.52	0.21	0.25	0.59	0.97	1.70	1.64	

The correlation between the two datasets is based on the Pearson correlation coefficient

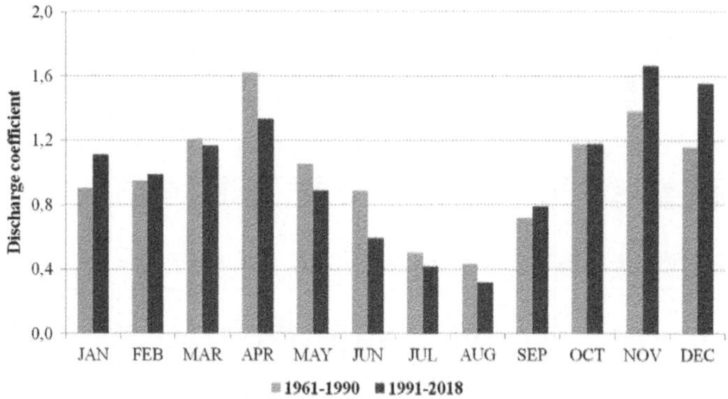

Fig. 9 Changes in the discharge regime of the Hubelj river at the Ajdovščina gauging station between the 1961–1990 and 1991–2018 periods. *Source* authors

Changes in the monthly discharge coefficients during the three-decade periods of 1961–1990 and 1991–2018 (Table 12) are largely uniform for all the Slovenian Mediterranean rivers analyzed. They indicate a gradual decrease in the spring and summer discharges and an increase in the fall and, to a slightly smaller extent, winter discharges. The greatest decrease in the discharge occurs from April to June, and the greatest increase from October to December. In addition to increased late-spring and summer evapotranspiration, an important reason for these developments is the gradual temperature increase and the resulting thinner and briefer snow cover. In the past, a significant share of fall precipitation was in the form of snow, which caused snow retention on the rivers, and the thicker snow cover at higher elevations in the mountains and on karst plateaus continued melting until late spring and even early summer.

Table 12 Changes in the monthly discharge coefficients between the 1961–1990 and 1991–2018 periods (blue shading indicates decreasing ratios and red shading shows increasing ratios)

River	Gauging station	Jan	Feb	Mar	Apr	May	Jun	Jul	Aug	Sep	Oct	Nov	Dec
Koritnica	Kal-Koritnica	0.14	0.04	0.07	−0.15	−0.33	−0.35	−0.13	−0.07	0.04	0.31	0.30	0.13
Soča	Kršovec	0.17	0.04	0.11	0.01	−0.38	−0.54	−0.18	−0.14	0.04	0.32	0.31	0.23
Soča	Log Čezsoški	0.09	−0.05	0.07	0.02	−0.31	−0.46	−0.19	−0.16	0.04	0.33	0.39	0.21
Učja	Žaga	0.19	0.02	0.05	−0.29	−0.44	−0.33	−0.11	−0.15	0.00	0.39	0.39	0.27
Soča	Kobarid	0.12	0.02	0.06	−0.07	−0.33	−0.41	−0.15	−0.15	0.03	0.31	0.38	0.19
Tolminka	Tolmin	0.11	0.06	0.07	0.05	−0.26	−0.47	−0.16	−0.07	0.11	0.13	0.31	0.12
Bača	Bača pri Modreju	0.05	0.01	−0.02	−0.28	−0.22	−0.19	0.04	−0.02	0.06	0.19	0.21	0.17
Cerknica	Cerkno	0.08	0.08	−0.05	−0.24	−0.19	−0.22	−0.05	−0.16	0.05	0.04	0.21	0.44
Trebuša	Dolenja Trebuša	0.01	0.10	0.04	−0.15	−0.09	−0.15	−0.05	−0.15	0.04	0.01	0.14	0.24
Idrijca	Podroteja	−0.06	0.18	0.05	−0.28	−0.08	−0.22	−0.15	−0.15	0.12	0.03	0.16	0.41
Idrijca	Hotešk	0.00	0.08	−0.02	−0.32	−0.14	−0.25	−0.08	−0.06	0.11	0.10	0.26	0.32
Soča	Solkan	0.07	0.03	0.01	−0.17	−0.25	−0.33	−0.12	−0.14	0.04	0.20	0.36	0.29
Hubelj	Ajdovščina	0.21	0.04	−0.04	−0.29	−0.17	−0.30	−0.09	−0.12	0.07	0.00	0.28	0.40
Vipava	Vipava	0.08	0.09	−0.05	−0.31	−0.13	−0.22	−0.09	−0.11	0.11	0.16	0.25	0.24
Vipava	Miren	−0.03	0.10	−0.05	−0.28	−0.10	−0.27	−0.11	−0.13	0.07	0.01	0.44	0.34
Reka	Cerkvenikov mlin	0.07	0.09	−0.10	−0.31	−0.12	−0.22	−0.04	−0.03	0.01	−0.05	0.35	0.34
Rižana	Kubed	0.00	0.23	−0.03	−0.18	−0.02	−0.24	−0.05	−0.07	−0.08	0.05	0.17	0.21

4.4 Floods

Because of the predominantly dynamic terrain with a small share of lowlands, floods only pose a threat to a small portion of the Mediterranean part of Slovenia. Minor flood-prone areas along the Soča, Vipava, Reka, Rižana, and Dragonja can be affected by lowland floods, whereas flashfloods may also occur along its tributaries (Komac et al. 2008). These flashfloods develop not only in the narrow gorges of the mountains in the Julian Alps, but also in the prealpine hills and Mediterranean low hills. A spatial-temporal analysis covering all of Slovenia showed that over a third of all flashfloods developed in Mediterranean Slovenia. A special threat factor is the long-lasting and heavy downpours that are most common in late summer or early fall (Trobec 2016).

Some areas of the narrow belt along the Slovenian coast are also exposed to sea floods. The lower parts of the coast are flooded several times a year. Between 1961 and 2015, there were eight to a maximum of 31 floods per year. They cause some minor economic damage. The main cause of sea floods is the combination of high tide, low air pressure, and a relatively strong wind coming from the south and causing significant waves (Bernot 1983; Kolega 2006; Volčanšek 2021).

Floods are connected with the absolute maximum discharge. In this regard, the increasing trends in certain watercourses are the main cause for concern. Between 1961 and 2018, the absolute maximum annual discharge trends were decreasing for the Učja, Bača, Cerknica, Trebuša, Idrijca, lower Soča, Hubelj, Reka, and Rižana, and increasing for the Koritnica, upper Soča, Tolminka, and Vipava (Table 13, Fig. 10). The confidence level is low, not even reaching 90% at fourteen gauging stations. The only exceptions are the stations at Kobarid on the Soča (99%), Bača pri Modreju on the Bača (95%), and Kubed on the Rižana (95%). At ten gauging stations out of seventeen, the absolute maximum annual discharge in the period observed decreased or increased by less than 10%.

In terms of water share, the absolute maximum annual discharge decreased the most at Bača pri Modreju on the Bača (i.e., by 54.39 m^3/s or 34.20%) and at Kubed on the Rižana (i.e., by 15.71 m^3/s or 28.9%). On the other hand, it increased the most at Kobarid on the Soča (i.e., by 124.68 m^3/s or 31.5%), at Kršovec on the Soča (i.e., by 23.87 m^3/s or 18.5%), and at Miren on the Vipava (i.e., by 28.89 m^3/s or 11.5%).

In terms of flood occurrence, the ratios of variation in the absolute maximum discharge between the 1961–1980, 1981–2000 and 2001–2018 periods were also established. Specifically, the variation ratios show the dispersion of data: the higher the ratio, the greater the dispersion (Sagadin 2003). In terms of river discharge, the higher the variation ratio, the higher the probability of the occurrence of both minimum and maximum discharges.

The calculations showed that the variation ratios for nearly all rivers were the lowest between 1961 and 1980. For the Koritnica, Soča, Tolminka, Cerknica, and Vipava they were the highest between 1981 and 2000, and for the Učja, Bača, Trebuša, Idrijca, Hubelj, Reka, and Rižana they were the highest in the most recent period, from 2001 to 2018 (Table 14). Based on this, greater deviations in the maximum

Table 13 Absolute maximum annual discharge trends, 1961–2018

River	Gauging station	Mann–Kendall test, *Z*	Confidence level, %	Sen's slope, *Q*	1961 trend value, m^3/s	2018 trend value, m^3/s	1961–2018 trend difference, m^3/s	1961–2018 trend difference, %
Koritnica	Kal-Koritnica	0.48	<90.0	0.106	63.63	69.66	6.03	9.48
Soča	Kršovec	0.91	<90.0	0.419	128.81	152.68	23.87	18.53
Soča	Log Čezsoški	0.12	<90.0	0.108	324.10	330.28	6.18	1.91
Učja	Žaga	−0.58	<90.0	−0.181	113.64	103.34	−10.30	−9.06
Soča (Fig. 10)	Kobarid	2.65	99.0	2.188	396.41	521.09	124.68	31.45
Tolminka	Tolmin	0.51	<90.0	0.109	74.17	80.40	6.23	8.40
Bača	Bača pri Modreju	−2.13	95.0	−0.954	158.93	104.54	−54.39	−34.22
Cerknica	Cerkno	−0.39	<90.0	−0.022	23.57	22.30	−1.27	−5.38
Trebuša	Dolenja Trebuša	−0.88	<90.0	−0.108	42.06	35.89	−6.17	−14.67
Idrijca	Podroteja	−0.28	<90.0	−0.166	203.24	193.76	−9.48	−4.67
Idrijca	Hotešk	−0.83	<90.0	−0.999	528.16	471.22	−56.94	−10.78
Soča	Solkan	−0.08	<90.0	−0.160	1365.68	1356.55	−9.13	−0.67
Hubelj	Ajdovščina	−0.91	<90.0	−0.055	33.55	30.44	−3.11	−9.27
Vipava	Vipava	1.00	<90.0	0.074	50.02	54.27	4.25	8.49
Vipava	Miren	0.89	<90.0	0.489	241.89	269.78	27.89	11.53
Reka	Cerkvenikov mlin	−0.14	<90.0	−0.105	183.26	177.28	−5.98	−3.26
Rižana	Kubed	−2.11	95.0	−0.276	54.43	38.72	−15.71	−28.86

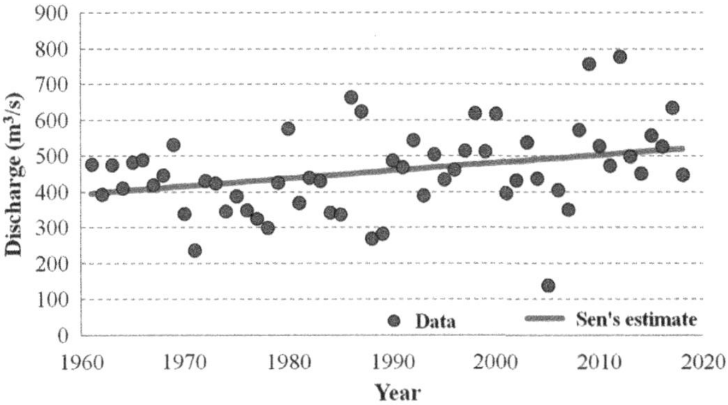

Fig. 10 Absolute maximum annual discharge trend of the Soča river at the Kobarid gauging station, 1961–2018. *Source* authors

Table 14 Absolute maximum annual discharge variation ratios, 1961–2018

River	Gauging station	1961–1980 ratio	1981–2000 ratio	2001–2018 ratio
Koritnica	Kal-Koritnica	23.05	30.75	25.82
Soča	Kršovec	21.87	34.39	28.46
Soča	Log Čezsoški	29.46	35.69	29.39
Učja	Žaga	32.06	35.72	41.06
Soča	Kobarid	28.27	37.25	34.00
Tolminka	Tolmin	21.89	36.46	32.43
Bača	Bača pri Modreju	20.91	34.22	34.28
Cerknica	Cerkno	21.07	38.81	37.79
Trebuša	Dolenja Trebuša	18.42	23.23	35.89
Idrijca	Podroteja	26.61	28.04	36.01
Idrijca	Hotešk	21.98	34.27	36.11
Soča	Solkan	24.82	43.20	37.01
Hubelj	Ajdovščina	23.15	26.69	31.19
Vipava	Vipava	20.78	25.59	25.09
Vipava	Miren	27.39	37.20	34.71
Reka	Cerkvenikov mlin	26.56	40.63	42.91
Rižana	Kubed	24.30	27.50	39.58

annual discharge, including "catastrophic" events, can be expected for these rivers. Variation ratios will continue to remain high in the future considering that these rivers also had high water levels in 2019 (Visoke vode 2020).

For example, in Fig. 10, which shows the absolute maximum annual discharge trend of the Soča at the Kobarid gauging station, the trend line varies between 396.4 m^3/s and 521.1 m^3/s, whereby certain annual values stand out significantly in both the downward and upward direction. For example, the 2005 absolute maximum discharge only reached 140.2 m^3/s, whereas the 2009 and 2012 ones exceeded 750 m^3/s, when the river also flooded (Polajnar 2010; Kobold 2013).

With regard to the potential occurrence of floods, the increasing fall maximum discharge should also be highlighted; it already exceeds the spring maximum for nearly all rivers because it is primarily the November and December water levels that are rising as a result of winter "running late." At the same time, the summer (August) minimum is falling increasingly (Tables 11 and 12). The former indicates an increased probability of high water and floods in the fall months and a smaller probability in the spring. The latter indicates the occurrence of summer droughts, which Mediterranean Slovenia is significantly exposed to (Zorn and Komac 2011; Cunja et al. 2019).

5 Discussion

The key findings about changes in the selected climate and hydrological variables between 1961 and 2018 largely overlap with the trends presented for Slovenia (e.g., Ulaga 2002; Uhan 2007; Frantar et al. 2008; Ulaga et al. 2008a, b, 2017b, 2018, 2020a, b; Kobold et al. 2012; Kovačič 2016; Kovačič et al. 2016; Makor 2016; Oblak 2017; Ravbar et al. 2017; Vertačnik and Bertalanič 2017; Vertačnik et al. 2018; Hrvatin and Zorn 2017a; Jelen et al. 2020) and the Mediterranean area (e.g., Hoerling et al. 2012; Flaounas et al. 2018) by other authors. In the Mediterranean area there has been a substantial drying since the beginning of the twentieth century, which is "consistent with an increase in surface air temperature and a decrease in precipitation" (Climate 2017: 301). These trends will continue in the future: compared to the period from 1961 to 1990, temperatures in the Mediterranean area are expected to rise by 1.5 °C in winter and by 2 °C in summer between 2021 and 2050, when annual precipitation is also expected to continue to decrease by approximately 5% (Climate 2017).

The average annual air temperature from 1961 to 2018 increased in the Mediterranean part of Slovenia by 0.029–0.042 °C a year at all temperature stations analyzed, which means that during that period the temperature increase was approximately 2 °C. Ravbar et al. (2017) established that during a slightly shorter period than the one studied here (i.e., from 1961 to 2013) the average annual air temperature in the southern part of the study area increased by 0.34 °C per decade, which is practically the same as the mean value of 0.35–0.36 °C per decade established here. Nearly the same results for the southern part of the study area were also reported by Kovačič (2016) and Kovačič et al. (2016). In addition, Hrvatin and Zorn (2017a) determined that from 1961 to 2010 the average annual air temperature in the central part of the study area increased by 0.34–0.35 °C per decade. The two authors also obtained similar results for the Slovenian Alps, which the northernmost part of the study area extends to (Hrvatin and Zorn 2017b, 2020a). For Slovenia overall, an increase in temperature of 0.15–0.29 °C per decade was reported for the period from 1951 to 2007 (De Luis et al. 2014), an increase of 0.23–0.46 °C per decade for the period from 1961 to 2011 (Vertačnik and Bertalanič 2017), and 0.3–0.5 °C per decade for the period from 1963 to 2014 (Milošević et al. 2017).

Annual precipitation in the period studied decreased at twenty-three out of twenty-five precipitation stations. Most differences in annual precipitation are moderate, not exceeding 20% at twenty-three stations. A 20% decrease in annual precipitation for western Slovenia between 1961 and 2011 is also reported by Tošić et al. (2016). Ravbar et al. (2017) mention decreasing annual precipitation of 4% per decade for the southern part of the study area between 1961 and 2013. Similar results are also reported by Kovačič (2016) and Kovačič et al. (2016). Hrvatin and Zorn (2017a) determined a decrease in annual precipitation of 10–20% in the central part of the study area between 1961 and 2010. For Slovenia in general, a decrease in annual precipitation of up to a few percentages per decade has been reported for the period from 1961 to 2011 (Vertačnik and Bertalanič 2017; Vertačnik et al. 2018).

In contrast to annual precipitation, the annual number of days with precipitation over 0.1 mm increased at sixteen out of twenty-five precipitation stations. At seventeen stations, the negative or positive trend deviations are smaller (i.e., below 10%). Similar results have also been reported by Hrvatin and Zorn (2017a, b, 2020a) for the central and northern parts of the study area.

The number of days with snow cover decreased significantly at all precipitation stations during the period studied, where it typically snows every year. At thirteen stations, the number decreased by 30–60%, or on average by thirty-six days. At the lowest precipitation stations, which can predominantly be found along the Adriatic coast (a total of seven), it snows only exceptionally and therefore it did not make sense to calculate the trends for them. Changes in snow cover are generally among the more evident indicators of climate change in Slovenia. This change is most prominent at lower elevations of the Alpine region, where the estimated mean snow cover trend for the period from 1961 to 2011 is as much as −20% per decade, which means that the snow cover halved during this period. The estimated mean snow cover trend for all of Slovenia is −16% per decade (Vertačnik and Bertalanič 2017; Vertačnik et al. 2018).

Climate change in the Mediterranean part of Slovenia is reflected in hydrological changes. In this context, we examined the river discharge trends and regimes from 1961 to 2018. Rivers in the study area show a decreasing minimum annual discharge trend (except for one) and a decreasing mean annual discharge trend, and nearly all of them also show a decreasing maximum annual discharge trend. In terms of these trends, most rivers recorded a relative decrease of 5–30%, but some also exceeded 40%. Very similar trends were also reported by Kovačič (2016), Kovačič et al. (2016), and Ravbar et al. (2017) for the southern part of the study area and by Hrvatin and Zorn (2017a, b, 2020a) for its central and northern parts. A decreasing mean annual discharge trend has been recorded for Slovenian rivers for quite some time now; for example, from 1955 to 1999 their discharges decreased by just over 18% on average (Ulaga 2002). A decrease of approximately 20% has also been observed in other parts of the Mediterranean area (Ducrocq et al. 2016).

The trends show that rivers in the Mediterranean part of Slovenia have a decreasing water quantities, which agrees with the fact mentioned at the beginning of this section–that is, that the Mediterranean area is affected by substantial drying (e.g., Ducrocq 2016; Schneider et al. 2013). More frequent droughts may cause problems for agriculture, affect public water supply, and lead to more frequent forest fires (Climate 2017). Predictions for the future are also very unpromising, indicating further decreases in the river discharges (Kobold 2007); in the northern part of the study area, they are even projected to decrease by up to 50% in the summer months by the end of the century (Janža 2013).

Despite the overall smaller volumes of water in the rivers and increasing aridity, an increase in the absolute maximum annual discharge, increasing discharge variability, and an increase in the fall discharge can imply a greater flood hazard in the Mediterranean part of Slovenia. Specifically, during the period studied the absolute maximum annual discharge trends increased at seven gauging stations out of seventeen. Similar trends in the absolute maximum annual discharge, discharge variability, and the fall discharge can also be observed in other parts of Slovenia (e.g., Hrvatin and Zorn 2020a, b, c). For southern Europe, Blöschl et al. (2019) report that a decreasing trend of river flood discharges can be observed at approximately three-quarters of the gauging stations. Mediterranean Europe is primarily characterized by flashfloods, which in certain parts of the Mediterranean area are more frequent than in the rest of Europe (Gaume et al. 2016). As already mentioned above, the spatial-temporal analysis covering all of Slovenia showed that over a third of all flashfloods developed in the Mediterranean part of Slovenia (Trobec 2016).

River discharge changes in the Mediterranean part of Slovenia influence the discharge regimes. A comparison of the discharge regimes based on the 1961–1990 data set and the based on the 1991–2018 data set shows that the impact of snow cover is decreasing rapidly. The impact of winter snow retention has decreased considerably and only remains significant in the northern (mountainous) part of the study area. The summer minimum discharge is becoming increasingly pronounced for all rivers, and the November and December water levels are increasing and, in many places, exceed the annual average, implying that winter is "running late." This is also characteristic of other parts of Slovenia (e.g., Hrvatin and Zorn 2020a, b, c), and at the national level such trends have already been reported by earlier studies (Frantar and Hrvatin 2005). Projections show that, in addition to the boreal climate zone, the Mediterranean area will experience the highest degree of discharge regime alteration in Europe (Schneider et al. 2013).

There are some limitations regarding our results because not only climate change is responsible for the changes in the hydrological regime. One must not forget the human impact on natural water flows, such as land-use changes, river training, water abstractions, barriers, and urbanization (e.g., Ulaga 2002; Bormann 2010; Zampieri et al. 2015; Goler et al. 2016; Šraj et al. 2016; Climate 2017). Another important factor to consider is the density of the temperature, precipitation, and gauging stations used, which is greater north of the Vipava Valley (Fig. 1)–that is, in the area that no longer belongs to Mediterranean Slovenia, according to Slovenia's natural-geography regionalization (Perko 1998; Perko and Ciglič 2020).

6 Conclusion

Climate and hydrological changes are correlated, and there is no doubt that the hydrological trends presented are also a reflection of the climate change described. This article shows that from 1961 to 2018 the average annual air temperature in the Mediterranean part of Slovenia increased by approximately 2 °C, alongside a reduction in annual precipitation (mostly not exceeding 20%), but not necessarily also a smaller number of days with precipitation. Climate change is the most evident in the number of days with snow cover, which in many places decreased by half, but in general by 30–60%. All of this is reflected in the river discharge trends and regimes. In the Mediterranean part of Slovenia, the (minimum, mean, and maximum) annual discharges are decreasing; in the period studied, they mostly decreased by 5–30% or more in some places, which indicates the area is more arid and has fewer water quantities. There will be even fewer of these in the future because the river discharges are forecast to decrease even further. Despite greater aridity, flood hazards cannot be expected to decrease because an increase in the absolute maximum annual discharge can be observed in some areas and rivers are characterized by significant discharge variability. Discharge changes cause changes in the discharge regimes. What is especially evident is that winter snow retention has decreased considerably, the summer minimum discharge is becoming increasingly pronounced for all rivers, and the November and December water levels are increasing and, in many places, exceed the annual average, implying that winter is "running late." It can be expected that the climate and hydrological changes will become even more pronounced in the future because the climate predictions for the Mediterranean area show further global warming and decrease in precipitation, which will also be expressed in the water budget (Lionello et al. 2012; Planton et al. 2012).

Acknowledgements This study was conducted as part of the research program The Geography of Slovenia (P6-0101), financed by the Slovenian Research Agency.

References

Arhiv hidroloških podatkov [Archive of hydrological data] (2020) Agencija Republike Slovenije za okolje, Ljubljana. Availabe at: http://vode.arso.gov.si/hidarhiv/; http://www.arso.gov.si/vode/podatki/arhiv/hidroloski_arhiv.html. Accessed 7 July 2020

Arhiv meteoroloških podatkov [Meteorological data archive] (2020) Agencija Republike Slovenije za okolje, Ljubljana. Availabe at: http://meteo.arso.gov.si/. Accessed 13 Aug 2020

Bard A, Renard B, Lang M, Giuntoli I, Korck J, Koboltschnig G, Janža M, d'Amico M, Volken D (2015) Trends in the hydrologic regime of Alpine rivers. J Hydrol 529:1823–1837. https://doi.org/10.1016/j.jhydrol.2015.07.052

Bernot F (1983) Vzroki in pogostost poplav ob slovenski obali [Causes and frequency of floods along the Slovenian coast]. In: Gams I (ed) Naravne nesreče v Sloveniji kot naša ogroženost. ZRC SAZU, Ljubljana, pp 50–53

Blöschl G, Hall J, Viglione A, Perdigão RAP, Parajka J, Merz B, Lun D, Arheimer B, Aronica GT, Bilibashi A, Boháč M, Bonacci O, Borga M, Čanjevac I, Castellarin A, Chirico GB, Claps P, Frolova N, Ganora D, Gorbachova L, Gül A, Hannaford J, Harrigan S, Kireeva M, Kiss A, Kjeldsen TR, Kohnová S, Koskela JJ, Ledvinka O, Macdonald N, Mavrova-Guirguinova M, Mediero L, Merz R, Molnar P, Montanari A, Murphy C, Osuch M, Ovcharuk V, Radevski I, Salinas JL, Sauquet E, Šraj M, Szolgay J, Volpi E, Wilson D, Zaimi K, Živković N (2019) Changing climate both increases and decreases European river floods. Nature 573:108–111. https://doi.org/10.1038/s41586-019-1495-6

Bormann H (2010) Runoff regime changes in German rivers due to climate change. Erdkunde 64(3):257–279. https://doi.org/10.3112/erdkunde.2010.03.04

Climate change, impacts and vulnerability in Europe 2016: An indicator-based report (2017) European Environment Agency, Copenhagen. https://doi.org/10.2800/534806

Cunja J, Kobold M, Šraj M (2019) Časovna in prostorska analiza največjih hidroloških suš v Sloveniji [Temporal and spatial analysis of the largest hydrological droughts in Slovenia]. Ujma 33:95–103

Čanjevac I, Orešić D (2018) Changes in discharge regimes of rivers in Croatia. Acta Geogr Slov 58(2):7–18. https://doi.org/10.3986/AGS.2004

da Silva RM, Santos CAG, Moreira M, Moreira M, Corte-Real J, Silva VCL, Medeiros IC (2015) Rainfall and river flow trends using Mann-Kendall and Sen's slope estimator statistical tests in the Cobres River basin. Nat Hazards 77:1205–1221. https://doi.org/10.1007/s11069-015-1644-7

De Luis M, Čufar K, Saz MA, Longares LA, Ceglar A, Kajfež-Bogataj L (2014) Trends in seasonal precipitation and temperature in Slovenia during 1951–2007. Reg Environ Change 14(5):1801–1810. https://doi.org/10.1007/s10113-012-0365-7

Ducrocq V (2016) Climate change in the Mediterranean region. In: Moatti J-P, Thiébault S (eds) The Mediterranean region under climate change: a scientific update. IRD Éditions, Marseille, pp 71–72. https://doi.org/10.4000/books.irdeditions.22908

Ducrocq V, Drobinski P, Gualdi S, Raimbault P (2016) The water cycle in the Mediterranean. In: Moatti J-P, Thiébault S (eds) The Mediterranean region under climate change: a scientific update. IRD Éditions, Marseille, pp 73–81. https://doi.org/10.4000/books.irdeditions.22908

Frantar P (2005) Pretočni režimi slovenskih rek in njihova spremenljivost [River flow regimes of slovene rivers and their fluctuation]. Ujma 19:145–153

Frantar P, Hrvatin M (2005) Pretočni režimi v Sloveniji med letoma 1971 in 2000 [Discharge regimes in Slovenia from 1971 to 2000]. Geografski Vestnik 77(2):115–127

Frantar P, Kobold M, Ulaga F (2008) Trend pretokov [Discharge trends]. In: Frantar P (ed) Vodna bilanca Slovenije 1971–2000. Agencija Republike Slovenije za okolje, Ljubljana, pp 50–60

Flaounas E, Kelemen FD, Wernli H, Gaertner MA, Reale M, Sanchez-Gomez E, Lionello P, Calmanti S, Podrascanin Z, Somot S, Akhtar N, Romera R, Conte D (2018) Assessment of an ensemble of ocean–atmosphere coupled and uncoupled regional climate models to reproduce the climatology of Mediterranean cyclones. Clim Dyn 51:1023–1040. https://doi.org/10.1007/s00382-016-3398-7

Gaume E, Borga M, Llasat MC, Maouche S, Lang M, Diakakis M (2016) Mediterranean extreme floods and flash floods. In: Moatti J-P, Thiébault S (eds) The Mediterranean region under climate change: a scientific update. IRD Éditions, Marseille, pp 133–144. https://doi.org/10.4000/books.irdeditions.22908

Gocic M, Trajkovic S (2013) Analysis of changes in meteorological variables using Mann-Kendall and Sen's slope estimator statistical tests in Serbia. Global Planet Change 100:172–182. https://doi.org/10.1016/j.gloplacha.2012.10.014

Goler RA, Frey S, Formayer H, Holzmann H (2016) Influence of climate change on river discharge in Austria. Meteorol Z 25(5):621–626. https://doi.org/10.1127/metz/2016/0562

Hamed KH (2008) Trend detection in hydrologic data: The Mann-Kendall trend test under the scaling hypothesis. J Hydrol 349(3–4):350–363. https://doi.org/10.1016/j.jhydrol.2007.11.009

Hartmann DL, Klein Tank AMG, Rusticucci M, Alexander LV, Brönnimann S, Charabi Y, Dentener FJ, Dlugokencky EJ, Easterling DR, Kaplan A, Soden BJ, Thorne PW, Wild M, Zhai PM (2013)

Observations: atmosphere and surface. In: Stocker TF, Qin D, Plattner G-K, Tignor M, Allen SK, Boschung J, Nauels A, Xia Y, Bex V, Midgley PM (eds) Climate change 2013: the physical science basis. Contribution of Working Group I to the Fifth Assessment Report of the Intergovernmental Panel on Climate Change. Cambridge University Press, Cambridge, pp 159–254

Hoerling M, Eischeid J, Perlwitz J, Quan X, Zhang T, Pegion P (2012) On the increased frequency of Mediterranean drought. J Clim 25(6):2146–2161. https://doi.org/10.1175/JCLI-D-11-00296.1

Hrvatin M (1998) Discharge regimes in Slovenia. Geografski zbornik 38:59–87

Hrvatin M, Komac B, Zorn M (2020) Water resources in Slovenia. In: Negm A, Romanescu G, Zelenakova M (eds) Water resources management in Balkan countries. Springer, Cham, pp 47–79. https://doi.org/10.1007/978-3-030-22468-4_3

Hrvatin M, Zorn M (2017) Trendi temperatur in padavin ter trendi pretokov rek v Idrijskem hribovju [Temperature and precipitation trends and river discharge trends in the Idrija Hills]. Geografski Vestnik 89(1):9–44. https://doi.org/10.3986/GV89101

Hrvatin M, Zorn M (2017) Trendi pretokov rek v slovenskih Alpah med letoma 1961 in 2010 [River discharge trends in Slovenian Alps between 1961 and 2010]. Geografski Vestnik 89(1):9–35. https://doi.org/10.3986/GV89201

Hrvatin M, Zorn M (2018) Recentne spremembe rečnih pretokov in rečnih režimov v Julijskih Alpah [Recent changes in river discharges and discharge regimes in the Julian Alps]. In: Zorn M, Mikša P, Lačen Benedičič I, Ogrin M, Marija Kunstelj A (eds) Triglav 240. Založba ZRC, Ljubljana, pp 107–129. https://doi.org/10.3986/9789610500841

Hrvatin M, Zorn M (2020a) Climate and hydrological changes in Slovenia's mountain regions between 1961 and 2018. Ekonomska i ekohistorija 16:201–218

Hrvatin M, Zorn M (2020b) Hidrološki odraz podnebnih sprememb v Podravju – bo poplav več ali manj [Hydrological reflection of climate change in the Podravje region – can we expected more or less floods]. In: Zorn M, Komac B, Ciglič R, Logar E (eds) Domači odzivi na globalne izzive. Naravne nesreče 5. Založba ZRC, Ljubljana, pp 105–126. https://doi.org/10.3986/NN0509

Hrvatin M, Zorn M (2020c) Climate and hydrological changes in Slovenia's Podravje region between 1961 and 2018. Podravina 19(33):161–180

Janža M (2013) Impact assessment of projected climate change on the hydrological regime in the SE Alps, Upper Soča River basin. Slovenia. Nat Hazards 67(3):1025–1043. https://doi.org/10.1007/s11069-011-9892-7

Jelen M, Mikoš M, Bezek N (2020) Kraški izviri v Sloveniji: analiza trendov [Karst springs in Slovenia: trend analysis]. Acta hydrotechnica 33(58):1–12. https://doi.org/10.15292/acta.hydro.2020.01

Kobold M (2007) Vpliv podnebnih sprememb na pretoke slovenskih rek [The impact of climate change on the discharge of Slovenian rivers]. 18. Mišičev vodarski dan:101–108

Kobold M (2013) Poplave konec oktobra in v začetku novembra 2012 [Floods at the end of October and in the beginning of November 2012]. Ujma 27:44–51

Kobold M, Dolinar M, Frantar P (2012) Spremembe vodnega režima zaradi podnebnih sprememb in drugih antropogenih vplivov [Water regime changes due to climate change and other anthropogenic impacts]. In: Brilly M (ed) 1. kongres o vodah Slovenije. Fakulteta za gradbeništvo in geodezijo, Ljubljana, pp 7–22

Kolega N (2006) Slovenian coast sea floods risk. Acta Geogr Slov 46(2):143–169. https://doi.org/10.3986/AGS46201

Komac B, Natek K, Zorn M (2008) Geografski vidiki poplav v Sloveniji [Geographical aspects of floods in Slovenia]. Geografija Slovenije 20. Založba ZRC, Ljubljana. https://doi.org/10.3986/9789612545451

Kovačič G (2016) Trendi pretokov rek jadranskega povodja v Sloveniji brez Posočja [Discharge trends of the Adriatic Sea basin rivers in Slovenia, excluding the Soča river basin]. Geografski Vestnik 88(2):9–29. https://doi.org/10.3986/GV88201

Kovačič G, Kolega N, Brečko Grubar V (2016) Vpliv podnebnih sprememb na količine vode in poplave morja v slovenski Istri [Climate change impacts on water quantities and sea flooding in Slovene Istria]. Geografski Vestnik 88(1):21–36. https://doi.org/10.3986/GV88102

Kraner Šumenjak T, Šuštar V (2011) Parametrični in neparametrični pristopi za odkrivanje trenda v časovnih vrstah [Parametric and nonparametric approach for trend detection in time series]. Acta Agriculturae Slovenica 97(3):305–312

Kundzewicz ZW (2008) Climate change impacts on the hydrological cycle. Ecohydrol Hydrobiol 8(2–4):195–203. https://doi.org/10.2478/v10104-009-0015-y

Lionello P, Abrantes F, Congedi L, Dulac F, Gacic M, Gomis D, Goodess C, Hoff H, Kutiel H, Luterbacher J, Planton S, Reale M, Schröder K, Struglia MV, Toretino Y, Tsimplis M, Ulbrich U, Xoplaki E (2012) Introduction: Mediterranean climate–background information. In: Lionello P (ed) The climate of the Mediterranean region: from the past to the future. Elsevier, Amsterdam, pp XXXV–XC. https://doi.org/10.1016/B978-0-12-416042-2.00012-4

Liu Z, Cuo L, Li Q, Liu X, Ma X, Liang L, Ding J (2020) Impacts of climate change and land use/cover change on streamflow in Beichuan River Basin in Qinghai Province, China. Water 12(4):1–26. https://doi.org/10.3390/w12041198

Ludwig W, Dumont E, Meybeck M, Heussner S (2009) River discharges of water and nutrients to the Mediterranean and Black Sea: major drivers for ecosystem changes during past and future decades. Prog Oceanogr 80(3–4):199–217. https://doi.org/10.1016/j.pocean.2009.02.001

Makor S (2016) Trendi spreminjanja pretokov rek v Sloveniji [Trends of river discharge changes in Slovenia]. BSc Thesis. Fakulteta za gradbeništvo in geodezijo, Ljubljana

Masseroni D, Camici S, Cislaghi A, Vacchiano G, Massari C, Brocca L (2020) 65-year changes of annual streamflow volumes across Europe with a focus on the Mediterranean basin. Hydrology and Earth System Sciences Discussion [preprint]. https://doi.org/10.5194/hess-2020-21

Milošević DD, Savić SM, Stankov U, Žiberna I, Pantelić MM, Dolinaj D, Leščešen I (2017) Maximum temperatures over Slovenia and their relationship with atmospheric circulation patterns. Geografie 122(1):1–20. https://doi.org/10.37040/geografie2017122010001

Oblak J (2017) Analiza sezonske spremenljivosti pretokov rek v Sloveniji [Analysis of the seasonal streamflow variation in Slovenian rivers]. MSc Thesis. Fakulteta za gradbeništvo in geodezijo, Ljubljana

Ogrin D (1996) Podnebni tipi v Sloveniji [The climate types in Slovenia]. Geografski Vestnik 68:39–56

Planton S, Lionello P, Artale V, Aznar R, Carrillo A, Colin J, Congedi L, Dubois C, Elizalde A, Gualdi S, Hertig E, Jacobeit J, Jordà G, Li L, Mariotti A, Piani C, Ruti P, Sanchez-Gomez E, Sannino G, Sevault F, Somot S, Tsimplis M (2012) The climate of the Mediterranean region in future climate projections. In: Lionello P (ed) The climate of the Mediterranean region: from the past to the future. Elsevier, Amsterdam, pp 449–502. https://doi.org/10.1016/B978-0-12-416 042-2.00008-2

Perko D (1998) The regionalization of Slovenia. Geografski Zbornik 38:11–57

Perko D, Ciglič R (2020) Slovenia's landscapes. In: Perko D, Ciglič R, Zorn M (eds) The geography of Slovenia: small but diverse. Springer, Cham, pp 211–225. https://doi.org/10.1007/978-3-030-14066-3_14

Perko D, Ciglič R, Zorn M (2020) Slovenia: A European landscape hotspot. In: Perko D, Ciglič R, Zorn M (eds) The geography of Slovenia: small but diverse. Springer, Cham, pp 1–20. https://doi.org/10.1007/978-3-030-14066-3_1

Polajnar J (2010) Visoke vode v Sloveniji leta 2009 [High waters in Slovenia in 2009]. Ujma 24:25–29

Ravbar N, Kovačič G, Petrič M, Kogovšek J, Brun C, Koželj A (2017) Climatological trends and anticipated karst spring quantity and quality: case study of the Slovene Istria. Geological Society, London, Special Publications 466:295–305. https://doi.org/10.1144/SP466.19

Sagadin J (2003) Statistične metode za pedagoge [Statistical methods for educators]. Obzorja, Maribor

Salmi T, Määttä A, Anttila P, Ruoho-Airola T, Amnell T (2002) Detecting trends of annual values of atmospheric pollutants by the Mann-Kendall test and Sen's slope estimates – the excel template application MAKESENS. Publications on Air Quality No. 31. Finnish Meteorological Institute, Helsinki

Schneider C, Laizé CLR, Acreman MC, Flörke M (2013) How will climate change modify river flow regimes in Europe? Hydrol Earth Syst Sci 17:325–339. https://doi.org/10.5194/hess-17-325-2013

Stagl J, Mayr E, Koch H, Hattermann FF, Huang S (2014) Effects of climate change on the hydrological cycle in Central and Eastern Europe. In: Rannow S, Neubert M (eds) Managing protected areas in Central and Eastern Europe under climate change. Advances in Global Change Research 58. Springer, Dordrecht, pp 31–43. https://doi.org/10.1007/978-94-007-7960-0_3

Staut M, Kovačič G, Ogrin D (2007) The spatial cognition of Mediterranean in Slovenia: (In) consistency between perception and physical definitions. Acta Geogr Slov 47(1):105–131. https://doi.org/10.3986/AGS47105

Šraj M, Menih M, Bezjak N (2016) Climate variability impact assessment on the flood risk in Slovenia. Phys Geogr 37(1):73–87. https://doi.org/10.1080/02723646.2016.1155389

Tošić I, Zorn M, Ortar J, Unkašević M, Gavrilov MB, Marković SB (2016) Annual and seasonal variability of precipitation and temperatures in Slovenia from 1961 to 2011. Atmos Res 168:220–233. https://doi.org/10.1016/j.atmosres.2015.09.014

Trobec T (2016) Spatio-temporal distribution of flash floods in Slovenia. Dela 46:21–39. https://doi.org/10.4312/dela.46.1.5-39

Uhan J (2007) Trendi velikih in malih pretokov rek v Slovenji [Trends in high and low river flows in Slovenia]. Ujma 21:164–166

Ulaga F (2002) Trendi spreminjanja pretokov slovenskih rek [Trends in discharge changes of Slovenian rivers]. Dela 18:93–114. https://doi.org/10.4312/dela.18.93-114

Ulaga F, Kobold M, Frantar P (2008a) Trends of river discharges in Slovenia. IOP conference series: earth and environmental science 4(1). https://doi.org/10.1088/1755-1307/4/1/012030

Ulaga F, Kobold M, Frantar P (2008b) Analiza časovnih sprememb vodnih količin slovenskih rek [Analysis of time changes of water quantities of Slovenian rivers]. 19. Mišičev vodarski dan: 9–18

Vertačnik G, Bertalanič R (2017) Podnebna spremenljivost Slovenije v obdobju 1961–2011, 3: Značilnosti podnebja v Sloveniji [Climate variability of Slovenia in the period 1961–2011, 3: Climate characteristics in Slovenia]. Agencija Republike Slovenije za okolje, Ljubljana. Availabe at: https://meteo.arso.gov.si/uploads/probase/www/climate/text/sl/publications/Znacilnosti%20podnebja%20splet.pdf. Accessed 19 Sep 2020

Vertačnik G, Bertalanič R, Draksler A, Dolinar M, Vlahović Ž, Frantar P (2018) Podnebna spremenljivost Slovenije v obdobju 1961–2011: povzetek [Climate variability of Slovenia in the period 1961–2011: summary]. Agencija Republike Slovenije za okolje, Ljubljana. Availabe at: https://meteo.arso.gov.si/uploads/probase/www/climate/text/sl/publications/PSSbrosura_spread_SLO.pdf. Accessed 19 Sep 2020

Visoke vode in poplave rek od 1. do 5. februarja 2019 [High waters and river floods from 1 to 5 February 2019] (2020) Agencija Republike Slovenije za okolje, Ljubljana. Availabe at: http://www.arso.si/vode/poro%C4%8Dila%20in%20publikacije/Visoke%20vode%20in%20poplave%20rek%201.%20do%205.%20februar%202019.pdf. Accessed 19 Sep 2020

Volčanšek M (2021) Dvig morske gladine v prihodnosti bo močno prizadel slovensko Istro [Rising sea levels in the future will severely affect Slovenian Istria] STA znanost. Availabe at: http://znanost.sta.si/2853068/dvig-morske-gladine-v-prihodnosti-bo-mocno-prizadel-slovensko-istro. Accessed 10 Jan 2021

Yue S, Wang CY (2004) The Mann-Kendall Test modified by effective sample size to detect trend in serially correlated hydrological series. Water Resour Manage 18:201–218. https://doi.org/10.1023/B:WARM.0000043140.61082.60

Zampieri M, Scoccimarro E, Gualdi S, Navarra A (2015) Observed shift towards earlier spring discharge in the main Alpine rivers. Sci Total Environ 503–504:222–232. https://doi.org/10.1016/j.scitotenv.2014.06.036

Zorn M, Komac B (2011) Damage caused by natural disasters in Slovenia and globally between 1995 and 2010. Acta Geogr Slov 51(1):7–41. https://doi.org/10.3986/AGS51101

Zorn M, Komac B, Carey AE, Hrvatin M, Ciglič R, Lyons WB (2020) The disappearing cryosphere in the southeastern Alps: introduction to special issue. Acta Geogr Slov 60(2):109–124. https://doi.org/10.3986/AGS.9396

Zorn M, Razpotnik Visković N, Repolusk P, Ferk M (2013) Prostorski in regionalni razvoj Sredozemlja - enotni pristop in izbrana orodja [Spatial and regional development of the Mediterranean: A unified approach and selected tools]. Založba ZRC, Ljubljana. https://doi.org/10.3986/9789610503514

Climate Change Impacts and the Role of Forestry: Insights from the Mediterranean Region

Veronika Andrea

Abstract The Mediterranean region is characterized by a great variety of forest ecosystems, biomass and soil properties. Devastating phenomena including extended forest fires as well as other natural disasters such as floods have been recorded in broader areas over recent decades. These results constitute strong evidence that climate-related natural disasters are a severe threat to the Mediterranean. To this end, forest policy should be replanned to meet current needs in the era of climate mitigation and resilience. The current chapter surveys the recent literature on the role of forestry in climate mitigation in line with the Paris Agreement and aims to set the primary goals for effective forest policy in Mediterranean countries. The findings indicate that forest industry models addressing climate neutrality are not yet well established in the Mediterranean, although they are considered to be of utmost importance in terms of sustainable development and climate mitigation alternatives. The establishment of a Mediterranean forest inventory and innovative monitoring systems, the enhancement of bioeconomy models, synergies in the forest industry, and the use of wood for housing and energy efficiency in buildings are some of the proposed measures, strategies and policies.

1 Introduction

As part of the history of the twenty-first century, climate change effects are now more evident than ever. Predictions of climate impacts in the 2050s and 2080s are raising concerns about the secure future of humankind (Harrison et al. 2018). Unless we limit emissions in line with the Paris Agreement (PA), there will be a catastrophic scenario for 2100 with a potential mean global temperature rise of over 5 °C (Fuss et al. 2014). Empirical studies have shown that climate-related disasters affect various strategic sectors that are closely affiliated with local, national and international economies,

V. Andrea (✉)
Department of Forestry and Management of the Environment and Natural Resources, Democritus University of Thrace, 193 Pantazidou Street, 68200 Orestiada, Greece
e-mail: vandrea@fmenr.duth.gr

W. Leal Filho and E. Manolas (eds.), *Climate Change in the Mediterranean and Middle Eastern Region*, Climate Change Management,
https://doi.org/10.1007/978-3-030-78566-6_5

multilevel governance, migration, food security, poverty and human health (Sait 2020; Kaczan and Orgill-Meyer 2020). In addition to members of the European Union, other countries play an important role in global climate policy and have pledged to the commitments of the PA. Mexico and China are striving for an ambitious climate-smart policy by substantially reducing deforestation over the next decades (Hernandez 2021).

The changing climate conditions play a critical role in the Mediterranean region due to its high levels of vulnerability (Baraldi et al. 2019). Experiential insights into the broader geographical area of the Mediterranean provide alarming estimates of the likelihood of temperature rise, drought phenomena, sea level rises, floods and other climatic disaster incidents (Cardoso et al. 2018; Soares et al. 2017; Turco et al. 2015; Argüeso et al. 2012). According to Bettini (2017), the poorest countries have suffered a large number of human and property losses, which can be attributed to the lack of sufficient climate-related disaster management or their inability to adapt to climate impacts.

Why the Mediterranean? The issue of the high climatic risk that affects this specific area is a historical concern for the research community. The relentless onslaught of climate change may be explained by the special characteristics and complexity of Mediterranean ecosystems. To deepen understanding in this field, special environmental components should be further investigated. To bridge the gap between policy and practice, land use change, particularly forest cover, should be monitored by the use of modern technology (Janus et al. 2021).

In particular, erosion phenomena lead to habitat loss and consequently endanger endemic flora and fauna species that serve as a driving force of the biodiversity and resilience of ecosystems (Hampe and Petit 2005). Another major point involves the wide range of forestland, biomass and soil properties in the broader area, which seriously affect vulnerability to climate impacts. However, it should be noted that the composition and outstanding ecological value of Mediterranean forests create great potential for carbon sequestration (Doukalianou et al. 2019).

There is a strong record in the last decades of devastating forest fires and catastrophic floods along with other natural disasters that pose evidence-based hazards in the Mediterranean. Forest fires constitute a major risk in this area as human-induced fires are the main factor leading to deforestation and the forest degradation of Mediterranean forestland (Gonçalves and Sousa 2017; Tampakis et al. 2005). Buras and Menzel (2019) argue that French, Spanish and Italian forests should receive special attention as they are subject to a high risk of forestland loss and insecurity of their silvicultural dynamics.

The major role of forests in climate mitigation was recently acknowledged on a global level by the PA. Notably, the forest biomass sector has great potential for the production of bioenergy, especially in Europe (Sikkema et al. 2021). Thus, the role of the forestry sector should be prioritized by decision makers to provide innovative and coordinated configurations for effective forest policy towards climate action.

This paper aims to investigate certain policy proposals established for Mediterranean countries after the PA that consider the crucial role of forestry in attaining climate mitigation targets. This chapter constitutes a conceptual study and uses a methodology of searching, organizing, analysing and synthesizing the data acquired from interdisciplinary studies. The main findings are presented in four subsections on the investigated literature that are separately described and further discussed. The main conclusions are presented, including recent challenges to humanity. Limitations of the study are also highlighted.

2 Methods

This paper involved qualitative research and the method used was theory synthesis. The study was initiated in August 2020 and lasted four months. The framework in which the study was conducted included an analysis of trends concerning the forestry sector in Mediterranean countries after the PA. Scopus and Google Scholar were the databases used for the literature search. Special limits were set as the relevant interdisciplinary literature addressed PA and beyond. A content analysis followed, and four subcategories were created. A qualitative synthesis took place for each subcategory. Elaborative discussion followed to generate practical insights and suggest potential effective policy measures indicative of the Mediterranean forestry sector. The methodological steps are illustrated in Fig. 1.

3 Results and Discussion

The results of the reviewed literature concerning forest policies followed by Mediterranean countries to advance forestry as a means of climate mitigation include a division into four subsections:

(1) Subsection 1 is titled "Paris agreement—a cleaner path- through article 5", involves a presentation of the measures that have been initiated and others that are expected to occur in the near future through the confirmation of article 5 of the PA, which acknowledges the role of forests in climate mitigation.
(2) Subsection 2, named "Climate-based forest policy in Mediterranean countries", provides a holistic view of the policies implemented to date in Mediterranean countries to confront climatic effects.
(3) Subsection 3, called "How much do we know about the contribution of forestry to global warming?", presents current trends and the value of monitoring tools and cross-cutting technologies as a vital part of up-to-date forest policy and forest governance towards climate mitigation.
(4) Subsection 4 is titled "Forest industry, a brother in arms towards climate neutrality" and mainly includes initiatives and measures that should be taken into consideration for an innovative, profitable and sustainable forest industry as a tool for climate resilience.

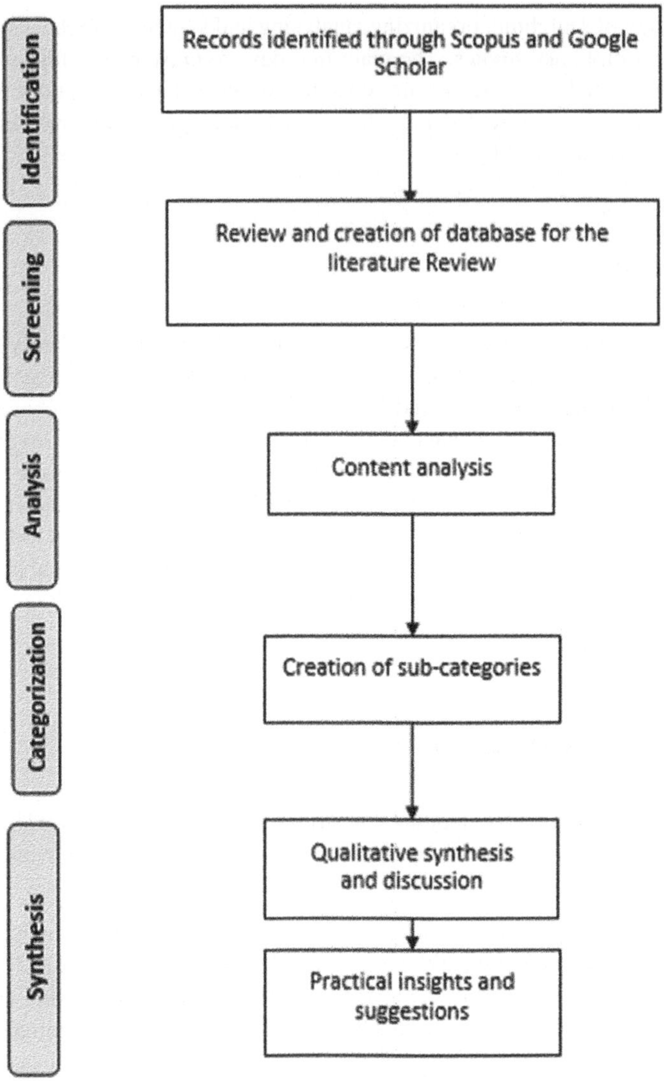

Fig. 1 Methodology flow chart

Each subsection is characterized by generalized policies, strategies and special measures that were implemented to harmonize with the pledged PA. The main findings are presented and discussed below.

3.1 *Paris Agreement—A Cleaner Path—Through Article 5*

The role of forestry in climate mitigation was ascertained by rational conceptions accepted by all co-signatories following Article 5 of the PA. In 2015, the United Nations called all nations to take emergency actions to address the threatening impacts of climate change. Based on the Framework Convention on Climate Change, the United Nations has committed 189 of 197 Parties to the Convention to ratify and pledge the Paris Agreement. The major goal of this agreement is presented in Article 2 as follows: "*The Paris Agreement central aim is to strengthen the global response to the threat of climate change by keeping a global temperature rise this century well below 2 °C above pre-industrial levels and to pursue efforts to limit the temperature increase even further to 1.5 °C*" (Paris Agreement 2015). In addition, it was strongly emphasized that the efficiency of natural sinks is indisputable and should be utilized more effectively to mitigate emissions and enhance absorptions.

Heck et al. (2018) identified sustainable forest management as a means to achieve ambitious climate targets and meet the determined commitments arising from the PA. These authors also note the development of new incentives that would sustain forestry through the potential of bioenergy. In fact, Article 5 indicates that the replanning of forest policy is critical to alleviate carbon drawbacks related to deforestation and forest degradation, which could be realized by providing motivations for sustainable forest management. Notably, the prioritization of forestry through a cleaner path, such as the forest-based scenario, is expected to satisfy the targets set by the PA within a long-term framework (Grassi et al. 2018).

Forest dynamics should be combined with national, regional and local special economic and social characteristics to form a case-based scenario that can serve as the driver for reducing current emissions by more than 25%. This means that national forest policies should establish a basis for innovation and new investments that would boost forestry. The establishment of incentives for entrepreneurship and the forest industry presupposes that Mediterranean governments will succeed in creating value for this market through the facilitation of strategic interaction between stakeholders (Grassi et al. 2017). However, it is vital to engage new tax-based motivations and raise awareness of both the role of forests in the carbon cycle and their vulnerability to changing climatic conditions.

Until recently, there was a "misconception" regarding the accounts of greenhouse gas emissions and removals per country. This misconception was confronted by the so-called "Land Use/Land Use Change/Forestry—LULUCF regulation", which has been the landmark for the incorporation of greenhouse gas emissions and removals from land use, land use change and forestry (Nabuurs et al. 2018). LULUCF regulation was adopted three years after the PA, in 2018, by the European Parliament and Council and commits all EU members to an offset of at least an equivalent removal of CO_2 from the atmosphere within the 2021–2030 period (European Commission 2018). This constitutes strong evidence that the European forest policy will seriously consider emissions and removals from forestry, which will be recorded in the next 10 years. In line with pledges through the adoption of Article 5 of the PA, the introduction of state-of-the-art land-use strategies involves the support and finance of forest businesses, including the promotion of forest biomass production and use.

3.2 *Climate-Based Forest Policy in Mediterranean Countries*

Mediterranean forests that include important mature wood for forestry mainly consist of broad-leaved, sclerophyllous species and dominant Quercus ilex and other species, such as Pinus halepensis and Pinus brutia (Amici et al. 2013; Zavala and Zea 2004). Although the broader basin has been characterized by land use changes and severe deforestation phenomena, there is currently an increase in biomass and forest continuity (Bedia et al. 2014). In parallel, the extended arid periods and the warmer and drier climate—as a consequence of climate change—have importantly altered resilience regimes and enhanced natural and human hazards such as wildfires (Hinojosa et al. 2021; Turco et al. 2018). Moreover, urbanization has strongly affected the current forest ecosystem formation as it has led to desertification due to rural land abandonment. These features have played a key role in the existing conditions that set the framework of the special identity of Mediterranean forest ecosystems (Chergui et al. 2018). There is also wide heterogeneity in the quality and status of Mediterranean soils (Hinojosa et al. 2021). Thus, a clear understanding of the dominant drivers that affect these ecosystems and their vulnerability to climate change is a structural component of efficient policy planning (Baudena et al. 2019).

The unique nature of Mediterranean forests is a challenging topic in setting the agenda for effective policies, drastic strategies and compelling measures towards climate mitigation. In fact, the employment of forests as carbon reservoirs through their vegetation and soils is a demanding task for Mediterranean countries. However, it is expected to lead to a new age of climate neutrality. Forest policies that were designed by Mediterranean countries in line with Article 5 of the PA focused on three strategic pillars, namely, land use change, forestry and the utilization of forests as carbon sinks.

With the aim of accelerating climate-based forest policies in Mediterranean countries, certain issues must be clarified, such as those arising from the implementation of LULUCF regulations. Specifically, emissions from agriculture and forestry performance, along with those that account for forest fires, hinder the effective interpretation of soil emissions, while certain measures are acknowledged as essential to trigger new investments in clean energy (Camargo et al. 2020).

Another basic policy pillar is the prevention of forestland loss due to agricultural purposes. Therefore, lessons learned by international programmes, such as REDD +, should be integrated in an effort to balance conflicts among stakeholders (Carter et al. 2017). To this end, stakeholder engagement and encouragement to participate in decision making are of utmost importance to enhance the quality of environmental decisions (Tubiello et al. 2015).

In some cases, such as Italy, which has a strong record of olive harvesting, policy indications are oriented towards agroforestry practices, which enhance soil carbon storage capacity (Bateni et al. 2019). Similarly, it is likely that Greece could also adopt such practices as it also holds an important share of olive cultivation. As a policy measure in the Mediterranean, agroforestry systems have been perceived as a common practice to reduce the hazard of wildfires and enrich protection schemes in complex ecosystems. According to Damianidis et al. (2020), agroforestry should be enhanced with certain policies that address Mediterranean ecosystems as they provide a useful method to address the devastating effects of climate change, such as wildfires. These authors also argue that if properly performed, agroforestry systems could offer sustained land management and could serve to reduce machinery fuel consumption and limit greenhouse gas emissions.

In addition, there is a policy targeting biodiversity preservation that addresses the protection and conservation of forestland comprising the pivotal places of the species' habitats. However, there are certain limitations due to the manifold nature of these dynamics, which involve a multilevel interaction due to the changing climate, bird demography, dispersal and relationships with other species (Martínez-Jauregui et. 2018).

Ultimately, there is consensus among stakeholders and decision makers that involves the promotion of both timber and non-timber forest products and services that support the bioeconomy (Malek et al. 2018; Verkerk et al. 2018). In fact, the potential of forest recreation and tourism, including nature-based and mountain tourism, is widely acknowledged as a growth prospect for these unique forest ecosystems that geographically belong to the Mediterranean region. They offer a mosaic of various ecosystem types that positively affect human well-being (Tampakis et al. 2019).

3.3 How Much do we Know About the Contribution of Forestry to Global Warming?

International organizations such as the Food and Agriculture Organization of the United Nations (FAO), which play a fundamental role in the formation of climate-based forest policies on a global level, underline the need to engage forest monitoring systems. According to a recent release presenting REDD + achievements, the FAO insists on the employment and operation of cross-cutting technologies that are technologically advanced and can represent the current status of forest-related emissions to provide accurate and accessible data (FAOa 2020).

The level of transparency, especially concerning LULUF emissions, could function as a milestone for transformational changes in forest policies aligned with the PA. These changes pertain to institutional, governmental, or legislative frameworks. In addition, unhindered access to information could catalyse community empowerment towards greener lifestyles. It would also motivate citizens to demand high-level political commitments for immediate, low-carbon and climate-resilient action through the capitalisation of forest resources.

To this end, the provision of reliable forest resource information is a way to acquire principal knowledge on forest-related activities that aggravate the problem of global warming. Information and communication technologies (ICTs) are currently sufficiently advanced that they can provide tangible support to researchers, practitioners, experts and decision makers and pave the way for climate-smart forest policy (Böcher, 2020). To date, forest monitoring tools include various systems that use the latest scientific and technological trends, such as remote-sensing systems and real-time transmission of spatial data, LiDAR, mapping, image analysis, optical radar and spatial resolution (Abad-Segura et al. 2020). Effective strategies for climate-based and sustained forestry could be facilitated by the provision of big data. Their parameterization is considered a valuable source of information on forest productivity and on prediction scenarios regarding the changes in forest carbon dynamics (Boisvenue and White 2019).

In line with these scenarios and with the aim of accurate measurements of forest conditions, tenure regimes and land extension may serve as operational guidelines for the improvement of harvesting procedures, biomass management and the preservation of forest carbon sinks (Kauranne et al. 2017). If forest-related data are properly analysed, they can provide a special field for case-based evidence on long-term patterns. The interpretation of this kind of data and emerging models—via forest monitoring tools—would intrinsically reinforce policy-making procedures in the field of climate-based forestry (Tabor and Connell 2019).

Sustainable forestry in REDD+ projects is delivered by national web geoportals that have already been developed or that will be developed in the future. Within REDD + national programmes, the contribution of national forest inventories along with other inventories that record emissions and removals has proven to be essential for the interpretation of forest carbon models. The latter are endorsed by satellite land monitoring systems and support decision making and efficient management of the forestry sector. They provide secure estimates of land use and land use change status, emission factors and forest carbon cycle dynamics (FAOb 2020). Coordinated efforts in the Mediterranean have proven that ICT integration can foster the role of forests in the era of climate neutrality and reduce emissions from deforestation and forest degradation (Özcan et al. 2018).

3.4 *Forest Industry, a Brother in Arms Towards Climate Neutrality*

Climate-smart forestry is one of the most critical sectors for the implementation of forest-based scenarios in mitigation strategies. Bowditch et al. (2020) define it as a substantial element for implementing "climate-smartness", which is an integral part of efficient and sustainable forest management, accompanied by social inclusion, adaptation and mitigation of climate effects.

Afforestation and carbon sequestration are establishing forestry as the prop of the PA (Bastin et al. 2019). The creation of new forests will bring Mediterranean countries closer to the commitments of the PA, although forest policy replanning is of utmost importance. Effective policy making should entail new models that revitalize forest economies by enhancing the bioeconomy, creating synergies in the forest industry and using wood for housing as well as improving energy efficiency in buildings.

Therefore, locality should be a priority in policy and decision making because case-based performance is more accurate and effective. In addition, it involves the specific indicators and characteristics of a certain area, such as the economic and social environment and ecosystem services. Furthermore, it includes numerous existing natural features, such as species mixtures, the productivity of soils and their capacity for new products, and other assets that constitute the local forest dynamics.

Predictions for Mediterranean models list forests at high risk of droughts and heat waves, where trees are exposed to water stress (Ogaya and Peñuelas 2020). However, carbon production seems to be increasing in the next years. These seasonal parameters negatively affect productivity levels in Mediterranean forests. Furthermore, there is a special status that raises additional concerns mainly in southern Europe, where there are large acreages of young and unmanaged forests. As claimed by Simioni et al. (2020), these special characteristics of Mediterranean forests emerge as a threat to the limited carbon sequestration capacity of forests. Timber is also affected as forest trees tend to appear with stem growth, defoliation, and high levels of mortality; ultimately, forests are severely degraded (Liu et al. 2018; Galiano et al. 2012).

Nabuurs et al. (2018) underscore that the implementation of sustainable forest management practices in Europe could substantially increase wood productivity. Forest monitoring is acknowledged as the technological keystone that will provide the necessary information for decision and policy makers and, in turn, for forest managers to engage with climate-smart forestry, adopt innovative measures for wood production and better promote them to new markets.

Another part of the solution is to reinforce certification processes that ensure quality and control in wood products. Kleindl et al. (2018) state that sustainable forestry can be attained if there is consensus among multiple stakeholders under a holistic approach that counteracts changing conditions and integrates wood business, related markets, resources, society and ecosystems. Timber and non-timber product certification ensures sustained production from responsibly managed forests accompanied by environmental, social and economic welfare. Wong et al. (2020) describe certification as a "win–win–win" effort to establish the basis for sustainable forest management that meets climate mitigation efforts. They also claim that it supports the forest industry and performs "social forestry" as its principles presuppose the safeguarding of local communities' rights to forest resources. To maintain ecosystem integrity and efficiently balance community-based measures, replanning forest policies in the Mediterranean region must entail coordinated efforts and innovative practices through the sustainable exploitation of forest resources. New models are required that are linked to market needs and attract private sector investments.

The next generation of forest bioeconomy intends to replace fossil-based products and non-renewable raw materials and create added value products such as biofuels, textiles, or even bioproducts with properties similar to chemicals and plastics (Verkerk et al. 2018). Although there is a need to institutionalize new climate-smart forest industry models that will strive for climate neutrality, it is obvious that there are not yet well-established models in the Mediterranean area. Painter (2020) highlights the way investors' decisions are influenced by changing trends, which could be utilized to introduce new circular models that can counteract climate change effects and achieve bioeconomic security. These new concepts and approaches will contribute to a more sustainable development of the forestry sector. Therefore, there is an emerging need to directly create incentives for businesses to invest in Mediterranean forest-based value chains and in the development of appropriate infrastructures, such as biorefineries (Koukios et al. 2017).

4 Conclusions

Extensive and unsustainable human activity negatively affects climate resilience. Consequently, Mediterranean forest ecosystems and their functions are facing a dearth of stability. Currently, humanity must deal with the outbreak of COVID 19, a deadly new pandemic. Politicians and decision makers are called to rethink measures to ensure stability in ecosystems in response to this and to any future health hazards as various new pandemics are expected to threaten humankind.

Based on the content analysis and thematic subsections examined in this conceptual paper, it is apparent that the forestry sector in Mediterranean countries offers great potential for sustainable forest management. However, an important part of forestland still lacks efficient management. To align their pledges with the PA, Mediterranean countries are at the forefront of delivering new policies addressing forest-based solutions to enhance their carbon reservoirs and mitigate emissions.

Some of these policies should focus on field practices such as agroforestry, afforestation, the prevention of wildfires and the preservation of biodiversity. However, others need preparatory activities for the better promotion of wood products as biofuels and construction materials. These also include the development of infrastructures, the implementation of certification processes and the creation of a new economic environment able to support bioeconomy products based on timber.

The PA is a great opportunity for policy makers in Mediterranean countries to design new policies for the forest industry by replacing fossil fuels used for energy production with forest biomass and non-renewable construction materials with wood. Synergies are welcomed in this scenario as the forest industry remains a challenging and high-risk business sector; thus, the engagement of multilevel stakeholders could make investment plans sufficiently realistic to implement. The support of the forestry sector should be at the top of the agendas in the national forest policies of all Mediterranean countries.

The executive added value of forest resources, including their unique services, are vital for human well-being, such as forest recreation and mountain tourism. Climate-smart policies that capitalize on the multifunctionality of Mediterranean forests demand information sharing, unhindered data flow and transparency. Mediterranean forest inventories as well as the establishment and operation of innovative monitoring systems are imperative if policy makers truly aim to transition to climate neutrality. The next generation of forest policy should be adapted to the challenges of the twenty-first century.

5 Limitations

The study was limited to literature selected after 2015 according to the Paris Agreement and does not provide an integrated view of historical comparisons. Furthermore, this study investigated the role of forestry in climate mitigation in Mediterranean countries, but it was not possible to deepen the analysis to the specific forest policies implemented per country and provide a wide scope analysis and possible comparisons. These two limitations provide suggestions for future research.

References

Abad-Segura E, González-Zamar M-D, Vázquez-Cano E, López-Meneses E (2020) Remote sensing applied in forest management to optimize ecosystem services: advances in research. Forests 11:969

Amici V, Santi E, Filibeck G, Diekmann M, Geri F, Landi S, Scoppola A, Chiarucci A (2013) Influence of secondary forest succession on plant diversity patterns in a Mediterranean landscape. J Biogeogr 40:2335–2347

Argüeso D, Hidalgo-Muñoz J, Gámiz-Fortis S, Esteban-Parra M, Castro-Díez Y (2012) High-resolution projections of mean and extreme precipitation over Spain using the WRF model (2070–2099 versus 1970–1999). J Geophys Res 117:D12108

Baraldi R, Chieco C, Neri L, Facini O, Rapparini F, Morrone L, Rotondi A, Carriero G (2019) An integrated study on air mitigation potential of urban vegetation: from a multi-trait approach to modeling. Urban Forest Urban Greening 41:127–138

Bastin JF, Finegold Y, Garcia C, Mollicone D, Rezende M, Routh D, Zohner CM, Crowther TW (2019) The global tree restoration potential. Science 365:76–79

Bateni C, Ventura M, Tonon G, Pisanelli A (2019) Soil carbon stock in olive groves agroforestry systems under different management and soil characteristics. Agroforestry Systems. Retrieved 19 June 2020, from https://doi.org/10.1007/s10457-019-00367-7

Baudena M, Santana VM, Baeza MJ, Bautista S, Eppinga MB, Hemerik L, Mayor AG, Rodriguez F, Valdecantos A, Vallejo VR, Vasques A, Rietkerk M (2019) In-creased aridity drives post-fire recovery of Mediterranean forests towards open shrublands. New Phytol 225(4):1500–1515

Bedia J, Herrera S, Camia A, Moreno JM, Gutiérrez JM (2014) Forest fire danger projections in the Mediterranean using ENSEMBLES regional climate change scenarios. Clim Change 122:185–199

Bettini G (2017) Where next? Climate change, migration, and the (Bio) politics of adaptation. Global Pol 8(S1):33–39

Böcher M (2020) Research trends: advanced approaches for a better understanding of scientific knowledge transfer in forest and forest-related policy. Forest Policy Econ 114:102165

Boisvenue C, White JC (2019) Information needs of next-generation forest carbon models: opportunities for remote sensing science. Remote Sens 11:463

Bowditch E, Santopuoli G, Binder F, del Río M, La Porta N, Kluvankova T, Lesinski J, Motta R, Pach M, Panzacchi P, Pretzsch H, Temperli C, Tonon G, Smith M, Velikova V, Weatherall A, Tognetti R (2020) What is climate-smart forestry? A definition from a multinational collaborative process focused on mountain regions of Europe. Ecosyst Serv 43:101113

Buras A, Menzel A, (2019). Projecting tree species composition changes of European forests for 2061–2090 under RCP 4.5 and RCP 8.5 scenarios. Front Plant Sci 9:1986

Camargo J, Barcena I, Soares PM, Schmidt L, Andaluz J (2020) Mind the climate policy gaps: climate change public policy and reality in Portugal, Spain and Morocco. Clim Change 161:151–169. https://doi.org/10.1007/s10584-019-02646-9

Cardoso R, Soares P, Lima D, Miranda P (2018) Mean and extreme temperatures in a warming climate: EURO CORDEX and WRF regional climate high-resolution projections for Portugal. Clim Dyn 52:129–157

Carter S, Manceur AM, Seppelt R, Hermans-Neumann K, Herold M, Verchot L (2017) Large scale land acquisitions and REDD+: A synthesis of conflicts and opportunities. Environ Res Lett 12(3):035010

Chergui B, Fahd S, Santos X, Pausas JG (2018) Socioeconomic factors drive fire-regime variability in the Mediterranean Basin. Ecosystems 21:619–628

Damianidis C, Santiago-Freijanes JJ, den Herder M (2020) Agroforestry as a sustainable land use option to reduce wildfires risk in European Mediterranean areas. Agroforestry Systems. Retrieved 10 November 2020, from https://doi.org/10.1007/s10457-020-00482-w

Doukalianou F, Radoglou K, Agnelli AE, Kitikidou K, Milios E, Orfanoudakis M, Lagomarsino A (2019) Annual greenhouse-gas emissions from forest soil of a peri-urban conifer forest in Greece under terms and conditions privacy policy different thinning intensities and their climate-change mitigation potential. Forest Sci 65(4):387–400

European Commission (2018) Land use and forestry regulation for 2021–2030. Available at: https://ec.europa.eu/clima/policies/forests/lulucf_en. Retrieved 01 June 2020

Galiano L, Martinez-Vilalta J, Sabaté S, Lloret F (2012) Determinants of drought effects on crown condition and their relationship with depletion of carbon reserves in a Mediterranean holm oak forest. Tree Physiol 32(4):478–489

Gonçalves AC, Sousa AMO (2017). The Fire in the Mediterranean Region: A Case Study of Forest Fires in Portugal. In: Fuerst-Bjelis B (ed) Mediterranean identities - environment, society. culture, IntechOpen. Available at: https://www.intechopen.com/books/mediterranean-identities-environment-society-culture/the-fire-in-the-mediterranean-region-a-case-study-of-forest-fires-in-portugal. Retrieved 19 Sept 2020

Grassi G, House J, Kurz WA, Cescatti A, Houghton RA, Peters GP, Sanz MJ, Viñas RA, Alkama R, Arneth A, Bondeau A, Dentener F, Fader M, Federici S, Friedlingstein P, Jain AK, Kato E, Koven CD, Lee D, Nabel JEMS, Nassikas AA, Perugini L, Rossi S, Sitch S, Viovy N, Wiltshire A, Zaehle S (2018) Reconciling global-model estimates and country reporting of anthropogenic forest CO2 sinks. Nat Clim Chang 8(10):914–920

Grassi G, House J, Dentener F, Federici S, Den Elzen M, Penman J (2017) The key role of forests in meeting climate targets requires science for credible mitigation. Nat Clim Chang 7(3):220–226

Hampe A, Petit RJ (2005) Conserving biodiversity under climate change: the rear edge matters. Ecol Lett 8(5):461–467

Harrison PA, Dunford RW, Holman IA, Cojocaru G, Madsen MS, Chen SP, Sandars D (2018) Differences between low-end and high-end climate change impacts in Europe across multiple sectors. Reg Environ Change 19:695–709

Heck V, Gerten D, Lucht W, Popp A (2018) Biomass-based negative emissions difficult to reconcile with planetary boundaries. Nat Clim Chang 8(2):151–155

Hernandez AM (2021) Mexico and China—sustainable, low-carbon transformation processes in democratic and authoritarian regimes. In: Taming the big green elephant. Globale Gesellschaft und internationale Beziehungen, Springer VS, Wiesbaden

Hinojosa MB, Albert-Belda E, Gómez-Muñoz B, Moreno JM (2021) High fire frequency reduces soil fertility underneath woody plant canopies of Mediterranean ecosystems. Sci Total Environ 752:141877

Janus J, Bożek P, Mitka B, Taszakowski J, Doroż A (2021) Long-term forest cover and height changes on abandoned agricultural land: an assessment based on historical stereometric images and airborne laser scanning data. Ecol Indic 120:106904

FAOa (Food and Agriculture Organization of the United Nations) (2020) From reference levels to results reporting: REDD+ under the United Nations framework convention on climate change. 2020 update. Rome, FAO. Retrieved 01 December from https://doi.org/10.4060/cb1635en

FAOb (Food and Agriculture Organization of the United Nations) (2020) REDD+ Reducing emissions from deforestation and forest degradation. FAO. Available at: http://www.fao.org/redd/areas-of-work/national-forest-monitoring-system/en/. Retrieved 07 December

Fuss S, Canadell JG, Peters GP, Tavoni M, Andrew RM, Ciais P, Jackson RB, Jones CD, Kraxner F, Nakicenovic N, Le Quéré C, Raupach MR, Sharifi A, Smith P, Yamagata Y (2014) Betting on negative emissions. Nat Clim Chang 4(10):850–853

Kaczan DJ, Orgill-Meyer J (2020) The impact of climate change on migration: a synthesis of recent empirical insights. Clim Change 158:281–330

Kauranne T, Joshi A, Gautam B, Manandhar U, Nepal S, Peuhkurinen J, Hämäläinen J, Junttila V, Gunia K, Latva-Käyrä P, Kolesnikov A, Tegel K, Leppänen V (2017) LiDAR-assisted multi-source program (LAMP) for measuring above ground biomass and forest carbon. Remote Sens 9(2):154

Kleindl WJ, Stoy PC, Binford MW, Desai AR, Dietze MC, Schultz CA, Starr G, Staudhammer CL, Wood DJA (2018) Toward a social-ecological theory of forest macrosystems for improved ecosystem management. Forests 9(4):200

Koukios E, Monteleone M, Texeira Carrondo MJ, Charalambous A, Girio F, Hernández EL, Mannelli S, Parajó JC, Polycarpou P, Zabaniotou A (2017) Targeting sustainable bioeconomy: a new development strategy for Southern European countries. The manifesto of the European Mezzogiorno. J Clean Prod 172:3931–3941

Liu D, Ogaya R, Barbeta A, Yang X, Peñuelas J (2018) Long-term experimental drought combined with natural extremes accelerate vegetation shift in a Mediterranean holm oak forest. Environ Exp Bot 151:1–11

Malek Ž, Verburg PH, Geijzendorffer IR, Bondeau A, Cramer W (2018) Global change effects on land management in the Mediterranean region. Glob Environ Chang 50:238–254

Martínez-Jauregui M, Serra-Varela MJ, Díaz M, Soliño M (2018) Mitigation strategies for conserving bird diversity under climate change scenarios in Europe: the role of forest naturalization. PLoS ONE 13(8):e0202009

Nabuurs GJ, Arets EJMM, Schelhaas MJ (2018) Understanding the implications of the EU-LULUCF regulation for the wood supply from EU forests to the EU. Carbon Balance Manage 13(18):1–10

Ogaya R, Peñuelas J (2020) Wood vs. Canopy allocation of aboveground net primary productivity in a Mediterranean Forest during 21 years of experimental rainfall exclusion. Forests 11(10):1094

Özcan O, Musaoğlu N, Türkeş M (2018) Assessing vulnerability of a forest ecosystem to climate change and variability in the western Mediterranean sub-region of Turkey. Journal of Forestry Research 29(3):709–725

Painter M (2020) An inconvenient cost: the effects of climate change on municipal bonds. J Financ Econ 135(2):468–482

Paris Agreement (2015) UNITED NATIONS. Available at: https://unfccc.int/process-and-meetings/the-paris-agreement/the-paris-agreement. Retrieved 13 May 2020

Sait MS (2020) Should Monrovian communities agree to voluntary slum relocations: land, gender and Urban Governance. In: Home R (ed) Land issues for Urban Governance in Sub-Saharan

Africa. Local and Urban Governance. Springer, Cham. https://doi.org/10.1007/978-3-030-52504-0_21

Simioni G, Marie G, Davi H, Martin-St Paul N, Huc R (2020) Natural forest dynamics have more influence than climate change on the net ecosystem production of a mixed Mediterranean forest. Ecol Model 416: 108921

Sikkema R, Proskurina S, Banja M, Vakkilainen E (2021) How can solid biomass contribute to the EU's renewable energy targets in 2020, 2030 and what are the GHG drivers and safeguards in energy- and forestry sectors? Renewable Energy 165(1):758–772

Soares P, Cardoso R, Lima D, Miranda P (2017) Future precipitation in Portugal: high-resolution projections using WRF model and EURO-CORDEX multi-model ensembles. Clim Dyn 49:2503–2530

Tabor K, Connell L (2019) Applications of forest monitoring tools for development projects. Bank Information Center, Washington. Available at: https://bankinformationcenter.cdn.prismic.io/bankinformationcenter%2F0627aed1-14a2-4e99-97f6-163b541d02aa_forest+monitoring+tools+report_may6.pdf. Retrieved 01 December

Tampakis S, Papageorgiou A, Karanikola P, Arabatzis G, Tsantopoulos G (2005) The forest fires in the Mediterranean from a policy point of view. New Medit, Mediterr J Econ Agric Environ 4(3):47–51

Tampakis S, Andrea V, Karanikola P, Pailas I (2019) The growth of mountain tourism in a traditional forest area of Greece. Forests 10(11):1022

Tubiello FN, Salvatore M, Ferrara AF, House J, Federici S, Rossi S, Biancalani R, Condor Golec RD, Jacobs H, Flammini A (2015) The contribution of agriculture, forestry and other land use activities to global warming, 1990–2012. Glob Change Biol 21:2655–2660

Turco M, Palazzi E, Von Hardenberg J, Provenzale A (2015) Observed climate change hotspots. Geophys Res Lett 42(9):3521–3528

Turco M, Rosa-Cánovas JJ, Bedia J, Jerez S, Montávez JP, Llasat MC, Provenzale A (2018) Exacerbated fires in Mediterranean Europe due to anthropogenic warming projected with non-stationary climate-fire models. Nat Commun 9:3821

Verkerk PJ, de Arano IM, Palahı M (2018) The bioeconomy as an opportunity to tackle wildfires in Mediterranean forest ecosystems. Forest Policy Econ 86:1–3

Wong GY, Moeliono M, Bong IW, Pham TT, Sahide MAK, Naito D, Brockhaus M (2020) Social forestry in Southeast Asia: evolving interests, discourses and the many notions of equity. Geoforum 117:246–258

Zavala MA, Zea E (2004) Mechanisms maintaining biodiversity in Mediterranean pine-oak forests: insights from a spatial simulation model. Plant Ecol 171:197–207

Energy Impacts on Climate Change: Issues, Challenges and Solutions with Clean Conversion Technology

Uttara Das, Champa Nandi, Somudeep Bhattacharjee, and Sarbani Mandal

Abstract Middle East countries are the front line sufferers in case of climate change. The Middle East has generally winter and summer weather with little rainfall. Due to climate change; lakes are running dry, the temperature hits higher level in the gulf, sandstorms occur in the cities, cyclones are waving in islands. Resulting, power generation through conventional resources faces high risk due to extreme climate change. Therefore, the use of renewable resources for power generation is one of the best ways to mitigate global warming. Renewable resources like stored water, solar, wind, biomass, etc. can easily solve the power generation problem, as conventional resources like coal, diesel, etc. will completely extinct after some years due to climate change. So, we should maintain our aim to use renewable resources in the case of power generation. Since the individual renewable resource is unable to generate a stable amount of power, the hybridization is used for maintaining continue power generation. Thus, this book chapter clarifies the cause of climate change in the Middle East, the impact of climate change, the effect of conventional and renewable sources on power generation. Mainly, this book chapter examines pollution-free hybrid system in the Middle East through model simulation and case study analysis.

1 Introduction

The climate of the Middle East consists of mainly hot and dry weather with mild winter and little rain. The desert of the north has an extremely high temperature and rain during winter and spring. In the other areas rainfall occurs during March and November. In the Middle East, summer temperature is around 85 F and it reaches up to even 100 F. In recent days, global warming is a major problem for the whole world which is one of the main reasons for climate change. The rising temperature of the air is the reason for the melting of ice for which sea level is increasing day by day (Nandi et al. 2019). The rising sea level puts many coastal cities of Middle East Countries at risk. People of the Middle East are experiencing the impacts of climate

U. Das · C. Nandi (✉) · S. Bhattacharjee · S. Mandal
Department of Electrical Engineering, Tripura University, Agartala, Tripura, India

W. Leal Filho and E. Manolas (eds.), *Climate Change in the Mediterranean and Middle Eastern Region*, Climate Change Management,
https://doi.org/10.1007/978-3-030-78566-6_6

change like floods in Jeddah and the effect of rising sea levels in the Mediterranean. On the Mediterranean coast of Egypt, Alexandria, because of the high level of water, basements of buildings are flooded which leads to fatal collapses. Also, the impact of climate change is the reason for extreme heat and drought as well as the flood in the desert. In June 2017, the temperature of Sweihan, Abu Dhabi is reached 50.4 °C and in Dubai, several cars caught fire due to extreme heat (Broom 2019). The reason for global warming is the increase of greenhouse gases in the environment. Emission of carbon is one of the primary causes behind climate change (Bhattacharjee et al. 2020a). Energy sectors dependent on fossil fuel are one of the main sources of carbon emission. The power generation of Middle East countries is mainly dependent on fossil fuel. In MENA (Middle East and North African) countries, 97% of the total energy is from oil, coal, and natural gas. Whereas Saudi Arabia and Iran are the 9 and 7th largest CO_2 emitters in the world and about 40% of CO_2 is emitted from these regions in 2018 (Nandi et al. 2019; Climate Change in the Middle East and North Africa 2020). From the energy sectors of the MENA region emission of CO_2 is about 38% (Timmerberg et al. 2019). Whereas, in Sudan emission of CO_2 from fossil fuel was 13,294,106 tons in 2016 among its 10.6% is from power industries and 52% is from transport (Worldometer 2020). So, it needs to reduce the carbon emission from fossil fuel generation plants. Hence, to get rid of these environmental hazards due to carbon emission, it is necessary to find alternative and clean sources of energy. Renewable energy sources like solar, wind, hydro, etc. are the best alternative way of energy generation which is needed to be increased (Bhattacharjee and Nandi 2020c). Different countries of the MENA region use fewer amounts of renewable energy sources. In 2017 contribution of renewable energy in Morocco, Egypt, Jordan is 16, 8, and 7% respectively which are the net-exporter of fossil energy. Net-exporter countries (Saudi Arabia, Libya, Tunisia, Algeria) of fossil fuel shows very small shares of renewable energy, in Tunisia shares of renewable energy is 3 and 1% in other countries. In 2017 installed capacity of the renewable energy-based power plants was 5% across the country (Timmerberg et al. 2019). Hence, the shares of renewable sources for energy generation have to increase in these regions. Also,the use of the electric vehicle for transportation can reduce the carbon emission in the environment (Bhattacharjee and Nandi 2020b). Whereas, in renewable energy-based power plant use of water is less thus water quality and availability can be improved in this region. A report of the IRENA (International Renewable Energy Agency) says that countries belong to North Africa have the great potential of solar and wind energy generation than all other parts of the continent (Climate Change in the Middle East and North Africa 2020). In Sudan radiation of solar exceeds 2000 kWh/m^2 and wind speed is about 5.1–7.1 m/s across the country. The largest on-going project of renewable energy in Sudan was approved by GEF (Global Environment Facility) (Team 2020). The potential of solar PV (photovoltaic) is comprehensive in the region and possible to tap into both utility and household levels. The intensity of solar irradiation in the desert region of North Africa is among the highest all over the world. Wind power is also has a huge potential in North Africa. Between 2010 and 2015, the installed capacity of wind power is reached 1942 MW from 857 MW in the countries of North Africa. Where at that time period in Morocco installed capacity of wind is increased from 253 to 934 MW (Tagliapietra 2019). Quarzazate Solar Power Station is the largest concentrated solar power plant in the world which is situated in the Draa-Tafilalet region of Morocco and it is expected that the plant capacity will be reached 2000 MW at the end of 2020 and will reduce 7×10^5 tons of carbon emission in every year. Also, the largest hydropower plants are installed in Egypt, and Morocco and the capacity of those plants are 2800 and 1770 MW respectively. Morocco has set a target of installation capacity of 52% of renewable energy by 2050 and renewable energy share is reached 28% in 2018 (Timmerberg et al. 2019; Climate Change in the Middle East and North Africa 2020).

Although influenced by climate change, (Nguyen et al. 2020) proposed a wind-hydro-thermal system to reduce the fuel cost of thermal power plant. By using rejected wind energy, (Nikolaou et al. 2020) designed pumped hydro energy storage system. This wind energy does not have any interconnection with electrical power system. It is preferably easy to get constantly same output when any inputs data are constant. (Huang and Bao 2016; Nandi and Chakraborty 2012) have proposed a wind based hydro power plant where constant wind speed is considered. Mihut et al. (2018) proposed an islanded wind, solar and hydro hybrid system in a location where no power supply is available. A hybrid wind and hydroelectric power plant is proposed by Opan et al. (2019) where, they have taken a factor when wind energy becomes zero.

However, renewable energy sources are variable in nature. Therefore, hybrid energy sources (a combination of two or more renewable energy sources) has to use for energy generation to get a continuous supply of energy (Bhattacharjee et al. 2020b). The aim of this chapter is not only to investigate the relation of CO_2 emissions, energy consumption, and economic growth but also to propose a new hybrid energy approach that will be reliable for the coming future. Thus, this book chapter consists of the following sections apart from the introduction in the first section. This book chapter illustrates different reasons, impacts of climate change, and the effect of conventional and renewable resources on power generation in the Middle East and North African countries in section two. Section three discussed about the proposed hybrid system methodology with different real-time case study data. To reduce environmental pollution and to reduce the cost of energy generation, section four illustrates the result with description. Thus, section five concluded the chapter. Mainly, a hybrid wind-based pumped hydro storage simulation model has been proposed to analyze the performance of the location with respect to load.

2 Climatic Overview in the Middle East and North Africa

The quality of any location critically depends on the climatic condition. It is required to understand that quality is closely connected with human activities. Due to the enhanced demand for human lifestyle, this globe wound face increasingly interconnected challenges.

2.1 Cause of Climate Change

Basically, the main reason for climate change is the imbalance activities between humans and nature. The basic reason for climate change in any area is firstly, the variation of the Sun energy reaching to the particular area, secondly, the Earth's surface and atmosphere's reflection change in that area, and lastly, greenhouse gasses action in the area for trapping heat in the earth surface. Actually, climate change is not a treat in theory or not a future prediction by scientists; it is the result of numerous decades of pollution which actually faced the Earth. Although, the Middle East and North African regions are one of the climate change 'hot-spot' over the world. Low and middle-income countries of MENA are the most vulnerable to the direct or indirect effect of climate change. Its region is affected by limited groundwater, extremely high temperature, lack of rainfalls, and deficiency of agriculture, and arable land, simply, harsh climate conditions. MENA is hosting the rapid growth of industrialization and the fastest urbanization in the world. Some important reason behind climate change in MENA are summarised below:

- Citizen conflict among most of the societies (particularly, Syria, Libya, Iraq, and Yemen) is an enormous problem in this region. Uses of different explosive types of equipment during conflict create the question about greenhouse gas emission. Thus, MENA has faced the world's fastest pollution rate. More population means more variety of lifestyles which directly hampers the climatic scenario. From 110 million in 1950, MENA reached 569 million of the population in 2017 (Lange 2019).
- Most of the people of low and middle-income countries are basically follow the rural lifestyle. Thus, the improper use of technologies in every sector and random, traditional, unsustainable agriculture is another cause of climate change in MENA which is increasing day by day. The use of underground freshwater in farming made a critical question in the drinking water security of MENA.
- Another concentrative one is, the MENA heavily depends on conventional resources to produce electricity. In 2018, it produced 8.7% of global greenhouse gases with 3.2 billion of carbon that is mostly from the energy sector (Karim 2018). The average 38% of MENA's CO_2 is emitted by the energy sector, followed by an average of 25% comes from the transportation sector.

2.2 Impact of Climate Change

Due to climate change, MENA is facing extreme temperature rise, low precipitation with huge depreciation in pursuance of scientific prognostic. From 1961 to 1990, MENA has faced a 0.2° increase in temperature per year and now it is facing a faster rate of increasing temperature than before. Thence, drought will be longer and frequent. The accumulative effects of climate change are visible on food storage, drinking water availability, and the economic structure of these countries. Most out the area of these countries are depending on the dams for the water in the field of irrigation and drinking. People of MENA are already started facing drinking water scarcity. Hither, most of the lands of MENA start receiving less than 300 mm of annual rainfall. This rainfall rate will decline 20–40% in the upcoming years in this area. Sandstorm is a long-live problem in MENA. But the sea-level rise in the Mediterranean areas of these countries throws a challenge for the coming future. The study said, it is recorded via tide gauges that averagely 1.1–1.3 mm sea level water is increasing per year. However, MENA has faced its highest recorded temperature of 54 °C in Kuwait in 2016 (Rignall 2019). None the all, this region is predicting the drying future as the temperature is increasing at the fastest rate than the global land temperature.

The impact of climate change is also visible in the public health of this region. High temperature and water-related diseases like cholera, dry condition and poor sanitization related disease like trachoma, a skin disease namely leishmaniasis, malaria-like vector-borne disease are the focusable effects of climate change already started spreading. These diseases are not only hampering the public health of MENA, it also the reason for the economic degradation of this region (Waha et al. 2017).

2.3 *Effect of Conventional and Renewable Resources on Power Generation*

Due to the high population and developing technologies, energy demand in the MENA region is increasing rapidly. Almost 75% of global electricity comes from fossil fuel resources or conventional resources. MENA is the global contributor of crude oil and natural gas as it has a high production capacity. But this strong pride makes the region so much vulnerable to global climate change. But, the extreme use of conventional resources can be the reason for the reduction of the storage of natural reserves. Conventional resources are responsible for ozone depletion, acid rain, and simply, global warming. Global warming is the impact of greenhouse gases emission. The greenhouse gasses like CO_2, CH_4, CFCs, N_2O, halon, ozone, etc. are produced for the use of conventional resources. For electricity and heat production by conventional resources, this region has faced almost 51% of carbon emission in 2014. However, the energy sectors of the MENA are the main contributor to air pollution.

On the other side, renewable resources based energy generation technologies are highly matured as they are abundantly available in nature (Bhattacharjee and Nandi 2020b; Nandi et al. 2018). Renewable sources such as hydro, wind, solar, geothermal, biomass are the largest power contributor in the world. To enhance the socio-economic development of any country, renewable resources give a chance to increase national income and job opportunities. In some countries of MENA, hydro is a prime renewable resource to produce power. But, it is remarkable that only 0.4% of total primary energy is supplied by the renewable resources in MENA. But, these

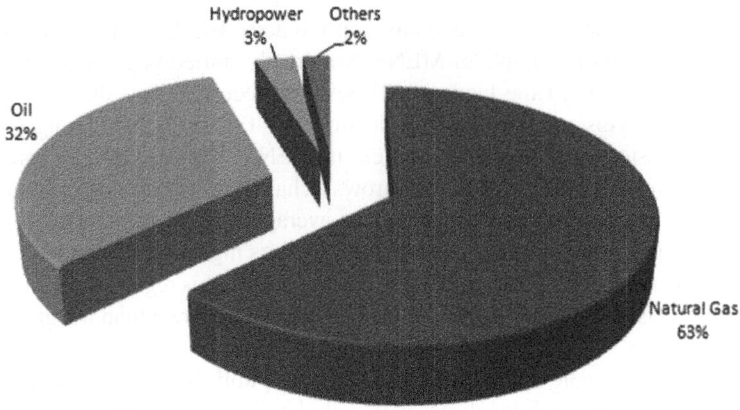

Fig. 1 Total energy generation scenario of MENA. *Source* Aghahosseini et al. (2020)

regions have the highest wind and solar energy potential over the world. Figure 1 indicates the total energy generation scenario in 2012 of MENA. Unfortunately, without Morocco, Tunisia, Yemen, Lebanon, and the State of Palestine, most of the countries of MENA are agree to use 100% renewable resources for energy generation by 2050 (Aghahosseini et al. 2020).Thence, renewable resources are only one solution to protect the MENA from the climate change issue. It is highlighted that the sustainability of renewable resources fully depends on the variation of climatic conditions (Elkadeem et al. 2019).

In MENA, Sudan, Oman, and Bahrain are the only countries that are practicing energy efficiency techniques by the use of renewable energies. Moreover, Sudan is one of the poorest countries in the world. But, Sudan is highly rich with renewable resource materials. So, as per the economic scenario, we proposed a renewable resource-based hybrid system to reduce the country's investment on conventional resources based energy sector. Table 1 shows some collected information about our focused location, Sudan (NASA Power 2018). This country has faced the drought in several times, but the snow melting water makes the rivers of Sudan always active.

3 Methodology

3.1 Location Analysis

Sudan is a North African country and it is the third-largest country of the Arab world. Sudan is a country that is really enriched by renewable resources. The Blue Nile is a state of the Republic of Sudan whose capital city is Ad-Damazin. In Ad-Damazin, a dam was constructed on the river Blue Nile in 1966 in the upstream of the one town of ErRoseires. Thence, the location implies the dam name as Roseires Dam

Table 1 Some information about selected location

Months	Air temperature (°C)	Relative humidity (%)	Earth temperature (°C)	Atmospheric pressure (kPa)
Jan	26.6	36.9	28.0	95.2
Feb	28.3	31.4	29.8	95.1
Mar	30.6	29.2	32.3	94.9
Apr	32.2	33.2	34.0	94.8
May	30.9	48.7	32.4	95.0
Jun	28.8	59.5	30.0	95.1
Jul	26.5	70.4	27.1	95.2
Aug	25.9	75.9	26.4	95.2
Sep	26.8	71.8	27.3	95.2
Oct	28.4	61.5	29.0	95.1
Nov	28.7	46.5	29.6	95.1
Dec	27.1	41.5	28.2	95.2

Source NASA Power (2018)

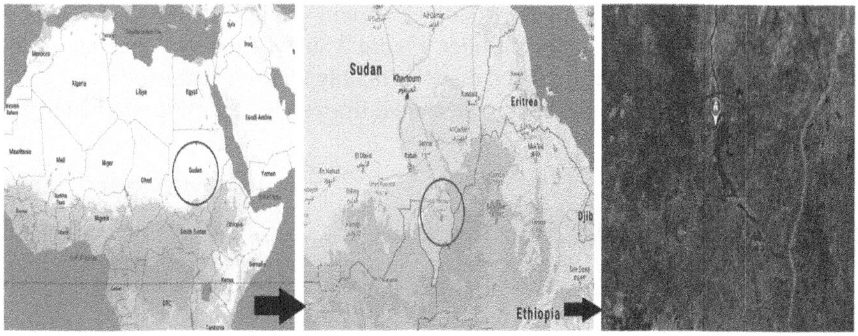

Fig. 2 Arial view of Roseires Dam. *Source* NASA Power (2018)

(Abuelnuor et al. 2020). Figure 2 displays an aerial view of the Dam (NASA Power 2018). This dam has various units to continue the service to generate electricity. Here, we consider only one unit of the Roseires hydroelectric plant to get a more efficient outage for continuous services.

3.2 *Block Diagram Analysis*

Renewable energy resources have the capability to give cheap energy and comfort economy. But, sometimes, they bring insufficient and low-security issues when they work individually to supply the power as per load demand (Bhattacharjee and Nandi

2020a). In this context, to find a cost-effective and clean solution for technical, financial, and environmental problems in the case of electricity generation, a hybrid renewable energy system is only one solution. Therefore, in this book chapter, we have proposed a wind power pumped hydro storage system to utilize wind power, hydropower, as well as the storage facility of hydro energy.

This hybrid model is proposed for only one unit of the Roseires hydroelectric power plant. In this modification, we consider a wind powered pumped hydro storage system to reuse the water after the first time generation. Generally, water from the dam or main reservoir comes to the upper reservoir, and then it falls down through the penstock to the river. Thus, the kinetic energy of water falling energy is converted to electrical energy by using the turbine-generator set. Here, we want to reuse the fallen water from the upper reservoir to lower reservoir. Therefore, wind energy is used to operate the pump during every season to pumping the water, otherwise, excess wind energy is directly sent to the grid for utility services. Therefore, an intelligent controller is used to control the operation of two different sources of energy. It supplies energy to the pumped storage system as per requirement and helps to maintain uninterrupted power to the load by taking the decision of how much energy is required from the wind turbine to operate the pump. Hence, the proposed modified system consists of three reservoirs (main, upper, and lower), penstock, turbine-generator set, pump, energy storage system (pumped hydro storage), end-user (load), and an intelligent controller. Figure 3 shows the block diagram of the modified power plant. This modified model is mainly focused on the following cases of operation. In the first case, energy generated from the windmill is checked by the

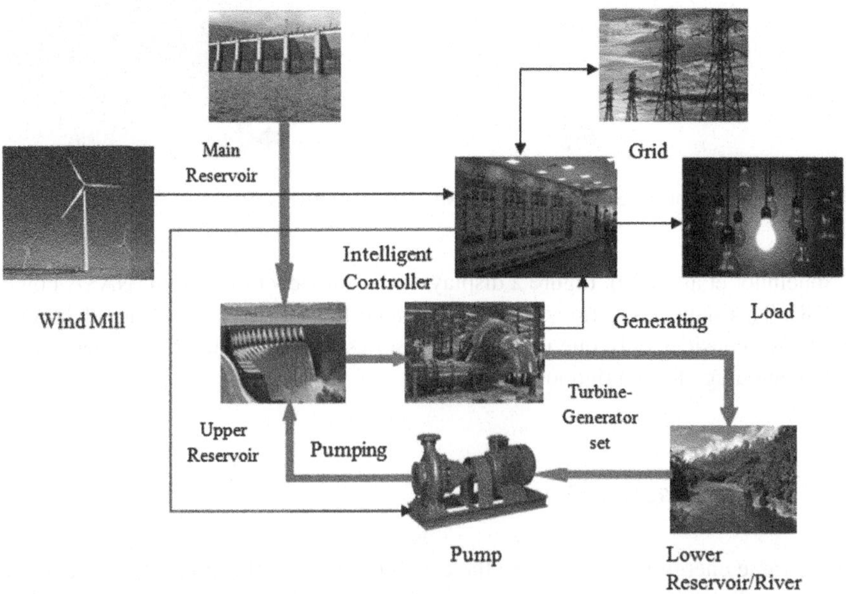

Fig. 3 Block diagram of the proposed hybrid system

intelligent controller. Secondly, the intelligent controller checks the water availability status of the main, upper and lower reservoirs and decides the pump needs to run or not. Thirdly, if the controller checked that water is available in the upper reservoir, which means no need to run the pump, and then the generated energy from the windmill is directly sent to the grid. Otherwise, wind energy is used to run the pump and excess wind energy after use in the pump is sent to the grid. In the fourth case, the generated energy from the pumped hydropower plant is directly sent to the grid to fulfil the load. At last, this intelligent controller checks, the generated energy from the modified model is sufficient or energy required taking from the grid to fulfil the load demand.

3.3 Analysis of Case Study Data

The case study data include average monthly water availability of that particular reservoir, wind speed of that location, locality load data, and the ambient temperature. These data are obtained from the references (Abuelnuor et al. 2020; NASA Power 2018). The latitude and the longitude of this particular location is $11^{\circ}8'$ and $34^{\circ}3'$ respectively.

3.3.1 Water Availability

Electricity generation from the hydroelectric plant depends on the water head and flow rate of water from the upper reservoir to the hydro-turbine. These water flow rate data are taken from (Abuelnuor et al. 2020).Water head means the vertical distance between water discharge points to headstream. Simply, the height of the water is water head. The maximum head of water refers to the maximum flow rate of water. Although, the hydro flow rate is defined as the rate of flow of water in litre per second. The turbine rotates most rapidly to generate more electricity. Figure 4 indicates the monthly average head and flow rate of water.

3.3.2 Wind Energy Resource

Figure 5 presents the average monthly wind speed of the location. The wind speed of any area depends on the difference in pressure of that particular area due to a change in temperature. Thus, the temperate of any area has a direct relation to wind speed. The maximum average wind speed of the Ad-Damazin area is found 4.9 m/s in June and the minimum wind speed is found 2.4 m/s in October. The speed of the wind is taken 50 m above the ground surface. The wind speed data is taken from the reference (NASA Power 2018). The average annual wind speed is 3.8 m/s for this particular area.

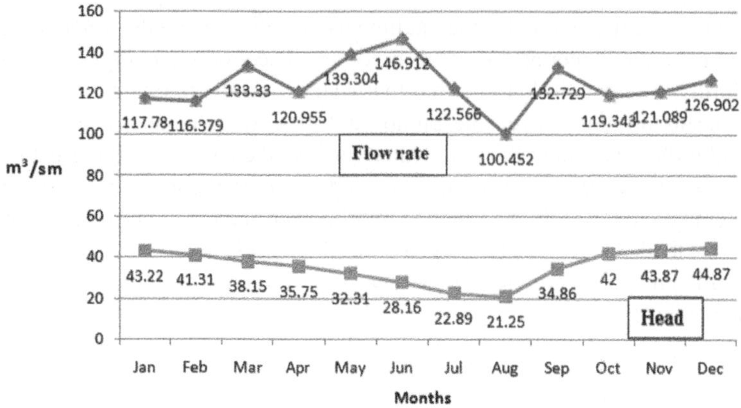

Fig. 4 The monthly average data of head in (m) and water flow in (m^3/s). *Source* Abuelnuor et al. (2020)

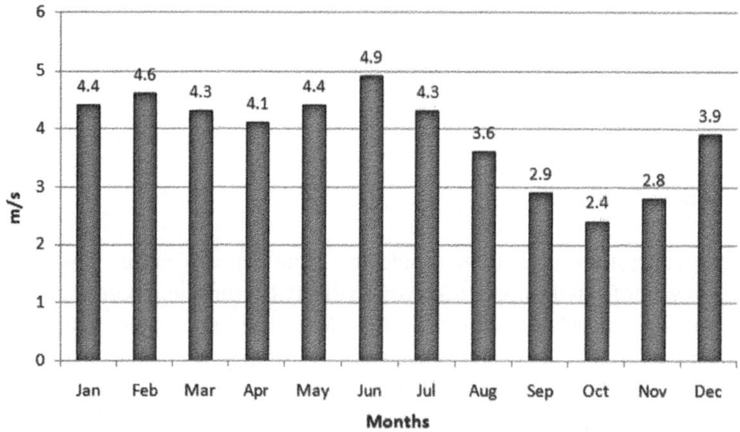

Fig. 5 The monthly average data of wind speed in (m/s). *Source* NASA Power (2018)

3.3.3 Load Profile

It is considered that the proposed hybrid system will operate with a single unit of the El Roseires hydropower plant in standalone mode to accomplish the load demand of a particular area of Ad Damazin, Sudan. The load profile of the system is considered as a summation of domestic load, small scale industries, and community load, etc. In order to place a feasible load profile, day to day variation of 15% and time-step variation of 20% are considered. In the proposed system, the load is taken as constant in all the months of a year. The load data are taken from the reference (Rabah et al. 2016).

3.4 Simulation Model Configuration

Agrid-connected hybrid system is consisting of two main sources of power namely hydro and wind. This proposed simulation model consists of a windmill, pumped hydro storage system, intelligent controller, and load. Figure 6 represents the simulation model of the proposed hybrid system.

3.4.1 Grid and Windmill

In the windmill, the wind turbine is connected to the induction generator to produce electricity. Here, we prefer an induction generator because of its excellent performance during variable wind speed. Induction generators are able to produce suitable power at different rotor speeds (Induction generator is more suitable for windmill 2010). Herein, a source or grid is connected to the whole hybrid system via windmill. The average hourly output of the wind farm is 3.4 MW. Therefore, the output of the wind farm is sent to the intelligent controller.

3.4.2 Intelligent Controller

The intelligent controller is the main operator of this proposed hybrid system. The output of the windmill has come to this controller and it sends power to the pump of the pumped hydro storage system. If the windmill power is not sufficient to run the pump, it takes power from the grid. On the other hand, this controller sends power directly to the grid to fulfils the load demand if the generated wind power is more than the requirement. Intelligent controller checks the water levels of the upper and the lower reservoirs of the pumped hydro storage system for deciding its actions.

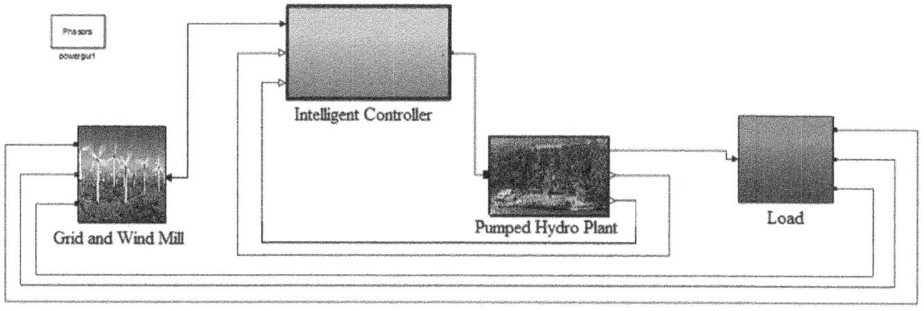

Fig. 6 The simulation model of the proposed hybrid system

3.4.3 Pumped Hydro Storage System

The pumped hydro storage system consists of reservoirs, penstocks, pumps, and turbine-generation units. In this storage system, the pump is operated by windmill power. A synchronous generator is used here for its high efficiency, good power factor characteristics. The output of this system sends it to the load via the grid. The average hourly output of the pumped hydro storage system is 36 MW.

4 Result and Discussion

This section reveals all possible system output that is generated in the software during the operation of the proposed hybrid system.

4.1 Production Analysis

We have seen in the case study data section that the water flow rate and head of water become low in August due to the climatic effect of this area. During this time of the year, windmill power and pumped hydro storage systems help to fulfil the load demand. Figure 7 shows the average monthly power output based on real-time data of the hydropower plant. Here, we can see that the maximum average monthly output of 35.31 MW is found in December and the minimum average monthly output of 16.11 MW is found in August.

Figure 8 indicates the monthly average power output of the windmill based on

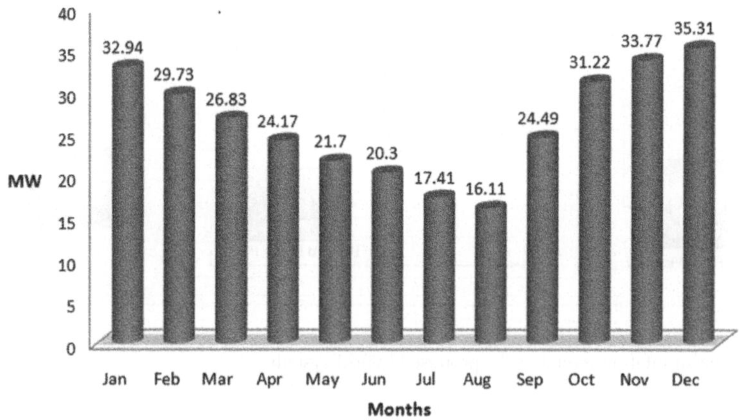

Fig. 7 The average monthly power output of the hydro power plantin MW. *Source* Authors' graph based on data from

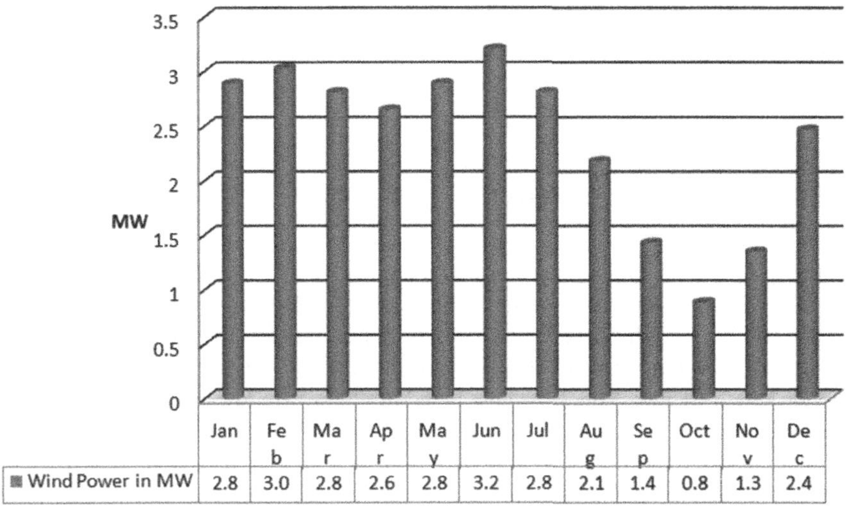

Fig. 8 The average monthly power output of the windmill in MW. *Source* Authors' graph based on data from

real-time data of the particular location. Here, we can see that the maximum average monthly output of 3.2 MW is found in June where the minimum average monthly output of 0.8 MW is found in October. The power production of the windmill absolutely depends on the speed of the wind. Table 2 shows the total energy generation and consumption scenario based on real-time data of water flow rate, water head, and wind speed. The total generated energy per year is 248993.4775 MWh and the total primary load is only 87,600 MWh per year. From Figs. 7 and 8 and Table 2, it is clear that the renewable power generation is quite good enough to maintain the continuity of power supply and successfully fulfill the load demand. Moreover, it sells a good amount of excess energy to the grid. Consequently, it is clear that the proposed hybrid

Table 2 Total energy generation and consumption scenario on a yearly basis

Component	Generation(MWh/year)	Percentage (%)
Energy generation scenario of the hybrid energy system		
Pumped hydro system	228,206.6909	91.65167
Wind mill	20,786.7866	8.34833
Energy from the grid	0	0
Total	248,993.4775	100
Energy consumption scenario of the hybrid energy system		
Primary load	87,600	35.18171
Total excess energy sending to the grid	161,393.4775	64.8184
Total	248,993.4775	100

system is profitable on the basis of all results. As the hybrid system is designed by using hydro and wind power sources, so zero greenhouse gas emissions. Therefore, we can tell this hybrid system is environment friendly for North African city.

4.2 Performance Analysis

Is this proposed hybrid system profitable or not? How profitable it is? These are the most important portions of this analysis. Figure 9 illustrates the energy sends to the grid per month of the proposed hybrid system. Now, if you focus on Fig. 9, it is noticeable that energies are available to send in every month of a year. For this proposed hybrid system, August is the minimum energy sending month where December is the maximum energy sending month of a year. Thus, this proposed system doesn't require taking energy from the grid. It can work standalone to fulfil the load demand. Attention, study says, water is almost available in the selected location to generate electricity. But, we know, due to the climate change effect, the desert areas are facing flood and the rainiest areas are facing high draught. Therefore, focusing on the climate change scenario, we proposed a wind-powered pumped hydro storage system for the upcoming future.

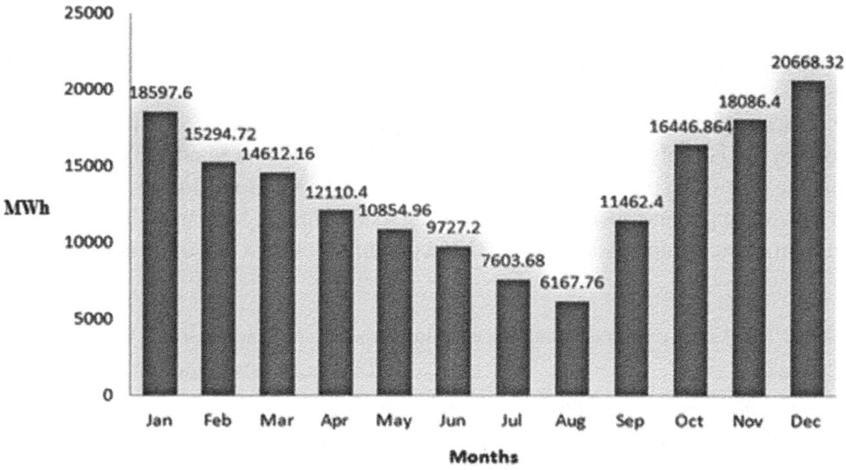

Fig. 9 Energy sends to the grid per month of the proposed hybrid system. *Source* Authors' graph based on data from

5 Conclusion and Recommendations

In the present scenario, climate change is rising like a curse for the whole world including Middle East countries. Due to rising impacts of the climate change, conventional energy sources are become endangered to found. So, we humans need to think about renewable energy sources for our better future. Here, focusing on the climate change scenario, we prefer renewable resources for power generation. To get clean and continuous energy, the hybridization of renewable resources is the best option. In this chapter, we proposed a hybrid system that incorporates hydropower and wind power together. By using real-time case study data of water flow and wind speed, this modified model is proposed for a single unit of the Roseires hydroelectric plant. This system is operated for uninterrupted service by no battery storage or no other conventional resources. By think about battery storage expenses, a pumped hydro storage system is taken as a backup of the proposed hybrid system. The following points are the key outputs of the analysis:

- The average annual wind speed is 3.8 m/s and the average annual water head is 35.72. This result shows the proposed system is perfectly suitable for the selected location.
- The proposed hybrid system doesn't purchase any energy from the grid and on the other hand, the proposed hybrid system sells 64.8184% of the total generated energy.
- 91.65167% of total energy is generated from pumped hydro system, which is 228206.6909 MWh/year and wind mill produces 8.34833% of total energy, numerically 20,786.7866 MWh/year.
- The proposed hybrid system produces no emissions of greenhouses gases.
- 35.18171% of the total generated energy is used to full the load demand.

These points show the proposed hybrid system is a preferable solution for the upcoming future in the Middle East. It successfully fulfils the load demand without affecting the environment. Moreover, it sends a good amount of energy to the grid, which increases the profits of the hybrid energy system. It is mentionable that wind energy helps to reduce the energy cost of the pump in the pumped hydro storage plant in our proposed model. Thus, the proposed hybrid system uses as a leading option for pollution-free economic growth.

In the proposed model, a grid-connected windmill is used to supply power for the operation of the pump in the pumped hydro storage system. Now, if some disturbances occur in the future and the grid or the windmill becomes unavailable, at that time who will supply power to the pump. This is one of the main limitations of the proposed hybrid system. To resolve this type of issue in the future, hydro energy could integrate with solar, biomass, etc. for better performance and continuity of power supply. Therefore, the future prospect of the proposed hybrid system is to integrate it with a solar power plant or biomass power plant for maintaining the power supply continuity during emergency situations.

References

Abuelnuor A, Abuelnuor A, Ahmed K, Saqr KM, Nogoud YAM, Babiker MEAM (2020) Energy analysis of large and impounded hydropower plants: case study El Roseires Dam (280 MW). Environ Prog Sustain Energy. 39:13362. https://doi.org/10.1002/ep.13362

Aghahosseini A, Bogdanov D, Breyer C (2020) Towards sustainable development in the MENA region: analysing the feasibility of a 100% renewable electricity system in 2030. Energy Strategy Rev 28:100466. https://doi.org/10.1016/j.esr.2020.100466

Alternative Energy Tutorials (2016) Home of alternative and renewable energy tutorials, induction generator. Available at: https://www.alternative-energy-tutorials.com/wind-energy/induction-generator.html. Retrieved Aug 2020

Bhattacharjee S, Das U, Chowdhury M, Nandi C (2020b) Role of hybrid energy system in reducing effects of climate change. In: Qudrat-Ullah H, Asif M (eds) Dynamics of energy, environment and economy, lecture notes in energy, vol 77. Springer, pp 115–138. https://doi.org/10.1007/978-3-030-43578-3_6

Bhattacharjee S, Nandi C (2020a) Design of an industrial internet of things-enabled energy management system of grid-connected solar-wind hybrid system-based battery swapping charging station for electric vehicle. In: Mandal J, Mukhopadhyay S, Roy A (eds) Application of internet of things, lecture notes in network and systems, vol 137. Springer, Singapore, pp 1–14. https://doi.org/10.1007/978-981-15-6198-6_1

Bhattacharjee S, Nandi C (2020b) Design of a smart energy management controller for hybrid energy system to promote clean energy. In: Bhoi A, Sherpa K, Kalam A, Chae GS (eds) Advances in greener energy technologies, green energy and technology. Springer, Singapore, pp 527–563. https://doi.org/10.1007/978-981-15-4246-6_31

Bhattacharjee S, Nandi C (2020c) Technical, economic, feasibility and comparative analysis of three different configurations of energy system to control intermittency of renewable energy. In: Proceedings of the 2nd international conference on information system & management science (ISMS) 2019, Tripura, India, SSRN electronic journal, Elsevier. https://doi.org/10.2139/ssrn.3512405

Bhattacharjee S, Chakraborty S, Nandi C (2020a) An optimization case study of hybrid energy system in four different regions of India. In: Bhoi A, Sherpa K, Kalam A, Chae GS (eds) Advances in greener energy technologies, green energy and technology. Springer, Singapore pp 399–437. https://doi.org/10.1007/978-981-15-4246-6_23

Broom D (2019) How the Middle East is suffering on the front lines of climate change. World economic forum, 05 April 2012. Available at: https://www.weforum.org/agenda/2019/04/middle-east-front-lines-climate-change-mena/. Retrieved 16 Aug 2020

Climate Change in the Middle East and North Africa (2020) Available at: https://en.wikipedia.org/wiki/Climate_change_in_the_Middle_East_and_North_Africa#Mitigation_and_adaptation. Retrieved 16 Aug 2020

Elkadeem MR, Wang S, Sharshir SW, Atia EG (2019) Feasibility analysis and techno-economic design of grid-isolated hybrid renewable energy system for electrification of agriculture and irrigation area: a case study in Dongola Sudan. Energy Convers Manag 196:1453–1478. https://doi.org/10.1016/j.enconman.2019.06.085

Huang XJ, Bao NS (2016) Modeling and simulation analysis of wind—hydro hybrid power plant. In: 2nd annual international conference on energy, environmental & sustainable ecosystem development (EESED 2016); Advances in Engineering Research, Atlantic Press, pp 169–179. https://doi.org/10.2991/eesed-16.2017.22

Karim ME (2018) Exploring the impact of climate change on the outbreak of early twenty-first-century violence in the Middle East and North Africa and the potential of permaculture as an effective adaptation. Weather, Clim Soc 10(1):179–186. https://doi.org/10.1175/WCAS-D-16-0130.1

Lange MA (2019) Impacts of climate change on the Eastern Mediterranean and the Middle East and North Africa Region and the water-energy nexus. Atmosphere 10(8):455. https://doi.org/10.3390/atmos10080455

Mihut VV, Baciu I, Panoiu C, Panoiu M (2018) Hybrid power generator based on wind, hydro and solar for use as an alternative solution for power supply. In: IOP conference series: materials science and engineering, international conference on applied sciences, vol 477. Banja Luke, Bosnia and Herzegovina, IOP Publishing limited, pp 9–11. https://doi.org/10.1088/1757-899X/477/1/012014.

NASA Power website (2018) Available at: https://power.larc.nasa.gov/. Retrieved July 2020

Nandi C, Bhattacharjee S, Reang S (2018) An optimization case study of hybrid energy system based charging station for electric vehicle on Mettur, Tamil Nadu. Int J Adv Sci Res Manage 3(11):225–231

Nandi C, Chakraborty AK (2012) Wind power plants with VSC based STATCOM in PSCAD/EMTDC environment. In: 2nd international conference on power and energy systems (ICPES), IPCSIT, vol 56. IACSIT Press, Singapore, pp 7–11. Available at: http://www.ipcsit.com/vol56/002-ICPES2012-PS022.pdf

Nandi C, Bhattacharjee S, Chakraborty S (2019) Climate change and energy dynamics with solutions: a case study in Egypt. In: Qudrat-Ullah H, Kayal A (eds) Climate change and energy dynamics in Middle East, understanding complex system, Springer, pp 225–257. https://doi.org/10.1007/978-3-030-11202-8_8

Nguyen TT, Phan V, Dinh BH, Phan TM, Nguyen TT (2020) Optimal power generation for wind-hydro-thermal system using meta-heuristic algorithms. Int J Electr Comput Eng 10(5):5123–5130. https://doi.org/10.11591/ijece.v10i5.pp5123-5130.

Nikolaou T, Stavrakakis GS, Tsamoudalis K (2020) Modeling and optimal dimensioning of a pumped hydro energy storage system for the exploitation of the rejected wind energy in the non-interconnected electrical power system of the Crete Island, Greece. Energies 13(11):2705. https://doi.org/10.3390/en13112705

Opan M, Unlu M, Ozkale C, Celik C, Saeac HI (2019) Optimal energy production from wind and hydroelectric power plants. Energy Sources, Part A 41(18):2219–2232. https://doi.org/10.1080/15567036.2018.1555626

Rabah AA, Nimer HB, Doud KR, Ahmed QA (2016) Modelling of Sudan's Energy Supply, transformation, and demand. J Energy 2016:5082678. https://doi.org/10.1155/2016/5082678

Rignall K (2019) Living climate change in the Middle East and North Africa. Int J Middle East Stud 51(4):629–632. https://doi.org/10.1017/S0020743819000710

Tagliapietra S (2019) Energy in North Africa: challenges and opportunities. Atlantic Community, 04 March 2019. Available at: https://atlantic-community.org/energy-in-north-africa-challenges-and-opportunities/. Retrieved 26 Aug 2020

Team R (2020) Renewable energy in Sudan: status and potential—Part 1. Renewables in Africa, 06 Jan 2020. Retrieved 26 Aug 2020. Available at: https://www.renewablesinafrica.com/renewable-energy-in-sudan-status-and-potential-part-1/#:~:text=The%20technical%20potential%20for%20renewable,be%20among%20the%20highest%20globally

Timmerberg S, Sanna A, Kaltschmitt M, Finkbeiner M (2019) Renewable electricity targets in selected MENA countries—assessment of available resources, generation costs and GHG emissions. Energy Rep 5:1470–1487. https://doi.org/10.1016/j.egyr.2019.10.003

Waha K, Krummenauer L, Adams S, Aich V, Baarsch F, Coumou D, Fader M, Hoff H, Jobbins G, Marcus R, Menge M, Otto IM, Perrette M, Rocha M, Robinson A, Schleussner CF (2017) Climate change impacts in the Middle East and Northern Africa (MENA) region and their implications for vulnerable population groups. Regional Environ Change 17:1623–1638

Worldometer (2020) Sudan CO2 emission. Available at: https://www.worldometers.info/co2-emissions/sudan-co2-emissions/. Retrieved 26 Aug 2020

Montenegro and World: Climate Change and Biodiversity Conservation

Mira Šorović

Abstract Last year, the fires of the Amazon rainforest have globally re-opened the discussion on Climate Change with closely related issues of Biodiversity conservation and global warming problem, thanks to several demonstrations around the world organized by nongovernmental environmental organizations and civil groups, led by a girl by name Greta Thunberg. Therefore, this article will present the main problems which the modern world and humankind have faced (negatively affecting Climate Change and Biodiversity conservation), but with special emphasis on the situation in Montenegro, as a good example of the first ecological country in the world. Hence, Montenegro needs to fulfill all necessary standards in the field of environmental protection and preservation, to become a full member of the European Union. In that context, in this article will be briefly presented the main projects that Montenegro has implemented, in order to preserve nature and the environment, supported by statistics and concrete facts from this area of activities. Thus, at the end of this paper, the main obstacles and the prospects for further progress and improvement of the preservation and protection of the living environment at the global level will be highlighted.

1 Introduction

Over the past years, the world has faced many different problems in the field of Climate Change and Biodiversity caused and made by humans. The damage is a real problem and represents a real threat to the Earth and its habitants. As a result, the current data advert that we need to decrease emissions of the greenhouse-gas by at least 80% till 2050: firstly, developed countries and later, less developed countries and areas, in order to have a possibility to stay below at an average temperature rise of over 2 °C (Woods, Williams, Hughes, Black and Murphy 2010).

Heavy industry and factory farming in less developed countries have a significant part in the greenhouse effect, especially great powers, like China and the United

M. Šorović (✉)
Humanistic Studies, University of Donja Gorica, Podgorica, Montenegro
e-mail: mira.sorovic@udg.edu.me

W. Leal Filho and E. Manolas (eds.), *Climate Change in the Mediterranean and Middle Eastern Region*, Climate Change Management,
https://doi.org/10.1007/978-3-030-78566-6_7

States of America which represent major contributors to the Climate Change and Biodiversity challenges, liberating in that manner enormous volumes of greenhouse gases, litter and waste of various origin. And that is not all. The gas processes in along with the pollution of the air, soil and water, in the recent decades, have had extremely negative effects on human health and Climate Change in many different shapes and ways. In accordance with the abovementioned, it is important to note that millions of people across the world are victims of the catastrophic effects caused by worsening of Climate Change. Because of that, it has increased speed of melting snow and glaciers in the Northern part of the world hemisphere, especially in the Arctic, to prolonging droughts in sub-Saharan Africa and destructive tropical storms and floods wide-ranging across the Pacific, Southeast Asia and the Caribbean Sea, while on the opposite side—the Southern part of the hemisphere, world is facing a different kind of the natural disasters, such as storms, fires and heatwaves during summer months. Despite all previously stated, we still do not understand the real consequences which may cause Climate Change processes on the Biodiversity. This is just the beginning of the end. In addition, if the devastation of the environment continues, the depth and the spread of its impacts will have effects on our natural world, creating in that manner the ruined and toxic environment for future generations. Hence, the interesting question is: Why do world leaders not pay attention to these crucial problems, like Climate Change and Biodiversity? (AMNESTY International 2020). Time and future generations will provide all answers. In the meantime, it is evident that many endemic species are in danger of the extinction, day by day. Hence, it is the right time that the world community take responsibility for its acts and actively prevent the further destruction of the environment and Biodiversity at the global level. In contrast, with such irresponsible behavior, the end of the human race is approaching.

In the second part of this paper, there will be a few words about Montenegro. Although Montenegro has a small surface area of 13.812 km^2, the fact is that it lies within the global Biodiversity hotspot of the Mediterranean Basin (Groombridge & Jenkins 2002, Ieronymidou et al. 2015, in Rubinić et al. 2019). It bounds "a rich array of geologic formations, climatic zones and natural and cultural landscapes. Following to its geographic position in the west of the Balkan Peninsula and its vertical extension from sea-level at the shores of the Adriatic Sea up to more than 2500 m a.s.l. in the Durmitor Massif and Prokletije Mountains, it harbors a particularly high habitat diversity. From alpine grasslands on the higher mountain ridges, together with a vast and extraordinarily preserved belt of mountain forests in the north and west of the country, to dry lowdown grasslands, sub-Mediterranean evergreen forests, and extensive coastal and marine ecosystems, Montenegro encompasses all major habitats of the Western Balkans and Eastern Europe" (Ministry of Spatial Planning and Environment 2010 in Rubinić et al. 2019: 305).

Also, it is important to underline that the territory of Montenegro has two basic bio-geographical regions: Mediterranean and Alpine. Although it is a small country surface, it encircles a wide range of habitat types and ecosystems. In accordance with the *National Biodiversity Strategy and Action Plan*, forest, dry grassland, alpine, freshwater and marine ecosystems are found in Montenegro.[1] Thanks to their specificity, types of habitats and geological structures, such as: karst, canyons, coastal habitats and caves, also have an important place in Biodiversity protection, as an addition to the main ecosystems (Ministry of Sustainable Development and Tourism 2014).

Montenegro as the first ecological country in the world has a clear aim to become a full member of the European Union. Therefore, it needs to fulfill all necessary standards in the field of environmental protection and preservation. In that context, it needs to update national legislation with the *EU acquis*, as a significant step towards country accession, but, in accordance with the technical assistance and advice from the Union. Hence, this article will briefly present the main projects that Montenegro has implemented in the last period, in order to preserve nature and environment, supported by statistics and concrete facts from this area of activities.

Despite many projects which support and preserve nature and the environment, Montenegro needs to face many threat categories in this field. They are identified in the *National Biodiversity Planning to Support the implementation of the CBD 2011—2020 Strategic Plan in Montenegro* (GEF 2019), as: uncontrolled urban and tourism development, changes in land use practices, unsustainable and illegal use of natural resources, habitat loss, poisoning and persecution of raptors, unsustainable hunting of game species and illegal hunting of protected species, etc. All those threats are connected especially with the bird species in Montenegro (Rubinić et al. 2019), but it could be connected and have some links with the other species and environment, generally speaking. Therefore, it is necessary to have a quick reaction and improvement of the current situation, both in theory and practice. Hence, in order to be acted and influenced at a global level, it is important to have all tasks and obligations completed and implemented at home, at the national level.

2 Aims and Objectives

This chapter aims to highlight the most important global problems in the world, in the field of Climate Change and Biodiversity, but with a special focus on Montenegro. Hence, there will be a few words in this article about the projects which have been implemented in Montenegro throughout the years and which are in harmony with nature and sustainable development, although they have roots in the field of Climate Change and Biodiversity.

[1] Montenegro is rich with floristic diversity. It is among the highest in the Western Balkans region (3250 plant species, while 223 endemic plant species and subspecies have been registered). Thus, the territory of Montenegro represents one of the 153 important floristic Biodiversity centers at global level (Ministry of Sustainable Development and Tourism 2014).

3 Materials and Methods

The information needed to assess existing issues which are connected with the theme of this topic—Climate Change and Biodiversity from the aspect of the world and Montenegro, were obtained from electronic databases such as Encyclopedia Britannica, UNESCO, GEF, NASA Global Climate Change, Montenegrin Ministry of Sustainable Development and Tourism and many other websites. The keywords such as: Climate Change, Biodiversity, Montenegro and world issues, were used isolated or combined, provided new and interesting information and facts in the field of this research. Also, the published journal articles on related topics in Montenegrin and English language were used for this article, with no year limit. The search generated a lot of articles and abstracts. Thus, these published sources were further filtered based on relevance, so only the most appropriate were selected and used for this chapter. The current situation in the field of Climate Change and Biodiversity was taken from reports of the Ministry of Sustainable Development and Tourism, the UNDP and other international bodies and many other institutions and is used for the purpose of this chapter. Hence, the chapter was based on research facts, statistics and scientific predictions as an important instrument for collection and analysis of a fresh field data, with the aim of further understanding crucial world problems.

4 World: Climate Change and Biodiversity

By definition, "Climate Change is a periodic modification of Earth's climate due to changes in the atmosphere and interactions between the atmosphere and other geological, chemical, biological and geographic factors within the Earth system" (Rafferty 2020). Thanks to the Earth-orbiting satellites and other advances of technology, throughout many years, scientists have collected various types of information about the Earth and its Climate Change on a global scale (NASA Global Climate Change). To sum up, in the last century, the average global temperature has significantly risen, while rainfall patterns have changed and affected the increased number of the extreme events. It is a fact that the Earth is in the phase of changing and getting warmer. Because of that, there is a reasonable need that global warming could be limited to 1.5 °C, but achieving this aim, we must reduce carbon emissions by 45% by 2030, which means reaching near-zero emissions by around 2050. Also, outputs of global warming are visible now, however catastrophic consequences are predicted in the near future. The global warming impacts have already been felt around the world, with declines in Biodiversity, crop yields, Arctic sea ice, coral reefs and the increased number of the heavy rainfalls and heatwaves. Additionally, in the last decade, sea levels have risen by 40.5 mm and they are predicted to continue rising through decades, even if all emissions of greenhouse gases were reduced to zero, starting from this moment (Keaveny 2018).

There is a need for urgent reaction. It is important to underline that our planet continues to warm, at the current rate of 0.2 °C per decade, while according to some calculations and predictions, this number will reach 1.5 °C of warming around 2040. Because of that, the Earth needs the same great changes. Thus, one of the best solutions at this moment is putting a price on global emissions of methane, carbon dioxide and other greenhouses in order to achieve the goal—reach net zero globally, by around 2050. This is, by opinion of most economists, the most efficient manner to save Biodiversity and natural environment. And that is not all. The emphasis will be put on renewable energy sources. Between 70 and 85% of the electricity at the global level will need to be provided by renewables. Also, there will be a need for investing in energy-efficient and low carbon technologies, in order to double them. On the other hand, investing in fossil-fuel extraction will need to be reduced by around a quarter. Hence, there will be a need for carbon dioxide removal technology as a real need for removing greenhouses from the atmosphere. But, the *Intergovernmental Panel on Climate Change* gives an alert because of the possible future risk—carbon dioxide removal has not been used on such a huge scale ever before. Hence, it could be used as an extra step against global warming, not just as an excuse to keep emitting greenhouse gasses in the atmosphere. Also, there would be a need for reduction of production, consumption and lifestyle choices (*Ibidem*) of the Western civilization primary. Well, it will be necessary to change the consciousness of society, at the global level. The consumer society must become more economical and rational. It must think beyond the current needs and satisfy its own desires. Humans have become selfish and egocentric. They forgot that they should live in harmony with the nature. Human beings, regardless of the fact that they build bridges, skyscrapers and railways, primarily need clean water, land, food and air, as necessary conditions for a healthy and balanced life. Therefore, they cannot think about short-term benefits, because, once we destroy and contaminate all available natural resources, banknotes will mean nothing to us—they will have no use value. In similar manner, this way of thinking is pictorially presented in the Cree Indian proverb: "Only when the last tree has died and the last river been poisoned and the last fish been caught will we realize we cannot eat money." In this sentence, there is all the wisdom and meaning of human existence and life. Additionally, an interesting thought about this theme—Climate Change and Biodiversity essentials, was given by Professor A. Hoffman (2015), who said that "Healthy ecosystems and rich Biodiversity are fundamental to life on our planet".

During last year, a fire has spread rapidly and has engulfed the Amazon tropical rainforest. This Biodiversity problem has globally re-opened the question of Climate Change, Biodiversity and global warming issues, thanks to several demonstrations around the world, organized by nongovernmental organizations and civil groups, led by a girl by name Greta Thunberg[2] (Šorović 2019). But, why has so much dust been raised about this problem, when day by day the collapse of the Biodiversity

[2] Greta Thunberg is the Swedish teenager girl who skipped school classes every Friday and protested because of Climate Change issues, outside the Swedish Parliament building. In that way, she inspired millions of people all around the world and shared her concerns about Climate Changes. She calls attention to individual action, sending a clear message about global warming and the global Climate

and environment is evident, as well as other natural disasters caused by man? The answer to this question is really simple. The Amazon Rainforest involves the drainage basin of the Amazon River and its tributaries in the northern part of South America. Thus, it covers a tropical area of 6,000,000 km^2 and represents around 40% of total area of Brazilian territory. Thus, the Amazon Rainforest is the richest and most-varied biological reservoir in the world. It contains several million species of plants, birds, insects and other life forms, many still unrecorded in science (Encyclopedia Britannica 2020). And that is not all. The Amazon Rainforest represents the most significant and vital climate stabilizer in the world, which means production of 20% of the world oxygen. Hence, this is the main reason why this tropical Rainforest is so precious for the whole world and with the right reason it got the name "the lungs of the world" (Šorović 2019). Additionally, the Amazon Rainforest, as previously pointed out, has an essential role in balancing the world's climate, while at the same time, it is helping in combating the Climate Change issue and global warming problem. But, at the moment, the Amazon Rainforest is suffering from the Coronavirus pandemic effects. It is important to underline that deforestation increased by 55% in the first four months of 2020, compared with the same period last year, which represents horrible data, because this kind of action needs to be stopped as soon as possible. However, predominantly poor people used this difficult situation in the world to illegally cut down the trees in the Rainforest. Beyond the deforestation process, there are visible other processes used in the last decade, such as illegal mining, soil destruction and rainforest fires, etc. Thus, scientists warn that we are approaching a point of no return at an accelerated pace, after which the Amazon Rainforest will no longer function properly, nor will the world (BBC News na srpskom 2020). In addition, new research published in *Nature Communications* warns that huge parts of the Amazon area have gotten so dry that scientists think that the Amazon tropical rainforest is on the brink of transitioning into an entirely different kind of ecosystem—savanna. Hence, this data research imputes that 40% of Amazon is in the transitioning phase. This is nothing new, bearing in the mind that this rainforest ecosystem is very sensitive to changes in precipitation (the wildlife-sustaining foliage depends on moisture receiving) (Noor 2020).

Such illegal actions and disruption of the balance of the environment, represents a real example of human violation of the multilateral agreements on Climate Change, Biodiversity and environment. Hence, it is my obligation to mention two other environmental disasters which have influenced at global level, while the extent of the negative impact on the Environment and the living world in it is still evident, even today.

In April 2010, an explosion was caused on the Gulf of Mexico oil spill, (also called Deepwater Horizon oil rig[3]) located in the Gulf of Mexico. It started with deadly force explosion, while the consequences of this event, according to many researches and

crisis. Because of that, Greta soon became a heroine who fights for reducing the harmful impact on the Biodiversity and global Environment.

[3] It operated and owned by offshore-oil-drilling company Transocean and leased by BP oil company—located in the Macondo oil prospect in the Mississippi Canyon (Pallardy 2020).

scientists are immeasurable, even today, ten years after this catastrophe that affected the entire world Biodiversity. The crew on the Deepwater Horizon oil rig "worked to close up an exploratory oil well deep under the Gulf of Mexico, a pulse of gas shot up, buckling the drill pipe. The emergency valve designed to cap well in case of an accident 'blowout protector', failed, and the gas reached the drill rig, triggering an explosion that killed 11 crewmembers" (Borunda 2020). In the next three months, more than 300 Olympic-sized swimming pools of the oil leaked into the waters of Gulf, making it the biggest oil spill in the history of the United States. Thus, the volume of oil escaping from the damaged well was originally estimated by BP to be around 1000 barrels per day, while the American officials have estimated at more than 60,000 barrels per day. To make matters worse, in first few months, claims by several research groups that had detected the subsurface plumes of dispersed hydrocarbons were dismissed by the National Oceanic and Atmospheric Administration (NOAA) and by BP, however they were verified the next month, as the plumes were in fact from the Deepwater spill. However, the effects of the microscopic oil droplets on the Biodiversity and ecosystem were unknown, although their presence was spread into the sea, seafloor and at the landfall of the beaches. In all, an estimated 1100 miles (1770 km) of shoreline were polluted (Alabama, Florida and Mississippi beaches). A group of government agencies of the United States of America, like the U.S. Coast Guard and the Environmental Protection Agency, coordinated by the National Response Team made effort to mitigate damage, however they did not succeed as the damage has been fatal for the Environment (Pallardy 2020). However, it was not just an ordinary nature pollution, but the destruction of the environment and its Biodiversity. The oil companies were responsible for unforeseeable destruction of flora and fauna, as well as natural habitat of many species, worth billions and billions of American dollars. Ten years after this horror, the marine ecosystem is slowly recovering, although the consequences of this unpleasant event are visible even today on very sensitive species such as birds, fishes, coral reefs, etc.

As a second example, it is important to mention the Doñana National Park in Andalusia, Spain. It is important to underline that this National Park is a wetland of international importance under the *Ramsar Convention* and a part of UNESCO's global network of Biosphere Reserves. On April 25, 1998, the worst environmental disaster in Europe happened. A supporting wall of the reservoir containing the toxic wastes of the Aznalcollar mine burst, thus releasing 5,000,000 m^3 of acidic water and toxic mud into the surrounding National Park landscape. The toxic waste entered into the Agrio River, a tributary of the Guadiamar River, which is situated within the Doñana National Park (UNESCO 2020). The spillage of toxic sludge in combination with acidic waters destroyed the land of the National Park and its ecosystem. Regardless of the huge financial investments in cleaning and removing the toxic sludge (around 400 tons), only a minimal progress has been made. Thus, this land needs forty years to be completely recovered.

Because of all previously stated, individual action has enormous importance for the future of our planet. Therefore, we all together need to be responsible and active, in order for those kinds of the environmental disasters to never happen again. Thus, it is of great importance that every citizen and every institution in the world put an effort in rising awareness and take real action to tackle the Climate Change and Biodiversity issues, thus accelerating the full implementation of the *Paris Agreement* and many other important documents. Also, at this moment, around one million species are facing extinction, many within the next few decades. The scientists and experts are warning that the Nature crisis could be as dangerous as the climate crisis, because in nature everything is connected. Improving further actions, we need to prevent the extinction of more species. During XX and first part of XXI century, many species do not exist anymore, such as: the Sicilian wolf and the Paradise parrot (1920s), the Tasmanian tiger (1930s), the Xerces blue (1940s), the Japanese sea lion, the Crescent nail-tail wallaby and the Bubal hartebeest (1950s), the St. Helena Darter and the Guam flying fox (1960s), the Caspian tiger, the Siamese flat-barbelled catfish and the Yunnan lake newt (1970s), the Golden toad (1980s), the Rotund rocksnail (1990s), the Pyrenean ibex and the Pinta giant tortoise (2000s), the West African black rhinoceros (2010s), etc. (Greenpeace 2020). Owing to the previous facts, during last year, the *United Nations Report* found that more than one million species around the world are threatened with extinction. In accordance with that, more than 65 states signed a global pledge to reverse the losses in the nature by 2030 (Briggs 2020). Nevertheless, it is not obvious that this pledge will be fulfilled, because much more important agreements have not been implemented.

Looking all that from a time distance, today's problems of global warming and human violation of international protocols and conventions, led by many different actors and large world economy polluters, need to be reduced and put on the renewable energy sources. Thus, we need to act on a local level, in order to introduce changes on a global scale. United in fighting against Climate Changes and Biodiversity conservation, everything is possible. It was particularly seen during the lockdown period due to coronavirus, when the factories were closed and the traffic was suspended. In that period, nature began to recover, although a new problem has emerged now—the excessive use of protective sleeves and masks which will soon affect the environment and it would have a negative effect on the world flora and fauna.

5 Montenegro: Brief Information

Montenegro is a small, but very beautiful Southeastern European country at the Adriatic coast. Although, its size and population is not that large, Montenegro has a great geographical position and natural richness of the water resources and forests at the northern part of the country, while in its south has the sea and mountains coast around 3000 km long. Bearing all that in mind, Montenegro has an interesting land configuration, which is completed with a favorable climate geo-diversity and vicinity of the sea and natural features (which are attractive for valorization in tourism and many activities based on the values of nature—attractive canyons, rivers, lakes, speleological localities and national parks for hikers, mountaineers, cyclists etc. (Government of Montenegro 2011).

Montenegro has rich biological diversity with a large number of endemic and relic species of fauna and flora of international importance; a big number of different types of the ecosystems diversity and a high concentration of ecosystems and species in one small place (*Ibidem*). In accordance with that, Montenegro[4] has nationally protected areas, which cover nearly 125,000 ha or 9.05% of the entire territory of the country (Ministry of Sustainable Development and Tourism 2014). In addition, it is important to say that Montenegro has five national parks: Durmitor, Biogradska gora, Prokletije, Lovćen and Skadar Lake, all protected by *Law on Nature Protection* (*Official Gazette of Montenegro*, 51/08) and *Law on National Parks* (*Official Gazette of Montenegro*, 56/09) (Parks.it n.d.). It is important to mention that on September 20, 1991 in Žabljak, the Montenegrin Parliament adopted the *Declaration on Montenegro* as an ecological country—the first country in the world, which stated the following: "Bearing in the mind our debt to nature, a source of health and our inspiration for freedom and culture, we are devoting ourselves to its protection for the sake of our survival and the future of our posterity" (Government of Montenegro 2011, 5).

In the next part of this paper presented are the main projects in Montenegro in the field of Climate Change and Biodiversity conservation.

In Montenegro, in October 2016 the ***Identification of Important Plant Areas (IPAs) and new Key Biodiversity Areas*** was completed within the Mediterranean Basin Hotspot. During that research, which was terminated through several projects, the results were harmonized with the new Key Biodiversity Areas (KBAs) standard. This enormous work was completed through the national validation and regional workshop with plant specialists. This regional workshop authorizes IPAs to be examined against new KBAs data and criteria on plant trigger species to be systemized. Also, it is important to underline that the project ***Fresh Water Biodiversity Assessment and Conservation Priorities for the Mediterranean Basin Hotspot***, implemented by the International Union for Conservation of Nature, with co-funding from the MAVA Foundation (in French: MAVA, Foundation pour la Nature), derived in complete Red List assessments of major groups of freshwater species in the area of the Mediterranean Basin. At the same time, this was the first-ever published list of the KBAs for the hotspot. To be more precise, a total number of described species in that moment was 1236, while one-third of it are not found nowhere else on the Earth, but that term involved mapping and their global threat status assessed (BirdLife International 2017). Of course, some projects are focused on monitoring water birds—unique habitats important for migratory bird species: Bojana River Delta and Skadar

[4] Five national parks in Montenegro encompasses approximately 80% of the total nationally protected areas, while the rest of it—the remaining 20% refers to the more than 45 sites created as areas of special natural characteristics, monuments of Nature and (specific and general) nature reserves. In addition, several sites hold important international designations: National Park Durmitor and Tara River. Thus, National Park Durmitor has been on UNESCO's World Natural List since 1980, while Tara River canyon is a part of the UNESCO's network of Man and Biosphere reserves since 1977. Also, it is important to mention that Kotorsko-risanski Bay is on the World's Natural and Cultural Heritage site, while National Park Skadar Lake and Tivat Saltpans are recognised as Ramsar sites (Ministry of Sustainable Development and Tourism 2014).

Lake,[5] as well as other interesting sites like: Bjelasica Mountain, Bjelopavlići Valley, Bratogošt, Bukovica Valley and Vojnik, Buljarica, Ćemovsko polje, Durmitor, Hajla, Komovi Mountains, Kovač, Gradina, Crni vrh, Kuči and Cijevna Canyon, Lim Valley and Plavsko Lake, Ljubišnja, Lovćen and Paštrovići, Matruge, Morača and Mala Rijeka Canyons, Moračke Mountains and Maganik, Nikšićko polje, North-Western Garač Mountain, Orjen Mountain, Pešter, Piva, etc. (Rubinić et al. 2019).

Also, during 2016 two interesting and at the same time important projects were initiated with the aim of improving the Montenegrin path in the European integration processes: ***Minamata Initial Assessment for Montenegro*** and ***Review and update of the national implementation plan for the Stockholm Convention on Persistent Organic Pollutants (POPs).***

The Minamata project (2016–2018) in Montenegro was developed by the United Nations Development Programme (UNDP). But, before describing the Minamata project, it is important to explain what really Mercury represents in practise. Mercury (Hg) is a chemical element, liquid metal of Group 12 (II b—Zinc group) of the Periodic table system. It is also called quicksilver. At an early date, Mercury is recognized as a toxic metal that is liquid at room temperature (melts at about 28.5 °C). Inspite of all those characteristics, its main use was for medical purposes (Encyclopedia Britannica 2020a). The main objective of the Minamata project was strengthening national decision in order to ratify the *Minamata Convention* and after that, to build up the capacities for implementation of future provisions. In the end, the Ministry for Sustainable Development and Tourism has undertaken the necessary activities for creating and enabling environment for decision-making of the ratification of the *Minamata Convention* and in developing a National Mercury Profile and Mercury Initial Assessment Report. Additionally, it is important to mention that prior to this Montenegro, became a signatory to the *Minamata Convention on Mercury* on September 24, 2014. This Convention for its creation has an aim to systematically reduce emissions and releases of Mercury to water and land, and to phase out the Mercury use where alternative solutions exist (UNDP 2018). This project had an important impact which enabled the Government of Montenegro to determine the national requirements and needs on the path of the European integrations. The similar project, in Chemicals and Waste focal areas is POPs project.

The POPs project (2016–2018) in Montenegro was developed by the United Nations Environment Programme (UNEP). Before describing this project, it is important to explain what POPs chemicals mean in practise and real life. POPs chemicals are organic compounds, registered as toxic for human and other living organisms. Also, they are bioaccumulative and persistent in the environment. They are able to

[5] The Skadar Lake and the Bojana River represents a wider wetland complex, which lies along the border between Montenegro and Albania. At the same time, it is an important wintering and resting site for migratory waterbirds. "In winter the Skadar Lake regulary harbours 40,000 – 80,000 waterbirds, with *Common Coot Fulica atra, Great Cormorant Phalacrocorax carbo, Pygmy Cormorant* and *Common Pochard Aythya ferina* as dominant species. The respective numbers of waterbird populations in the Ulcinj Salina fluctuate between 9,000 – 19,000 birds in winter and from fewer than 2000 up to more than 32,000 waterbirds during autumn and spring migration " (Schneider-Jacoby et al. 2006, Stumberger et al. 2008 in Rubinić et al. 2019: 306).

travel a long distance from the point of the origin. POPs chemicals have low water solubility and high lipid solubility. Therefore, they are easily transported through phospholipids in biological membranes, later deposited in adipose tissue and other tissues with high lipid levels. Thus, all those characteristics cause POPs chemicals to be widespread in the environment, even in the regions and areas where they have never been used earlier. Therefore, this issue requires a global solution, which is defined in the *Stockholm Convention* which entered into force in 2004. Montenegro is a signatory country to the *Stockholm Convention on Persistent Organic Pollutants* since 2011. The objective of this Convention is to protect environment and human health from those chemicals. Keeping all that in mind, the Center for Ecotoxicological Research in cooperation with the UNEP and the Ministry of Sustainable Development and Tourism had to update the old POPs inventory, develop the new POPs inventory and at the same time develop the Action plans for reduction/elimination risks and exposure to POPs, raising the awareness of target groups on the risk of exposure to POPs chemicals (CETI 2018).

The aim of ***Sustainable Urban Mobility Plan for the Boka Bay area and the Old Royal Capital Cetinje in Montenegro*** is to promote sustainable tourism, thus minimizing Montenegro's carbon footprint and at the same time, promote sustainable transport options, in cooperation with the municipalities in the area of the Kotor Bay (Kotor, Tivat and Herceg Novi) as a Poly-centric Sustainable Urban Mobility Plan (Poly-SUMP) for the region. This project represents an ambitious plan for low-carbon or simply said green development. Thus, it contributes to resource efficiency and sustainable Biodiversity and nature conservation for future generations. In that context, it boosted the connection between the southern and northern part of the country. Also, the low-carbon tourism in the last few years has become a new mode of travel, offering rich experiences for tourists through reducing the carbon footprint[6] – CO_2 emissions during all sorts of tourism activities (Šorović 2019).

The Poly-SUMP project facilitates a safer, secure, reliable urban transport system for this area, which will be environmentally friendly, bearing in mind the specific characteristics of the region. This plan is focused on maritime and land transport (provided solutions for non-motorized and motorized transport modes). To be more precise, the main challenge is bringing together different stakeholders and municipalities, in order to develop a common approach for addressing specific transport needs. It will promote sustainable transport options for the region, as a great opportunity to overcome any barriers and reach a common vision for the future area in terms of urban mobility accessibility (PolySUMP 2015).

During October 2020 online education will start as well as research under the theme *Massive Open Online Course (MOOC)*, organized by the United Nations Environment Programme (UNEP) and the Open University from Netherland and Belgium, as a part of the UNEP *Clean Seas* initiative, aim of which is to fight against

[6] To make this kind of tourism real, there was developed an online carbon footprint calculator (You may find it on: www.izračunajCO2). During 2017, it was established in many hotels along the Montenegrin coastline and local tourism organizations as great project support with the aim of reducing CO_2 emissions, providing on that way active and interesting vacation for tourists (Šorović 2019).

marine plastic pollution. Thus, this initiative—course will give contribution to the goals of the Global Partnership on Marine Litter, in order to change the standards, habits, practices and policies at the global level and at the same time to decrease litter in marine and its negative effect on the marine environment and human health. Also, it is important to underline that the MOOC on marine litter will represent an interesting initiative that is useful for all of us to have insight in the different tools and options which are necessary in the research and learning process (University of Montenegro 2020).

To be more precise, there are many other projects at the national level that are being implemented, in order to achieve significant progress in the field of Environment and Climate Change, which is primarily related to the Chap. 27. It is important to mention that Montenegro is the European Union candidate country. This Chapter is the most difficult and expensive one (opened during 2018). Therefore, Montenegro needs its lawyers, politicians, all relevant country institutions and Montenegrin citizens to work hard and take enormous responsibility to harmonize the domestic legislation with the EU *acquis* (Šorović 2019).

6 Results and Discussion

There is a lot of evidence that shows the negative sides of Climate Change and its effects on Biodiversity. Because of that, in accordance with the *Millennium Ecosystem Assessment*, Climate Change will become one of the most significant drivers which will cause enormous Biodiversity loss by the end of our century (Convention on Biological Diversity 2010). Thus, "Climate Change is already forcing Biodiversity to adapt either through shifting habitat, changing life cycles, or the development of new physical traits" (*Ibidem*).

As an example of the negative impact of the man on the environment and the changes that occur in nature, followed by the effects of Climate Change, are also visible in the Arctic area. The aim of The *Dark Snow Project*[7] is "to enable the general public to fund Earth science research and to provide educational insight into the current health of our planet. Research projects are primarily focused on Climate Change and natural hazard monitoring (i.e. earthquakes, tsunamis, landslides, etc.)" (Leonardo DiCaprio Foundation 2020). Well, simply said, this project tries to explain and quantifies "the distant snow/ice melting impact of industrial and wildfire black carbon soot; mineral dust, each melt factor having some human driven enhancement" (*Ibidem*).

[7] The Dark Snow represents an organization, a group of scientists, led by Spanish polar explorer R. Larramendi. This project for its aim has "the collection of black carbon snow samples to investigate spatial variability and its relationship to snow melt and to assess the WindSled capacity for future green science" (Edwards 2017). Their idea was gathering measurements from the surface of ice in Greenland, in advancing the understanding of the melting of the Arctic which has a crucial importance for further investigations and in the same time to inform the global public what is happening with the Earth (*Ibidem*).

On the other hand, as a good example of ecosystem conservation is the protection and the planting of mangrove forest in Mexico. It is important to mention that, in the last 40 years, Mexico has lost 9% of its mangrove forest due to deforestation and pollution impacts. The remaining part of the mangrove forest on the Baja California Peninsula and in the Gulf of California represents essential habitat for diverse species: many migratory birds, sea turtles and marine mammals. Also, those trees are effective "at atmospheric carbon intake and sequester more carbon than most plant species on Earth. They also provide protections against storm surges, which will be increasingly more important as Eastern Pacific tropical hurricanes continue to intensify and become more frequent, as well as sea level rise" (Lopez 2020). It is estimated that mangrove forest in Mexico will prevent the emission of at least 70.8 million tons of carbon dioxide. In the next period (next four to five years), Mexican Federal agencies in cooperation with WILDCOAST[8] will protect and preserve a total of 71,374 acres of mangrove trees in Northwest Mexico. In accordance with that, preliminary research indicates that this area would prevent emissions equivalent to half of Baja California Sur's carbon dioxide emissions from transportation and industrial activities in the same period, around 53 million tons of carbon dioxide. This process of Biodiversity conservation could be replicated all around the world (*Ibidem*).

When it comes to Montenegro, during 2019, the process of updating the *Unified Single List of Priority Infrastructure Projects* has been completed, in accordance with the recommendations of the European Commission. This process refers to the coordination, collection and data processing, related to projects in the sector covered by the Single List like: Transport, Energy, Environment, Digital infrastructure, Social activities (education, justice and health), etc. There are no significant changes in the sectorial list. Priorities at the sector level remained largely the same, although there was a significant increase in costs for individual projects, while, on the other hand, a number of new projects were added (Sekretarijat za razvojne projekte 2019). Also, it is important to record a significant increase in costs, mostly related to projects in the field of Wastewater, such as: construction of wastewater treatment plants and sewage network in the municipality of Rožaje; construction of a wastewater treatment system in Cetinje; construction of wastewater treatment plants and sewerage network segments in the municipality of Kolašin; improvement of the wastewater management system in municipality of Mojkovac; and construction of missing and reconstruction of the existing fecal and atmospheric sewerage networks of Cetinje (II phase)—secondary collectors in the settlement of Donji kraj (*Ibidem*).

In addition to all above, the fact is that Chap. 27 is one of the most demanding in the negotiation process and it already requires, according to some projections, enormous financial allocations (exceeding the amount of one billion euros), but also, the construction of competent administrative capacity. Therefore, it is necessary to work on strengthening the capacity in the executive institutional bodies and in some judicial instances. Because of all of that, it is of great importance to implement

[8] It represents an international team which aims to conserve marine and coastal ecosystems and addresses Climate Change through natural solutions (WILDCOAST 2020).

projects financed by the European Union, and thus, improve the environment and Biodiversity issues, but, at the same time implement the *acquis communautaire* in Montenegro.

Thus, Montenegro, despite the Coronavirus pandemic, has made progress in several areas. It was stated at the *13th Subcommittee Meeting on Transport, Environment, Energetics and Regional Development*, between Montenegro and the European Commission, which was held online, June 9, 2020. Hence, it is of great importance to underline that most of activities on the transposition of the *acquis communautaire* into the national legislative framework have been successfully completed or they are in the final stage. On the other hand, Montenegro proceeds with the implementation of the obligations that require strong administrative capacities in all fields covered by this Subcommittee and great financial resources, primarily needed for the infrastructure buildings. In addition, the importance of the *European Green Deal*, which represents a new growth strategy, created by the European Commission. It aims to turn climate and environmental challenges into opportunities. With the appearance of Coronavirus, the world economy has collapsed, while nature starts to renew. Hence, this is a challenging moment for the economy. It creates conditions for a green economic recovery and a new socio-economic model that would be much more efficient in terms of resources, more resistant to health, environmental and economic threats (Eu.me 2020).

7 Conclusions

> "We only have one planet. Humankind must become accountable on a massive scale for the wanton destruction of our collective home. Protecting our future on this planet depends on the conscious evolution of our species."
>
> "This is the most urgent of times, and the most urgent of messages."
>
> Leonardo DiCaprio.
>
> (To the United Nations in 2014.)

Climate Change is a real problem. Problem for all of us. We need to work together in order to prevent global chaos. Thus, Climate Change means more weather disasters and problems every year. Hence, year by year, nature and Biodiversity have been destroyed, while over a million plant and animal species are endangered (in the phase of extinction). Because of that, there is a need for urgent action by reaching near-zero emissions in the atmosphere. Therefore, the reaction of the leading state leaders is crucial in this issue. Only in such manner, it is still possible to stop the destruction of nature and natural resources, thus preserving environment and Biodiversity conservation. On that path, Montenegro as the first ecological country in the world, needs to find its importance and role. Despite the implementation of numerous directives and regulations of the European Union, accompanied by many projects, Montenegro needs to work hard for a long time and to be committed to preserving Biodiversity and the environment. It seems pretty simple, but practice in some aspects of this area is slowing down.

In accordance with the European Commission *White Paper on Adapting to Climate Change—Towards a European Framework Action* and the *EU Strategy on Adaptation to Climate Change*, "both recognized the importance of ecosystems in tackling Climate Change. The White Paper encouraged the development of 'measures which address Biodiversity loss and Climate Change in an integrated manner to fully exploit co-benefits and avoid ecosystem feedbacks that accelerate global warming'" (European Commission 2020). But, this is not enough. We need more politicians and other officials who will make their own contribution in raising the awareness among the people, in order to avoid unforeseeable consequences for the world, nature and humanity. A good example of such behavior and actions was given to the world by the famous actor, Leonardo DiCaprio, who fights day by day, with all his available tools, for a healthy Environment, planet and better quality of our lives. Hence, I will conclude with his very inspirational speech at the United Nations. "No more talk, no more excuses, no more 10-year studies. This is the body that can do what is needed, all of you sitting in this very hall. The world is now watching. You will either be lauded by future generations or vilified by them" (McCarthy 2017).

8 Recommendations

- Each country needs to preserve human health and the quality of the ecosystem. In accordance with that, Montenegro needs to protect the species of flora and fauna, through the implementation of the *acquis* communautaire;
- Conserving freshwater, natural terrestrial and marine ecosystems;
- Preserving the balance and ecological stability of nature, through the rational and adequate use of natural resources;
- Reducing carbon emissions by 45% by 2030 and reaching near-zero emissions by around 2050;
- Use of renewable energy sources (gradual disposal of petroleum products and plastics from use);
- Preventing illegal logging and mining (deforestation and forest degradation) and planting trees in those affected areas;
- Establishing diverse agroforestry and storing seeds systems (for cereal crops) to cope with increasing risk from changed climatic conditions.

References

AMNESTY International (2020) Available at: https://www.amnesty.org/en/what-we-do/climate-change/?utm_source=google&utm_medium=cpc&gclid=EAIaIQobChMI18XXwK-B7AIVBd-yCh1EvAH8EAAYAyAAEgInaPD_BwE. Retrieved Sept 11 2020

BBC News na srpskom (2020) Amazonija u opasnosti: Požari, seča šuma i korona virus. Available at: https://www.bbc.com/serbian/lat/svet-52850902. Retrieved Sept 25 2020

BirdLife International (2017) Ecosystem profile: Mediterranean Basin Biodiversity hotspot. Available at: https://www.cepf.net/sites/default/files/mediterranean-basin-2017-ecosystem-profile-english_0.pdf. Retrieved Sept 29 2020

Borunda A (2020) National geographic. We still don't know the full impacts of the BP oil spill, 10 years later. Available at: https://www.nationalgeographic.com/science/2020/04/bp-oil-spill-still-dont-know-effects-decade-later/. Retrieved Sept 23 2020

Briggs H (2020) Sir David Attenborough warns world leaders over extinction crisis. BBC. Available at: https://www.bbc.com/news/science-environment-54329813. Retrieved Sept 29 2020

CETI (2018) Available at: http://eng.ceti.me/?page_id=4136. Retrieved Sept 30 2020

Convention on biological diversity (2010) Climate Change and Biodiversity. Available at: https://www.cbd.int/climate/intro.shtml. Retrieved Nov 16 2020

Edwards R (2017) 2017 WindSled traverse – green science. Dark Snow. Available at: https://www.darksnow.org/. Retrieved Nov 17 2020

Encyclopaedia Britannica (2020) Amazon rainforest. Available at: https://www.britannica.com/place/Amazon-Rainforest. Retrieved July 30 2020

Encyclopedia Britannica (2020a) Available at: https://www.britannica.com/science/mercury-chemical-element. Retrieved July 30 2020

European Commission (2020) Environment: Biodiversity and Climate Change. Available at: https://ec.europa.eu/environment/nature/climatechange/index_en.htm. Retrieved Nov 17 2020

Eu.me (2020) Crnogorski napredak vidljiv u oblastima saobraćaja, energetike, životne sredine i regionalnog razvoja. Available at: http://www.eu.me/mn/press/saopstenja/pregovori-o-pristupanju/item/4236-crnogorski-napredak-vidljiv-u-oblastima-saobracaja-energetike-zivotne-sredine-i-regionalnog-razvoja. Retireved Nov 23 2020

GEF (2019), National Biodiversity planning to support the Implementation of the CBD 2011 – 2020 strategic plan in Montenegro. Available at: https://www.thegef.org/project/national-biodiversity-planning-support-implementation-cbd-2011-2020-strategic-plan-6. Retrieved Sept 11 2020

Government of Montenegro, National Council for Sustainable Development (2011) Analysis of achievements and challenges of the Ecological State: 20 years of the Ecological State of Montenegro, Podgorica

Greenpeace (2020) Available at: https://twitter.com/Greenpeace. Retrieved Sept 18 2020

Hoffman A (2015) Climate Change and Biodiversity. Australian Academy of Science. Available at: https://www.science.org.au/curious/earth-environment/climate-change-and-biodiversity. Retrieved Sept 27 2020

Keaveny P (2018) New UN report outlines urgent, transformational change needed to hold global warning to 1.5 °C. Available at: https://theconversation.com/new-un-report-outlines-urgent-transformational-change-needed-to-hold-global-warming-to-1-5-c-103237. Retrieved Sept 24 2020

Leonardo DiCaprio Foundation (2019) Dark snow arctic research. Available at: https://www.leonardodicaprio.org/dark-snow-arctic-research/. Retrieved Nov 17 2020

Lopez G (2020) Ecosystem based Climate Change adaptation through Mangrove conservation. Leonardo DiCaprio Foundation. Available at: https://www.leonardodicaprio.org/ecosystem-based-climate-change-adaptation-through-mangrove-conservation/. Retrieved 17 Nov 2020

McCarthy J (2017) 10 Quotes from Leonardo DiCaprio, environmental champion. Global Citizen. Available at: https://www.globalcitizen.org/en/content/leonardo-dicaprio-quotes-on-climate-change-environ/. Retrieved Nov 23 2020

Ministry of Sustainable Development and Tourism (2014) The fifth national report to the United Nations convention on biological diversity

NASA Global Climate Change (2020) Climate Change: how do we know? Available at: https://climate.nasa.gov/evidence/. Retrieved Aug 31 2020

Noor D (2020) 40 Percent of the Amazon is on the brink of transitioning to Savanna. Gimodo. Available at: https://earther.gizmodo.com/40-percent-of-the-amazon-is-on-the-brink-of-transitioni-1845274803?utm_medium=sharefromsite&utm_source=_twitter. Retrieved Aug 31 2020

Pallardy R (2020) Deepwater horizon oil spill. Encyclopaedia Britannica. Available at: https://www.britannica.com/event/Deepwater-Horizon-oil-spill. Retrieved Sept 23 2020

Parks.it (n.d.) Available at: http://www.parks.it/world/YU/Eindex.html. Retrieved July 28 2020

PolySUMP (2015) Sustainable urban mobility. Available at: http://www.poly-sump.eu/. Retrieved Sept 30 2020

Rafferty JP (2020) Is Climate Change real? Britannica. Available at: https://www.britannica.com/story/is-climate-change-real. Retrieved Sept 20 2020

Rubinić B, Sackl P, Gamatikov M (2019) Conserving wild birds in Montenegro: a first inventory of potential special protection areas. Agency for Nature and Environmental Protection, Montenegro. ISBN: 978-9940-9924-1-5. COBISS: CG-ID: 38345488. Available at: https://rsis.ramsar.org/RISapp/files/46601290/documents/ME2399_lit190509.%20Conserving%20Wild%20Birds%20of%20Montenegro%20-%20potential%20SPA%20Natura%202000.pdf

Sekretarijat za razvojne projekte (2019) Polugodišnji izvještaj o radu Nacionalne investicione komisije (za period od 01. januara do 30. juna 2019. godine) sa ažuriranom Jedinstvenom listom prioritetnih infrastrukturnih projekata. Available at: www.srp.gov.me. Retrieved Nov 23 2020

Šorović, M (2019) Montenegro and Climate Change. The Balkan Forum. Available at: https://thebalkanforum.org/en/montenegro-and-the-climate-change. Retrieved Sept 29 2020

UNDP (2018) Final report: Minamata initial assessment for Montenegro

UNESCO (2020) Available at: https://whc.unesco.org/en/news/144. Retrieved Oct 14 2020

University of Montenegro (2020) Massive Open online course on marine litter (MOOC). Available at: https://www.ucg.ac.me/objava/blog/21920/objava/84988-masovni-otvoreni-online-kurs-o-morskom-otpadu-mooc. Retrieved Oct 3 2020

WILDCOAST (2020) A voice for the coast and the ocean. Available at: https://wildcoast.org/. Retrieved Nov 17 2020

Woods J, Williams A, Hughes JK, Black M, Murphy R (2010) Energy and the food system. The Royal Society. Retrieved July 28 2020, from https://doi.org/10.1098/rstb.2010.0172

Portrayals of Climate Change and Drought in the Politically Oriented Turkish Press: Socialist, Islamist, and Nationalist Accounts of Extreme Weather in 2007 and 2014

Mehmet Ali Üzelgün

Abstract Although the basic link between the global climate and local weather conditions is obvious, the extent to which anthropogenic pressures on the climate system produce a particular extreme weather event is a complex and contentious matter. This highlights the framing and sense making processes that are involved in communicating extreme weather events. This study takes the case of the 2007 and 2014 droughts in Turkey to examine the press portrayals of, and connections drawn between, global warming and drought. The analysis focuses on newspapers paralleling political parties and ideologies—socialist, islamist, nationalist—and examine a corpus of articles from 6 dailies (N = 285). Both quantitative and qualitative findings point out considerable differences between 2007 and 2014 in how global warming features in the Turkish media. The study also suggests that political ideology has a significant bearing upon the reporting of drought and its connection to global warming. The paper briefly discusses the three core findings concerning the differences (1) between the reporting context of 2007 and 2014, (2) among the coverage of socialist, islamist, nationalist newspapers, and (3) in making attribution links between drought and global warming in newspapers of different ideological orientation.

1 Introduction

Global climate change, or global warming[1] (GW), is a slow-paced phenomenon unfolding on a different temporal and spatial scale than that of human perception.

[1] To match the vocabulary of the press articles examined in this chapter, I prefer to use the term global warming, especially to refer to the geophysical phenomenon. The use of the term climate change will thus be limited to the contexts of international diplomacy and policy.

M. A. Üzelgün (✉)
Instituto de Filosofia da Nova, Universidade Nova de Lisboa, Colégio Almada Negreiros, Campolide, 1099-032 Lisboa, Portugal
e-mail: uzelgun@fcsh.unl.pt

Munazara and Argumentation Ethics Research Center, Ibn Haldun University, Istanbul, Turkey

W. Leal Filho and E. Manolas (eds.), *Climate Change in the Mediterranean and Middle Eastern Region*, Climate Change Management,
https://doi.org/10.1007/978-3-030-78566-6_8

As a problem difficult to objectify and visualize, GW features in mass media often through the "events" it is linked to through scientific knowledge. This primacy of impacts and hazards associated with GW is confirmed for the press of industrializing countries (Üzelgün and Castro 2015; Takahashi and Meisner 2012; Gordon et al. 2010), as well as industrialized ones (Lyytimäki and Tapio 2009; Smith and Joffe 2009).

By making GW visible and newsworthy, extreme weather events can be said to take part in occasioning "teachable moments" (Leiserowitz 2004). These moments of sporadic media attention and social scare are not simple consequences of extreme weather. Instead, they are occasioned through the concatenation of different sources of information consistently with each other, and with local weather extremes such as heat waves and droughts (Ungar 2014). In these critical episodes of GW coverage, the dangers and risks usually represented at temporal and spatial distances suddenly appear here and now (Poumadere et al. 2005). In a way, the local bad weather serves as a sort of sensorial proof of the reality of the otherwise abstract global risks and dangers. In accordance, the public "discovery" of the phenomenon of GW, for instance in 1988 in the US (Ungar 1992), and in 2007 in Turkey (Üzelgün and Castro 2015), seems to have required dramatic local weather extremes: "what rendered 1988 so extraordinary was concatenating physical impacts felt by the person in the street" (Ungar 1992, p. 490). On the other hand, the public salience of GW and extreme weather events owe a great extent to media portrayals that saturate these with reality, and make them act as the "harbingers of things to come" (Lieserowitz 2004, p. 56; see also Ungar 2014).

Turkey is one of the most industrialized countries in the Middle East with high economic ambitions and rapidly increasing emissions. Like other countries in the East Mediterranean, it is considered as highly vulnerable to GW, particularly to increased droughts (Türkeş et al. 2020; Şen et al. 2012). Yet, too little is known about the public portrayals of both GW and droughts in Turkey, as well as their connection in the media and public imagination. There are also no studies on the reception of these phenomena by different segments of the society, such as in different ideological cultures. The goal of this study is to examine the portrayals of GW and drought in the Turkish press. To achieve this, articles published in the drought conditions of 2007 and 2014 in six Turkish newspapers oriented to different political ideologies are examined. To further specify the general goal and provide the socio-political background, in the following subsections I briefly introduce the relevant geographical and political context. After setting forth the specific research questions and the method, I report the differences in GW coverage of the socialist, islamist, and nationalist newspapers, discussing also the differences in their coverage between 2007 and 2014.

1.1 Droughts in Turkey

Located in the eastern part of the Mediterranean basin, Turkey is one of the countries that are highly prone to drought and desertification. Except for its northern coast,

Turkey's weather conditions are shaped by the dry-summer subtropical microclimate belt associated with diverse climate change impacts (Türkeş 2020). In line with the IPCC (2012), numerous regional climate projections (e.g., Türkeş et al. 2020; Şen et al. 2012) anticipate significant precipitation decreases, especially for the western and southern parts of Turkey, particularly during the winter months. These projections also anticipate more frequent and more intense drought events, as well as water shortages and loss in agricultural production.

In the last forty years the most severe and widespread droughts in Turkey have occurred in 1983–1984, 1989–1990, 1996, 2001, 2007–2008, 2013–2014 (Şahin and Kurnaz 2014; Türkeş and Erlat 2003). This study focuses on the two most recent of these (2007–2008 and 2013–2014), both of which have converted from meteorological to agricultural and hydrological droughts, with extensive socioeconomic impacts (Şahin and Kurnaz 2014; Türkeş 2014). Due to significantly less snow and rainfall in the three preceding autumn and winter months, the impacts of the 2007–2008 drought were already being felt in late 2006. The drought extended widely across the country, resulting in significant loss in agricultural yield (Dellal and McCari 2010), decline in water bodies of all sorts, and severe water cuts in cities like Istanbul and Ankara (Şahin and Kurnaz 2014). Although the country received above long-term average rainfall in the years following the drought, from 2012 onwards meteorological drought conditions prevailed again in the middle and eastern regions. Having received 37% less than long-term average rainfall in the winter months of 2013–2014, by mid-2014 the regional conditions once again converted to agricultural and hydrological drought (Türkeş 2014). Severe water rationing and interbasin diversion measures were applied in many cities during both droughts.

In short, Turkey is "extremely vulnerable to climate change, particularly to increased droughts" (Türkeş et al. 2020, p. 487). The basic link between GW and local weather conditions is obvious in this quotation, as well as being a matter of consensus in the branch of climate science called Extreme Event Attribution (EEA). As the studies on EEA show, however, the extent to which anthropogenic climate change induces a particular extreme weather event is still a complex and contentious matter (Osaka and Bellamy 2020; Osaka et al. 2020; Schiermeier 2018). Arguably, EEA science becoming part of media reporting across the world would significantly shift the dynamics of public reception of extreme weather events, their linkage to GW, as well as public understanding of climate change.

1.2 Drought and Global Warming in the Media

Local weather extremes such as heat waves and droughts have the potential—if communicated in ways that are conducive—to increase public concern and engagement with climate change. Indeed, meteorological events are known to occupy a significant share of GW coverage across countries (Areia et al. 2019; Üzelgün and Castro 2015; Dotson et al. 2012; Takahashi and Meisner 2012; Gordon et al. 2010). An important question then is the extent to which the media associates the coverage of local weather extremes with GW. Of the few studies addressing this question, Osaka et al. (2020) report that 66% of the mainstream media articles in the US that covered the California drought of 2011–2017 involve attribution links to GW. Another recent study of the coverage of 2012 Midwestern US drought (Church et al. 2020) reports the exploration of the causes of the drought as very low overall. Both studies register some heterogeneity—if not divergence—among the newspapers in attributing the droughts to GW. Although sample sizes were small, it seems that politically conservative newspapers both explore the cause of the droughts to a lesser extent, and report fewer causal links to GW than politically liberal newspapers.

This relates to a broader literature on a key source of heterogeneity in mass media reporting of GW: ideology, or ideological cultures (Carvalho 2007), which refer to systems of goals, values, preferences, and actions—vis-à-vis a given political order. Ideologies seem to have a bearing on how the press in large industrialized countries such as the UK (Doulton and Brown 2009; Carvalho 2007) and France (Dirikx and Gelders 2010) portray the scientific knowledge on GW, as well as the necessity and urgency of mitigation actions. Likewise, there appears a consistent ideological divide between left- and right-leaning citizens of industrialized countries, where those on the left are more likely to believe GW is occurring, consider it a serious problem, and support policies to reduce carbon emissions (McRight et al. 2016). A similar divide also seems to emerge in the press of industrializing countries, such as Chile (Dotson et al. 2012), and Greece (Gkiouzepas and Botetzagias 2017), where the left-leaning newspapers report GW with higher frequency, longer articles, and more pro-climate action frames.

1.3 Climate Change Policy and Communication in Turkey

Turkey is one of the most industrialized countries in the Middle East and East Mediterranean. Its geopolitical position contributed to Turkey's becoming part of the Western Bloc, a NATO-state, and a founding member of the OECD. However, when classified with other OECD countries and listed in Annex II of the UNFCCC in 1992, Turkey pleaded its relatively low carbon emissions in comparison to other OECD member states and refused to ratify the Convention (Üzelgün and Şahin 2017). After a decade of negotiations Turkey was removed from the Annex II of the UNFCCC in recognition of its "special circumstances", and became a party to the Convention in 2004 as an Annex I country (Turhan et al. 2016).

This delayed entry into the global climate change regime was also partly due to the political turmoil of the 1990s, which led to Justice and Development Party's (AKP) political hegemony establishing a neo-liberal agenda of intense privatization wedded with Islamic conservatism. Erdoğan's AKP has regarded climate change primarily as a matter of foreign policy, thus a matter of national interest (Cerit Mazlum 2009). Pursuing an increasingly proactive and exasperated political agenda in the Middle-East and East-Mediterranean, Erdoğan focused on procuring a triad of geopolitical, security, and economic advantages in the region, a triad that also constitutes the very foundation of the country's energy policy (İşeri and Bartan 2019). Accordingly, in 2013 Turkey recorded the highest rate of increase (110.4%) in GHG emissions among the above-mentioned Annex I countries (Turhan et al. 2016).

A nation's energy and climate change policy is commonly regarded in a dynamic relationship with the public concern and media reporting (Boykoff and Rajan 2007). On this account, it is not surprising that media attention to GW was close to non-existent in Turkey until 2006, when it rocketed in the context of the 2006–2007 drought (Üzelgün and Castro 2015; Şen 2013). With the alleviating effect of the reduction of domestic political conflict and terror that usually overwhelm Turkish media, the year 2007 became special in many ways for public concern with and understanding of GW in Turkey (Şen 2013). Yet the mainstream media's dramatized reporting of the impacts and risks associated with GW in this period was oriented more towards consuming a spectacle than covering a complex phenomenon (Üzelgün and Castro 2015). The representation of climate science as certain was accompanied by its simplification and obscuration (Üzelgün and Castro 2014) and this arguably did not help public understanding. The securitization of GW in the mainstream media both accentuated the existential threat and circumscribed the political responses (Günay et al. 2018). While the political responses were typically represented outside the country's national borders (Üzelgün and Castro 2015), alternative media provided a forum for the circulation of more diverse and critical discourses implied by securitization (Günay et al. 2018).

Since 2007 domestic political conflict augmented, and the Turkish media landscape has become increasingly polarized (Somer 2018). For instance, the discussion of the benefits and risks of nuclear power now clearly distinguishes and polarizes the pro-government and anti-government media (Ersoy and İşeri 2020). Of course, the recent polarization only builds upon historical camps and cleavages, e.g., in the case of the coverage of the 2007 drought (see Ergül et al. 2010). As the government's pressure over the media intensified and generalized, even those issues previously with low political traction have started to separate the politically oriented press. In this regard, the years 2007 and 2014 represent two very different settings in terms of both the socio-political agenda and the journalistic context.

2 Aims and Objectives

The study aims to examine the portrayals of drought and climate change in the drought conditions of 2007 and 2014 in six Turkish newspapers with national distribution. In scrutinizing the politically oriented press, namely the socialist, islamist, and nationalist newspapers, it seeks to answer the following research questions:

Q1. What are the differences in GW coverage between 2007 and 2014? Do these differences involve connecting GW to local weather extremes?

Q2. What are the differences among the socialist, islamist, and nationalist newspapers in reporting GW?

Q3. Does newspaper ideology have a bearing on the connections drawn between GW and drought conditions? If so, how?

3 The Material and Method

To build the corpus for the analysis of GW coverage in six Turkish newspapers in the drought conditions of 2007 and 2014, I used the print media archive of the National Library at Ankara. Systematic sampling strategies would be extremely time-costly due to the lack of a digital database for the newspapers. I thus opted for all January and February issues—of both 2007 and 2014—of the six newspapers[2] in order to probe the above-mentioned differences. A total of four months of content for each newspaper was scanned page by page for articles that report on climate change, global warming, greenhouse gases, fossil fuels, drought, or water scarcity. Relevant articles were digitized and classified according to the year and newspaper.

The newspapers were selected due to their politically differentiated readership. *Birgün* and *Evrensel* are two socialist newpapers linked to left parties, while *Yeniçağ* and *Ortadoğu* are linked to nationalist-chauvinistic parties, and *Milli Gazete* and *Yeni Akit* to two strands of the islamist movement.[3] Having a differentiated readership and relatively low circulation figures,[4] the newspapers cannot be categorized as part of the mainstream; indeed at many instances they bear antagonistic relations to what can be called mainstream, striving to pull the political spectrum towards their respective political standpoints. For the lack of a better word, a decision was taken to call them ideological newspapers.

A systematization procedure was carried out, coding for whether the article (1) mentions any of the keywords: climate change, global warming, greenhouse gas, fossil fuel; (2) takes GW or drought as its central issue; (3) is published in the front/back covers or inside pages; (4) is a news item, op-ed, or part of a dossier; (5) is short or long; (6) involves some form of discussion-conflict, or just diffuses simple images; (7) uses outside sources: scientific, government, civil society; (8) incorporates images, as well as for (9) the primary, and (10) secondary issues: drought, water scarcity, water policy, agriculture, local extremes, GW impacts, GW risks, GW mitigation, GW causes/responsibility, environmental pollution, industry, economy. Further emergent variation was registered as notes.

[2] The rationale in this decision was to avoid heat-relevant—combined—coverage and probe into the reporting of winter-droughts, as mentioned in section Droughts in Turkey.

[3] Although none of the papers are party organs—i.e., they are independent as organizations—they are well-known to portray the figures and voices of certain marginal political parties and leaders that do not as saliently and dominantly appear in other—commercial—media organs. Such political parallelism does not concern only the newspapers in our focus, it is a historical characteristic of Turkish media (Kaya and Çakmur 2010).

[4] In the first week of February 2014 Birgün and Evrensel (socialist) sold on average about 12.000 and 6.500 respectively, Milli Gazete and Yeni Akit (islamist) about 25.000 and 30.000, and Yeniçağ and Ortadoğu (nationalist) about 50.000 and 7.700.

4 Results

Starting with the first research question that concerned the differences in GW coverage between 2007 and 2014, the quantitative findings point to a contrast. While in the first two months of 2007 the six newspapers produced a total of N = 136 articles that mention GW in one way or another, in the same months of 2014 the total sum was N = 36 (see Fig. 1 for the distribution of these values among ideologies). This great difference indicates that while in 2007 a considerable attention, perhaps a 'teachable moment' (Lieserowitz 2004), was created with the concurrence of the release of the IPCC AR4 summary, Al Gore's movie An Inconvenient Truth, and other international events, in 2014 such a momentum was not created in a way to interpret and resignify the local and regional drought conditions.

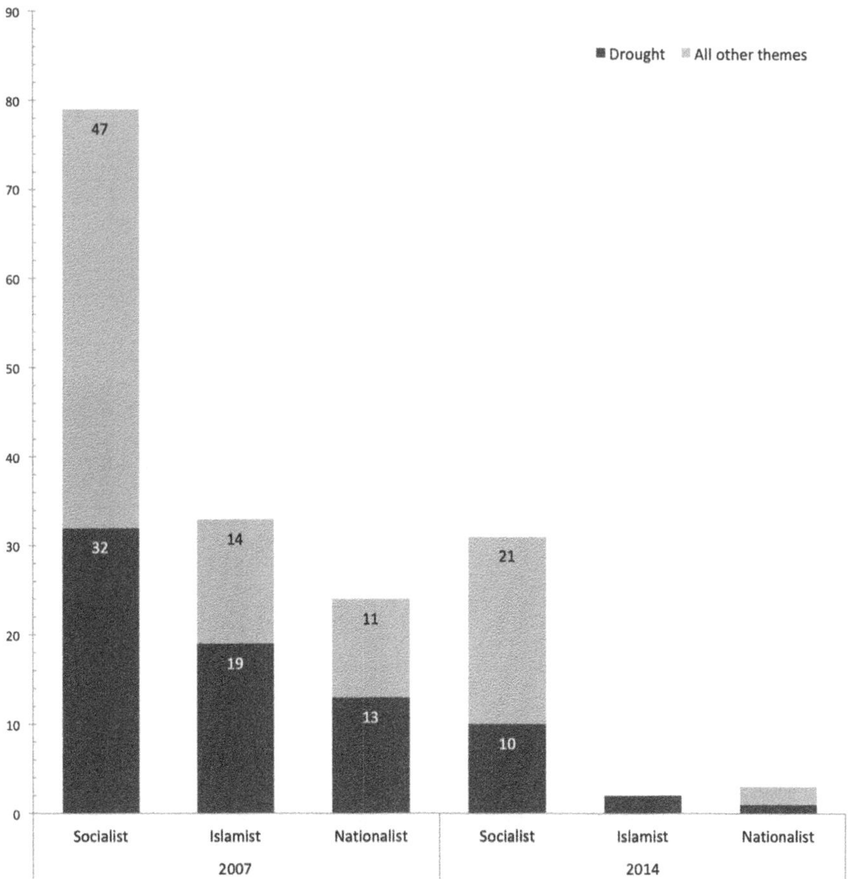

Fig. 1 Articles that mention GW in the first two months of 2007 and 2014, stacked as those that report on the droughts, and those on all other themes

Seeing the fourfold fall in coverage volume overall, one could doubt whether these two drought events were geophysically and hydrologically comparable. Or perhaps the social and economic impacts of the drought in 2014 were not felt as much as in 2007, at least in the winter months? The number of articles that report on the droughts without any mention of GW responds to such doubts. This figure for the six newspapers in the first two months of 2007 is N = 25, while in 2014 the same total sum is N = 88 (see Fig. 2 for the distribution of these values among ideologies). This means that all six newspapers did cover the 2014 drought considerably, but without connecting it to GW, as they did in 2007.

In sum, there emerges a substantial difference in the coverage of GW between the years 2007 and 2014. This difference seems to outweigh other differences in reporting, such as those between different ideologies. In other words, the decline in the number of articles from 2007 to 2014 seems to concern all newspapers and ideologies. I will examine the further qualitative differences in GW coverage between

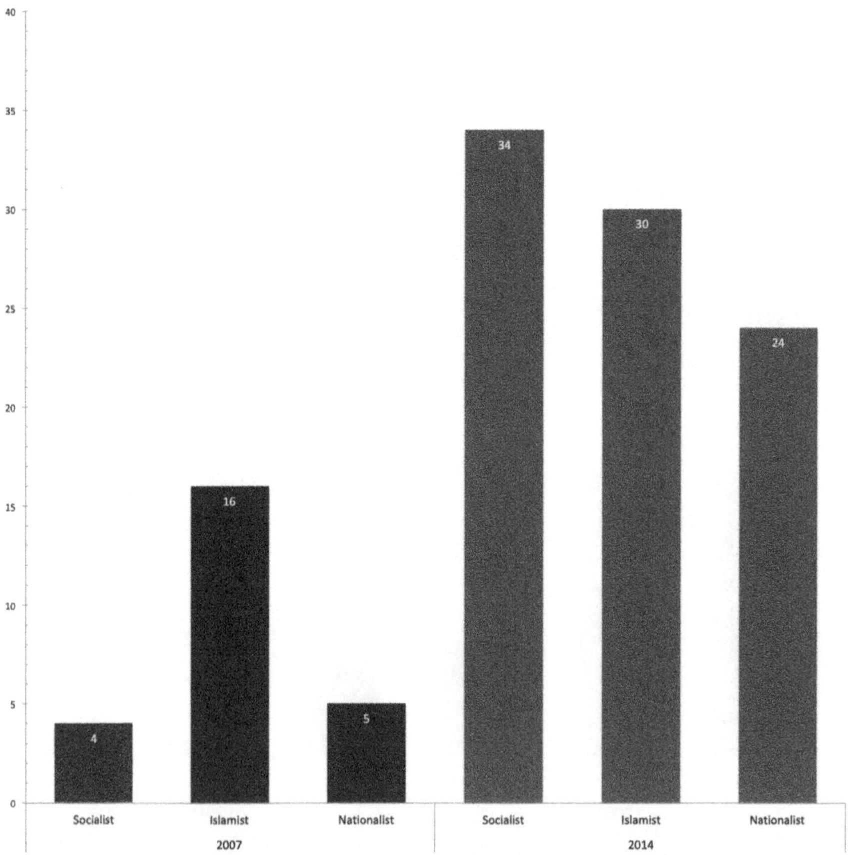

Fig. 2 Articles in the first two months of 2007 and 2014 that report local droughts without mention of GW

2007 and 2014 in the following sections that are organized based on the differences between ideologies.

4.1 Ideology and Global Warming Coverage

The second research question concerned the differences between the socialist, islamist, and nationalist newspapers in reporting GW. The quantitative findings indicate that newspaper ideology has an important bearing upon GW coverage volumes. Overall, the socialist newspapers report on GW more (N = 79 in 2007, N = 31 in 2014) than the islamist (N = 33 in 2007 and N = 2 in 2014) and the nationalist newspapers (N = 24 in 2007, N = 3 in 2014).

Unsurprisingly, a significant part of this coverage is concerned with local droughts and water scarcity. For instance, in January–February 2007, N = 32 of the total of 79 articles on GW published in the socialist newspapers (40%) were concerned with local droughts and water scarcity (see Fig. 1). This ratio of the articles on GW that focus on drought to the total number of articles on GW differs from the islamist (57%) and nationalist (54%) newspapers. In other words, while the reporting of the right-wing newspapers in 2007 is characterized strongly by the alarming local events, the socialist papers seem to have opened a wider space for reporting GW beyond the immediate—drought-related—context.

The openness of the socialist press to climate relevant issues other than the local catastrophe becomes more evident in 2014—when the connections between GW and local weather extremes were somehow lost, the coverage of GW in the islamist and nationalist papers diminished almost completely (no articles and N = 2, respectively). The socialist papers however still published a total of 21 GW news pieces that focused on issues such as coal-powered plant constructions, Turkey's GHG emissions, biodiversity loss, and aviation. This suggests that a sustained attention to GW—rather than sporadic spikes—and its causes is preserved only by the socialist press.

Another important finding concerns the length or depth of the articles. Overall, socialist newspapers have published more longform articles (N = 68), sometimes reserving a whole page to interpret various aspects of the problem, locate responsibilities, and criticize the government. The nationalist and islamist newspapers had relatively fewer longform articles (N = 19 and N = 15 respectively), oftentimes utilising simple images of strange weather.

To address the third research question that concerned the connections drawn between GW and drought by different newspapers, a comparison between Figs. 1 and 2 will be most helpful. For instance, a comparison of the number of articles that cover the 2007 drought by attributing it to GW (Socialist: 32, Islamist: 19, Nationalist: 13), and those that cover the same drought without any attribution or mention to GW (Socialist: 4, Islamist: 16, Nationalist: 5) is indicative of a remarkable difference. That is, ideology seems to have a significant effect on the linkage of local weather phenomena to the global climate, as socialist newspapers seem to link the local droughts with GW in many more articles than the islamist newspapers. If the 2014 figures do not conform to this pattern, it is mainly because the right-wing newspapers in this period seem to have given up the mention of GW altogether. The more subtle differences between the coverage of the newspapers with different ideologies will be reported in the three subsections that follow.

4.1.1 Socialist Newspapers

"Disaster knocks the door, governments say we are not home".

In January–February 2007, Birgün and Evrensel together published 79 articles on GW, and another 4 articles that report the drought conditions without mentioning GW. Among these articles, 58 were accompanied by at least one image, 26 were front or—mostly—back cover headlines, and 47 were longform articles. The two socialist papers not only have given a good deal of space to GW, but also discussed it at depth. A salient feature of these articles is what may be called the locus of responsibility. Unlike other articles of this corpus, and unlike the mainstream press reporting in the same period (Üzelgün and Castro 2015), carbon emissions, the fossil fuel industry, the rich, the government, and capitalism are figures repeating in the articles, many of which are oriented to distinguish the culprit. The frontpage headline of Evrensel on February 3rd depicts this tendency—accompanying a large image combining a cyclone and a chimney, the top headline reads: "Nature's Rebellion against Capitalism".

In January–February 2014, Birgün and Evrensel published 31 articles on GW, and another 34 articles that discuss the droughts without mentioning GW. As these figures suggest, while their coverage of droughts did not decline at all in 2014, the socialist papers did not attribute the extreme local events to GW as much as in 2007. To cover the impacts of drought across the country both newspapers mainly used local NGOs, academics, and spokespersons of chambers of agriculture as sources. A distinguishing characteristic of the news pieces in 2014 is a decided focus on thermal power plants: when GW coverage completely diminished in other newspapers of the corpus, the socialist newspapers featured local resistance movements against the coal industry and the government's deeds. While a similar opposition also emerges at the global level between the oil, coal and nuclear industries and nature (to be specific, e.g., polar bears, barn owls, Antarctica), the conflict is more manifest and politically charged at the national level. In this regard both newspapers have frequently resorted to irony and ridicule. A key figure, the minister of forestry and water works, has been repeatedly ridiculed for not expressing any plan against the drought. For instance, in an article dated January 22nd, accompanying the image of an imam and dozens of men praying in the fields, Birgün quotes the minister calling for prayers, contrasting his ridiculed appeal with three experts who call for serious measures.

Among the two socialist newspapers, Birgün should be marked as the paper that has created the remarkable difference among ideologies. When in 2014 GW coverage in other papers ceased almost completely, and when Evrensel's coverage also retreated to 7 articles in the two months, Birgün maintained the reporting of

GW with 24 articles. That is besides the articles that reported on droughts without attributing to GW. This was probably partly due to an editorial decision to reserve, on a daily basis, the complete back cover of the paper to environmental issues and struggles. 21 of the just mentioned 24 GW news were thus staged on the back cover, with numerous colorful visuals, and in-depth analyses. In turn, the editorial decision and Birgün's persistence in reporting GW related phenomena could be interpreted in the context of the newspaper's organic connections to the green-left movements.

4.1.2 Islamist Newspapers

"My Allah give us rain!".

In January–February 2007, Milli Gazete and Vakit together published 33 articles on GW, and another 16 articles that report on the drought conditions without alluding to GW. The majority of the 33 articles on GW also covered drought and water scarcity. This means that the islamist papers almost levelled with the socialist ones in covering the extreme drought, though with lesser attributions to GW. In GW coverage, Milli Gazete and Vakit published both basic informative articles, aiming at raising awareness, and oppositional ones, aiming at exposing the blameworthy. While the first are oriented merely to diffuse the risks and dangers associated with GW, the second type issues a conflict, again, between the culprit and the victim. The opposed sides however differ from those in the socialist press: unlike an opposition between capitalism and the planet/the people, in the islamist newspapers the line seems to have drawn mainly between the industrial west—primarily the US—and the rest of the world.

In January–February 2014, the islamist newspapers published only two articles that mention GW. The only article in Milli Gazete, entitled "Drought danger in Turkey", accompanied an image of a landscape experiencing desertification, and referred to a Turkish scientist who warned against an intensifying and serious drought. The only article in Yeni Akit,[5] entitled "There will be no water shortage", featured the Minister of Forestry and water works—with a smiling picture of him hugging the Istanbul Mayor—depicting him in control of the situation. All other 30 articles that reported on the drought and water scarcity did not mention GW at all. This radical difference in GW coverage—from 33 to 2 articles in two comparable time frames—requires a better understanding and further focus, but is indicative of the attention GW received after its alarmed discovery in 2007.

An associated difference concerns the figure of the government. While the socialist press was very vocal in criticizing the government over the lack of both drought and climate change policies, the religious papers were somewhat mute in this regard. Detecting no mention of Turkey's soaring emissions, this can be said especially for climate policy. While there were a few critical points raised against the government

[5] The newspaper has changed its name from Vakit to Yeni Akit in 2010. The apparent shift in positioning vis-à-vis the government mentioned above could be related to a shift in editorial policy associated with the change of newspaper's name.

in January–February 2007, by 2014 both Milli Gazete and Yeni Akit seem to have shifted their position vis-à-vis the government's measures. The shift is most evident in Yeni Akit's reporting, which, rather than having a critical engagement, seems more about depicting the government in control of the situation concerning potential water-related stresses. Four of the (N = 8) articles Yeni Akit published in this period were oriented to pass on the government's messages, exemplified by the headlines "We will not let Istanbul be without water" and "The capital will not suffer from water scarcity".

A further distinguishing characteristic of religious press reporting of GW concerns the space devoted to rain prayers. In six articles of Milli Gazete and four articles of Vakit, either the article's own voice or the sources quoted, e.g., members of the cabinet, announce calls for prayers or report collective invocation ceremonies, almost always accompanying images of collective prayer. This connection with the divine (resolution) is common in the islamist media in Turkey, and at times employed with political purposes across a variety of disasters (Ergül et al. 2010).

4.1.3 Nationalist Newspapers

"Spring arrived on a winter day".

In January–February 2007, Yeniçağ and Ortadoğu together published 24 articles on GW, and another 5 articles that report the droughts without any link to GW. Almost half of these articles were on the back cover, accompanying a colorful photo, and many times calling attention to an imminent danger. Very much like the mainstream press in Turkey in that period (Üzelgün and Castro 2014, 2015), the two nationalist newspapers focused in a rather alarming and dramatizing manner on impacts that act as proof of GW occurring, and risks such as desertification and sea-level rise. Still, the majority of the news was permeated by the local drought conditions. Unlike a risk reporting characterized by temporal and spatial distance, the danger was already "here and now", due to the alarming local conditions. An example is Yeniçağ's back cover of February 14th that shows a near-empty dam reservoir with the shocking title "Global warming has hit Istanbul".

In January–February 2014, Ortadoğu published no articles that mention GW, while Yeniçağ published only 3. The two newspapers together published 24 articles on the drought conditions without any mention of GW. Of the 3 GW articles in Yeniçağ, one was on the impacts of local droughts on agriculture, one on the dangers awaiting island states, and one on the "World's climate getting out of balance". Of the 24 articles reporting the drought, the vast majority were concerned with the water levels in dams and lakes. Using the local NGOs, mayors, experts, the national meteorological agency, and the minister of forestry and water himself as sources, an alarming state of affairs was depicted, albeit without any link whatsoever to the global climate.

Overall, it is not possible to speak about a manifest conflict in the nationalist press reporting of GW. Without an explicit critical stance or flattery vis-à-vis both the government's measures and the global climate change regime, the nationalist news items were rather devoid of a political conflict, perspective, or subject. It may simply be the case that in 2014 GW was still not conceived by the Turkish extreme nationalist milieu as a matter of politics. This confirms, in turn, that the more preoccupied a culture is with the national interests and conflicts, the less open to the cosmopolitan proposals concerning a world which is readily considered to be facing huge risks.

5 Conclusion

This study demonstrates three noteworthy differences in the three questions it posed. The first difference concerns the coverage of GW in 2007 and 2014: while all six newspapers dedicated considerable attention to GW in the first two months of 2007, their attention had ceased almost completely in the first two months of 2014. This means that the first attention cycle that occurred around the 2006–2007 drought in Turkish media (Üzelgün and Castro 2015; Şen 2013) was not followed by a second cycle in 2014. One limitation of these results is the partial representation of coverage—the January and February months only—instead of a systematic sampling of articles from 2007 and 2014. For instance, one may argue that while the Summary for Policymakers of the IPCC AR4 WG1 was released in February 2007, no IPCC reports were released in the first two months of 2014, and that creates a biased comparison. Yet, the difference in coverage volumes is so large that it cannot be accounted for by individual events. Instead, the radical difference in reporting indicates different social and political contexts within which developments and events related to GW are interpreted. Those differences among the contexts of disaster coverage, and the factors contributing to them, require further scholarly attention and discussion.

The second difference emerges between the reporting of socialist, islamist, and nationalist newspapers, confirming that newspaper ideology has an important bearing upon GW coverage. This study attests that this difference can be more nuanced than left-wing or liberal media devoting more attention and space to environmental issues, as compared to right-wing or conservative ones. In the Turkish context, reporting of the extreme nationalist newspapers bears some similarities to the mainstream press (Üzelgün and Castro 2015) in depicting a spectacle of a dramatic risk in the absence of a political subject or conflict. Unlike such depoliticized coverage, in the socialist and islamist newspapers conflicts were rather explicit, albeit in different ways. In the socialist press, the dividing line is drawn between the oil, coal, nuclear industries, the government, or capitalism as a whole system, and—the people striving to defend—nature. In the islamist newspapers, the main dividing line is drawn between the industrial West and the rest of the world. Furthermore, while the political subject is rather explicit in the socialist newspapers (local inhabitants, activists, NGOs), in the islamist newspapers God makes a surprising entry, as invoked by both politicians and the local population. This adverse finding for the EEA literature requires closer attention, and Turkish media appears as one of the sites that deserve that attention (see Ergül et al. 2010).

The third difference concerns the extent to which the droughts were attributed to GW by newspapers of different ideologies. This question too, like the first one, distinguishes the socialist newspapers from what may be called the right-wing press. Firstly, overall the socialist newspapers reported on the droughts by alluding to GW more often than the other newspapers. Furthermore, besides having almost completely ceased reporting on GW in 2014, the nationalist and islamist newspapers also ceased reporting on the drought events by alluding to GW in this period. This tendency to dissociate the local extremes form GW could be identified in one of the socialist newspapers too, but the socialist papers in 2014 have partly maintained the links established in 2007 between local droughts and GW. In the main, attribution of local disasters to GW should be less of an ideological issue and be regarded as a specific mode of science reporting. With the development of EEA science, media coverage of local weather extremes and its relation to public understanding would probably change considerably, a task for the future.

Overall, the difference between the socio-political contexts of 2007 and 2014 seems to be remarkable, as its effects are noticeable across other research questions—ideologies and attribution. One of the factors contributing to this difference could be the scientific, diplomatic, and cultural events taking place beyond the national borders, and the coincidence of such cosmopolitan events with the "GW impacts" felt on local water systems and agriculture. Another factor could be the national agenda and its relation to geopolitical and local grassroots movements. Ultimately, the media representation and public attention to GW could be contemplated with regard to how well "the symptom" fares against other problems competing for attention (Ungar 2014). Considering that the news agenda in Turkey has been increasingly loaded with political conflict and polarization since 2007, it may be concluded that the cosmopolitan events and ideas had a harder time to propagate in 2014. In this sense, the study highlights the "collateral damage" of more tangible and short-term conflicts over the public understanding and engagement with more complex and long-term problems such as climate change.

References

Areia NP, Intrigliolo D, Tavares A, Manuel Mendes J, Sequeira MD (2019) The role of media between expert and lay knowledge: a study of Iberian media coverage on climate change. Sci Total Environ 682:291–300. https://doi.org/10.1016/j.scitotenv.2019.05.191

Boykoff MT, Rajan SR (2007) Signals and noise: mass-media coverage of climate change in the USA and the UK. EMBO Rep 8:207–211

Carvalho A (2007) Ideological cultures and media discourses on scientific knowledge: re-reading news on climate change. Public Underst Sci 16:223–243

Cerit Mazlum S (2009) Turkey's foreign policy on global atmospheric commons: climate change and ozone depletion. In: Harris PG (ed) Climate change and foreign policy: case studies from east to west. Routledge, London, pp 68–84

Church SP, Bentlage B, Weiner R, Babin N, Bulla BR, Fagan K, Haigh T, Stuart Carlton J, Prokopy LS (2020) National print media versus agricultural trade publications: communicating the 2012 Midwestern US drought. Climat Change 161:43–63

Dellal I, McCari BA (2010) The economic impacts of drought on agriculture: the case of Turkey. Options Méditérr 95:169–174

Dirikx A, Gelders D (2010) Ideologies overruled? An explorative study of the link between ideology and climate change reporting in Dutch and French newspapers, Environ Commun 4(2):190–205

Dotson DM, Jacobson SK, Kaid LL, Carlton JS (2012) Media coverage of climate change in Chile: a content analysis of conservative and liberal newspapers. Environ Commun 6:64–81. https://doi.org/10.1080/17524032.2011.642078

Ergül H, Gökalp E, Cangöz I (2010) From the wrath of Allah to divine judgment: natural disasters in the Islamist press in Turkey. J Multicultural Disc 5:1–20. https://doi.org/10.1080/17447140903197233

Ersoy M, İşeri E (2020) Framing environmental debates over nuclear energy in Turkey's polarized media system. Turk Stud. https://doi.org/10.1080/14683849.2020.1746908

Gordon JC, Deines T, Havice J (2010) Global warming coverage in the media: trends in a Mexico City newspaper. Sci Commun 32:143–170

Gkiouzepas G, Botetzagias I (2017) Climate change coverage in Greek newspapers: 2001–2008. Environ Commun 11:490–514. https://doi.org/10.1080/17524032.2015.1047888

Günay D, İşeri E, Ersoy M (2018) Alternative media and the securitization of climate change in Turkey. Alternatives 43:96–114

IPCC (2012) Managing the risks of extreme events and disasters to advance climate change adaptation: special report of the intergovernmental panel on climate change. Field CB, Barros V, Stocker TF, Dahe Q (Eds) Cambridge University Press

İşeri E, Bartan AÇ (2019) Turkey's geostrategic vision and energy concerns in the Eastern Mediterranean security architecture: a view from Ankara. Tziarras Z (Ed) The new geopolitics of the Eastern Mediterranean: trilateral partnerships and regional security (pp 111–124) Prio

Kaya R, Çakmur B (2010) Politics and the mass media in Turkey. Turk Stud 11:521–537. https://doi.org/10.1080/14683849.2010.540112

Leiserowitz, A (2004) Before and after the day after tomorrow: A U.S. study of climate change risk perception. Environment 46(9):22–37

Lyytimäki J, Tapio P (2009) Climate change as reported in the press of Finland: from screaming headlines to penetrating background noise. Int J Environ Stud 66:723–735

McCright AM, Dunlap RE, Marquart-Pyatt ST (2016) Political ideology and views about climate change in the European Union. Environ Polit 25(2):338–358

Osaka S, Bellamy R (2020) Natural variability or climate change? Stakeholder and citizen perceptions of extreme event attribution. Glob Environ Change 62:102070. https://doi.org/10.1016/j.gloenvcha.2020.102070

Osaka S, Painter J, Walton P, Halperin A (2020) Media representation of extreme event attribution: a case study of the 2011–17 California Drought. Weather, Climate, Soc 12(4):847–862

Poumadere M, Mays C, Le Mer S, Blong R (2005) The 2003 heat wave in France: dangerous climate change here and now. Risk Anal 25(6):1483–1494

Schiermeier Q (2018) Droughts, heatwaves and floods: How to tell when climate change is to blame. Nature 560(7717):20–23

Smith N, Joffe H (2009) Climate change in the British press: the role of the visual. J Risk Res 12:647–663. https://doi.org/10.1080/13669870802586512

Somer M (2018) Turkey: the slippery slope from reformist to revolutionary polarization and democratic breakdown. Ann Am Acad Pol Soc Sci 681:42–61. https://doi.org/10.1177/0002716218818056

Şahin Ü, Kurnaz L (2014) Climate change and drought. IPC-Mercator Policy Brief. (In Turkish) Available at: https://ipc.sabanciuniv.edu/Content/Images/CKeditorImages/20200326-02030608.pdf. Retrieved Oct 28 2020

Şen OL (2013) Media coverage of climate change: The world versus Turkey. IPC-Mercator Policy Brief. Available at: https://ipc.sabanciuniv.edu.tr/wp-content/uploads/2013/07/Omer_Lutfi_Sen_Policy_Brief_Final.pdf. Retrieved Sep 22 2020

Şen B, Topçu S, Türkeş M, Şen B, Warner JF (2012) Projecting climate change, drought conditions and crop productivity in Turkey. Clim Res 52:175–191

Takahashi B, Meisner M (2012) Climate change in Peruvian newspapers: the role of foreign voices in a context of vulnerability. Public Underst Sci 22:427–442

Turhan E, Cerit Mazlum, S, Şahin Ü, Şorman AH, Gündoğan AC (2016) Beyond special circumstances: climate change policy in Turkey 1992–2015. Wiley Interdisciplinary Reviews: Climate Change 7(3):448-460

Türkeş M (2020) Climate and drought in Turkey. In: Harmancioglu N, Altinbilek D (Eds) Water resources of Turkey. World Water Resources, vol 2. Springer, Cham

Türkeş M (2014) Türkiye'deki 2013–2014 Kuraklığının ve Iklim Değişikliği ve Klimatolojik/Meteorolojik Nedenlerinin Çözümlenmesi. Konya Toprak Su Dergisi 20–34. Available at: https://arastirma.tarimorman.gov.tr/konyatopraksu/Belgeler/Toprak%20Su%20Dergisi/Toprak_Su_Dergisi_Sayi_2.pdf. Retrieved Nov 3 2020

Türkeş M, Erlat E (2003) Precipitation changes and variability in Turkey linked to the North Atlantic Oscillation during the period 1930–2000. Int J Climatol 23:1771–1796

Türkeş M, Turp MT, An N, Ozturk T, Kurnaz ML (2020) Impacts of climate change on precipitation climatology and variability in Turkey. In: Harmancioglu N., Altinbilek D (eds) Water Resources of Turkey. World Water Resources, vol 2. Springer, Cham https://doi.org/10.1007/978-3-030-11729-0_14

Ungar S (1992) The rise and (relative) decline of global warming as a social problem. Sociolog Quart 33:483–501

Ungar S (2014) Media context and reporting opportunities on climate change: 2012 versus 1988. Environ Commun 8(2):233–248. https://doi.org/10.1080/17524032.2014.907193

Üzelgün MA, Castro P (2014) The voice of science on climate change in the mainstream Turkish press. Environ Commun 8:326–344. https://doi.org/10.1080/17524032.2014.898674

Üzelgün MA, Castro P (2015) Climate change in the mainstream Turkish press: Coverage trends and meaning dimensions in the first attention cycle. Mass Commun Soc 18:730–752

Üzelgün MA, Şahin Ü (2017) Climate change communication in Turkey. Climate Science: Oxford Research Encyclopedias. https://doi.org/10.1093/acrefore/9780190228620.013.466

Climate Changes and Insolation in the Mediterranean Basin: The Case of Montenegro

Dragan Burić and Miroslav Doderović

Abstract The paper analyses the changes in insolation in Montenegro from 1961 to 2018. Data from 8 main meteorological stations were used where the homogeneity of time series was examined using MASH v3.0 methods. Calculations were made on a monthly, seasonal, and annual basis. During the period 1961–2018, there was a trend of increasing duration of sunshine for 5, and decreasing for 3 stations. We need to adapt to current climate change, and the primary way to mitigate further anthropogenic temperature rises is to reduce the use of fossil fuels and switch to clean energy sources. The results for Montenegro show that insolation is high (most of the country has an annual average of 2200–2600 h) and quite stable. Therefore the Mediterranean region has good potential for the development of solar energy, in general. This is supported by the fact that the period after 1990 (1991–2018) in most parts of Montenegro is sunnier than the period 1961–1990.

1 Introduction

Solar radiation is most often shown in the form of intensity (J/m^2/s or W/m^2) of a given component (direct, diffuse, global ... radiation), which is denoted as radiation and over the duration of sunshine (hour), also known as insolation. This is one of the most important climatic elements, because solar energy is the basic driver of almost all processes that occur in the atmosphere, the basic factor in climate formation, the most important source of heat on the Earth and the cause of many phenomena and processes in the atmosphere, hydrosphere and biosphere. Temperatures of the soil and air depend on the amount of sunshine, and thus the values of other meteorological elements that are correlated with the temperature. The sums of sunshine

D. Burić (✉) · M. Doderović
Faculty of Philosophy Nikšić, University of Montenegro, Danila Bojovića bb, Nikšic, Montenegro
e-mail: dragan.buric@meteo.co.me

M. Doderović
e-mail: dodemir@t-com.me

W. Leal Filho and E. Manolas (eds.), *Climate Change in the Mediterranean and Middle Eastern Region*, Climate Change Management,
https://doi.org/10.1007/978-3-030-78566-6_9

are proportional to the size of the angle at which rays of the Sun fall, and they are inversely related to the clouds.

Solar energy intensity and temperature are directly correlated (Nullet and Ekern 1988; Erlykin et al. 2009). During the geological history of our planet, the intensity of solar radiation changed in cycles that lasted from several 10 years to several 100 and even 1000 years (Berger et al. 1993; Huybers 2006; Drysdale et al. 2009; Lohmann 2017). In the past, variations in solar energy led to climate changes, namely glacial and intergacial shifts (Lorenz and Lohmann 2004; Lohmann et al. 2013). The relationship between the variability of solar radiation and cloud cover in the United States was discussed in the paper by Hanrahan et al. (2017). Based on data from 31 meteorological stations for 1971–2018, it was noted that there is a statistically significant trend in the duration of sunshine in Poland from April to September (Bartoszek et al. 2020). The authors obtained a statistically significant negative correlation of insolation with the total amount of cloudiness, while phase synchrony is present with pressure variations.

Power (2003) examined the trend of sunshine duration, then global, direct and diffuse radiation in Germany based on data from 13 meteorological stations, where time series varied between 11 and 48 years. The author noted that there were spatial differences in terms of sign and trend intensity of the considered radiation components. His research has shown that long-term aerosol reduction is the most likely cause of changes in certain radiation components. For one station (Freiburg), the author linked the increase in direct and diffuse radiation with the increase in the duration of sunshine. Spatial differences in the solar trend for the period 1936–2013 were also observed in Italy (Manara et al. 2015). Studying the patterns of climate change in Belgrade (Serbia) for 1888–2013, Vujović et al. (2018) point out: "The underlying structure of the data was analyzed by principal component analysis, which detected the following four easily interpreted dimensions: More sunshine-Higher temperature, Precipitation, Extreme heats, and Changeable summer".

In large cities, the intensity of radiation is reduced, as well as the duration of sunshine, due to its urbanization and high pollution (Omran 2000). In Greece, there was a negative correlation between the duration of sunshine (SDU) and total cloud cover (TCC). However, since the mid-1980s almost all meteorological stations in that state have shown a trend of increasing annual insolation, but this increase is not accompanied by a synchronous decrease of TCC, possibly pointing to the impact of aerosol loads on solar radiation and SDU fluctuations (Founda et al. 2017).

It is interesting to see which element can be viewed as a potential factor in solar fluctuations. The relationship between temperature and the duration of sunshine can be seen as a positive feedback mechanism. Under other same conditions, longer sun exposure = higher temperature, and vice versa. During the previous 6–7 decades, there is a significant trend of increasing air temperature in Montenegro (Burić et al. 2019; Burić and Doderović 2021). A logical question arises: can longer insolation be considered as a factor of heating amplification, and vice versa. However, changes in the length of insolation cannot directly answer whether this also affects the change (increase) in temperature, because the relationship between these two elements (temperature and insolation) is far more complex. Other indicators (e.g.

number of sunspots, magnetic activity indices, etc.) and more complex statistical-mathematical apparatus are used to determine the impact of changes in solar energy on temperature. On the other hand, insolation is closely related to cloud cover. Therefore, the correlation of insolation with the total cloud cover was examined.

2 Aims and Objectives

The main goal of this paper is to analyze the trend of insolation in Montenegro in 1961–2018. Due to the current climate change, it is important to reduce the use of fossil fuels and adjust to the use of alternative, i.e. clean energy sources. Accordingly, not only will anthropogenic impact on the global temperature be mitigated, but also the quality of the environment will be improved. Montenegro is a part of the Mediterranean, which is one of the sunniest regions in the world. Therefore, this paper has the following tasks: to examine whether there are significant changes in insolation in Montenegro in the instrumental period then which climate element may be considered as a potential factor in insolation fluctuations and also to accentuate to decision makers that Montenegro has good potential for solar energy development.

3 Research Area, Data and Methodology

The study covers the area of Montenegro i.e. an area of 13,812 km^2. The country is located in the southwest of the Balkan Peninsula and faces the Adriatic Sea in the length of about 100 km. The relief is dissected, mostly mountainous, with few plains to the south. The altitude of the terrain ranges from 0 (Adriatic Sea) to 2543 m (the top of Zla Kolat, the mountain range of Prokletije). According to Kepen's classification of climate, most of Montenegro has the characteristics of the Mediterranean climate—*Csa, Csb* and *Cfb*, which is characterized by warm, in the south and hot summers and mild and rainy winters. In the higher mountain areas in the north, above 1000–1200 m, a moderately cold climate is present—*Dfb* and *Dfc* (Burić et al. 2013, 2014).

Solar and cloudiness data from 8 main meteorological stations, relatively homogeneously distributed on the territory of Montenegro, obtained from the Institute of Hydrometeorology and Seismology of Montenegro—IHMSM were used (see Fig. 1): 3 stations are coastal (Bar, Herceg Novi and Ulcinj), 2 from the central region (Podgorica and Nikšić) and 3 from the mountainous north of the country (Kolašin, Žabljak and Pljevlja). Cloud cover is determined visually, while the duration of sunshine at IHMSM stations is traditionally measured using a Campbell-Stokes heliograph. For the last few years, the duration of insolation at the main stations in Montenegro has been measured using automatic meteorological stations (sensors), but in order to use this data, a statistical procedure is needed to harmonize with data obtained using Campbell-Stokes heliograph, as done in the Czech Republic (Valík et al. 2019). Therefore, only Campbell-Stokes heliograph data were

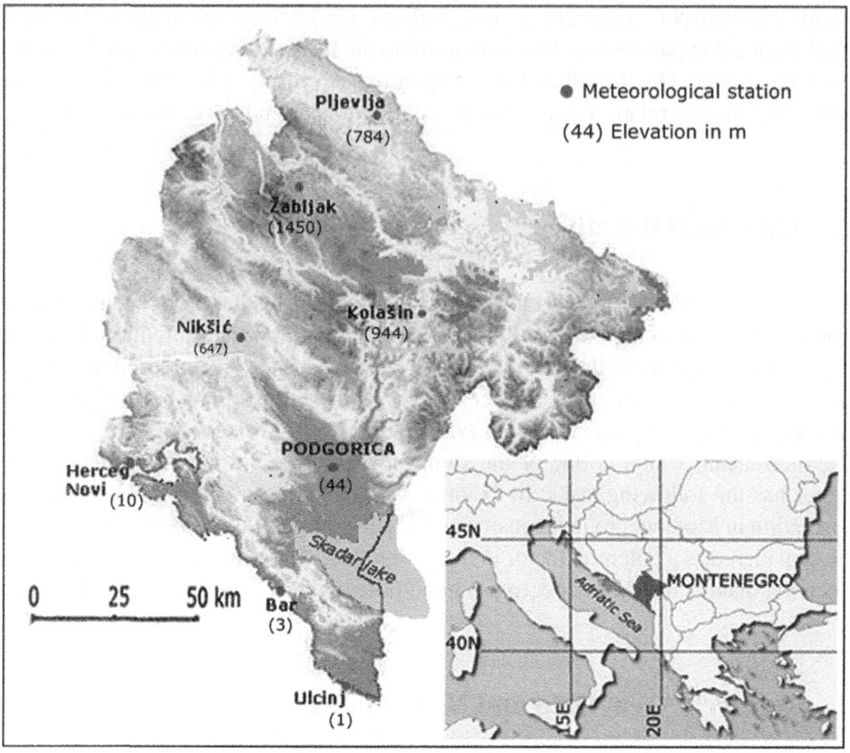

Fig. 1 Locations and altitude of meteorological stations and the position of Montenegro in the Eastern Mediterranean

used for this paper. For the observed 58-year period (1961–2018), there was 3–5% of missing data for insolation and cloud cover, depending on the time series of individual stations. The Multiple Analysis of Series for Homogenization (MASHv3.02) method, developed by the Hungarian Meteorological Service (Szentimrey 2007), was used to test the homogeneity of the datasets and to evaluate missing data in sequences. Testing showed that there were up to 1% inhomogeneous data in all-time series. These data were excluded from the time series and then estimated using the MASHv3.02 method. Initially, data from 10 meteorological stations were used, but two time series for insolation did not meet the conditions of homogeneity, so these two stations were excluded from the analysis.

For the purposes of the paper, the trend method and the correlation method were used. The linear regression method starts from the assumption that the distribution is normal (McBean and Motiee 2008). However, it was found that meteorological and hydrological variables do not meet all the criteria for normal distribution in a series (Viessman et al. 1989), especially when it comes to quantitative elements (precipitation, insolation, evaporation, etc.). Therefore, the authors emphasize the advantage of using nonparametric tests for detection and significance of the trend of a given time series, primarily Sen's slope estimates and Mann–Kendall (e.g. Šumenjak and Schuster 2011; Mondal et al. 2012), which was used in this paper. It is a non-parametric test, which is also recommended by the World Meteorological Organization (WMO 1966). The significance of the trend was examined for significance levels $p <$ of: 0.001, 0.01, 0.05 and 0.1 (the degree of accuracy of the hypothesis of 99.9, 99, 95 and 90%). The correlation between insolation and cloudiness was calculated using the Pearson coefficient, and the significance of the relationship was checked by t-test at the 95% ($p < 0.05$) and 99% ($p < 0.01$) level of hypothesis acceptance.

4 Results and Discussion

Average sunshine sums were given first for the standard climate period 1961–1990 and then for the period 1991–2018. In such a way, changes of desiccation could be observed by simple difference.

On a monthly basis, the shortest sunshine was in winter, with a minimum in December, whose sums ranged from 36 h (h) in Pljevlja to 114 h in Ulcinj. From winter to summer, the length of insolation increased permanently. The sunniest month was July from 230 to 234 h in Bijelo Polje and Pljevlja to as much as 351 h in Ulcinj (see Table 1). During the month of July, the sun shone a little longer than 10 h/day in Nikšić (10.1 h/day) and Podgorica and along the coast around 11 h/day. Žabljak, at an altitude of 1450 m, had an average of about 1928 h of sunshine per year. Isolation was shorter in Pljevlja, Bijelo Polje and Kolašin, because the mentioned places are located in valleys, which are often foggy in the colder part of the year. Apart from the topography of the terrain, the data clearly indicated the dependence of the sun on the length of the day, but also the cloud cover, which was greatest in the colder part of the year. Sun exposure was in direct correlation with the length of the day, and in inverse correlation with cloudiness.

Based on available data, the average annual amount of sunshine was the lowest in Pljevlja, 1631 h or 4.5 h per day. The sunniest place in Montenegro was Ulcinj, with

Table 1 Mean monthly and annual sunshine amounts (in hours) in Montenegro (period 1961–1990)

Station	Jan	Feb	Mar	Apr	May	June	July	Aug	Sep	Oct	Nov	Dec	Year
Ulcinj	126	130	171	204	269	299	351	323	254	199	134	114	2574
Bar	123	123	171	199	261	297	351	318	249	198	125	112	2527
H.Novi	111	118	160	192	254	288	343	318	248	190	114	102	2438
Podgorica	121	130	170	194	251	276	340	314	250	201	127	110	2484
Nikšić	115	120	151	168	215	240	312	285	227	188	117	108	2246
Kolašin	78	92	129	148	187	193	247	238	183	152	92	67	1806
Žabljak	91	100	139	159	191	204	257	236	195	162	106	88	1928
Pljevlja	51	81	126	147	175	180	234	224	170	134	73	36	1631

the average annual sunshine of 2574 h, i.e. the sun shone an average of 7.1 h/day. Coastal places, then Podgorica and Nikšić had annual sun exposure for more than 2200 h (see Fig. 2). There was a lot of sunshine in Podgorica, 6.8 h/day, which was an annual amount of 2484 h. Thus, places on the Montenegrin coast, as well as the central part of the country (Podgorica, Danilovgrad and Nikšić), were at the very top in terms of sunshine in the Balkans and Europe.

Since changes in the duration of sunshine can directly or indirectly indicate anomalies of some other elements, such as e.g. cloudiness, humidity, temperature, etc., it is interesting to see what happened to the insolation after 1990, whether it increased or decreased. Therefore, the sunshine amounts for the period 1991–2018 were calculated and compared with the corresponding climatic norms of the period 1961–1990. When it comes to the differences between the two mentioned periods (1991/2018–1961/1990), the results indicated that Ulcinj, Bar, Nikšić, Pljevlja and Žabljak were sunnier in the last 28-year period (1991–2018). At the annual level, the average insolation in the mentioned places was longer by 70 h in Žabljak to 92 h in Ulcinj compared to the climate period 1961–1990 (see Table 2). On the other hand, in Herceg Novi, Kolašin and Podgorica, the average annual insolation was shorter in the second (1991–2018) than in the first period (1961–1990), which was not to be

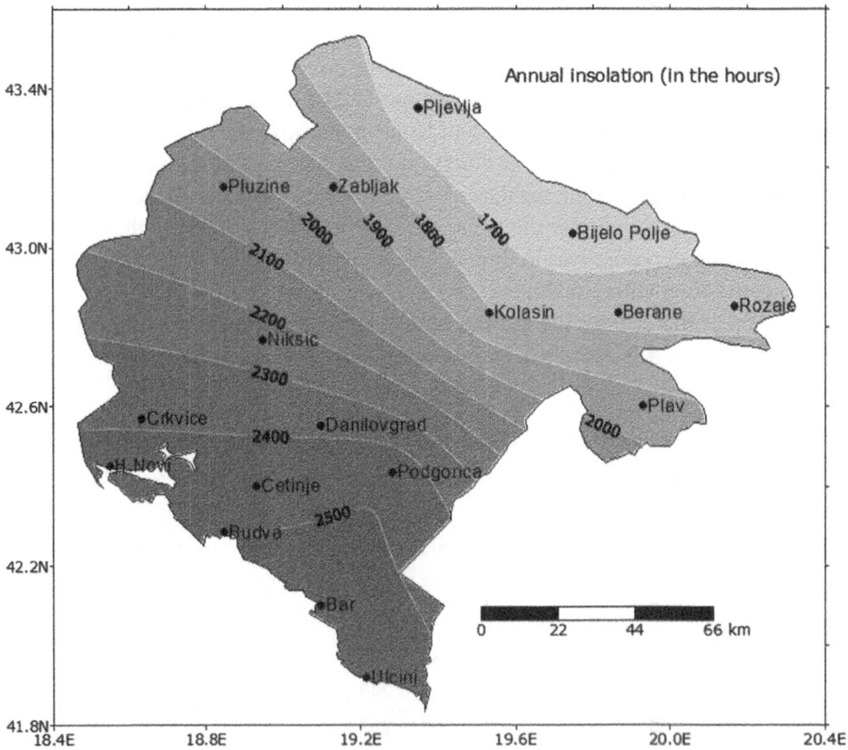

Fig. 2 Annual sunshine sums in Montenegro on the horizontal surface period 1961–1990

Table 2 The difference between the mean monthly and annual insolation between 1991 and 2018 (A) and 1961–1990 (B)

Station	Difference	Jan	Feb	Mar	Apr	May	June	July	Aug	Sep	Oct	Nov	Dec	Year
Ulcinj	A-B	14	24	21	10	−1	5	−7	7	−7	5	8	13	92
Bar	A-B	6	16	15	6	9	18	4	8	−7	−3	4	9	85
H.Novi	A-B	−2	15	11	−4	−7	4	−16	−3	−20	−16	−3	−5	−46
Podgorica	A-B	−2	2	5	−1	−1	20	−3	4	−24	−23	−10	−8	−41
Nikšić	A-B	11	14	16	5	12	34	4	15	−20	−16	7	9	91
Kolašin	A-B	−22	−9	−6	−7	1	27	8	−2	−24	−27	−26	−30	−112
Žabljak	A-B	13	19	12	1	4	18	3	17	−15	−6	1	4	70
Pljevlja	A-B	8	13	6	1	2	35	15	14	−12	−5	2	9	87

+ Longer insolation
− Shorter insolation

expected. Namely, in the conditions of modern climate change, i.e. global warming, in the area of the Western Balkans, to which Montenegro also belongs, a smaller amount of precipitation and cloudiness is expected (IPCC 2014), so the logical conclusion is that the solar system is increasing.

On a monthly basis, certain regularity was observed for June and September. In September, all stations showed a decrease in sunshine, while June was sunnier in all the considered places in the last almost three decades (1991–2018) compared to the climatic norm (1961–1990). In October, it was sunnier only in Ulcinj, while in other places the duration of sunshine decreased. In general, it can be concluded that in Herceg Novi, Kolašin and Podgorica, the reduction of monthly sunshine amounts dominated, while in other places it was sunnier.

The basic direction of movement of a climate element showed the trend of components. In the period 1961–2018 in Bar, the trend of monthly sunshine amounts was negative only in September and October, while during other months the duration of sunshine increased. However, the trend of changes in sunshine in Bar, during the observed 58-year period, was insignificant in all months. In Herceg Novi, there was a significant decrease in insolation in July (−3.8 h/decade), September (−4.4 h/decade) and October (−4.8 h/decade), but the tendency was significant only at the lowest risk level of accepting the hypothesis of 0.10 (the lowest level of confidence of 90%). In other months, the trend of sunshine in Herceg Novi was mostly positive, but the trend was insignificant (see Table 3). In Podgorica, there was a significant decrease in sunshine in September and October, while in Ulcinj all months showed a trend of increasing the length of sunshine, and it was significant in April, November and December. The Nikšić station mainly recorded an increasing trend of monthly sunshine sums, but it was significant only in November (5.7 h/decade) at a 95% confidence level.

In Kolašin, the three summer months were getting sunnier, and the trend of increasing sunshine in June was significant at the 95% level of acceptance of the hypothesis. However, in the period from September to May in Kolašin, the insolation decreases, and the negative trend was statistically significant in the period October-February. Apart from the amount of clouds, a possible cause of the decrease in sunshine in the colder part of the year in Kolašin could be an increase in the frequency of the day with fog, but this requires a more detailed analysis. Pljevlja also showed insignificant changes in sunshine on a monthly basis, except in December when the trend was positive (2.7 h/decade) and significant at 95% confidence level. In Žabljak, the changes in sunshine were insignificant in all months.

The trend was calculated on both seasonal and annual levels. During the winter, the insolation decreased significantly in Kolašin, and increased in Ulcinj, Bar, Pljevlja and Žabljak. In spring, almost all analyzed stations showed a trend of increasing insolation, however, significant only in Ulcinj (12.9 h/decade). Also, there was a trend of increasing summer sunshine at almost all stations. It was significant in Nikšić (10.8 h/decade) and Bar (8.6 h/decade). On the other hand, autumn sunspots had a negative momentum, which was significant in Herceg Novi, Kolašin and Podgorica (see Table 4).

Table 3 Trend of monthly sunshine sums (h/decade) for the period 1961–2018

Station	Jan	Feb	Mar	Apr	May	June	July	Aug	Sep	Oct	Nov	Dec
Ulcinj	2.4	3.5	4.4	5.6^{+}	0.7	0.2	1.4	2.8	0.7	3.5	5.0*	6.5^{+}
Bar	1.4	1.7	1.1	3.0	2.3	2.3	2.8	2.6	−1.3	−0.4	3.3	5.1
H.Novi	−2.1	0.6	0.5	0.8	−1.3	−2.0	-3.8^{+}	0.8	-4.4^{+}	-4.8^{+}	0.4	0.2
Podgorica	−2.9	−4.0	−1.7	2.1	0.1	2.9	0.5	1.3	-5.7^{+}	−7.7*	−1.9	−1.4
Nikšić	2.0	0.1	0.8	2.8	3.0	4.8	1.8	4.7	−3.7	−2.5	5.7*	4.2
Kolašin	−7.4***	-4.7^{+}	−2.9	−1.0	−0.1	6.0*	2.9	1.0	-5.4^{+}	−7.1*	−5.5**	−8.0***
Žabljak	3.9	3.4	2.2	1.0	0.1	−0.2	0.1	5.7	−2.6	−1.2	3.3	1.5
Pljevlja	1.8	2.7	1.9	−1.0	0.5	5.4	3.9	2.8	−3.0	−0.9	1.4	2.7*

Significance of the trend at the level: $^{***}p < 0.001$, $^{**}p < 0.01$, $^{*}p < 0.05$ and $^{+}p < 0.10$

Table 4 Trend of annual and seasonal sunshine sums (h/10 years) for the period 1961–2018

Station	Winter	Spring	Summer	Autumn
Ulcinj	14.3*	12.9*	1.4	7.3
Bar	10.6⁺	7.5	8.6⁺	1.4
H.Novi	−0.8	0.1	−6.9	−8.9⁺
Podgorica	−5.4	−0.3	4.7	−15.7**
Nikšić	7.1	5.3	10.8*	−2.4
Kolašin	−18.5***	5.0	7.6	−21.6***
Žabljak	13.0*	5.5	3.5	−0.1
Pljevlja	7.5⁺	3.2	10.6	−1.1

Significance of the trend at the level: ***$p < 0.001$, **$p < 0.01$, *$p < 0.05$ and $^{+}p < 0.10$

At the annual level, a larger number of stations showed an increase in the length of insolation, and it was significant in Bar, Nikšić and Ulcinj. In the period 1961–2018, in the area of Ulcinj, the sun shone longer by 37.3 h/decade and the value of the trend was significant at 99% confidence level. A significant increase in sunshine was also registered at the station in Bar.

In Herceg Novi, Kolašin and Podgorica, the trend of sunshine was negative, but it was significant in Kolašin (−35.0 h/decade). However, for the last 28-year period (1991–2018), an increasing trend of insolation was registered in Kolašin (see Fig. 3). Let us remind you, the average annual insolation in Kolašin for the period 1991–2018 was 112 h less compared to the period 1961–1990. But, in the period 1991–2018, there was a trend of increasing insolation in Kolašin. This contradiction is of a methodological nature—there are two different methods (difference and trend).

The changes in sunshine were most pronounced during the winter and spring, and the least in the summer at most stations. The annual course of insolation is inversely related to cloud cover—inverse correlation. In this regard, a correlation was calculated between the monthly values of insolation and cloud cover for the period 1961–2018. Calculations showed that the relationship between the monthly values of insolation and cloud cover was mostly over 0.9 (see Table 5) at 99% confidence level, which was to be expected.

Thus, it was found that there was a strong relationship between sunshine and clouds, statistically very significant. The results obtained in this paper are in line with the research conducted by Burić and Stanojević (2020). Specifically, for 1961–2017, the authors acknowledged the opposite trend of total cloud cover (TCC) in Montenegro in relation to insolation. During winter, spring and summer, the trend of decreasing TCC dominates, while the autumn season in most parts of Montenegro registers a tendency to increase TCC. Future research should address the factors that had the strongest impact on cloud cover, and thus indirectly on insolation.

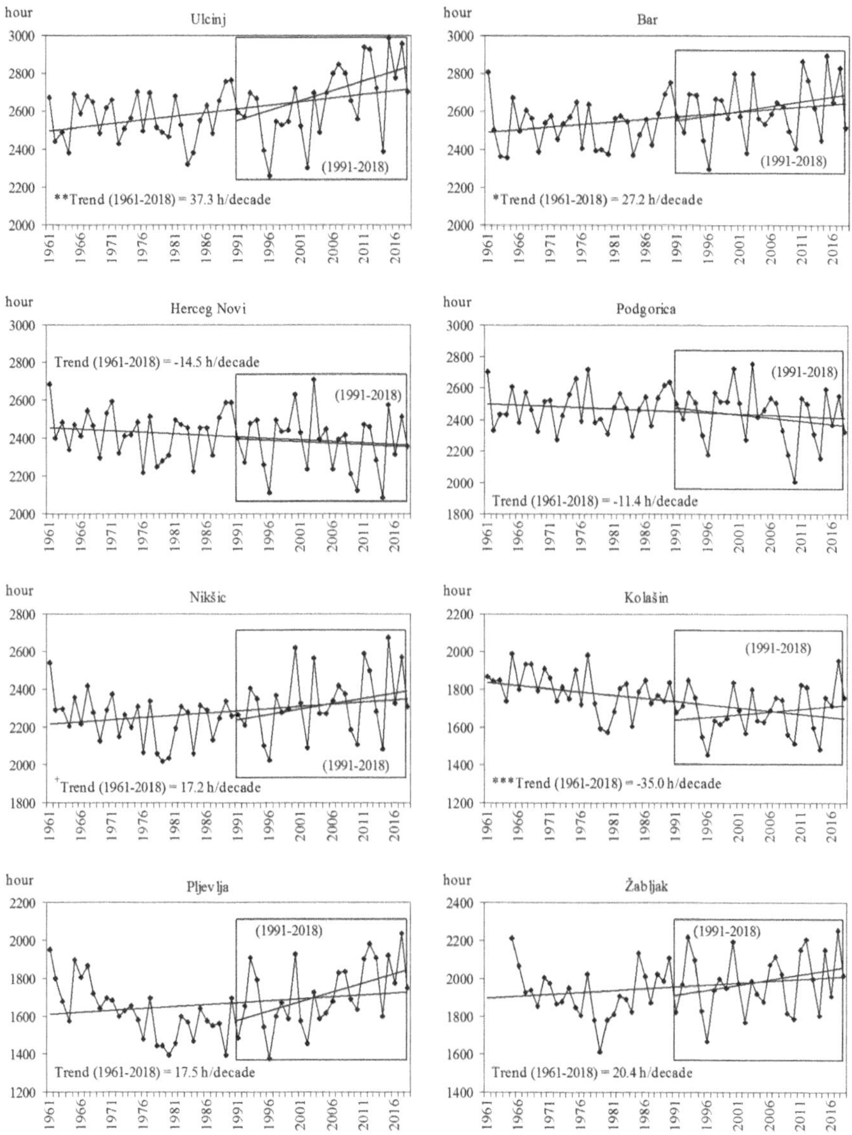

Fig. 3 Trend of annual suns in the period 1961–2018 (significant: ***$p < 0.001$, **$p < 0.01$, *$p < 0.05$ and $^{+}p < 0.10$)

5 Conclusion

In this paper, the changes of insolation in the area of Montenegro were analysed and the connection of this element with cloudiness was examined. Data from 8 meteorological stations, relatively homogeneously distributed in Montenegro, were

Table 5 Matrix of correlations between insolation and cloud cover in Montenegro (1961–2018)

Station	Jan	Feb	Mar	Apr	May	June	July	Aug	Sep	Oct	Nov	Dec
Ulcinj	−0.93	−0.95	−0.90	−0.87	−0.76	−0.55	−0.52	−0.54	−0.81	−0.88	−0.91	−0.95
Bar	−0.97	−0.96	−0.95	−0.91	−0.91	−0.88	−0.82	−0.90	−0.91	−0.95	−0.95	−0.95
H.Novi	−0.94	−0.96	−0.94	−0.75	−0.86	−0.72	−0.61	−0.89	−0.92	−0.93	−0.86	−0.86
Podgorica	−0.93	−0.93	−0.90	−0.84	−0.82	−0.81	−0.86	−0.89	−0.91	−0.91	−0.85	−0.87
Nikšić	−0.94	−0.96	−0.93	−0.90	−0.85	−0.84	−0.83	−0.89	−0.94	−0.95	−0.93	−0.96
Kolašin	−0.78	−0.84	−0.90	−0.90	−0.79	−0.78	−0.66	−0.80	−0.82	−0.89	−0.67	−0.54
Žabljak	−0.86	−0.89	−0.86	−0.85	−0.57	−0.80	−0.88	−0.85	−0.90	−0.89	−0.91	−0.85
Pljevlja	−0.74	−0.90	−0.89	−0.86	−0.82	−0.81	−0.91	−0.93	−0.91	−0.92	−0.83	−0.72

In all cases, there is a significance of correlation at the level of risk $p < 0.01$

used, and the period 1961–2018 was observed. Research showed that the period after 1990 (1991–2018) in most parts of Montenegro was sunnier than the period 1961–1990. It was further determined that in the period 1961–2018, there was a trend of increasing duration of sunshine to 5, and decreasing for 3 stations, and that was the general picture on a monthly, seasonal and annual level. The significance of the trend was halved—at 4 stations the changes in annual insolation were statistically significant, and at the same number the tendency was insignificant. In the last 28-year period (1991–2018), the trend of increasing insolation was positive at 6 stations and negative at 2. The fact that insolation was a function of cloudiness was confirmed, because the correlation calculations between these two elements showed a strong relationship, mostly with coefficient values >0.9.

During the previous 6–7 decades, there has been a significant trend of rising temperatures in Montenegro, but this cannot be related to the change in sunshine. This does not mean that changes in solar activity do not affect the temperature, on the contrary. To examine the relationship between changes in solar activity and temperature, it is necessary to use other indicators (number of sunspots, AA index, etc.). Further research is needed to address the possible causes of the rise in temperature in this country, primarily consideration of changes in solar activity, the impact of the anthropogenic greenhouse effect, and variations in atmospheric and oceanic oscillations. Finally, the results show that Montenegro, especially the coastal and central part, where the annual insolation is 2200–2600 h, has a good potential for the development of solar energy, which can be used for heating water and residential and other buildings, as well as for electricity production.

Recommendations

- To form a database for the basic parameters of solar radiation: insolation, direct and diffuse, i.e. global radiation.
- In order to identify the most suitable locations for the installation of solar systems it is necessary to map the parameters of solar radiation for the entire territory of Montenegro in the form of high-resolution thematic maps.
- Even though it is the most efficient and cleanest source of energy, the solar potential in Montenegro is minimally used. Through appropriate development plans and strategies, decision-makers (government and ministries) should promote and encourage subsidies for the mass use of solar energy, not only for space and water heating, but also for electricity production. Hence the emission of harmful gases caused by the burning of fossil fuels will be reduced, and Montenegro will give its significant contribution in the fields of mitigating the anthropogenic greenhouse effect and improving the quality of the environment.

References

Berger A, Loutre MF, Tricot C (1993) Insolation and Earth's orbital periods. J Geophys Res 98(D6):10341–10362. https://doi.org/10.1029/93JD00222

Bartoszek K, Matuszko D, Węglarczyk S (2020) Trends in sunshine duration in Poland (1971–2018). Int J Climatol (accepted for Publication). https://doi.org/10.1002/joc.6609

Burić D, Ducić V, Mihajlović J (2013) The climate of Montenegro: modificators and types—part one. Bull Serbian Geogr Soc 93(4):83–102. https://doi.org/10.2298/GSGD1304083B

Burić D, Ducić V, Mihajlović J (2014) The climate of Montenegro: modificators and types—part two. Bull Serbian Geogr Soc 94(1):73–90. https://doi.org/10.2298/GSGD1401073B

Burić D, Dragojlović J, Penjišević-Sočanac I, Luković J, Doderović M (2019) Relationship between atmospheric circulation and temperature extremes in Montenegro in the period 1951–2010. In: Leal Filho W, Trbic G, Filipovic D. (eds) Climate change adaptation in Eastern Europe. Climate change management. Springer, Cham, pp 29–42. https://doi.org/10.1007/978-3-030-03383-5_3

Burić D, Doderović M (2021) Changes in temperature and precipitation in the instrumental period (1951–2018) and projections up to 2100 in Podgorica (Montenegro). Int J Climatol 41(S1):E133–E149. https://doi.org/10.1002/joc.6671

Burić D, Stanojević G (2020) Trends and possible causes of cloudiness variability in Montenegro in the period 1961–2017. Clim Res 81:187–205. https://doi.org/10.3354/cr01615

Drysdale RN, Hellstrom JC, Zanchetta G, Fallick AE, Sanchez Gni MF, Couchoud I, McDonald J, Maas R, Lohmann G, Isola I (2009) Evidence for obliquity forcing of glacial Termination II. Science 325(5947):1527–1531. https://doi.org/10.1126/science.1170371

Erlykin AD, Sloan T, Wolfendale AW (2009) Solar activity and the mean global temperature. Environ Res Lett 4(1):014006 (6 pp). https://doi.org/10.1088/1748-9326/4/1/014006

Founda D, Pierros F, Sarantopoulos A (2017) Evidence of dimming/brightening over greece from long-term observations of sunshine duration and cloud cover. In: Karacostas T, Bais A, Nastos P (eds) Perspectives on atmospheric sciences. Springer Atmospheric Sciences. Springer, Cham, pp 753–758. https://doi.org/10.1007/978-3-319-35095-0_108

Hanrahan J, Maynard A, Murphy SZ, Zercher C, Fitzpatrick A (2017) Examining the Climatology of Shortwave Radiation in the Northeastern United States. J Appl Meteorol Climatol 56(10):2869–2881. https://doi.org/10.1175/JAMC-D-16-0420.1

Huybers PJ (2006) Early Pleistocene glacial cycles and the integrated summer insolation forcing. Science 313(5786):508–511. https://doi.org/10.1126/science.1170371

Lohmann G, Pfeiffer M, Laepple T, Leduc G, Kim JH (2013) A model-data comparison of the Holocene global sea surface temperature evolution. Clim past 9:1807–1839. https://doi.org/10.5194/cp-9-1807-2013

Lohmann G (2017) Atmospheric bridge on orbital time scales. Theor Appl Climatol 128:709–718. https://doi.org/10.1007/s00704-015-1725-2

Lorenz SJ, Lohmann G (2004) Acceleration technique for Milankovitch type forcing in a coupled atmosphere-ocean circulation model: method and application for the Holocene. Climate Dyn 23:727–743. https://doi.org/10.1007/s00382-004-0469-y

Manara V, Beltrano MC, Brunetti M, Maugeri M, Sanchez-Lorenzo A, Simolo C, Sorrenti S (2015) Sunshine duration variability and trends in Italy from homogenized instrumental time series (1936–2013). J Geophys Res Atmospheres 120(9):3622–3641. https://doi.org/10.1002/2014JD022560

McBean E, Motiee H (2008) Assessment of impact of climate change on water resources: a long term analysis of the Great Lakes of North America. Hydrol Earth Syst Sci 12(1):239–255. https://doi.org/10.5194/hess-12-239-2008

Mondal A, Kundu S, Mukhopadhyay A (2012) Rainfall trend analysis by Mann-Kendall test: a case study of north-eastern part of Cuttack district, Orissa. Int J Geol Earth Environ Sci 2 (1):70–78. Available at: https://www.researchgate.net/publication/268438767

Nullet D, Ekern PC (1988) Temperature and insolation trends in Hawaii. Theor Appl Climatol 39:90–92. https://doi.org/10.1007/BF00866393

Omran MA (2000) Analysis of solar radiation over Egypt. Theor Appl Climatol 67:225–240. https://doi.org/10.1007/s007040070011
Power HC (2003) Trends in solar radiation over Germany and an assessment of the role of aerosols and sunshine duration. Theor Appl Climatol 76:47–63. https://doi.org/10.1007/s00704-003-0005-8
Szentimrey T (2007) Manual of homogenization software MASHv3.02. Hung Meteorol Serv 61
Šumenjak K, Šuster V (2011) Parametrični in neparametrični pristopi za odkrivanje trenda v časovnih vrstah. Acta agriculturae Slovenica 97(3):305–312. Available at: http://www.dlib.si/?URN=URN:NBN:SI:doc-QIDKTLWM
Valík A, Brázdil R, Zahradníček P, Tolasz R, Možný M, Řezníčková L (2019) Measurements of sunshine duration by automatic sensors and their effects on the homogeneity of long-term series in the Czech Republic. Clim Res 78(1):83–101. https://doi.org/10.3354/cr01564
Viessman W, Knapp JW, Lewis GL (1989) Introduction to hydrology. Harper and Row Publishers. New York, pp. 149–355. ISBN 006046822X
Vujović D, Todorović N, Paskota M (2018) Summer weather characteristics and periodicity observed over the period 1888–2013 in the region of Belgrade, Serbia. Theor Appl Climatol 132:103–114. https://doi.org/10.1007/s00704-017-2075-z
WMO (1966) Climatic change. Tech Note No 79, WMO, Geneva, p 79

[illegible] Wilks DS (2011) Statistical methods in the atmospheric sciences [illegible] ISBN [illegible]

[illegible]

The Conservation Challenge of Traditional Agroecosystems in Morocco: The Case Study of Six Oases Agroecosystems

Mhammad Houssni, Jalal Kassout, Abdeltif El Ouahrani, Mohammed El Mahroussi, Vladimiro Boselli, Khalil Kadaoui, Abdelouahab Sahli, Mohamed Kadiri, and Mohammed Ater

Abstract Global changes, caused by anthropogenic activities, are imposing serious and growing challenges for both natural and human systems. For instance, climate change and the emergence of global diseases, as the recent pandemic of coronavirus (Covid-19), greatly affecting economic and social stability of many vulnerable communities worldwide. Traditional agroecosystems (TAE) are remarkable adaptations of crop production systems that ensure the balance of human-nature interactions. However, the abovementioned challenges are already affecting the basic foundation of crop production and rural livelihoods in TAE particularly within arid African and MENA regions. In this context, we hypothesize that these agroecosystems are playing a key role in maintaining and strengthening the resilience of human-nature systems. To verify this later, prospective surveys were conducted in six major traditional oases agroecosystems (TOAE) in southern Morocco (Alnif, Zagora, Aoufous, Rich, Guelmim and Tata oases) to describe and analyse the status of their agro-diversity considering climatic trends and characteristics. The primary results revealed that the studied TOAE are structured on a multi-strata cropping system embracing more than 180 different crops including palm trees (50), cereals (29), legumes and vegetables (42). Almond, olive and fig trees are major elements of arboreal strata. In addition, the results show that TOAE are cornerstones for maintaining both genetic resources conservation, as well as an important driver for local socio-economic resilience. Nevertheless, the vulnerable aspects of TOAE in the current context of climate change is undeniable. This study represents a basic initiative to value many neglected and poorly studied components of TAE and provides potential head-points for future

M. Houssni (✉) · J. Kassout · A. E. Ouahrani · M. E. Mahroussi · K. Kadaoui · A. Sahli · M. Kadiri · M. Ater (✉)
Department of Biology, Bio-Agrobiodiversity Team, Faculty of Sciences, Abdelmalek Essadi University, Tetouan, Morocco

J. Kassout · M. Ater
Associated International Laboratory (IRP), EVOLEA, INEE-CNRS, CNRST, Montpellier, France

V. Boselli
Dipartimento di Ingegneria Civile, Architettura, Territorio, Ambiente e di Matematica, Università degli Studi di Brescia, Brescia, Italy

W. Leal Filho and E. Manolas (eds.), *Climate Change in the Mediterranean and Middle Eastern Region*, Climate Change Management,
https://doi.org/10.1007/978-3-030-78566-6_10

agricultural adaptation programmes to counter global change impacts and ensure a sustainable development of such arid agro-ecosystems.

1 Introduction

Along with their historical presence on earth, humanity has transformed nature. Today, human-induced impacts took a planetary dimension on the earth system generating unprecedented global changes (Vitousek et al. 1997; Steffen et al. 2007, 2015). The ongoing climate change and anthropogenic pressures are influencing worldwide terrestrial landscapes integrity, services and sustainability (Parmesan 2006; Steffen et al. 2007; Ellis 2011; Newbold et al. 2015). Deforestation, habitat disturbance, overexploitation of natural resources and increasing of population density has certainly led to complex challenges (McMichael et al. 2003; Goudie 2013; Newbold et al. 2015). The emergence of global disease, as evidenced by the current coronavirus pandemic which rapidly escalated across the world (WHO 2020; Manzanedo and Manning 2020), and food security issues (Godfray et al. 2010; Shaheen et al. 2017; Campbell-Lendrum et al. 2018) are actually the most accurate challenges threatening all aspects of global society and the planet we live on. Furthermore, the fast growing rate of world population, estimated at nine billion by the year 2050, accompanied by land degradation, soil erosion, and water resources depletion, impose serious pressures on agricultural systems to produce more food on limited arable land (Tilman et al. 2011). Moreover, the projected increase of global temperature will affect water reservoirs and supplies, crop yields—in general-agroecosystems and ultimately agroecosystems performance (Godfray et al. 2010; IPCC 2014). For instance, climate change scenarios expect a drier and warmer condition for Mediterranean region (Giorgi and Lionello 2008; IPCC 2014). This will negatively affect crop water demands and agroecosystems yields productivity (Fuhrer 2002; Saadi et al. 2014; Malek et al. 2018; Madani et al. 2018), and consequently threaten food security, mainly for vulnerable communities and remote arid localities in Middle East and North Africa (MENA) regions.

To address these challenges, the United Nations has adopted the 2030 development agenda, based on 17 Sustainable Development Goals (SDGs; UN 2015). These goals aim to protect global biodiversity, search for new innovative and sustainable solutions for human prosperity, to end poverty and hunger by ensuring food security through agricultural sustainable systems (ICSU 2015; FAO 2016). However, critical trade-offs may occur between these goals. For example increased agricultural and energy production can compete for the same resources and consequently affect the implementation of SDGs (Weitz et al. 2014). Actually, the challenge for the agriculture sector is to provide enough food to the growing population with less resources, such as water and farmland (Godfray et al. 2010; FAO 2015). Ultimately, there is a need for an effective sustainable use and management of the existing agricultural systems. This requires more attention to small-scale agricultural production systems (Rapsomanikis 2015), such as the case for traditional agroecosystems in the

Rif and Atlas mountains of Morocco (Hmimsa and Ater 2008; Ater and Hmimsa 2013; Ziyadi et al. 2019; Boselli et al. 2020). We argue here that maintaining such traditional agricultural systems is crucial to meet the SDGs.

Traditional agroecosystems (TAE) may present a key component to solving some of the global food challenges. This is factual for at least two main raison: (i) TAE are characterized by a significant diversity in crop varieties and local landraces representing a significant raw material for future crop selection and amelioration programs, and (ii) an important traditional knowledge holding by the local farmers through generations (Ater and Hmimsa 2008; Achtak et al. 2010; Lauri et al. 2019; Boselli et al. 2020). For instance, oases agroecosystems represent an effective example of TAE associated with a sustainable agriculture model, which can provide solutions to many issues related to food security challenges. In fact, Traditional oases agroecosystems (TOAE) are holding an enormous diversity of date palm with high genetic diversity (Terral et al. 2012; Gros-Balthazard 2013). These TOAE are often accompanied with an important diversity of arboreal species, vegetable plants and seasonal corps (Boselli et al. 2020; Houssni et al. 2020).

In Morocco, oases are distributed from the north-east region of the country to the south through the Anti-atlas region (Kabiri 2014) within an area of 48,000 ha holding 1.7 million inhabitants (Sbaï 2011). These oases are characterized by important phytogenetic resources (Belarbi et al. 2004), promoted to be incorporated in the Moroccan oasis biosphere reserve of the east (RBOSM) recognized by the UNISCO in 2000. Furthermore, TOAE could be seen as 'biodiversity hot-spot' by conserving in situ local crops and landraces (Brush 1995; Achtak et al. 2010).

Though, TOAE in Morocco, as in North Africa and the Middle East, are undergoing socioeconomic, physical and ecological changes (Lightfoot 1996; Belarbi et al. 2004; Hamidi 2005). The rapid rate of climate change may represent the most important threat to these agroecosystems sustainability (Dixon 2012; Gill et al. 2014). In fact, the overexploitation and depletion of underground water, using modern irrigation techniques, has led to massive rural exodus events, land abandonments and consequently degradation and disappearance of agrobiodiversity within TOAE (Battesti 2012; Boselli et al. 2020). The decline of oasis agrobiodiversity has serious implications for species, ecosystem services and local farmers who depend on environmental and natural resources for their livelihood (Battesti 2012; Gill et al. 2014). Despite the fact that Morocco had developed the 'Morocco Green Plan' (MGP), which aims to modernize the agriculture sector to become more competitive and more integrated in the global market, it has neglected the importance of traditional agroecosystems as a source of genetic resources and sustainable agriculture models.

In this context, the present study aims to explore the main traditional oases agroecosystems in Morocco as a resilient approach to counter global changes impacts, including climate change, and ensure sustainable development of these arid agroecosystems and food security for local populations. To this end, we present a guideline to the valorisation of many neglected and poorly studied components of traditional oases agroecosystems, which can provide potential head-points for future agricultural programmes.

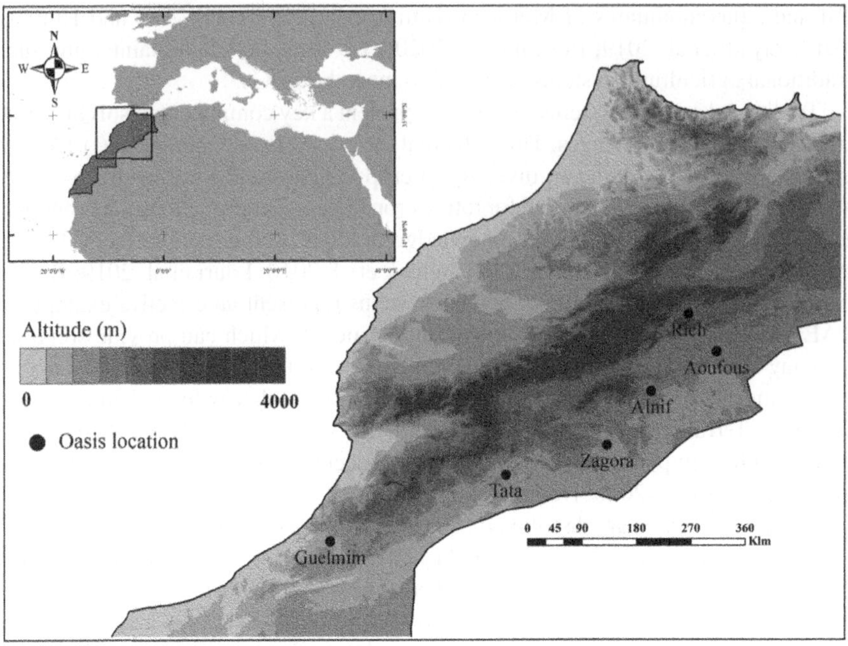

Fig. 1 Geographical localization of the six studied oases of Rich, Aoufous, Alnif, Zagora, Tata and Guelmim

2 Methodology

2.1 Study Area

The study area is ranged from the Anti-Atlas to the southern part of the country along 700 km transect (Fig. 1), covering Drâa-Tafilalet and Sous-Massa regions. The study investigates six oases considered as the largest in Morocco: Alnif, Zagora, Aoufous, Rich, Guelmim and Tata. The investigated oases show different geographic and climatic characteristics (Table 1). The agroecological and socioeconomic features, along with cultural symbolism, make these oases an ideal traditional oases agro-ecosystems (TOAE) case study, where the conservation of valuable landraces and crops varieties in the current context of global changes is a real challenge.

2.2 Agro-Diversity Surveys

To collect data, we adopted semi-structured surveys. The farmers were chosen at random in different angles of each surveyed oasis. This technique allows to collect maximum qualitative and quantitative data from the local population (Benaoun et al.

Table 1 Climatic and geographic characteristics of the six studied oases

Oasis	Latitude	Longitude	PP (mm/year)	T_{max} (°C)	T_{min} (°C)	T_{mean} (°C)	Altitude (m)
Alnif	31.1166	−5.1664	145.29	26.29	11.78	18.52	1011
Zagora	30.3367	−5.8254	85.72	29.64	14.54	21.71	735
Aoufous	31.6871	−4.1915	203.80	26.43	11.76	18.65	909
Rich	32.2368	−4.6074	363.36	22.34	8.24	14.63	1433
Guelmim	28.9420	−9.9382	154.58	28.80	14.52	20.71	211
Tata	29.9007	−7.3204	135.52	30.71	15.39	22.66	831

Climatic variables are abbreviated as follows: PP (mm/year), mean annual precipitation; T_{max} (°C), mean maximum temperature; T_{min} (°C), mean minimum temperature; T_{mean} (°C), mean annual temperature

2014; Houssni et al. 2020). To this end, five *ksours*[1] (plural of *ksar)* from each oases were selected for this study, and 20 farmers per each *ksar* were interviewed. In total, 100 farmers were surveyed per oasis (Table 2). The questionnaires were designed to collect information on different farming layers and present a comprehensive inventory of the traditional crops (cereals, pulses, local landraces and varieties), as well as introduced varieties. This approach has already been applied in previous similar studies on agroecosystems in the Rif (Hmimsa and Ater 2008; Ater and Hmimsa 2013; Barontini et al. 2017), Anti-Atlas (Ziyadi et al. 2019; Boselli et al. 2020) and in the south of Morocco (Ait Hmida 2003; Houssni et al. 2020). To assess possible similarities between the studied oases, we calculated, hierarchical Cluster Analyses (HCA) and Correspondent Analysis (CA) to explore the structure of TOAE components and to analyse the questionnaires responses. These analyses were conducted using R Software (R Development Core Team 2015).

2.3 *Climate Change and Aridity Impacts on Traditional Oases Agroecosystems (TOAE)*

We conducted a fractal analysis of wet and dry periods according to the methodology developed by Bazuhair et al. (1997). Data, representing 37-year monthly meteorological data, were retrieved from NASA Power (Earthdata 2020). This approach allows identifying the periods of consecutive days without rain and to examine pluviometric information and highlight possible patterns of arid periods that can occur at each oasis (Boselli et al. 2020). In addition, using monthly average temperatures and monthly average precipitations, climagraphs were used to describe local climate in each oasis. The long-term climate change impact on the studied area was then analysed, using Köppen-Geiger climate type classification change (Kottek et al. 2006). Observed

[1] Represent agglomerative area inside the oasis.

Table 2 Investigated *Ksours* and number of questionnaires per each oasis

Oasis	*Ksar* name	Number of questionnaires
Alnif	Ammar	20
	Tizi	20
	Alnif	20
	Achbarou	20
	Ait Zeggane	20
Zagora	Asrir N ilemchane	20
	Tansita	20
	Amazrou	20
	Sart	20
	Aghla Oudrar	20
Aoufous	Takhyamte	20
	Rbiet	20
	Zrigate	20
	Lamâarka	20
	Zaouia Jdida	20
Rich	Ait Moussa Oali	20
	Balite	20
	Zaouia Sidi Boukil	20
	M'zizel	20
	Tamagourte	20
Guelmim	Tighmert	20
	Ifrane	20
	Fask	20
	Tagante	20
	Taghjijt	20
Tata	Tissint	20
	Akka	20
	Ait oubelli	20
	Foum zguid	20
	Foum lahcen	20

(1976–2000) and projected (2051–2075 and 2076–2100) climatic conditions were extracted from World Maps of Köppen-Geiger climate classification, with a resolution of 0.5° latitude and longitude, then projected over the study area using QGIS software (Quantum Gis Development Team 2015). The different classes are coded according to the Köppen-Geiger climate classification. The climagraphs are obtained according to the method of Bagnouls and Gaussen (1953). The IPCC business as usual (BAU) scenario (A1FI) was adopted to estimate the worst climate change scenario until 2100 (IPCC 2007; Ziyadi et al. 2019).

3 Results

3.1 Composition of Traditional Oases Agroecosystems (TOAE) in Morocco

The conducted surveys in the six investigated oases (Alnif, Zagora, Aoufous, Rich, Guelmim and Tata) resulted in 183 different types of crops including landraces and traditional varieties (Fig. 3). These oases were also characterized by an important specific richness in crops with 53 different types of cultures. For instance, we recorded the presence of 43 cultures in Alnif, 40 in Tata, 36 in Guelmim, 30 in Rich and 28 in Zagora. From agrodiversity perspective, Tata presents the highest number with 106 varieties, while Rich oasis shows the lowest number with only 40 varieties. Meanwhile, Guelmim, Alnif, Aoufous and Zagora present 80, 73, 58 and 50 different varieties respectively (Fig. 3a). For detail purposes, we divided the inventoried crops into six different categories: fruit trees, vegetables, legumes, cereals, fodder plants and date palms. Fruit trees were found to be the most diverse category ranging from eight varieties in Zagora to 35 in Guelmim. Vegetables come in 2nd place with 14 categories found in Rich to 23 in Alnif. Date palms ranged from nine different varieties in Alnif to 27 in Tata. For cereals, Tata is leading with 17 varieties, followed by Alnif with 16 varieties, Guelmim with nine varieties and five different varieties each found in Aoufous, Zagora and Rich (Fig. 3b).

3.2 Oases Cropping System Characteristics

The surveys presented a comprehensive inventory of the traditional crops (cereals, pulses, local landraces and varieties) as well as introduced varieties. The study allows exploring and recognizing three main farming layers within TOAE; namely date palm layer, arboreal layer and herbaceous layer (Figs. 2 and 3). However, Rich oasis shows the existence of only two layers; the arboreal and herbaceous layer, with absence of date palm layer due to altitude effect which create colder winters.

Date palm layer represents the main and superior tree layer in the investigated oases (except for Rich) and is mainly composed of palm trees (*Phoenix dactylifera*). Our results revealed the existence of 50 different varieties of palm tree in the six studied oases with the presence of high economic value varieties such as '*Majhoul*' and '*Boufggouss*'. Moroccan oases are well known for their more than 220 known varieties of palm tree with high genetic diversity (Toutain et al. 1971; Sedra 2003; El Houmaizi et al. 2006; Bodian et al. 2012; Harrak et al. 2018).

In fact, *Phoenix dactylifera* is the emblematic tree of the oases agro-systems (Terral et al. 2012; Abul-Saod et al. 2017). It is highly appreciated by local farmers. This is due to at least three main reasons: (i) its fruit, or dates, which are highly appreciated for their nutritive value and economic revenue during production season; (ii)

Fig. 2 General overview of the studied oases of Alnif **a**, Aoufous **b**, Rich **c**, Zagora **d**, Tata **e** and Guelmim **f**

provides shade for seasonal and herbaceous plants and consequently prevent excessive desiccation of crops under high temperatures; and (iii) used as soil fixator against erosion. Additionally, several studies have shown that bacterial communities associated with root palm system is important and diverse (Ferjani et al. 2015; Mosqueira et al. 2019), which contribute largely to maintain soil viability and consequently positively affects agroecosystems productions (Chao and Krueger 2007).

The secondary arboreal layer is composed by fruit trees, an important component of the traditional agro-ecosystems (Ater and Hmimsa 2008; Boselli et al. 2020). According to our investigations, 54 different varieties representing 15 different

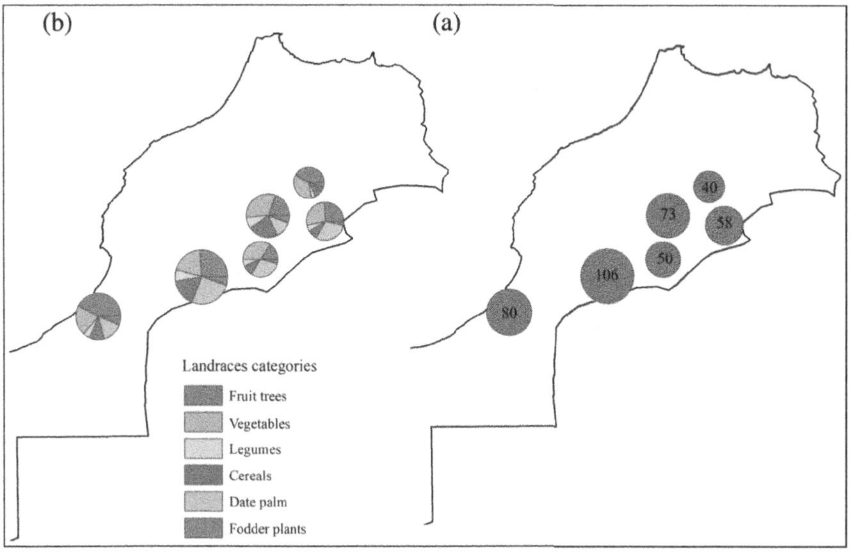

Fig. 3 Number of total crop varieties **a** and crop types by oasis **b**

species of fruit trees were identified. It includes fig (*Ficus carica*), pomegranate (*Punica granatum*), grapevine (*Vitis vinifera*), olive (*Olea europaea*), almond (*Prunus dulcis*), apple (*Malus pumila*), apricot (*Prunus armeniaca*), orange (*Citrus sinensis*), carob (*Ceratonia siliqua*), prickly pear (*Opuntia ficus-indica*), lemon (*Citrus limon*), mandarin (*Citrus nobilis*), quince (*Cydonia oblonga*), walnut (*Juglans regia*) and common pear (*Pyrus communis*). The fig, apricot, almond and olive trees are the dominant species in the secondary layer. The olive tree, identified by the farmers under the denomination of '*Zitoune*' and previously identified by genetic markers as 'Picholine marocaine' (Khadari et al. 2008), is widely spread in Moroccan agroecosystems and largely appreciated by local farmers (Hmimsa and Ater 2013; Ater et al. 2016; Bosselli et al. 2020). The preference for olive, fig and almond trees, in the studied oases, is in agreement with what has been reported by Hmimsa and Ater (2013) in the Rif traditional agroecosystems and Bosselli et al. (2020) in Asrargh oasis in the Anti Atlas region.

Vegetables, legumes, cereals and fodder plants species compose the herbaceous layer. This third layer is particularly important for local communities, as it provides various vegetables for family consumption and fodder for animals (Santoro et al. 2020). This layer hosts 29 varieties of cereals, eight varieties of legumes, 35 varieties of vegetables and seven varieties of fodder plants represented mainly by alfalfa (*Medicago sativa*). For local farmers, alfalfa is highly demanded as forage crop for animals and its economic value (Sraïri et al. 2017). In fact, the existence of forage crops indicates that animal breeding is maintained as an essential component of TOAE (Benaoun et al. 2014). Our results indicate, as well, the presence of two important varieties of cereals: millet and sorghum. Millet is found only in Tata oasis and the sorghum is generally rare in Moroccan agroecosystems (Ater and Hmimsa 2008). The association of herbaceous layer varieties alongside with other crops are found to be beneficial for sustainable agricultural practices (Wezel et al. 2014). Therefore, small land plots with arid conditions have led local farmers towards more efficient agricultural system management, by applying the concept of Terraces Agro-Systems (TAS) combined with a polyculture crop production system which implies the association of seasonal and perennials crops (Ziadi et al. 2019).

3.3 The Structure of TOAE in Morocco

The structure of the investigated traditional agroecosystems, analysed by Hierarchical Cluster Analyses (HCA), shows a clear separation between the studied oases and their inventoried *ksours* (Fig. 4). Our analysis displays three main separated clusters; 1: include Rich and Aoufous, 2: include Alnif and Zagora, and 3: include Guelmim and Tata. The two first clusters composed by Rich, Aoufous, Alnif and Zagora were found to be more related, but separated from Guelmim and Tata oases. In fact, Alnif and Zagora show similar structures likely due to their similarity in climatic

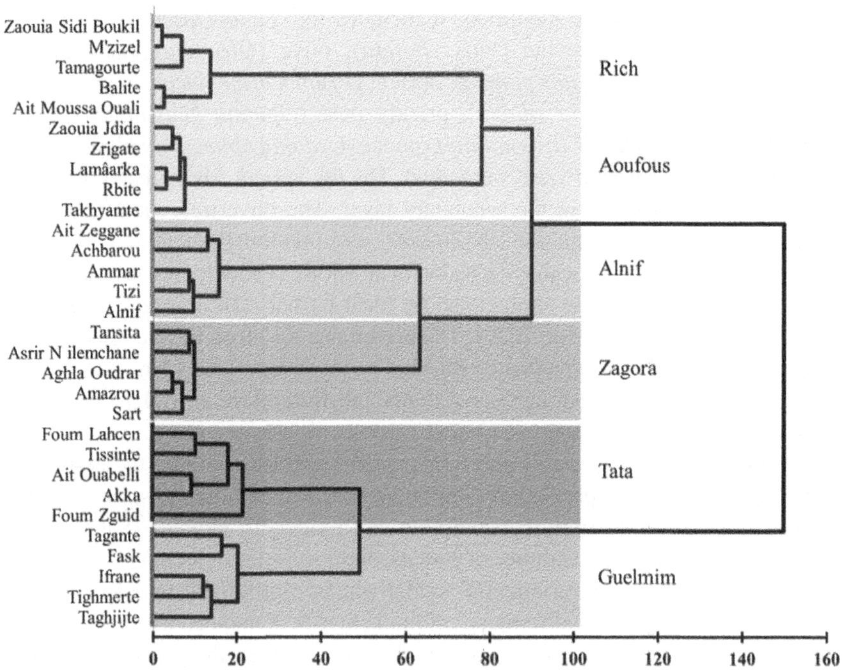

Fig. 4 Hierarchical cluster analysis of the studied oases and their *ksours*

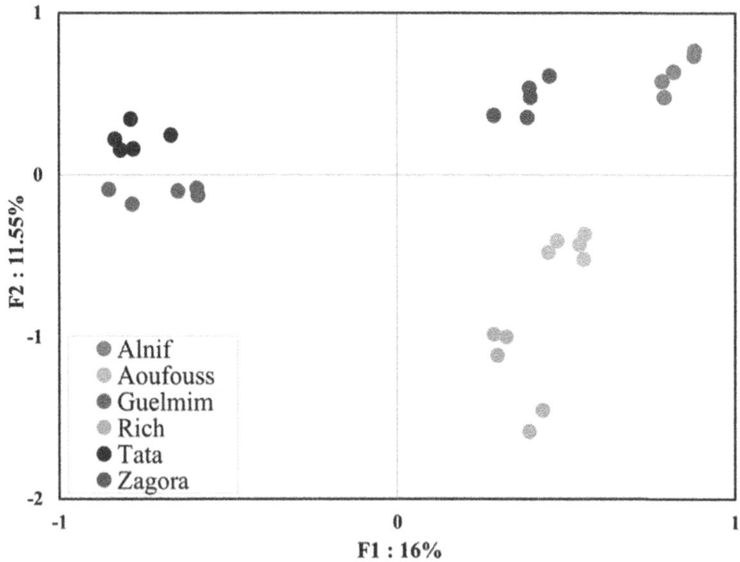

Fig. 5 Correspondent analysis (CA) of the studied oases

conditions and vegetable varieties. Furthermore, Alnif, Zagora, Rich and Aoufous are geographically close to each other compared to Tata and Guelmim.

Correspondent Analysis (CA) (Fig. 5) confirmed these results and shows a clear discrimination, along the first axis (16% of total variation), between Guelmim and Tata (negative side axis1) compared to other oases of Rich, Aoufous, Alnif and Zagora (positive side of axis1). The first axis was found mainly and positively correlated with fruit trees and negatively correlated with palm date varieties and cereals. Meanwhile, axis 2 (11.55% of total variation) shows a discrimination between Tata, Alnif and Zagora at the positive side and Guelmim, Aoufous and Rich at the negative side of axis 2 (Fig. 5). These results indicates that specific richness in fruit trees are important in Aoufous, Rich and Guelmim oases, which is related to suitable conditions and farmers choice for the same species as the case for some apple varieties and walnut trees found only in Rich and some palm and pomegranate varieties found only in Aoufous oasis. However, low fruits trees in the other studied oases are balanced by an important richness in vegetable and cereal crops. The analysis of oases agrodiversity reflects also a differentiation within the investigated *Ksours*.

3.4 *Drought Incidence and Climate Change on Traditional Oases Agroecosystems*

Analysis of oases climagraphs, developed from the 37-year monthly meteorological data of NASA Power, shows that Alnif, Zagora, Aoufous, Guelmim and Tata oases

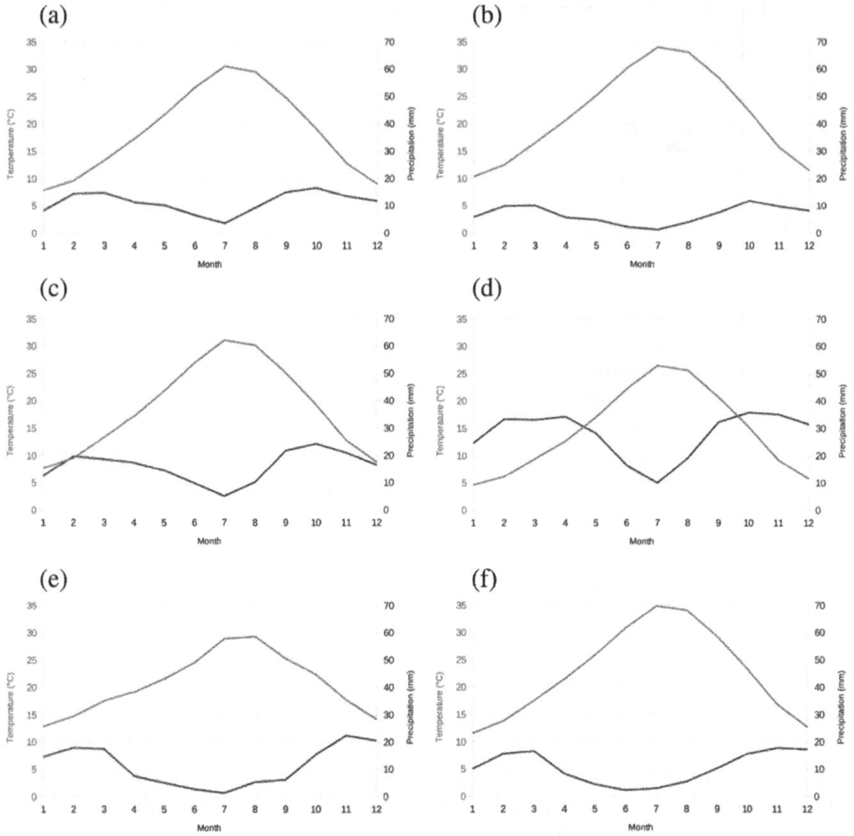

Fig. 6 Climographs of the studied oases of Alnif **a**, Zagora **b**, Aoufous **c**, Rich **d**, Guelmim **e** and Tata **f**

are characterized by a Saharian and hyper arid climate (Fig. 6). Only Rich oasis has a Mediterranean climate type with a wet season that develops from October to the end of March (Fig. 6d).

The fractal analysis of wet and dry periods (Fig. 7, Table 3) shows a pattern of arid and semi-arid climates. We observed a slight increase in the duration and persistence of non-rainy periods in the oases of Alnif, Zagora, Guelmim and Tata (Fig. 7a, b, e, f). This kind of analysis can be useful if cross-referred with questionnaires on the perception of rainy days and non-rainy days nowadays and in the past decades. Thus, our analysis highlights a decrease of rainy days, which can be considered a climate change signal (Table 3).

The climate type classification using Köppen-Geiger Climate Classification of the studied oases reveals a clear pattern of a drier future. In fact, the used climate data allows presentation of climate change patterns from observed to future estimates based on IPCC scenarios (Kottek et al. 2006). Our results (Fig. 8) show three maps

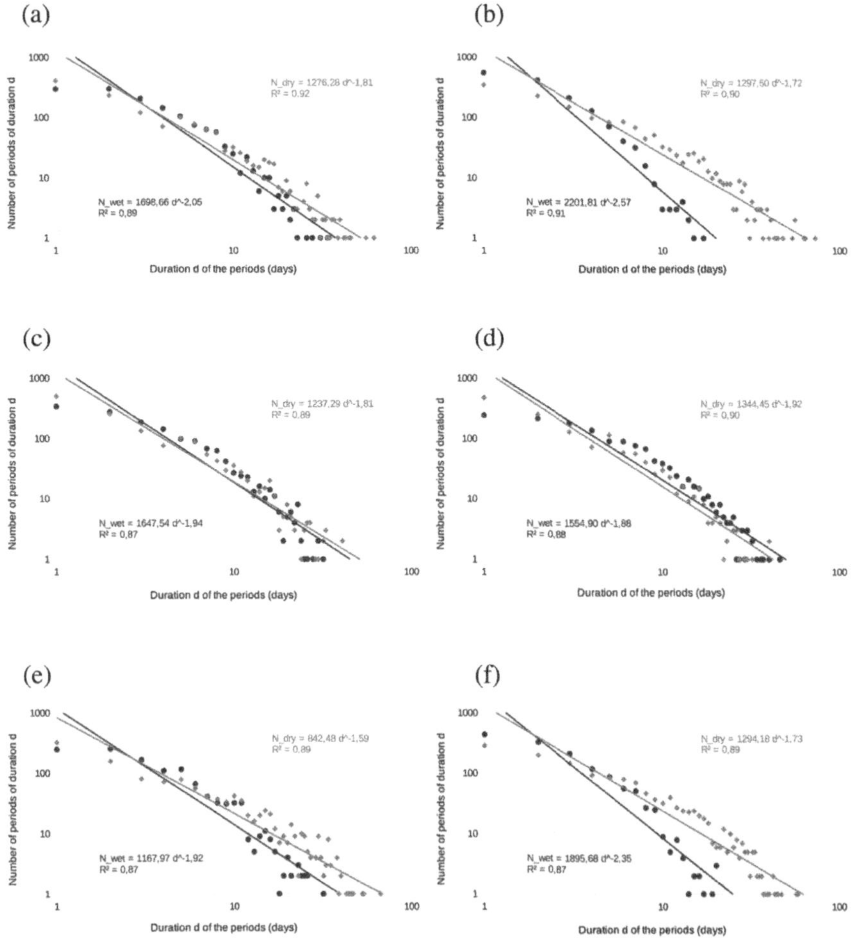

Fig. 7 Fractal analysis of wet (in blue) and dry (in red) periods of Alnif **a**, Zagora **b**, Aoufous **c**, Rich **d**, Guelmim **e** and Tata **f** Oases

representing observed (Ob 1967–2000) and projected (A1FI 2051–2075 and A1FI 2076–2100) data of 25 years each. Along from Rich to Guelmim oases, three climate types were identified, abbreviated as BWK, BWh and BSh. The first letter provides an indication about the main climates (B refers to arid climate), the second letter designates the main precipitation regime (W refers to desert and S refers to steppe), and the third letter designates a type of temperature regime (here: k refers to cold arid temperature and h to hot arid temperature). As observed in Fig. 8, under current climate scenario, Rich oasis is under arid climate with desert environment and cold arid temperature (BWk), Alnif is under arid climate with steppe environment and hot arid temperature (BSh), while, Aoufous, Zagora, Tata and Guelmim are under arid climate with desert environment and hot arid temperature (BWh). Projected

Table 3 The longest periods of consecutive dry and wet days of the studied oases

Oasis	Dry period			Wet period		
	Year	Duration in days	Last day of longest dry period	Year	Duration in days	Last day of longest wet period
Alnif	2002	62	28 Jul	1997	34	24 Apr
	2005	56	8 May	2018	31	10 Oct
	1995	46	8 Feb	1995	28	24 Apr
	2007	45	8 Jul	1990	26	1 Oct
	1992	43	29 Jan	2011	23	29 May
	2016	42	3 Jan	1987	22	4 Oct
Zagora	2018	73	1 Aug	2003	17	16 Aug
	2016	64	3 Jan	1988	15	5 Mar
	2002	63	28 Jul	1991	14	16 Dec
	2005	56	8 May	1996	14	28 Jan
	2019	56	18 Jan	1993	13	3 Mar
	2012	54	21 Jun	1996	13	19 Jun
				1997	13	23 Aug
				2005	13	24 Nov
Aoufous	1997	41	22 Mar	1984	32	24 May
	2016	41	3 Jan	1990	30	1 Oct
	1989	33	5 Jan	2009	30	29 Sep
	2015	33	17 May	1995	28	24 Apr
	2019	33	15 Jan	2002	26	20 Apr
	2000	32	29 Jul	1990	25	14 May
	2018	32	3 Jul			
Rich	2002	40	3 Jun	1984	21	21 Mar
	2005	40	4 Jun	2004	2	2 Jan
	1995	32	25 May	1985	14	14 Jan
	2007	30	9 Jun	2011	2	2 Aug
	1992	29	13 May	1993	28	28 Jul
	2016	27	30	2009	24	24 Feb
				2015	18	18 Jun
Guelmim	1995	67	7 Feb	2019	32	14 Aug
	2005	54	8 May	2015	26	31 Oct
	1998	53	27 Nov	2016	26	14 Jul
	1983	47	29 Jun	1985	25	3 Jun
	2013	46	16 Jan	1997	25	14 Apr
	1992	45	7 Oct	1982	24	18 Jun
				2004	24	25 May

(continued)

Table 3 (continued)

Oasis	Dry period			Wet period		
	Year	Duration in days	Last day of longest dry period	Year	Duration in days	Last day of longest wet period
Tata	2005	58	8 May	1993	20	15 Nov
	1997	57	25 Jul	1996	20	3 Feb
	2013	47	17 Jan	2018	20	29 Sep
	1992	45	30 Jan	2015	19	27 Oct
	2007	44	7 Jul	1993	17	7 Mar
	2012	43	17 Mar	1988	16	5 Mar
				2006	16	12 Sep

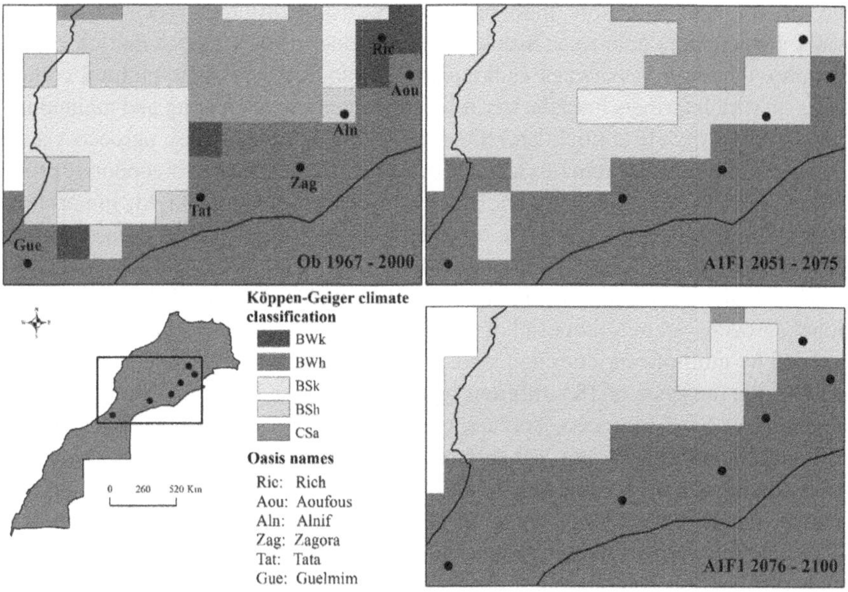

Fig. 8 Observed (Ob) and projected (A1FI) climate type shifts: Ob1967-2000, A1FI2051-2075, and A1FI2076-2100, based on the Köppen-Geiger climate classification world maps, using IPCC fossil intensive (A1FI IPCC) scenario. (here: B refers to arid climate and C to warm temperate climate), the 2nd letter designates the main precipitation regime (here: W refers to desert and S refers to steppe), and the 3rd letter designates a type of temperature regime (here: k refers to cold arid temperature, h to hot arid temperature and a to hot summer). BWk: desert environment and cold arid temperature; BWh: desert environment and hot arid temperature; BSk: steppe environment and cold arid temperature; BSh: steppe environment and hot arid temperature; CSa: steppe environment and hot summer temperature

climate of the study area shows a shift from steppe to desert environment. Thus, our results show a shift in climate types of Rich oasis from BWk to arid climate with steppe environment and cold arid temperature (BSk). The other studied oases show no change of climate types. However, the precipitation deficiency is expected to be more drastic, which puts the region under high risk of drought and desertification.

4 Discussion

4.1 Traditional Oases Agroecosystems

Agrodiversity represents the way in which farmers use the natural diversity of the environment to ensure high productivity, using different types of land and water management practices (Brookfield and Padoch 1994). This definition implies complex interactions between ecological, hydrological and socio-cultural components, in which farmers hold the key role in the dynamic of creating and maintaining this agrodiversity (Brookfield and Stocking 1999). In return, these agroecosystems provide valuable ecosystem services and bring back important socio-economic profits to their users. In this context, traditional oases agroecosystems are highly important as centers for agrodiversity conservation and traditional knowledge (Ater and Hmimsa 2008; Achtak et al. 2010; Boselli et al. 2020).

The obtained results show that the studied oases possess a well-diversified traditional agroecosystems composed of a considerable number of crop types and differentiated by multiple varieties and landraces. The present study makes it possible to identify the presence of 183 different types of crops, landraces and varieties within the main oases of Morocco. This quite high number of crops reflects a variety in soil and microclimate types and different preferences of local farmers. This richness is important also when compared to other oases of Tunisia, Algeria and Libya (Hammer and Perrino 1985; Allam et al. 2013; Benaoun et al. 2014; Aljane et al. 2017) and represents valuable genetic resources for local agroecosystems (Hammer et al. 1988). The small land plots drive farmers to apply the concept of agroforestry consisted in the association between seasonal crops as cereals and legumes and perennials ones as fig (*Figus carica*), almond (*Prunus dulcis*) and olive (*Olea auropaea*). This approach was previously reported in the Anti-Atlas agroecosystems (Ziyadi et al. 2019; Boselli et al. 2020) and may lead to the observed diversification in crop varieties and landraces of the studied traditional oases agroecosystems. Moreover, our results show the existence of a considerable number of crop varieties within date palm, fruit trees, vegetables, legumes, cereals and fodder plants species. This diversification can be seen as an efficient security food system, in which the practice of multiple crop types led to satisfy human and livestock needs, and thus, plays a key role in both the conservation of important genetic resources (Brush 1995; Achtak et al. 2010) and the adaptation to a changing climate. It is well known that vegetation plays an important role in maintaining agricultural stability in arid regions (Gaudin et al. 2015).

The present work contributes to the comprehension of traditional oases agroecosystems functioning and sustainability. It represents a first step toward assessing oases agrodiversity with regards to its richness and distribution and to monitor its variation facing global changes. In fact, monitoring agrodiversity changes and erosion through inventories is a key step toward biodiversity conservation efforts. However, studies on biodiversity inventories and assessment are still not as comprehensive as they should be, and lacking some precision (Hambler 2004). Therefore, land race and varietal inventories are required for conservation programmes and allow to detect changes among biodiversity components. Since the integration of domesticated plant species in diversity inventories, the number of inventoried species has increased (Khoshbakht and Hammer 2008; Myers et al. 2012), including crops, varieties and landraces used for food production.

4.2 Drought Incidence in Traditional Oases Agroecosystems (TOAE)

From a climatic and hydrological perspective, the studied oases are almost under arid climatic conditions, except for Rich oasis, which can be considered under a sub-humid regime. The structural water deficit of the environment is compensated at the very local oasis level by the water drained by the *wadi* and is often present in the form of a perennial hyperemic stream flow.

A general methodology to estimate the water availability and water deficit at the oasis level can be implemented on the basis of the area drained by the *Assif*,[2] its geomorphology and monthly precipitation and temperature. This simple data associated with models for the hydrological balance of the agricultural soil can be used to make a proper estimation of water surpluses and deficits at the oasis level, as described by Boselli et al. (2020). Generally, oases are affected by hydrological and climatic data scarcity since they are in areas which are mostly desert, relatively poor (Buytaert et al. 2012) and of scarce interest for extensive agricultural production. A fast analysis can be done by means of a climagraph to identify the months in which a natural plant water deficit is present (in our case only Rich presents a humid season from October to March). A climagraph should be used as an indicator of potential hydrological balance for agricultural soil. Another important factor in arid, semiarid and sub-humid climates is the temporal structure/aspect of the dry and wet periods (Şen 2008), since prolonged dry periods can affect agricultural production and water allocation. Fractal analysis is a good methodology proposed by Bazuhair et al. (1997) and developed by Şen (2008) for investigating *wadi* hydrology. We applied this analysis on rainy and non-rainy days, to the investigated oasis, and we observed a slight increase in the duration and persistence of non-rainy periods in the

[2] Represent the main river.

oases of Zagora, Alnif and Tata. This kind of analysis can be quite useful if cross-referred with questionnaires on farmers' perception of rainy and non-rainy days in the present and in the past decades.

4.3 Climate Change and the Challenge of Conservation

Our results, retrieved from the Köppen-Geiger Climate Classification, show a general pattern of changing climate types between the current time and the future scenarios. These changes are mainly linked to precipitation regime, which shifted from a steppe to a desert environment. Therefore, the region is expected to face an aridification featured with low precipitation and hot temperature in summer. In fact, the climate scenarios for the Mediterranean region predict an increase in temperature and a decrease in precipitation (Polade et al. 2014; IPCC 2014). Thus, the future climate over Morocco shows an important decrease in annual mean precipitation accompanied with a change in the precipitation distribution and in extreme events (Driouech et al. 2010). Similar previous findings have shown that agroecosystems of the Anti-Atlas region will face drastic conditions of aridity in the light of future climate scenarios (Ziyadi et al. 2019). These conditions will likely increase the incidence of drought and prolonged periods of water deficiency, which will create challenging conditions for many traditional crop varieties across the TOAE.

In this study, agrodiversity surveys coupled with the assessment of climate change impacts within Moroccan traditional oases agroecosystems make it possible to explore different components of these valuable ecosystems and show their vulnerability under future global changes. As discussed previously, species surveys and inventories are required to understand biodiversity structure and distributions, and to monitor its changes at multiple ecological scales under different anthropic pressures (Hambler 2004). The absence of detailed surveys of crop diversity and local landraces and varieties is a major concern, especially in the light of rapid biodiversity erosion (Newbold et al. 2015). Therefore, as in this study, agrodiversity needs to be assessed through long-term inventories among several socio-economic backgrounds to realise effective richness and dynamics of traditional agroecosystems. In addition, the lake of genetic characterization of local landraces and varieties could not reflect the real richness of traditional agroecosystems, since many local crops names were attributed by local farmers and do not necessarily reflect their genetic identities (Ater and Hmimsa 2008; Achtak et al. 2010). On the other hand, the assessment of climatic change incidence on oasis agroecosystems is limited by the complexity of interaction between factors acting at global scale and those acting at local scale. In fact, human, physical, biological and social variables at a local scale need to be integrated in models aimed at the assessments of climate change effects on agroecosystems, which can bring new insights on local and regional dynamics of traditional oases agroecosystems.

5 Conclusion

Throughout their historical existence, oases ecosystems have shown the capacity to resist and adapt to challenging environmental conditions. In the light of the current context of global changes, they represent a buffer zone and a relevant study model for climate-human landscape dynamics. Traditional oases agroecosystems play a crucial role in the development of arid zones. They constitute a refuge of agrodiversity under harsh climatic conditions. This agrodiversity represents a rolling-force of economic activities for local farmers as for the national economy. The present study allowed for characterising different components of TOAE. The study confirmed the aspect of polyculture practices in Moroccan oases and highlighted their relevant roles as holders of important genetic resources. In addition, the assessment of TOAE agrodiversity allows for featuring specific characteristics of each studied oases, which reflects differences in their local dynamics. Polyculture practices represent an essential pathway to preserve local landraces and varieties, and thus, to maintain sustainable land management. However, the ongoing aridification and desertification along with modernisation (changing lifestyle) practices of local farmers might lead to the disappearance of many valuable traditional knowledge and crops.

In fact, climate change has profoundly influenced the structure of TOAE. Water scarcity and food security emerge as major concerns within agroecosystems. Furthermore, low water availability will generate severe impacts on many crop types and consequently limits their resistance capacity to the predicted drought events. In addition, immigration movement of the younger generation towards cities has triggered the loss of local knowledge of elders, which represent a high threat for the maintenance of traditional crop varieties and landraces, and thus valuable genetic resources. Moreover, plot abandonment is likely to threaten the sustainability of traditional oases agroecosystems.

As a consequence of the current state, and as evidenced by the Covid-19 pandemic, the conservation of traditional oases agroecosystems is actually the most urgent action to be put at the high level agenda of local authorities. Such action should promote TOAE's traditional products as a key solution for economic, touristic and social prosperity, especially if women involvement is promoted. The integration of traditional varieties and local landraces should also be promoted in national biodiversity conservation programmes. These programmes should focus on genetic improvement of newly adapted crops. We believe that the above-suggested efforts could encounter the effect of climate change and desertification of TOAE, create job placement for local young generations, ensure the transfer of traditional farming knowledge and strengthen the sustainability of TOAE.

Acknowledgements Authors present their gratitude to Bio-Agrodiversity team members, and Pr. Younès Hmimsa and Pr. Salama Elfatehi for the valuable discussions and comments. In addition, authors thank all farmers and locals of the studied oases.

References

Abul-Soad AA, Jain SM, Jatoi MA (2017) Biodiversity and conservation of date palm. In: Ahuja MR, Jain SM (eds) Biodiversity and conservation of woody plants, vol 17. Springer International Publishing, Switzerland, pp. 313–353. Sustain Dev Biodivers

Achtak H, Ater M, Oukabli A, Santoni S, Kjellberg F, Khadari B (2010) Traditional agroecosystems as conservatories and incubators of cultivated plant varietal diversity: the case of Fig (*Ficus carica* L.) in Morocco. BMC Plant Biol 18:10–28

Ait Hmida A (2003) Système de production et stratégies des agriculteurs dans les oasis de la région d'Errachidia au Maroc. New Medit 2:37–43

Aljane F, Elbekkay M, Hichem Neily M, Hamza H (2017) Prospecting and inventory of agro-diversity in the traditional oases of Gafsa. J New Sci Agric Biotechnol CSIEA 10:2671–2677

Allam A, Tirichine A, Cheloufi H, Arif Y, Tama M, Mimouni A (2013) Etude de la diversité biologique des espèces maraichères cultivées dans les palmeraies de la vallé d'Oued Righ (Cas de la région de Tougourt). BioRessources 03:64–71

Ater M, Hmimsa Y (2008) Agriculture traditionnelle et agrodiversité dans le bassin versant de l'Oued Laou (Maroc). Travaux De L'institut Scientifique : Série Générale 5:107–115

Ater M, Hmimsa Y (2013) Agrodiversité des agroécosystèmes traditionnels du pays Jbala (rif, Maroc) et produit de terroirs. In: Ilbert H, Tekelioglu Y, Çagatay S, Tozanli S (eds) Options Méditerranéennes, A 104—Indications Géographiques, dynamiques socio-économiques et patrimoine bio-culturel en Turquie et dans les pays méditerranéens. CIHEAM, Montpellier, France, pp 197–208

Ater M, Barbara H, Kassout J (2016) Importance des variétés locales, de l'oléastre et des pratiques traditionnelles de l'oléiculture dans la région de Chefchaouen (Nord du Maroc). In: Ater M, Essalouh L, Ilbert H, Moukhli A, Khadari B (eds) L'oléiculture au Maroc de la préhistoire à nos jours: Pratiques, Diversité, Adaptation, Usages, Commerce et Politiques. Options Méditerranéennes A, vol 118. CIHEAM, Montpellier, France, pp 109–121

Bagnouls F, Gaussen H (1953) Saison sèche et indice xérothermique. Document Pour Les Cartes Des Prod Veget Serie: Généralité 1:1–49

Barontini S, Boselli V, Louki A, Ben Slima Z, Ghaouch F-E, Labaran R, Raffelli G, Peli M, Al Ani AM Vitale N et al (2017) Bridging mediterranean cultures in the international year of soils 2015: a documentary exhibition on irrigation techniques in water scarcity conditions. Hydrol Res 48:789–801. https://doi.org/10.2166/nh.2017.113

Battesti V (2012) The power of a disappearance: water in the Jerid region of Tunisia In: Johnston BR et al (eds) Water, cultural diversity, and global environmental change: emerging trends, sustainable futures? UNESCO/Springer, Paris, Jakarta, pp 77–96. Available at: http://hal.archivesouvertes.fr/hal-00569337

Bazuhair SA, Al-Gohani A, Şen Z (1997) Determination of monthly wet and dry periods in Saudi Arabia. Int J Climatol 17:303–311

Belarbi A, Bouayad A, Diaou M, Kaassis N, Tidjani Maliki M (2004) Agrobiodiversité et durabilité des systèmes de production oasiens dans la palmeraie d'Aoufouss. Errachidia-Maroc, IRCA/INRA-Maroc, p 167

Benaoun A, Elbakkay M, Ferchichi A (2014) Change of oases farming systems and their effects on vegetable species diversity: cas of oasian agro-systems of Nefzaoua (South of Tunisia). Sci Hortic 180:167–175

Bodian A, El Houmaizi MA, Ndoye Ndir K, Hasnaoui A, Nachtigall M, Wehling P (2012) Genetic diversity analysis of date palm (*Phoenix dactylifera* L.) cultivars from Figuig oasis (Morocco) using SSR markers. IJSAT 2(3):96–104

Boselli V, Ouallali A, Briak H, Houssni M, Kassout J, El Ouahrani A, Michailidi EM (2020) System dynamics applied to terraced agroecosystems: the case study of Assaragh (Anti-Atlas Mountains, Morocco). Water 12:1693

Brookfield H, Padoch C (1994) Appreciating agrodiversity: a look at the dynamism and diversity of indigenous farming practices. Environment 36(5):8–11

Brookfield H, Stocking M (1999) Agrodiversity: definition, description and design. Glob Environ Change 9(2):77–80
Brush SB (1995) (1995) In situ conservation of landraces in centers of crop diversity. Crop Sci 35:346–354
Buytaert W, Friesen J, Liebe J, Ludwig R (2012) Assessment and management of water resources in developing, semi-arid and arid regions. Water Resour Manag 26:841–844. https://doi.org/10.1007/s11269-012-9994-3
Chao CCT, Krueger RR (2007) The date palm (Phoenix dactylifera L.): overview of biology, uses, and cultivation. Hort Sci 42(5):1077–1082. https://doi.org/10.21273/HORTSCI.42.5.1077
Campbell-Lendrum D, Wheeler N, Maiero M, Villalobos Prats E, Nevelle T (2018) World Health Organization COP24 special report on health and climate change. World Health Organization. Available at: https://unfccc.int/sites/default/files/resource/WHO%20COP24%20Special%20Report_final.pdf
Dixon GR (2012) Climate change—impact on crop growth and food production, and plant pathogens. Can J Plant Path 34(3):362–379
Driouech F, Déqué M, Sanchez-Gomez E (2010) Weather regimes-Moroccan precipitation link in a regional climate change simulation. Global Planet Change 72:1–10
Earthdata (2020) Earthdata Search. Available at: https://earthdata.nasa.gov/. Accessed on 23 May 2020
Elhoumaizi MA, Devanand PS, Fang J, Chao CT (2006) Confirmation of 'Medjool' date as a landrace variety through genetic analysis of 'Medjool' accessions in Morocco. J Am Soc Hortic Sci 131:403–407. https://doi.org/10.21273/JASHS.131.3.403
Ellis EC (2011) Anthropogenic transformation of the terrestrial biosphere. Philos Trans Ser A Math Phys Eng Sci 369:1010–1035
Ferjani R, Marasco R, Rolli E, Cherif H, Cherif A, Gtari M, Boudabous A, Daffonchio D, Ouzari HI (2015) The date palm tree rhizosphere is a niche for plant growth promoting bacteria in the oasis ecosystem. BioMed Res Int 2015:153851 https://doi.org/10.1155/2015/153851
Food and Agriculture Organization (2015) FAO and the 17 Sustainable Development Goals Food and Agriculture Organisation of the United Nations: Rome, Italy
Food and Agriculture Organization (2016) Food and Agriculture: Key to Achieving the 2030 Agenda for Sustainable Development; Food and Agriculture Organization of the United Nations: Rome Italy
Fuhrer (2003) Review: agroecosystem responses to combinations of elevated CO_2, ozone, and global climate change. Agric Ecosyst Environ 97:1–20. https://doi.org/10.1016/S0167-8809(03)00125-7
Gaudin ACM, Tolhurst TN , Ker AP, Janovicek K , Tortora C, Martin RC, Deen W (2015) Increasing crop diversity mitigates weather variations and imporves yield stability. PLos ONE 10. https://doi.org/10.1371/journal.pone.0113261
Gill BS, Raupp WJ, Friebe B (2014) Dual threats of imperiled native agroecosystems and climate change to world food security: genomic perspectives. J Crop Improv 28:88–98
Giorgi F, Lionello P (2008) Climate change projections for the mediterranean region. Global Planet Change 63(2–3):90–104
Godfray HC, Beddington JR, Crute IR, Haddad L, Lawrence D et al (2010) Food security: the challenge of feeding 9 billion people. Science 327(5967):812–818
Goudie A (2013) In: Goudie AS (ed) The human impact on the natural environment: past, present and future, 7th ed. Wiley, p 376
Gros-Balthazard M (2013) Hybridization in the genus phoenix: a review. Emir J Food Agric 25:831–842
Hambler C (2004) Conservation. Cambridge University Press, Cambridge, p 14. ISBN 0-521-80190-7
Hamidi MH (2005) Dynamiques agraires et perspectives d'actions de développement rural des bassins versants des oasis de Tafilalet, province d'Errachidia, (Maroc). Mémoire d'Ingénieur des techniques agricoles de Clermont Ferrand, p 116

Hammer K, Perrino P (1985) A check-list of the cultivated plants of the Ghat oases. Kulturpflanze 33:269–286
Hammer K, Lehmann CO, Perrino P (1988) A check-list of the Libyan cultivated plants including an inventory of the germplasm collected in the years 1981, 1982 and 1983. Kulturpflanze 36:475–527
Harrak H, Hamouda A, Nadi M (2018) Évaluation et amélioration de la qualité des pâtes traditionnelles de dattes, produits du terroir des oasis. Cah Agric 27:15001. https://doi.org/10.1051/cagri/2017057
Hmimsa Y, Ater M (2008) Agrodiversity in the traditional agrosystems of the Rif mountains (North of Morocco). Biodiversity 9(1–2):78–81
Houssni M, El Mahroussi M, Ben Sbih H, Kadiri M, Ater M (2020) Agriculture traditionnelle et agrodiversité dans les oasis du Sud du Maroc: cas des oasis de la région Drâa-Tafilalet. In: El Moujabber M, Belhouchette H, Belkhodja M, Kalaitzis P (eds) Research and innovation as tools for sustainable agriculture food and nutrition security. Options méditerranéennes A, vol 124(2). CIHEAM, Bari, Italy, pp 82–88
ICSU (2015) Review of targets of the sustainable development goals: the science perspective. International Council for Science (ICSU), Paris, France, pp 31–34
IPCC (2007) Climate change 2007: synthesis report. In: Pachauri RK, Reisinger A (eds) Core writing team, contribution of working groups I, II and III to the fourth assessment report of the intergovernmental panel on climate change. IPCC Geneva, Switzerland, p 104
IPCC (2014) In: Pachauri RK, Meyer LA (eds) Core writing team, climate change 2014: synthesis report. Contribution of working groups I, II and III to the fifth assessment report of the intergovernmental panel on climate change. IPCC Geneva, Switzerland, p 151
Kabiri L (2014) Etude comparative des modes de gestion, de conservation et de valorisation des ressources naturelles des oasis et des zones désertiques des pays de MENA DELP. Projet MENA-DELP (Projet de coordination et de partage des connaissances sur les moyens de subsistance et les écosystèmes désertiques, au profit de l'Algérie, Egypte, Jordanie, Maroc et Tunisie), p 228
Khadari B, Charafi J, Moukhli A, Ater M (2008) Substantial genetic diversity in cultivated Moroccan olive despite a single major cultivar: a paradoxical situation evidenced by the use of SSR loci. Tree Genet Genomes 4:213–221. https://doi.org/10.1007/s11295-007-0102-4
Khoshbakht K, Hammer K (2008) How many plant species are cultivated? Genet Resour Crop Evol 55(7):925–928. https://doi.org/10.1007/s10722-008
Kottek M, Grieser J, Beck C, Rudolf B, Rubel F (2006) World map of the Köppen-Geiger climate classification updated. Meteorol Z 15:259–263. https://doi.org/10.1127/0941-2948/2006/0130
Lauri P, Barkaoui K, Ater M, Rosati A (2019) Agroforestry for fruit trees in the temperate Europe and dry Mediterranean. In: Mosquera-Losada MR, Prabhu R (eds) Agroforestry for sustainable agriculture. Burleigh Dodds Science Publishing, Cambridge, UK, pp 385–418. ISBN 978 1 78676 220 7. Available at: www.bdspublishing.com
Lightfoot DR (1996) Moroccan Khettara: traditional irrigation and progressive desiccation. Geoforum 27:261–273. https://doi.org/10.1016/0016-7185(96)00008-5
Madani N, Kimball JS, Ballantyne AP, Afleck DL, Bodegom PM, Reich PB (2018) Future global productivity will be affected by plant trait response to climate. Sci Rep 8:2870
Malek Ž, Verburg PH, Geijzendorffer R, Bondeau A, Cramer W (2018) Global change effects on land management in the Mediterranean region. Glob Environ Chang 50:238–254
Manzanedo RD, Manning P (2020) COVID-19: lessons for the climate change emergency. Sci Total Environ 742(10):140563
McMichael AJ, Campbell-Lendrum DH, Corvalan CF, Ebi KL, Githeko A, Scheraga JD, Woodward A (2003) Climate change and human health: risks and responses. World Health Organisation, p 322
Meyer RS, Du Val AE, Jensen HR (2012) Patterns and processes in crop domestication: an historical review and quantitative analysis of 203 global food crops. New Phytol 196:29–48
Mosqueira MJ, Marasco R, Fusi M, Michoud G, Merlino G, Cherif A, Daffonchio D (2019) Consistent bacterial selection by date palm root system across heterogeneous desert oasis agroecosystems. Sci Rep 9:1–12. https://doi.org/10.1038/s41598-019-40551-4

Newbold T, Hudson LN, Hill SLL, Contu S, Lysenko I, Senior RA, Börger L, Bennett DJ, Choimes A, Collen B, Day J, De Palma A, Díaz S, Echeverria-Londoño S, Edgar MJ, Feldman A, Garon M, Harrison MLK, Alhusseini T, Ingram DJ, Itescu Y, Kattge J, Kemp V, Kirkpatrick L, Kleyer M, Correia DLP, Martin CD, Meiri S, Novosolov M, Pan Y, Phillips HRP, Purves DW, Robinson A, Simpson J, Tuck SL, Weiher E, White HJ, Ewers RM, Mace GM, Scharlemann JPW, Purvis A (2015) Global effects of land use on local terrestrial biodiversity. Nature 520:45–50. https://doi.org/10.1038/nature14324

Polade SD, Pierce DW, Cayan DR, Gershunov A, Dettinger MD (2014) The key role of dry days in changing regional climate and precipitation regimes. Sci Rep 4(1):4364. https://doi.org/10.1038/srep04364

Parmesan C (2006) Ecological and evolutionary responses to recent climate change. Annu Rev Ecol Evol Syst 37:637–669

Persson GM, Ramanathan LM, Reyers V, Sorlin B (2015) Planetary boundaries: guiding human development on a changing planet. Science 347(6223):1259855

Quantum Gis Development Team (2015) Système d'Information Géographique Libre et Open Source. Available at: https://www.qgis.org/fr

Rapsomanikis G (2015) The economic lives of smallholder farmers—an analysis based on household data from nine countries. Organisation of the United Nations, Rome, Italy

R Development Core Team (2015) R: The R Project for Statistical Computing. Available at: https://www.r-project.org/

Saadi S, Todorovic M, Tansijevic L, Pereira LS, Pizzigali C, Lionello P (2014) Climate change and Mediterranean agriculture: Impacts on winter wheat and tomato crop evapotranspiration, irrigation requirements and yield. Agric Water Manag 147(1):103–115

Santoro A, Venturi M, Ben Maachia S, Benyahia F, Corrieri F, Piras F, Agnoletti M (2020) Agroforestry heritage systems as agrobiodiversity hotspots. The case of the mountain oases of Tunisia. Sustainability 12(4054):1–180. Available at: https://doi.org/10.3390/su12104054

Sbaï L (2011) Oasis du Maroc: Cadre Juridique et Institutionnel. IUCN, p 70

Shaheen S, Ahmad M, Haroon N (2017) Food security: a global problem. In: Edible wild plants: an alternative approach to food security. Springer, Cham. https://doi.org/10.1007/978-3-319-63037-3 1

Sedra MH (2003) Le Palmier Dattier base de la mise en valeur des oasis au Maroc Techniques phoénicicoles et Création d'oasis. Inra Eds., Rabat, Maroc, p 265

Şen Z (2008) Wadi Hydrology, 1st ed. CRC Press, Taylor & Francis Group, Boca Raton, FL, USA, p 347

Steffen W, Crutzen J, McNeill JR (2007) The Anthropocene: are humans now overwhelming the great forces of nature? AMBIO J Human Environ 36(8):614–621. https://doi.org/10.1579/0044-7447(2007)

Steffen W, Richardson K, Rockstrom J, Cornell SE, Fetzer I, Bennett EM, Biggs R, Carpenter SR, de Vries W, de Wit CA et al (2015) Planetary boundaries: guiding human development on a changing planet. Science 347(6223):1259855

Sraïri MT, M'ghar FA, Benidir M, Bengoumi M (2017) Analyse typologique de la diversité et des performances de l'élevage oasien. Cah Agric 26:15005

Terral J-F, Newton C, Ivorra S, Gros-Balthazard M, Tito de Morais C, Picq S, Tengberg M, Pintaud J-C (2012) First insights into the complex structure of date palm agrobiodiversity (*Phoenix dactylifera* L.) and history of ancient Egyptian cultivated forms assessed by geometric morphometrical analysis of modern and archaeological seeds. J Biogeogr 39(5):929–941

Tilman D, Balzer C, Hill J, Befort BL (2011) Global food demand and the sustainable intensification of agriculture. Proc Natl Acad Sci USA 108(50):20260–20264

Toutain G, Bachra A, Chari A (1971) Cartographie Variétale de la Palmeraie Marocaine. Technical Report; Direction de la Recherche Agronomique, Rabat, Maroc, p 242

United Nations (2015) Transforming our world: the 2030 agenda for sustainable development. In: A/RES/70/1. United Nations, New York, NY, USA

Vitousek PM, Mooney HA, Lubchenco J, Melillo JM (1997) Human domination of Earth's ecosystems. Science 277:494–499

Weitz N, Nilsson M, Davis M (2014) Anexus approach to the post-2015 agenda: formulating integrated water, energy and food SDGs. SAIS Rev Int Aff 34:37–50

Wezel A, Casagrande M, Celette F, Vian J, Ferrer A, Peigné J (2014) Agroecological practices for sustainable agriculture. A review. Agron Sustain Dev 34:1–20. https://doi.org/10.1007/s13593-013-0180-7

WHO (2020) World Health Organisation. Available at: https://www.who.int/docs/default-source/coronaviruse/situation-reports/20200211-sitrep-22-ncov.pdf?sfvrsn=fb6d49b1_2

Ziyadi M, Dahbi A, Aitlhaj A, El Ouahrani A, El Ouahidi A, Achtak H (2019) Terraced agroforestry systems in West Anti-Atlas (Morocco): incidence of climate change and prospects for sustainable development. In: Castro P, Azul AM, Filho WL, Azeiteiro UM (eds) Climate change-resilient agriculture and agroforestry. Springer, Cham, Switzerland, pp 1–19

Climate Change Adaptation and Resilience Initiatives

Enhancing the Resilience of Oasis Agrosystems to Climate Change in Morocco

Jannate Chehbouni, Latifa Elhadioui, Nour-Eddine Benaoda Tlemçani, Latifa Daadaoui, Salma Daoud, and Cherif Harrouni

Abstract Oases worldwide are unique agrosystems adapted to adverse climatic conditions. Morocco is known for having several oases constituting breathtaking landscapes, especially in the southern areas of the country. However, in recent years several oases have lost their attractiveness because of social factors. These agrosystems that were once productive under extreme climatic conditions are being neglected leading to an unpredictable future. If there are no attempts to restore them they will revert back to their initial desert state. This paper focuses on the importance of these agrosystems as among the best adapted to harsh climatic conditions, and therefore their conservation as sound and as sustainable as possible against climate change. Moroccan oases have developed water management practices, crops and animal husbandry that are adapted to local conditions and sociocultural behaviour. Water mobilization, distribution, and utilisation as well as traditionally cultivated crops are investigated in relation to their importance to local livelihood. Highlighting oasis assets and introducing better management practices will certainly contribute to their appreciation by the young generation and enhance the resilience of these agrosystems to climate change. It is well established that marginal lands are the ones that will be most affected by climate change.

1 Introduction

The Moroccan oases are located south of the Atlas Mountains (High Atlas and Anti-Atlas) along a line extending from West to South-East along the border with Algeria

J. Chehbouni · N.-E. B. Tlemçani · C. Harrouni (✉)
Hassan 2nd Institute of Agronomy and Veterinary Medicine, Rabat, Morocco
e-mail: c.harrouni@iav.ac.ma

L. Daadaoui
Mohamed 5th University, Rabat, Morocco

L. Elhadioui · S. Daoud
Faculty of Science, Ibn Zohr University, Agadir, Morocco

W. Leal Filho and E. Manolas (eds.), *Climate Change in the Mediterranean and Middle Eastern Region*, Climate Change Management,
https://doi.org/10.1007/978-3-030-78566-6_11

(Fig. 1). They benefit from a great variety of landscapes, plants and animals, land management techniques, customs, and social organizations. Moroccan oases are also known for their variety of water management techniques: harnessing, mobilization, and distribution. The reason for this diversification is due to their strategic position as gates to the Great Sahara or as old pre-Saharan roads and crossroads. A rich heritage resource (Senil and Landel 2016), that makes them an important asset for Morocco. Their topographic location is also an important asset as their morphologies change accordingly. These are oases of piedmont, flood plains and valleys but also mountain oases. It is important to point out that in the case of the latter, date palm trees may not exist if the climate is too cold (low winter temperatures). Moroccan oases represent

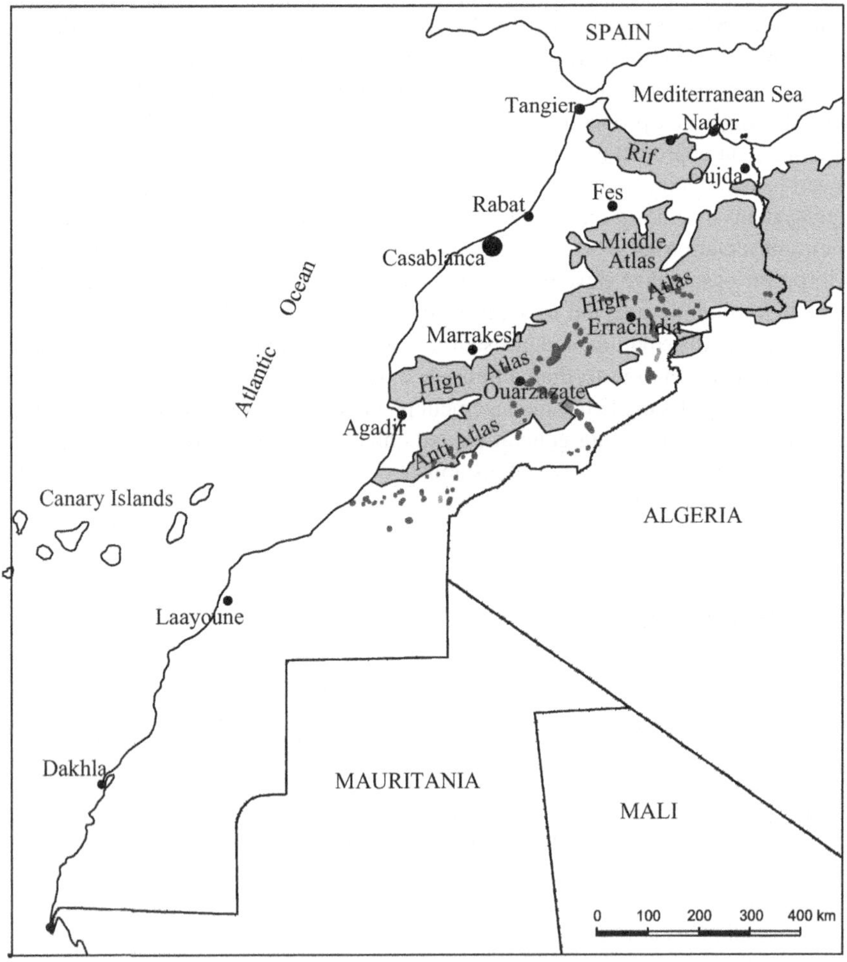

Fig. 1 Location of Moroccan oases (green and red areas)

1.5% of the total agricultural area, including 44 000 ha of date palm groves, 6.7% of the rural population and 8.9% of organized production farms (Anonymous 2007).

Oases are examples of man-made, productive agro-systems adapted to the constraints of extreme conditions: high temperatures, water scarcity and salinity, cultivable soil rarity and quality. They represent complex agro-ecosystems that have demonstrated their efficiency for centuries as among the most resilient productive systems (Cheneval and Burger 2015; Fassi 2017). Indeed, they enabled civilisations to rise, develop and thrive in spite of adverse climatic conditions. Oases are located in Saharan climates characterized by very low rainfalls and extreme temperatures. In these conditions agricultural production was made possible thanks to water mobilization and distribution and arable land management was made possible thanks to the special organization and social and cultural assets of the local populations. Understanding such community managed adaptations in land use, animal husbandry, cropping patterns, collective action over time is an interesting research topic as it strongly determines the resilience of oasis (Alemayehu et al. 2009; Al-Rawahi et al. 2010).

Unfortunately these agro-ecosystems which have proven to be resilient to different climatic and ecological limitations are being either neglected by local populations or exhausted by overexploitation of the limited resources thus threatening their existence. The threat is even greater because of water scarcity, desertification, unauthorized urbanization, and date palm susceptibility to *Bayoud*, a fungal disease caused by *Fusarium oxysporum* f. sp. *Albedinis*. All these may be aggravated by climate change as the case for oases around the world (Liu et al. 2018). Moreover, lack of field maintenance on the one hand and modern agriculture on the other, are also threats to the survival of oasis agrosystems (King and Thomas 2014; Wurl et al. 2018; Mokadem et al. 2015; Guo et al. 2019; Imaz-Lamadrid et al. 2020). It is therefore necessary to urgently take measures for the restoration of the Moroccan oases otherwise they will get back to their initial desert state.

2 Water Mobilization and Distribution in Oases

Water is a vital source of inestimable value especially in the intra-desertic regions where it is almost absent. Oasis civilizations have been able to identify water resources necessary for their subsistence and the practice of successful crop production. Once water was identified it was channelled from its source to where soil was available or made available by importation from alluvial sites or other fields. Water mobilization was achieved through different practices. Sometimes it was diverted from a perennial river upstream of the oasis, captured from a spring within or close to the oasis or drawn from a well using a pendulum system powered by animal or human force. However, the most challenging method of water mobilization is the horizontal system for groundwater abstraction consisting of a network of underground channels and vertical wells (Zella and Smadhi 2006; Antequera Fernández et al. 2014). This impressive network is named differently depending on the regions: *Falaj* in the Arabian Peninsula, *Foggara* in Algeria, *Galería* in Andalucía, *Khettara*

(Arabic) and *Tasfelt* (Berber) in Morocco and *Qanat* in Iran. These systems consisted of identifying a reliable source of underground water which was channelled towards the fields to be irrigated. To achieve this goal, wells were dug at a more or less regular spacing with the aim of aerating the underground galleries and retrieving the excavated earth during construction and maintenance. Such structures were built thanks to a strong sense of belonging to and solidarity with a community that kept its members particularly close-knit. This community feeling was most probably reinforced by religious convictions. Some of these kinds of systems date back to the Iron Age, over 3000 years ago (Al Tikriti 2011). However, many of them were disseminated and developed during the early Islamic era and in Islamic countries. Later on, they were transferred as far as Central and Southern America (Barnes and Fleming 1991; Hamidiana et al. 2015) and Japan (Al-Ghafri et al. 2003). The design of oasis hydraulic systems would be constituted of diversion dams, underground drainage galleries or wells with or without regulation storage reservoirs.

Once water was mobilized as explained above, it was carefully distributed among the entitled persons—that is the members of the community who contributed to the construction of the network who become holders of water rights. Given the sacred value of water, its management in oases was achieved through strict regulations accepted by the local population and applied by a trusted person designated for this purpose whose role was to ensure respect for the rules. This person was named differently depending on the location: *Arrif* (Arabic) in the United Arab Emirates and Oman (Al-Ghafri et al. 2003), *Kial l'ma* (Arabic) in Algeria (Zella and Smadhi 2006), *Alim* (Arabic) *Amghar n'Waman* (Skouran 2006) or *Amazzal* (Berber) in Morocco (Ouhajjou 1996).

Water distribution among rights holders was mainly based on their contribution to the construction of the structure and its control was governed by a general consensus but may also have been an expression of a balance of power within the population (Zella and Smadhi 2006; Al-Ghafri et al. 2003). However original water distribution was decided and rights holders had water shares expressed in terms of definite volumes (if water was stored in reservoirs) or periods of time during which they could use the valuable resource and manage as they thought suitable for the irrigation of their cultivation plots. Before the use of clocks, time periods were generally measured using a copper bowl (called *Tanast*) with a small hole in the bottom placed at the surface of a bucket full of water. This bowl acted like a sandglass as water filled it progressively until it sank indicating one share of water (Janty 2013). Water shares were measured by units taking different names depending on the location. In Figuig, the unit of water measurement is called *Kharruba* corresponding to a forty-five-minute share (Madani 2006); in Fam Al Hisn, it is called *Tanast* and corresponds to ten minutes (distributed in pairs). Rights holders receive their share of water over a given period of time: two to three weeks generally but it can be more if water is not abundant. Therefore each rights holder has to wait for their turn to benefit from the valuable resource. This is why we can see oases shrink or expand depending on the availability of water. Plenty of water means higher flow rate and shorter time for the irrigation of a given plot and hence the possibility to provide water to more land. On the other hand, less water means slower flow rate and longer time to irrigate. In this case farmers give priority to plots that are vital for their subsistence leading to the shrinking of the cultivated land.

The right to water is becoming more and more complex due to inheritance fragmentation and the ever-increasing water demand of farmers to irrigate their plots and meet the urgent requirements to maintain their activities.

The water shares were considered as private property and were therefore freely managed by rights holders who could rent, lend or sell some or all of their shares to obtain an immediate cash flow. This was among the reasons for social problems to start appearing as when water shares were sold, the new owners had more water and the previous ones had less water and eventually had no water at all. Land without water loses its value and land owners may end up selling it at low prices. This process brought poverty to certain families among the oasis community.

It is important to note that given the continuous water flow in the oases, it had to be used round the clock. Therefore the water share of a right holder may be due at any time of the day or night, depending on their turn for its use. This was among the constraints of the laborious work in the oases. However, water distribution in shares only applied to irrigation since water for human and animal consumption was considered as an inalienable right for all members of the community. The water is managed by a collective organization which ensures the cohesion within the oasis population. However, the separation of water ownership from land ownership makes governance quite complicated.

In any case, maintenance of this system required considerable human effort and a great mastery of water management. The practices adopted in oases involving water harnessing, mobilization, distribution and specific rules for its community sharing are considered as an intangible heritage. These practices came from experiences inherited from oasis populations' ancestors over several thousand years as witnessed by the thousands of kilometres of *Khettaras* that criss-cross Morocco and other oases all over the world (Charoy and Torrent 1990).

It is therefore well established that water is essential to the existence of oases agro-ecosystems. But the contribution of humans is also essential to the sustainability of such productive systems as they are man-made. Water scarcity due to low and irregular rainfall, combined with recurrent drought periods may be exacerbated by climate change in a context of ever-increasing demand.

3 Land Management and Agricultural Production in Oases

The agricultural production system of oases is complex but has enabled the locals to be self-sufficient in terms of food. It consists of irrigated micro-farms deploying a combination of crops with three layers: date palms as the top level surrounding the plot and providing shelter to the other crops, fruit trees in the second level and at the ground level, cereals, vegetable crops, and forage plants are cultivated. This farming system is complemented by adapted livestock (cattle, sheep, goats and camels). In the context of the oasis, women play an important role in preserving and improving agriculture and animal husbandry. According to Guerradi et al. (2004), the role of women is not limited only in the preservation of the varietal biodiversity of palm trees but also in field work, processing of palm grove products and by-products, breeding and decision-making in all that concerns the management of animal and plant diversity within domestic agriculture. Nowadays, the presence of women in the fields is becoming scarcer since agricultural work has become devalued. Most women, especially the young, are now either going to school with the aim of getting a more rewarding job or working in other sectors demanding less labour.

Date palm varieties such as *Mejhoul, Boufegous, Bouzekri, Jihel, Bouittob* in Moroccan oases are recognized for their quality. Barathon et al. (2005) mentioned that Morocco is ranked 3rd in the world for the quality of dates produced. Derouich (2013) stated that most dates sold in Europe come from Draa, Tafilalet and Bani, thanks to their good reputation. Unfortunately, date palm groves production and surface area are declining. Haddouch (1993) attributed the loss of 500,000 palm trees in Morocco to drought and Messar (1993) found that hundreds of hectares are abandoned in Tata province due to water scarcity and desertification. In terms of plant health, Mahé (2006) warned about the high infection of Guelmim-Assa palm grove with *Bayoud* disease. In Fam Al Hisn oasis, the population attributes the loss of the third of their palm trees-about 50,000 trees-to *Bayoud* disease. More recently, some oases have been subject to the practice of a new crop: the intensive cultivation of water melons and other water demanding cucurbits. These practices have been introduced by investors coming from other parts of the country in search of available groundwater and cheap labour. They generally cultivate land in the outskirts of oases using modern agricultural techniques without any consideration to sustainability and these crops will eventually lead to the depletion of groundwater.

Nowadays agriculture in these unique places is declining due to palm trees aging, *Bayoud* disease, increasing water scarcity, fragility of soils, fragmentation of farms and water rights (inheritance issue), water and wind erosion and the degradation of natural resources. As a result of agricultural deterioration, poverty among oasis populations has increased, there has been a loss of local pastoral and plant genetic resources and there has been a massive exodus of young members of the population to urban areas.

4 Moroccan Oases and Climate Change

In the context of climate change, it is important to examine how the performance of oases may be impacted in terms of productivity and ability to cope with harsher conditions. In the past these agro-ecosystems have shown their soundness as sustainable production systems under adverse climatic conditions. With the aim of analysing the variations of temperature and rainfall over time, three oases representing 3 distinct locations are studied in terms of climatic data between 1964 and 2018. We have chosen the oases of Erfoud from south-eastern Morocco; Zagora that represents a southern location, and Fam Al Hisn located in the south-west (Fig. 2). Climate data

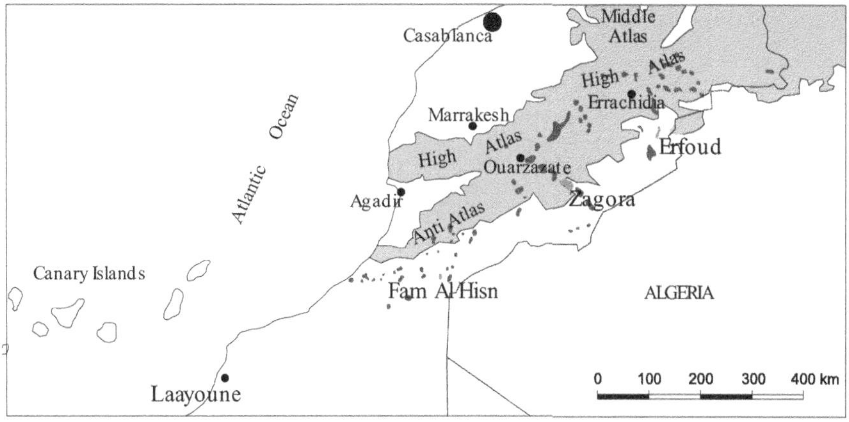

Fig. 2 Location of Moroccan oases chosen for climate change study (Erfoud, Zagora and Fam Al Hisn)

for the three oases were obtained from TerraClimate website (https://climate.northwestknowledge.net).

4.1 Climatic Characteristics of Moroccan Oases

4.1.1 Rainfall

Morocco is a country with a Mediterranean climate where rainfall occurs during the period from September to April (cool season). The hotter months receive little or no rain at all and when it occurs it is in the form of violent storms that bring damage to land, crops and houses. That is also the case for oases located south of the Atlas Mountains where rainfall is very low, irregular in time and space and occurs during the cool season. The dry season is longer than elsewhere and can extend to more than 8 months. Average annual rainfall for each location calculated over a period of 55 years is shown in Table 1.

Table 1 shows that rainfall is low in all locations with important variations. Maximum values were recorded for Erfoud and Zagora in 1965 (170 and 157 mm respectively) and for Fam Al Hisn in 1967 (230 mm). The distribution of rain through the year is also irregular. Table 2 gives the average monthly rainfall as calculated over the 55 years period. It shows that rain essentially falls from October to February. The

Table 1 Average annual rainfall in the oases of Erfoud, Zagora and Fam Al Hisn (1964–2018)

Oasis	Erfoud	Zagora	Fam Al Hisn
Annual Rainfall (mm)	113.8 ± 32	60.9 ± 37	114.8 ± 51.3

Table 2 Average monthly rainfall (mm) in the oases of Erfoud, Zagora and Fam Al Hisn (1964–2018)

	Jan	Feb	March	April	May	June	July	Aug	Sept	Oct	Nov	Dec
Erfoud	11.29 ± 9.89	14.16 ± 10.36	9.85 ± 6.7	8.51 ± 7.42	6.52 ± 5.43	3.71 ± 2.96	2.47 ± 2.31	4.07 ± 3.32	7.07 ± 4.88	14.93 ± 12.06	16.41 ± 11.89	14.78 ± 10.85
Zagora	5.02 ± 13.2	6.23 ± 11	3.64 ± 7.43	6.11 ± 14	2.07 ± 5.55	1.2 ± 3.83	0.47 ± 1.13	3.72 ± 6.32	4.57 ± 11.2	11.9 ± 21.1	7.26 ± 10.3	8.75 ± 14.4
Fam Al Hisn	12.5 ± 12.33	13 ± 13.65	12.7 ± 13.87	8.4 ± 11.69	2.7 ± 3.18	0.7 ± 0.95	0.8 ± 1.59	1.1 ± 1.87	8.1 ± 9.41	11.2 ± 10.76	21.9 ± 24.56	21.7 ± 19.99

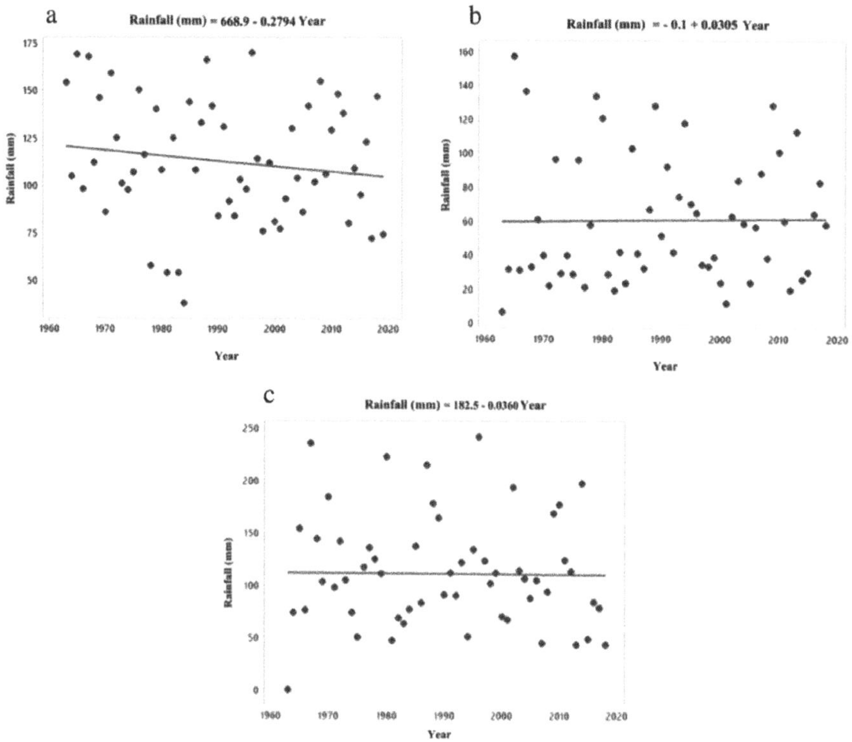

Fig. 3 Regression curves for rainfall trend between 1964 and 2018 (**a**: Erfoud; **b**: Zagora; **c**: Fam Al Hisn)

dry period lasts from May to August but the amount of rain is always low and any month can be completely dry. Therefore no crop can be grown without irrigation.

In order to determine the trend of rainfall between 1964 and 2018, regression curves were drawn for the annual precipitation using Minitab software (Fig. 3).

Figure 3 shows that rainfall did not seem to change much over time in Zagora and Fam Al Hisn but in the oasis of Erfoud the regression curve shows a decrease of 13% between 1964 and 2018. However, given the scattered data, the variation is high and the coefficient of regression r^2 is 2.1%.

4.1.2 Temperature

Regression curves for temperatures show different patterns (Figs. 4, 5, 6, 7 and 8). Mean annual temperatures increased by over 2 °C between 1964 and 2018 in Erfoud (Fig. 4). In Fam Al Hisn the increase in temperature was a little smaller as it did not exceed 0.8 °C. Quite surprisingly, mean annual temperature decreased in Zagora by 0.7 °C.

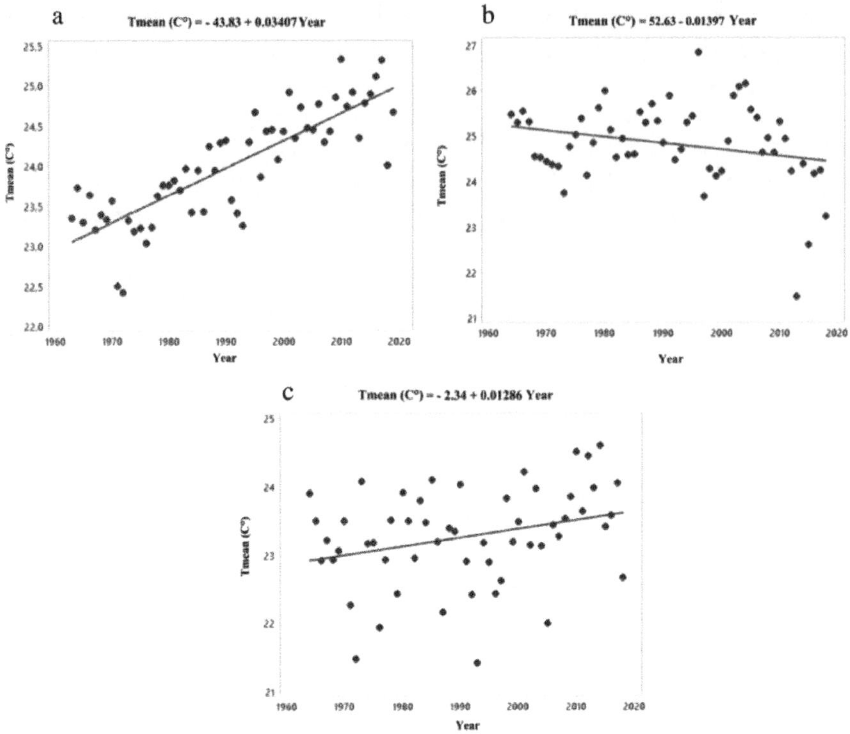

Fig. 4 Regression curves for the trend of mean annual temperature between 1964 and 2018. (**a**: Erfoud; **b**: Zagora; **c**: Fam Al Hisn)

As far as maximum temperatures are concerned, Zagora had the highest record with 46.8 °C followed by Erfoud and Fam Al Hisn (44.5 and 42.5 °C respectively) for the month of July being the hottest. Regarding minimum temperatures of the coldest month (January), Zagora also had the lowest record with −0.2 °C followed by Erfoud and Fam Al Hisn with 1.7 and 3.3 °C respectively. Regression curves of data for the hottest and the coldest months show that in Erfoud and Fam Al Hisn the trend is toward an increase of maximum temperatures of the coldest month (January) by 0.65 and 0.5 °C respectively (Fig. 5).

Regarding minimum temperatures in January, they did not change much in Erfoud, decreased in Zagora by 2 °C and increased by 1.3 °C in Fam Al Hisn (Fig. 6).

Average maximum temperatures of the hottest month (July) increased by 2.25 °C over the period in Erfoud (Fig. 7) and by 2 °C in Zagora whereas in Fam Al Hisn it decreased by 5.5 °C. Average minimum temperatures of the hottest month increased by 1.6 °C in Erfoud, 1.45 °C in Zagora and 0.3 °C in Fam Al Hisn between 1964 and 2018.

It can be concluded that climate change is affecting Moroccan oases in 2 ways: reduction of annual rainfall combined with an increase in temperature or increase of

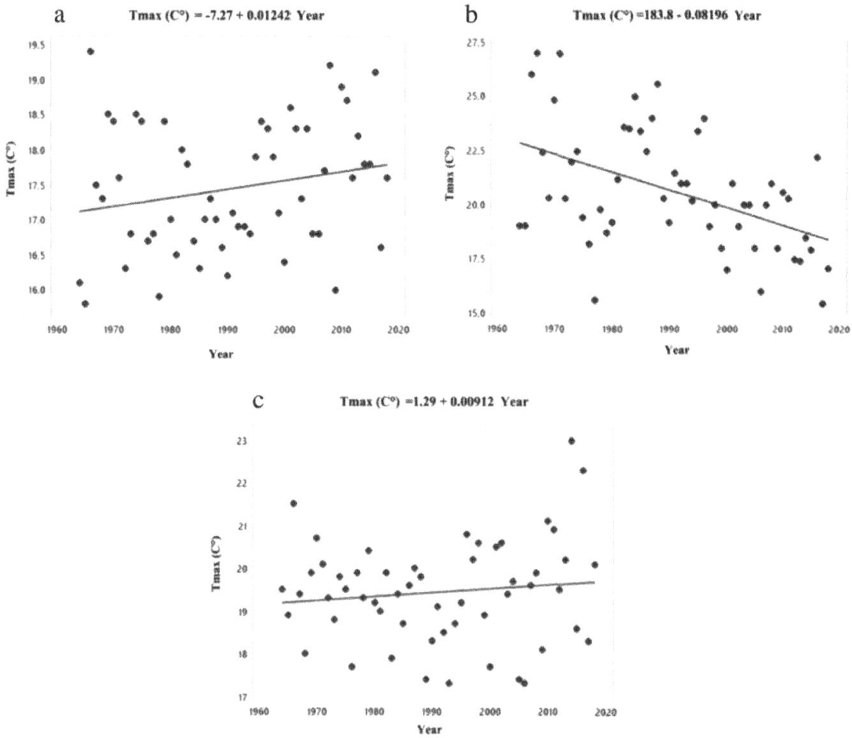

Fig. 5 Regression curves for mean maximum temperature of the coldest month (Jan.) between 1964 and 2018 (**a**: Erfoud; **b**: Zagora; **c**: Fam Al Hisn)

temperature with no effect on rainfall. Given existing low rainfall and high temperature conditions in oases such climatic changes mean it is vital to be creative in order to maintain populations in such environments.

5 Challenges Facing Moroccan Oases

Climate change is definitely affecting the local climate through the decrease of the already low rainfall as in the case of Erfoud and the increase of temperatures in most oases. Further reduction of rainfall and higher temperatures (higher evaporation rates) mean more difficulties for crop plant growth and the ability of animals to cope with the climate. Water supply is declining almost everywhere and especially in desert regions where it is less available in oases. In addition, underground channel systems (*Khettaras*) are drying up because of the decline in the water table, but mainly because of the lack of continuous maintenance required for the network to keep functioning. As explained above, building and maintaining these systems require strict discipline

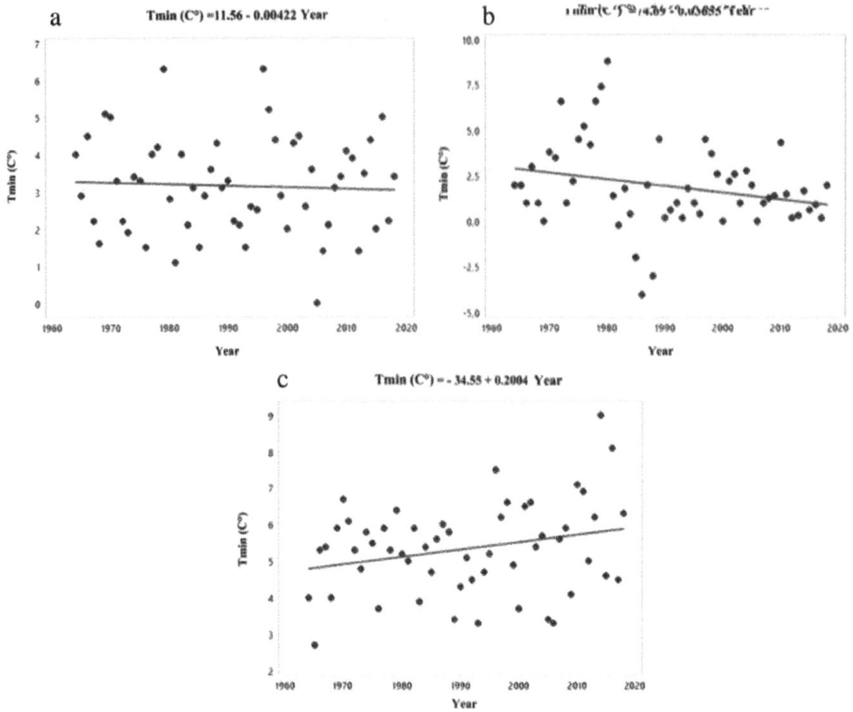

Fig. 6 Regression curves for mean minimum temperature of the coldest month (Jan.) between 1964 and 2018 (**a**: Erfoud; **b**: Zagora; **c**: Fam Al Hisn)

and a high sense of community ties. In the context of less and less bound societies, the maintenance of underground channels or any collective building is not considered a priority. The abandonment of such infrastructure is leading to a lower productivity and sometimes a reduction in the size of oases or even their total disappearance.

However, climate change is not the only factor contributing to oasis decline. Socioeconomic factors are also involved. Crops produced in oases have lost their economic value and do not provide enough income to local farmers. Some crops are facing competition with those produced in highly productive areas that are often sold cheaper. Animal husbandry is an important component in the oasis agro-ecosystem but keeping animals requires hard work and continuous care.

From a social point of view, this particular agro-ecosystem is losing attractiveness to the young population because of the evolution of society. Surveys conducted in oases reveal that young people (men and women) think that emancipation means neither having to work hard nor living under climatic constraints. They also think that going to school and getting a degree will exempt them from doing hard manual work. Generally, young people from marginal areas aspire to leave their villages and go to big cities. If this trend does not reverse, the oasis agro-ecosystem could degrade and the land would turn back to its desert origin.

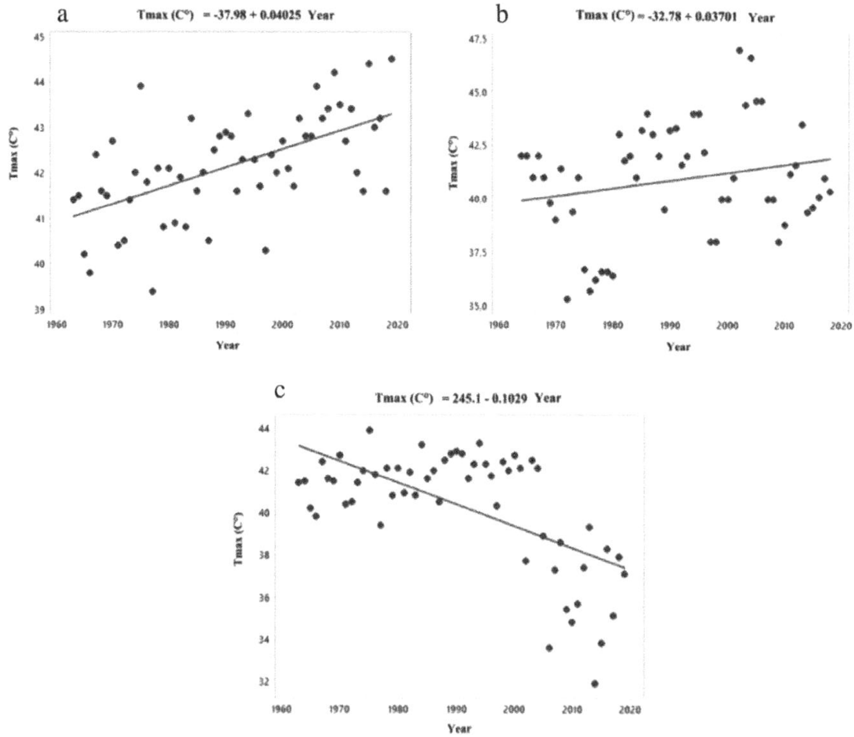

Fig. 7 Regression curves for mean maximum temperature of the hottest month (Jul.) between 1964 and 2018 (**a**: Erfoud; **b**: Zagora; **c**: Fam Al Hisn)

During the last decade investors attracted by the mildness of the climate have started crop production projects within oases or in their vicinity. Such projects require large amounts of water pumped from the underground resulting in overexploitation of the aquifer. Table 3 shows the importance of melon and water melon production in the region of Zagora. These crops represented up to 23% of the irrigated area depending on the agricultural season. Such crops are not suitable for the oasis ecosystem as they require huge amounts of water that cannot be compensated by rainfall but rather mined from the aquifer making them totally unsustainable.

Moroccan oases are subject to threats due to different factors. In order to face these threats, it is necessary to devise appropriate management measures to maintain the resilience of these man-made, centuries-old agro-ecosystems. For that it is essential to mobilize young people and interest them both economically and ecologically. They are more likely to be receptive to new technologies and new practices.

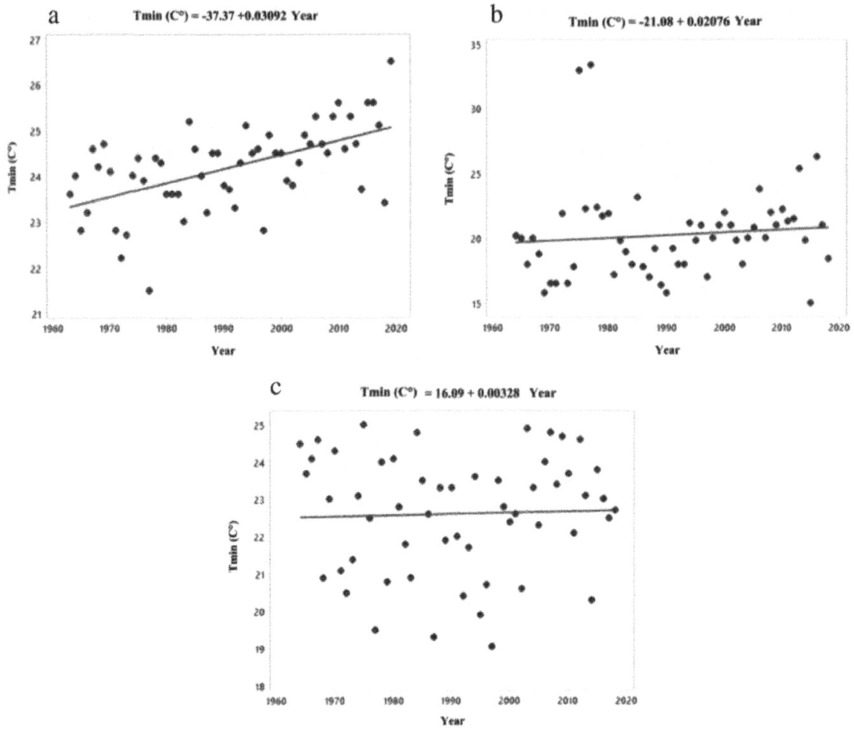

Fig. 8 Regression curves for mean minimum temperature of the hottest month (Jul.) between 1964 and 2018 (**a**: Erfoud; **b**: Zagora; **c**: Fam Al Hisn)

Table 3 Area dedicated to melon and water melon production in Zagora from 2014 to 2019

Agricultural season	2014–2015	2015–2016	2016–2017	2017–2018	2018–2019
Melon (ha)	–	400	760	310	400
Water melon (ha)	1515	2800	3200	1740	2250
Total irrigated area (ha)	21,605	19,500	17,050	14,650	15,800
Percentage	7	16.41	23.22	13.99	16.77

Data communicated by the Regional Office of Agricultural Development (ORMVA), Ouarzazate

6 Enhancing the Resilience of Moroccan Oases

Oasis agro-ecosystems have shown their soundness and sustainability through centuries if not millennia. Each oasis has specific assets. It is important to investigate how it has developed and thrived in a completely dry climate under adverse conditions. Understanding inner characteristics of the oasis and the management practices underlying its functioning may be of great importance for the maintenance of these resilient agro-productive systems and make them capable of coping with climate change. In the actual world where digital technology is overtaking all other practices and where globalization may give the impression that local actions may not be of any relevance, it is still important to draw lessons from sound productive systems, especially if these systems ensure organic crop production and waste recycling while making the best of the available water. These achievements have been possible only because of peculiar cultural assets specific to oasis inhabitants.

Encyclopaedia Britannica defines *ecological resilience* as "the ability of an ecosystem to maintain key patterns of nutrient cycling and biomass production after being subjected to damage caused by any disturbance. The term *resilience* is […] used interchangeably with *robustness* to describe the ability of a system to continue functioning under stresses or pressures". The UN Habitat stated that *human settlement resilience* is "the ability of any urban system to maintain continuity through all shocks and stresses while positively adapting and transforming towards sustainability".

Resilience can be maintained and even improved in oases through several actions that can be summarized as follows.

6.1 Better Water Mobilization

Water is the most limiting factor to life on Earth. Oases have developed thanks to water mobilization and channelling towards cropping plots. Water can be better mobilized by the maintenance of the water network and the restoration of collapsed galleries and their consolidation. It is highly recommended to ban any unsustainable crops that require water mining from the aquifer (cf. cucurbits production). Actions should be planned to enable aquifer recharge with better infiltration of runoff and improve water supply. Underground channels may also be expanded or deepened to reach more important sources of water. Technology nowadays can be very helpful in this respect. In 2008, an infiltration dam was built upstream the oasis of Fam Al Hisn to enable water to percolate and feed the aquifer downstream. As a result, water flow in the underground gallery (*Khettara*) has been substantially improved. Moreover, making impervious irrigation distribution canals saves water and increases its velocity.

6.2 Better Water Distribution and Use

Water distribution in oases is very complicated and requires a good knowledge of water shares and the transactions to which they have been subject. In most oases traditional water rights represent a hindrance to development since a substantial proportion of the population may not have any water shares and thus no access to water for irrigation while others may have water shares that they do not use for irrigation for one reason or another. As a vital source of life, water should be inalienable so that it benefits all members of the community, but its distribution should be proportional to the appropriateness of its use as a means of production. An authority emanating from representatives of local communities should be appointed for this purpose and vested with the required powers but subject to the control of water users for good governance.

Water must be optimally used. Therefore, water requirements of crops should be well known so that they are met without any waste. This way more land can be planted with the same amount of water. Crop water requirements should be assessed under the prevailing climatic conditions of each oasis. A peculiar aspect related to water quality must be addressed, namely salinity. In most oases the available water is saline. Nevertheless, crops grown in this context have adapted to salinity. Elhadioui (2017) collected several seeds of local crops in Zagora oasis and undertook germination experiments. She found that all crops were able to germinate and grow normally under salinity up to 8 dS/m.

6.3 Crops with High Value Added

It is important that the crops produced in the oasis ecosystem have a high value added so that they are not in competition with crops cultivated with conventional practices. This is why the organic and local produce characters should be emphasized. Dietary properties of peculiar oases crops should also be indicated in order to improve their marketing. Date palms are represented by numerous varieties. The best ones should be prioritised as they are more likely to improve the income of farmers. In combination with the best crops, fruit and vegetable processing and packing should be included in the chain of value. It is also urgent to identify crops specific to oases and establish gene banks with the aim of conserving local crops and maintaining this man-made biodiversity.

Other crops with high value added must be proposed to local farmers if they are more likely to improve water profitability and income.

6.4 Better Land Management

The soil is an important component in crop production, especially in oases. In such environments land ownership is regarded as a sign of wealth. Nevertheless, due to inheritance distribution, the basically small area of land becomes more and more fragmented ending up with plots too small to allow any substantial crop production. Consolidation is therefore an important aspect of land management for resilience maintenance. Soil amendment is a traditional practice in oases but it must be improved based on scientific evidence. Organic waste composting must become a regular practice. In order to stimulate better land management, it is necessary to encourage the invention of new organizational models adapted to the present situations ensuring the educated youth play a leading role for the promotion of innovative practice.

6.5 Better Animal Production

Animal breeding is a complementary activity in oases. Animals consume biomass and any leftovers from human consumption. Moreover, they represent a source of cash flow for farmers. The correct choice of animals adopted is essential as well as the way they are fed. People are more and more aware of the quality of meat and dairy products. Oasis farmers should be trained in the field of animal breeding and dairy product preparation under organic and health and safety conditions. Some local breeds of sheep are prolific and their breeding should be encouraged as they will improve the income of farmers while ensuring the recycling of organic waste.

6.6 Encouraging Ecotourism and Valorisation of Local Products

Tourists at the national and the international levels are nowadays interested in discovering other cultures. They also would like to experience ways of life different from those existing in big towns and cities. Ecotourism (ecological, agricultural and cultural tourism) is gaining in attractiveness. The oasis agro-ecosystem is very appealing to city dwellers that are in search for authenticity in terms of social relations and links to the land. Visitors will undoubtedly appreciate local products organically cultivated. Ecotourism can therefore be an important component of the oasis agro-ecosystem and will contribute to its resilience. Local products and their valorisation will certainly help diversifying the economic activities within oases.

6.7 Increasing Attractiveness

Oasis agro-ecosystems are appealing to agriculture and ecology specialists, but have little attraction to their own people, particularly the youth. Nowadays, oasis agro-ecosystems are in decline also because of the general exodus of people towards large towns and cities.

A societal approach of the oases agro-ecosystem is essential to understand the type of relationships that were at the basis of a close-knit society. Such an approach will make it possible to identify the factors that have transformed the inhabitants of oases from a communitarian society into an egocentric one no longer interested in communal affairs.

So far most educated youth originating from oases would like to leave them for big towns or cities where they think there are more job opportunities. Efforts should therefore be deployed to improve productivity of the system and living conditions for the local population with the aim of encouraging the youth to stay in their towns and optimally and sustainably exploit the available resources. As presented by Daoud et al. (2016) Morocco has a population dominated by the youth with a high potential for capacity building and innovation.

Improving and facilitating field work using new technologies, small agricultural machinery and providing technical support to optimise yield and quality of oasis products can all be incentives to the youth to stay and work in the oases. It is also important to encourage the diaspora to purchase oasis products. The oasis diaspora also represents a potential market in large towns and cities as most of them have a nostalgia for their home towns.

6.8 Improving Innovation Capacity

In order to improve resilience of oasis ecosystems to climate change, it is necessary to enhance the innovation capacity of the local population. Climate change projections predict situations of uncertainty in the future. Local authorities and populations must develop capacities of reaction to worse climatic conditions. Networking for exchange of information and mutual learning, improving community and stakeholders' awareness about sound management and exchange of best practices represent ways towards better resilience. Efforts must be undertaken for the consolidation and capitalization of knowledge about adaptation to climate change involving the policy makers. Finally, implementing monitoring and warning systems such as an *observatory* for oases could help supervising their vulnerability and promote adapted actions.

7 Conclusion

Oasis agro-ecosystems are in the frontline of desertification. They have been resilient for centuries thanks to specific management practices based on solidarity and sound governance. Societal changes have induced labour devaluation and globalization has changed economic considerations. In the absence of any actions to make life in oases more profitable, and relatively more comfortable, people may leave them and the eco-system will eventually degrade and revert to its desert origin. Local politicians are aware of the situation but require innovative approaches to initiate economic development in their territories.

The analysis of the climate in terms of rainfall and temperatures between 1964 and 2018 showed that oases are affected by climate change as the trend reveals a reduction in rainfall and an increase in temperatures with adverse effects on human wellbeing and crop production. To counter the adverse climatic conditions, it is vital to mobilize more water resources. This can only be achieved if governmental bodies undertake relatively big projects such as the construction of storage or infiltration dams upstream, the digging of underground galleries using modern technology or the nationalization of water ownership. These actions will make more water available to crop production but its distribution and use must be optimized through the choice of value added crops and efficient techniques of production.

Among the solutions envisaged to maintain the resilience of oases, the inclusion of the educated young generation is essential. To achieve this goal, it is necessary to offer better living conditions as the youth represent the guarantors of sustainability and resilience of these agro-ecosystems to climate change with adaptation to potentially more adverse climatic conditions. Advantage should be taken from the opportunities that science, technology and market can offer. Adapted innovative management approaches should be envisaged for the maintenance of the resilience of this particular farming system.

Although our knowledge of the oasis agro-ecosystems is relatively good thanks to several studies and surveys, a holistic approach is necessary to tackle the different problems facing them in the context of climate change. The scientific community should approach politicians and local representatives and explain ideas developed in this chapter with the aim of making them implemented. This is one of the major limitations of any scientific work, especially when it deals with innovation or the change of social behaviours. Daoud et al. (2014) have presented assets of the Moroccan agriculture, discussed opportunities and insisted on involving the youth for the development of adapted models.

Different oasis agro-ecosystems exist at the edge of the Sahara from the Atlantic Ocean to central Asia. Encouraging cooperation and exchange of management practices will enable stakeholders to compare their managerial choices and methods with the aim of improving oasis resilience.

References

Alemayehu F, Taha N, Nyssen J, Girma A, Zenebe A, Behailu M, Deckers S, Poesen J (2009) The impacts of watershed management on land use and land cover dynamics in Eastern Tigray (Ethiopia). Resour Conserv Recycl 53(4):192–198

Al-Ghafri A, Inoue T, Nagasawa T (2003) Daudi Aflaj: the Qanats of Oman. In: Proceedings of the third symposium on Xinjang Uyghor (China), Chiba University, Japan, pp 29–36. Available at: https://researchgate.net/publication/259931405

Al-Rawahi MN, Brinkmann K, Schlecht E, Buerkert A (2010) Effects of changing water availability on land use in irrigated mountain oases of Al Jabal Al Akhdar, northern Oman. J Geogr Soc Berlin

Al Tikriti WY (2011) Archaeology of the Aflaj. A field study of the ancient irrigation systems of the United Arab Emirates, Department of Historic Environment, Abu Dhabi Culture and Heritage

Anonymous (2007) Prospective Maroc 2030. Prospective énergétique du Maroc: enjeux et défis. Note sur le secteur de l'énergie au Maroc, Haut Commissariat au Plan, Royaume du Maroc.

Antequera Fernández M, Iranzo García E, Hermosilla Pla J (2014) Las galerías drenantes en España: cuantificación y clasificación tipológica de los sistemas horizontales de captación de aguas subsuperficiales. In Sanchis-Ibor C, Palau-Salvador G, Mangue Alférez I, Martínez-Sanmartín LP (eds) Irrigation, society, landscape. tribute to Thomas F. Glick, Universitat Politècnica de València

Barathon JJ, El Abbassi H, Lechevalier C (2005) Les oasis de la région de Tata (Maroc): abandon de la vie oasienne traditionnelle et adaptation à la vie urbaine. In: Annales de géographie (4). Armand Colin, pp 449–461

Barnes M, Fleming D (1991) Filtration-gallery irrigation in the Spanish New World. Lat Am Antiq 2(1):48–68
Charoy J, Torrent H (1990) Origine, gestion de l'eau, évaluation des aquifères dans les Oasis. Revue options méditerranéennes, CIHEAM, pp 229–235
Cheneval J-B, Burger P (2015) Adaptation of oasis ecosystems in Maghreb to climate change: how training and on farm based dissemination of agro ecological practices can help. In: Drynet, 2015, Climate change mitigation and adaptation in dry lands
Daoud S, Lyagoubi A, Harrouni C (2014) Moroccan agriculture facing climate change: adaptation and distribution of the value added. In: Behnassi M, Shahid SA, Habib-Mintz N (eds) Science, policy and politics of modern agricultural system. Springer Science+Business Media, Dordrecht, pp 83–95
Daoud S, El Brik K, Tachbibi N, Bouqbis L, Brakez M, Harrouni C (2016) The potential use of halophytes for the development of dry marginal areas in Morocco. In Khan MA, Ozturk M, Gul B, Ahmed MZ (eds) Halophytes for food security in dry lands. Elsevier Academic Press, pp 141–156
Derouich S (2013) Les Palmeraies du pays des Id Brahim et des Aït Herbil: un espace en crise du sud ouest marocain. Thèse de doctorat, Université de Lorraine, France
Elhadioui L (2017) Effet de la salinité sur la germination de quelques cultures de la région de Zagora (Maroc). Mémoire de Master, Faculté des Sciences d'Agadir, Université Ibn Zohr, Morocco, p 65
Fassi D (2017) Les oasis du Monde, carrefour des civilisations et modèle fondamental de durabilité. In: Cahiers agricultures, vol 26(4). Juillet-Août 2017 (Open Access)
Guerradi M, Outlioua K, Hamdouni N (2004) Rôle de la femme dans la gestion de la diversité génétique du palmier dattier dans les oasis du Maghreb. Revue des régions arides, Numéro spécial, pp 869–873
Guo X, Feng Q, Si J, Wei Y, Bao T, Xi H, Li Z (2019) Indentifying the origin of groundwater for water resources sustainable management, in an arid oasis, China. Hydrol Sci J 64(10):1253–1264. https://doi.org/10.1080/02626667.2019.161908011
Haddouch M (1993) Situation actuelle et perspectives de développement du palmier dattier au Maroc. In: Ferry M, Greiner D (eds) *Le palmier dattier dans l'agriculture d'oasis des pays méditerranéens*. CIHEAM, Options Méditerranéennes: Série A. Séminaires Méditerranéens, Zaragoza (28), pp 63–79
Hamidiana A, Ghorbanib M, Abdolshahnejad M, Abdolshahnejad A (2015) Qanat, traditional eco-technology for irrigation and water management. Agric Agric Sci Procedia 4(2015):119–125
Imaz-Lamadrid MA, Wurl J, Arizpe-Covarrubias O, Romero-Vadillo E (2020) Deterioration of oases subject to climate change and coastal development: the case of Todos Santos Oasis, Baja California Sur, Mexico. Groundwater Sustain Dev 11:100401
Janty G (2013) Capacité d'adaptation des pratiques traditionnelles de gestion et de partage de l'eau dans l'oasis de Figuig (Maroc). Available at: https://www.cairn.info/revue-autrepart-2013-2-page-129.htm
King C, Thomas DSG (2014) Monitoring environmental change and degradation in the irrigated oasis of the Northern Sahara. J Arid Env 103:36–45. https://doi.org/10.1016/j.jaridenv.2013.12.00920
Liu M, Jiang Y, Xu X, Huang Q, Huo Z, Huang G (2018) Long-term groundwater dynamics affected by intense agricultural activities in oasis areas of arid inland river basins, Northwest China. Agric Water Manag 203(2018):37–52. https://doi.org/10.1016/j.agwat.2018.02.028
Madani T (2006) Le partage de l'eau dans l'oasis de Figuig (Maroc oriental). Approche historique et archéologique. Mélanges de la Casa de Velázquez. Nouvelle série, (36-2), pp 61–81
Mahé E (2006): Programme de lutte contre la désertification et lutte contre la pauvreté par la sauvegarde et la valorisation des oasis, Composante secteur de Guelmim-Assa-Tata. Proposition de projet, Agence pour la promotion et le développement économique et social des provinces du sud du royaume-PNUD, p 72

Messar EM (1993) Le secteur phœnicicole algérien: Situation et perspectives à l'horizon 2010. In: Ferry M, Greiner D (éds) *Le palmier dattier dans l'agriculture d'oasis des pays méditerranéens*. Zaragoza: CIHEAM, Options Méditerranéennes: Série A. Séminaires Méditerranéens, vol 28, pp 23–44

Mokadem N, Hamed Y, Hfaid M, Ben DH (2015) Hydrogeochemical and isotope evidence of groundwater evolution in El Guettar Oasis area, Southwest Tunisia. Carbonates Evaporites 30:417–437

Ouhajou L (1996) Espace hydraulique et société au Maroc: Cas des systèmes d'irrigation dans la vallée du Dra. Editions de la Faculté des Lettres et des Sciences Humaines, Université Ibn Zohr, Agadir

Senil N, Landel PA (2016) De la ressource territoriale à la ressource patrimoniale. EricGlon; Bernard Pecqueur. Au cœur des territoires créatifs. Proximités et ressources territoriales. Presses Universitaires de Rennes, 978-2-7535-4951-7

Skouran H (2006) La gestion des eaux dans les oasis du sud marocain: cas de l'oasis de Ferkla, PFE Maîtrise Sciences et Techniques, Option: Protection de l'Environnement, FST d'Errachidia

Wurl J, Gámez AE, Ivanova A, Imaz-Lamadrid MA, Hernández-Morales P (2018) Socio-hydrological resilience of an arid aquifer system, subject to changing climate and inadequate agricultural management: a case study from the Valley of Santo Domingo, Mexico. J Hydrol 559:486–498. https://doi.org/10.1016/j.jhydrol.2018.02.050

Zella L, Smadhi D (2006) Gestion de l'eau dans les oasis algériennes. Larhyss J (5):149–156. ISSN 1112-3680

Climate Change and Agricultural Production in Algeria

Mohammed Bouznit, Mohamed Elaguab, Mohammed Mostefa Selt, Mohammed Himrane, and Rachida Aïssaoui

Abstract The agricultural sector plays an important role in Algeria's economic growth: in 2017, it contributed to more than 12% of the country's GDP and employed directly and indirectly around 13 million people. However, climate change's impacts on both the agricultural sector and food security pose a serious threat to sustainable development. The aim of this study is to assess the effects of climate change on agricultural production in Algeria during the period 1966–2016. Findings from a simultaneous equations model show that the agricultural yield is positively affected by rainfall and negatively impacted by CO_2 emissions. The latter are positively related to temperatures, which leads over time to aggravate the warming climate phenomenon. The results further show that GDP per capita indirectly lowers agricultural yield through its effects on CO_2 emissions. Therefore, promoting renewable energy to reduce CO_2 emissions provides the best option to fight climate change and reach food security in Algeria.

1 Introduction

Climate warming is a transnational phenomenon. Developed and developing countries must work together and combine their efforts to preserve our planet from the

M. Bouznit (✉)
Laboratoire Economie et Développement, Faculté des Sciences Economiques, Commerciales et des Sciences de Gestion, Université de Bejaia, Bejaia, Algeria

M. Elaguab · M. M. Selt
Laboratoire: Méthodes Quantitatives dans les Sciences Economiques et les Sciences Managériales et ses Applications pour le Développement Durable (MQEMADD), Faculté des Sciences Economiques, Commerciales et des Sciences de Gestion, Université de Djelfa, 17000 Djelfa, Algeria

M. Himrane
University Mohammed Seddik Ben Yahia, Jijel, Algeria

R. Aïssaoui
College of Business, Ohio University, Athens, USA

W. Leal Filho and E. Manolas (eds.), *Climate Change in the Mediterranean and Middle Eastern Region*, Climate Change Management,
https://doi.org/10.1007/978-3-030-78566-6_12

disastrous consequences of climate change and to achieve sustainable development. In 2015, 196 countries met at the United Nations Climate Change Conference in Paris (COP21), and all agreed to adopt the Paris Agreement. The latter aims at fighting climate change by limiting global warming to well-below 2 °C and pursuing efforts to limit it to 1.5 °C, as well as by strengthening countries' ability and resources to deal with the impacts of climate change (EC 2020). Algeria was one of the first developing countries to submit the Intended Nationally Determined Contribution (INDC) to the United Nations Framework Convention on Climate Change (UNFCCC) (Bouznit et al. 2020).

In fact, addressing climate change mitigation, whilst ensuring sustainable development, will be the new challenge for Algeria in the coming years. In that context, Algeria adopted the National Climate Plan (NCP) in 2019 which lays out the targets and actions for climate mitigation and adaptation to be achieved during the period 2020–2030. Among these goals is the reduction by 7% of greenhouse gas emissions (GHG) by 2030. It is even argued that GHG can be reduced by 22% with the support of developed countries, notably through financial support and technology transfer (INDC-Algeria 2015). According to Sahnoune and Imessad (2017), such support is critical to ensuring the effective implementation of programs of renewable energy and energy efficiency, which, in turn, will generate several positive effects on the environment. For instance, the amount of CO_2 reduction will approximate 348 $MtCO_2$, and the emissions will reach the turning point by 2024. To reach this objective, the mitigation and adaptation measures taken by Algeria to address the negative effects of climate change focus specifically on the sectors of energy, forests, housing, transport, industry, agriculture, and waste.

Climate change mitigation measures are critical given the effects of climate change on national economies, notably for developing countries. Indeed, Rezai et al. (2018) argue that climate change undermines the economy, by, for instance, reducing profitability, investment and potential income level, and reducing productivity in the short and long runs. Similarly, Fankhauser and Tol (2005) find that climate change tends to lower productivity outputs, which then creates obstacles to capital formation, production, and consumption per capita. Dell et al. (2008) further find these negative effects are more pronounced in poor countries than in developed countries. Specifically, the authors confirm that climate warming in poor countries seriously hinders sustainable development, as it reduces agricultural and industrial production, undermines investments in physical capital, and even stirs up political instability. Moreover, change in crop productivity is strongly linked to change in rainfall, and the spatial and temporal distribution of rainfall (Lobell and Burke 2008; Kumar et al. 2004). Porter and Semenov (2005) noted that climate and weather directly affect yield and quality of food crops, and showed that increasing temperature and precipitation variability will strongly impact the crop yields. Similar results are obtained in a large number of studies examining the effects of climate change on crops in countries around the world (Walthall et al. 2012; Mendelsohn 2014; Pathak et al. 2018; Chandio et al. 2020; Chen and Gong 2021; DePaula 2020).

Yet, the agricultural sector plays an important role in Algeria's economic growth. In 2017, it contributed to more than 12% to GDP, and improved the living conditions of 13 million people—about 30% of the national population–living in rural areas through direct and indirect employment. Moreover, the agricultural sector had experienced a high annual growth rate over the period 2004–2014; it was estimated at 7.02% compared to the national economic growth rate at 2.62% (Bessaoud et al. 2019). However, Algeria is particularly vulnerable to the multiform effects of climate change, including desertification, increasing of erosion, water scarcity, and degradation of the ecological and biodiversity system. Specifically, Algeria is suffering from the irregularity of the temporal and spatial rainfall (INDC-Algeria 2015). The average rainfall has declined by more than 30% over the past decade, and the droughts are becoming more persistent, leading to dire consequences on the levels of dams and ground water, agricultural production, and food insecurity. As a result, Algeria is classified as a poor country in term of water resources, as the Word Bank fixed 1000 m^3 per capita/year as a scarcity threshold and Algeria's water per capita is estimated at 500 m^3 per year (NCP-Algeria 2019). In addition, Sahnoune et al. (2013) note the several negative effects of a small rise of temperature on the country's development. For instance, the authors explain that temperatures in Algeria will increase by 2 °C by 2050, which will further aggravate water scarcity and the desertification of agricultural lands and reduce agricultural yields.

Chabane (2012) confirms Algeria's geographic challenges, noting the correlation between agricultural yield and climate change, notably the high variability of annual precipitations. The latter, the author indicates, tends to be more pronounced in the western parts of the country, and the difference between the southern and the northern regions is paramount with annual rainfalls amounting to 50 mm/year (in Sahara) and 1500 mm/year (in Jijel) respectively (Chabane 2012).

Moreover, the Algerian NCP identifies fives effects climate change has on the agricultural sector namely, desertification, increase of erosion, decrease in productivity due to the disruption of the plant cycle, scarcity of productive factors (water and land), and degradation of food security. This chapter seeks to further evaluate the effects of climate change on Algeria's agricultural production over the period 1966–2016. To do so; we use a simultaneous equations approach, which allows identifying both the direct and indirect channels through which climate change affects agricultural production. To the best of our knowledge, there is no study using such methodological approach to analyse the direct and indirect channels through which climate change affects agricultural production in Algeria. This study is therefore important because the Algerian food system relies heavily on cereal products which are highly sensitive to precipitations. It further suggests that mitigating warming climate and adapting to negative effects of climate change are necessary steps to reach both food security and sustainability development.

This chapter is structured as follows. Section 2 gives an overview of the energy and greenhouse gas emissions in Algeria. The Algerian National Climate Plan is presented in Sect. 3. The methodology used in the study is explained in Sect. 4. Data and descriptive analysis are provided in Sect. 5. In Sect. 6, the results are presented and discussed. Finally, the main conclusions are reported in Sect. 7.

2 Energy and Greenhouse Gas in Algeria

Energy consumption, environmental degradation and climate change are highly interdependent (Bouznit and Pablo-Romero 2016; Rehmen and Rashid 2017; Munir and Riaz, 2020), which is particularly relevant for Algeria whose economy heavily depends on hydrocarbon exports. Indeed, more than 97% of Algeria's exports are crude oil and gas products. Moreover, 99% of the country's energy consumption is produced from fossil fuels, which contributes to the greenhouse effect responsible for climate change (Bouznit et al. 2020).These data place Algeria as one of the first CO_2 emitters among developing countries and the third among African countries (Sahnoune et al. 2013). With an amount of 147Mt CO_2 emissions in 2014, Algeria ranked 34th in the world in fossil fuel emissions and 6th in CO_2 emissions from gas flaring (Olivier et al. 2015). In 2017, electricity consumption was estimated at 64.6 TWh, and was expected to reach 80 TWh in 2020 and 150 TWh in 2030, i.e., a consumption of more than 54 billion cubic feet and 102 billion cubic feet of natural gas respectively (Bouznit et al. 2020; Sahnoune and Imessad 2017).

2.1 Energy

The amount of electricity generated from fossil fuels has been increasing over time and is growing at a rapid rate, tripling between 2000 and 2016, from 24 billion kWh in 2000 to 43.3 billion kWh in 2010, and finally 67 billion kWh in 2016. Table 1 uses data from the U.S. Energy Information Administration database (U.S.-EIA 2019) to illustrate the trends in annual electricity consumption per capita for the World, Algeria, Morocco, and Tunisia for the period 2000–2016. On average, annual electricity consumption per capita in Algeria, Tunisia and Morocco are smaller than the global average: worldwide, average annual electricity consumption was 2401.78 kWh for the period 2000–2010, and 2830.41 kWh between 2011 and 2016; comparatively, Algeria's annual electricity consumption per capita was only 770.36 kWh and 1222.32 kWh, for the same periods. However, the growth rate of electricity

Table 1 Electricity consumption per capita: World, Algeria, Morocco, and Tunisia (2000–2016)

	Average electricity consumption per capita (kWh)		Annual growth rate (%)	
	2000–2010	2011–2016	2000–2010	2011–2016
World	2401.78	2830.41	2.43	1.28
Algeria	770.36	1222.32	4.68	6.79
Morocco	597.12	787.41	6.43	0.09
Tunisia	1080.88	1292.49	3.47	1.80

Source U.S. Energy Information Administration (U.S-EIA 2019)

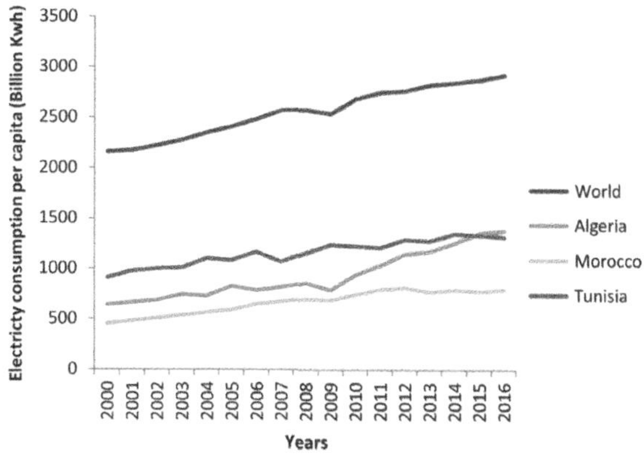

Fig. 1 Electricity consumption per capita: World, Algeria, Morocco, and Tunisia (2000–2016). *Source* U.S. Energy Information Administration (U.S-EIA 2019)

consumption per capita in Algeria was higher than that observed in the world over the two studied periods: 4.68% for Algeria and 2.43% worldwide in the first period, and 6.79% for Algeria compared to 1.28% worldwide in the second period. In addition, electricity consumption per capita in Algeria was higher than Morocco's, and it started to surpass Tunisia's in 2015 (see Fig. 1).

This substantial increase in electricity consumption for Algeria can be attributed to its growing economy and population (Belaid and Abderrahmani 2013). Moreover, in recent years, the residential sector has been using much more electricity than any other sector, representing a share of 63% of the total electricity consumption in 2014, compared to 35% for the industry sector. Bouznit and Pablo-Romero (2016) and Belaid and Abderrahmani (2013) explain that several factors, including life style changes, may be responsible for this trend. Together, electricity consumption is expected to double by 2030 (Nachmany et al. 2015).

With respect to the main sources of electricity generation, Fig. 2 shows the share of fossil fuels and renewable energies in electricity generation in the world and Algeria over the period 2000–2016. In Algeria, fossil fuels were the principal source of electricity generation (98%) throughout the period of 2000–2016. Before 2011, hydroelectricity was the only renewable source of electricity generation, representing 0.85% of the total electricity generation for the period 2000–2010, and even experiencing a downward trend to 0.45% during 2011–2016.

Since 2011, Algeria has also begun to generate electricity from solar energy (solar photovoltaic technology). However, despite Algeria's endowment in one of the highest solar potentials in the world (more than 3000 h of sunshine per year and 5 kWh of daily energy received on a horizontal surface in most of the country) (Ghezloun et al. 2015; INDC-Algeria, 2015), its share of renewable energies in electricity generation remains marginal, markedly with respect to both global trends and Algeria's goals in renewable energy development. To illustrate, photovoltaic electricity generation did not surpass 0.20 billion kWh during the period 2011–2016, which represents only 0.35% of the total electricity generation. This is in stark contrast with

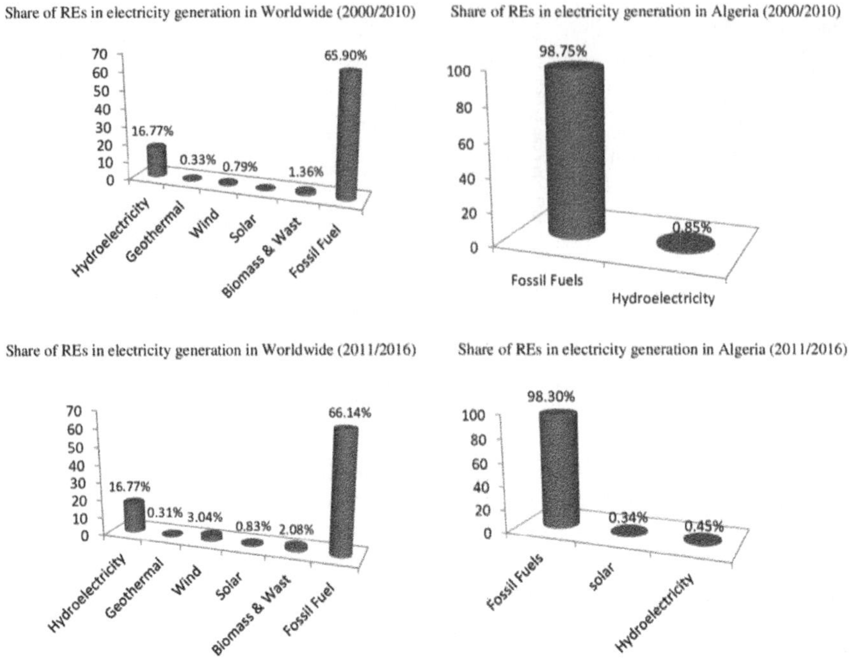

Fig. 2 Share of fossil fuels and renewable energies in electricity generation: World and Algeria, 2000–2016. *Source* U.S. Energy Information Administration (U.S-EIA 2019)

global trends whereby electricity generation from one of the five sources of renewable energy—hydroelectricity, solar, geothermal, wind and biomass and waste—has increased from 19.25% over the period 2000–2010 to 23.03% in 2011–2016. During these two periods, renewable energies have increased at an annual rate of 4.6% during the first period and 6.9% during the second period. The US-EIA (2019) suggests hydroelectricity remains the main source of renewable energy, contributing to an estimated 16.77% of total electricity generation. Meanwhile, wind power experienced a significant growth: its contribution was only 0.79% throughout the period 2000–2010, but rose to 3.04% during the period 2011–2016. Lastly, solar power, biomass & waste and geothermal represented 2.08%, 0.83% and 0.31% respectively of the electricity produced in 2011–2016.

2.2 *Greenhouse Gas Emissions*

The World Development Indicators suggest that Algeria's per capita CO_2 emissions have experienced a steep upward trend over the period 1966–2016, with an average annual growth rate of 9.32%. Comparatively, the rates for Morocco, Tunisia and the MENA region were estimated at 7.31%, 6.51% and 5.06%, respectively.

The upward trend of CO_2 emissions per capita in Algeria can be divided into three periods. During the first period from 1966 to 1978, the amount of CO_2 per capita emitted accelerated extremely rapidly, with the annual growth rate reaching its top level at 35.72%. Bouznit and Pablo-Romero (2016) explain this trend by the four government development plans implemented over the period 1967–1977. At the time, a substantial increase in revenues from oil exports allowed the Algerian government to fund its development strategy, which consisted primarily of boosting the accumulation of gross fixed capital formation in industries such as the heavy industry and the hydrocarbon sector. For instance, between 1973 and 1977, the equivalent of 61% of the country's GDP was invested in the industry sector, compared to 34% in the heavy infrastructure and 5% in the agricultural sector. As a result, gross fixed capital formation in the industry sector experienced incredible growth in a short period, from 7.56 billion Algerian dinars (DZD) in 1967 to 44.4 billion DZD in 1977, or an annual growth rate of 48.73% (Palloix 1980). In this context, large factories were built, such as El-Hadjar Steel Complex, which directly impacted electricity consumption from fossil fuels, and thus increased GHG emissions (Bouznit and Pablo-Romero 2016; De-Bernis 1971). Bouznit and Pablo-Romero (2016) find the annual growth rate of energy consumption was 7.52% in 1970–1974 and 10.37% in 1975–1979.

During the second period from 1979 to 1982, the emission of CO_2 per capita fell drastically from 3.45 to 1.92 metric tons, i.e., a 44.34% decrease. Later, though, CO_2 emissions per capita would regain its upward trend from 1982 to 1987 with an annual growth rate of 16.85%, which could have been higher had public investment in the industry sector not decreased significantly. Indeed, Algeria revised its development strategy from 1967 to 1977 to focus on meeting its population's human development needs by 1990 (Ollivier 1985). Investment in the industry sector thus decreased sharply to 40% of GDP compared to its previous levels (60% of GDP over the period 1967–1977).

During the third period 1987–2016, GHG emissions stabilized between 3.53 and 3.69 metric tons per capita, i.e., an annual growth rate of 0.15%. Since 1988, the Algerian economy has faced a serious crisis reflected in a slow economic development, and the deterioration of its industry. Bouznit and Pablo-Romero (2016) add that the Algerian economy had been suffering, since 1982, from a modest investment in the manufacturing sector, a rapid de-industrialization and an increased reliance on imports. For instance, the share of manufacturing in Gross Value Added decreased from 17% in 1970 to less than 5% in 2010, which led to a decrease in GHG emissions as manufacturing activities are highly energy intensive, accounting globally for one-third of energy use and 40% of CO_2 emissions (Brown et al. 2012) (Fig. 3).

3 Algeria's National Climate Plan

The final version of Algeria's NCP, adopted in 2019, defined the legal framework for the government's actions and programs to fight global warming (NCP-Algeria 2019). The sectors most affected by climate change are water, sea fishing, agriculture, and

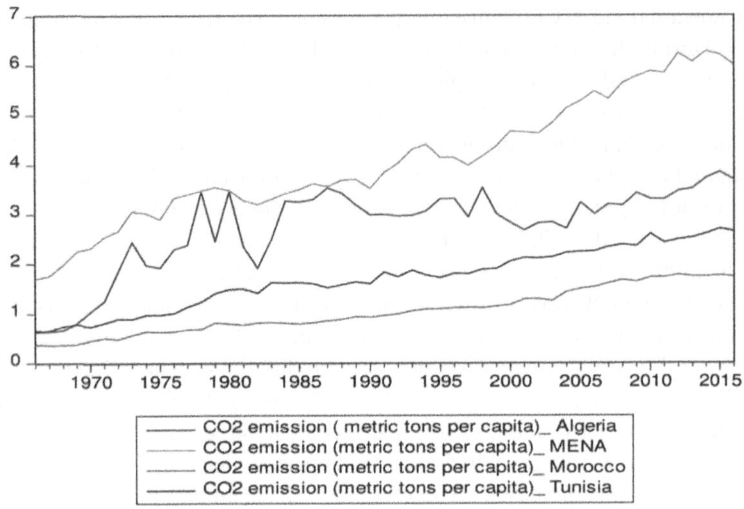

Fig. 3 Evolution of CO_2 emission per capita in Algeria, Morocco, Tunisia and the MENA region. *Source* WDI (2020)

industry. Stern (2006) estimates the annual cost of climate change by 2050 between 5 and 20% of the 2005 world GDP. However, measures to mitigate GHG are estimated at a 1% annually of the world GDP.

To meet its international commitments and sustainable development goals, Algeria's CPN defines the actions to mitigate GHG emissions and adapt the economic sectors to the negative effects of climate change. This plan set ambitious goals which it aims to achieve by mobilizing financial and human resources and combining the efforts of institutional and civil actors. The plan is to be implemented gradually during the period 2020–2030, under the supervision of the Ministry of Environment and with the support of the Algerian-Germany cooperation. In total, 155 actions have been selected, 63 adaptation actions, 76 mitigation actions and 16 transversal actions related to climate change governance, financial helps and sensitization.

3.1 Mitigation Actions

Climate change mitigation actions aim to reduce GHG emissions by decreasing significantly fossil fuels consumption. In this context, the Algerian's INDC reports the legal and institutional framework for the national policies and programs to fight global warming. The INDC will be implemented from 2021 to 2030 to reach the targets defined by the United Nations Framework Convention on Climate Change (UNFCCC). Algeria committed to spare no efforts to fight climate change and to implement the measures needed to mitigate and capture GHG emissions. These measures will affect various sectors, including energy, agriculture, forests, housing, transport, industry, and waste. Algeria set the overarching goal to reduce GHG emissions by 7% by 2030; however, this reduction could reach 22% if Algeria can garner the support from international partners in the forms of financial help and technology transfer.

A pilot version of the national program of renewable energy and energy efficiency was approved in 2011, and the final plan was revised and approved in 2015. This program seeks to address global warming and preserve fossil energy resources for future generation by developing solar energy production (Bouznit et al. 2020). Indeed, Algeria's solar potential is one of the highest in the world, with more than 3000 h of sunshine per year and 5 kWh of daily energy received on a horizontal surface in most part of the country (Ghezloun et al. 2015). This explains why Algeria has selected to intensify its investment in the mix energy photovoltaic solar energy, solar thermal energy and wind energy. Renewable energy is expected to contribute to 27% of the total electricity production by 2030 (Bouznit et al. 2020). To achieve this objective, Algeria plans to produce 22,000 MW of renewable energy by 2030; 12,000 MW will serve domestic needs and 10,000 MW will be exported (WEC 2019). The 22,000 MW will be generated from six renewable energy sources: 13,575 MW from solar photovoltaic; 5010 MW from wind; 2000 MW from solar thermal; 1000 MW from biomass; 400 MW from cogeneration; and 15 MW from geothermal (Bouznit et al. 2020; INDC-Algeria 2015). However, the national program of energy transition of 2020 defined a new objective of 16,000 MW of renewable energy from photovoltaic by 2035 (CEREFE 2020). In doing so, 300 billion cubic meters of natural gas will be saved, which represents a reduction of 348 $MtCO_2$ or the equivalent of CO_2 emissions for the year of 2014 (Sahnoune and Imessad 2017). Other measures have been designed to increase the absorption of CO_2 emissions, notably measures aimed at preserving the forest sector such as forestation, reforestation and forest and fires prevention.

3.2 Adaptation Actions

According to UNFCCC (2020), adaptation refers to adjustments of the ecological, social, or economic systems in response to actual or expected climatic stimuli and their effects. All countries have to find innovative solutions and implement suitable actions to be more resilient to the negative effects of climate change. For Algeria specifically, the sectors of water, sea fishing, agriculture and industry are all classified as the most vulnerable to climate change. In total, 64 adaptation actions have been selected, including 36 actions to be achieved in the short run and 27 actions in the medium term. Table 2 summarizes the adaptation actions related to the water and agricultural sectors.

Table 2 Adaptation actions in the sectors of water and agriculture

Sector	Short term actions (2020–2025)	Medium term actions (2020–2035)
Water	• Identification of solutions to the impacts of climate change on coastal aquifers and marine intrusion • Optimization and rehabilitation of the irrigation systems	• Valorization of non-conventional waters in arid zones • National assessment of the purified wastewater recycling • Recycling of dematerialized drainage water in agriculture: Oued Righ valley • Implementation of the project to fill a pilot aquifer using rainfall water • Assessment of the vulnerability of groundwater sources from climate change
Agriculture	• Valorization and duplication of the Oran's experience to produce organic compost as a natural fertilizer • Use and generalization of the geographic system to efficient management of agricultural droughts	• Improvement of agricultural practices to better use the rainfall waters in semi-arid zones • Introduction and valorization of the new adapted species to climate change of Algeria • Selection and genetic improvement of varieties of cereals, food legumes, fodder, vegetable corps and arboriculture • Development of marginal areas sensitive to erosion the techniques of direct sowing • Preservation of the apple orchards of Bouhmama against advanced drought and implementation of measures to guarantee a sustainable irrigation

Source National Climate Plan of Algeria (NCP-Algeria 2019), http://www.meer.gov.dz/a/?wpdmpro=plan-national-climat

4 Methodology

Due to its vulnerability to climate change, Algeria's agricultural production is expected to decrease over time, which compromises the country's food security. Although the agricultural sector in Algeria suffers from several extreme geographic features, Chabane (2012) finds the weak agricultural yield is strongly linked to physical and climatic factors, such as the high variability of annual precipitations. The latter tends to increase the more we move to the East, and the difference in annual rainfall is paramount between the southern and the northern region −50 mm/year in Sahara and 1500 mm/year in Jijel, respectively (Chabane 2012).

A simultaneous equations model was used to estimate the direct and indirect effects of climate change on cereal yields in Algeria over the period 1966–2016. First, cereal yield is a function of a set of explanatory variables, including rainfall, temperature, and CO_2 emissions. Second, temperature is also a function of the temperature for the previous year and CO_2 emissions. Third, CO_2 is a function of

GDP per capita. The functional form of the model is formulated as follows:

$$LnCY_t = C_1 + \beta_1 LnP_t + \beta_2 T_t + \beta_3 LnCO2_t + \beta_4 trend + \varepsilon_{1t} \quad (1)$$

$$LnT_t = C_2 + \beta_5 LnT_{t-1} + \beta_6 LnCO2_t + \varepsilon_{2t} \quad (2)$$

$$LnCO2_t = C_3 + \beta_7 LnGDPC_t + \varepsilon_{3t} \quad (3)$$

where CY is cereal yield measured in kg per hectare. P is the annual average in mm of rainfall during the growing season (from September to August). T is the annual average of the temperature in °C. CO_2 measures the per capita CO_2 emissions (tons metric). GDPC is the real gross domestic product per capita (constant 2010 US$). β_1, β_2, β_3, β_4, β_5, β_6 and β_7 are parameters to be estimated. ε_{1t}, ε_{2t} and ε_{3t} are white-noise error terms and *Ln* is the natural logarithm.The data are annual figures covering the period 1966–2016 and expressed in natural logarithm. Moreover, full information methods, Three-Stage Least Squares (3SLS) were used to estimate the simultaneous equations model (1–3) (Wooldridge 2010).

5 Data and Descriptive Analysis

The variables in the model are observed on the period of 1966–2016 and use data from the World Development Indicators database and the Climate Change Knowledge Portal database (CCKP). Table 3 reports the definition of selected variables and data sources; Table 4 provides descriptive statistics. Our choice of the period of 1966–2016 was motivated by two reasons: the independence of Algeria in 1962 and the availability of data.

Algeria can be divided into two main regions, the northern region located along the Mediterranean Sea, and the southern region located in the Sahara which forms a

Table 3 Definition of selected variables and data sources

Variables	Symbol	Measure	Data source
Cereal yield	CY	Cereal yield (in Kg per hectare)	WDI
Rainfall	P	Annual average of rainfall during the growing season (from September to August) (in mm)	CCKP
Temperature	T	Annual average of temperature (°C)	CCKP
CO_2	CO_2	CO_2 emissions (metric tons per capita)	WDI
Real GDP per capita	GDPC	Real gross domestic product per capita (at constant 2010 US$)	WDI

Notes WDI: World Development Indicators database
CCKP: Climate Change Knowledge Portal database

Table 4 Descriptive statistics

	LnCY	*LnP*	*LnT*	*LnCO2*	*LnGDPC*
Mean	6.76	4.39	3.13	0.93	8.17
Max	7.50	4.77	3.18	1.34	8.48
Min	6.00	3.95	3.08	-0.45	7.65
Std. Dev	0.39	0.16	0.02	0.46	0.19
Obs	51	51	51	51	51
	CY	*P*	*T*	*CO2*	*GDPC*
Mean	933.65	81.74	23.07	2.74	3624.12
Max	1813.10	118.58	24.08	3.85	4828.62
Min	403.90	52.29	21.92	0.63	2120.45
Std. Dev	379.34	13.25	0.54	0.85	672.07
Obs	51	51	51	51	51

large band of 350 km. The climate differs substantially between the two regions. It is either hyper arid, arid or semi-arid in nearly three-quarters of Algeria (Nouaceur et al. 2013). The nearest zone to the sea has a Mediterranean climate with a total rainfall of 2000 mm per year, while the climate of the Atlas-Tellien zone is arid and its nature is dry and tropical (NCP-Algeria 2019). In the southern region which includes the Sahara area with more than 2 million km^2, the climate is hot, sunny and arid. Figure 4 provides a detailed overview of the geographical distribution of average temperatures and rainfalls per month in Algeria (Fanack 2020).

Figures 5 and 6 further show the evolution of the annual average temperature and rainfall respectively in Algeria over the period 1966–2016. Temperatures have experienced an upward trend with an increase of 1.5 °C, between 1966 and 2016; and the global warming phenomenon is manifest in the acceleration of temperature increases as the same increase of 2 °C is expected to be reached by 2050. Moreover, the annual average of rainfall has experienced strong variations, but has generally tended to decrease. Nouaceur et al. (2013) reached a similar conclusion, finding that average temperatures in Algeria have changed substantially, notably from 1987 to the present time. The authors identified three periods, the wet period started from 1971 to 1986 followed by a strong drought period between 1987 and 2002. However, during the period 2003–2012, rainfalls regained previous levels, but their spatial and temporal distribution became irregular, causing floods on multiple occasions.

Stern (2006) defined five categories of climate change impacts along a scale of global warming degrees. Table 5 summarizes these categories.

Table 6 summarizes the specific effects of climate change on Algeria's water and agricultural sectors (NCP-Algeria 2019).

Furthermore, the Algerian agricultural policy focuses on developing cereal cultivation to secure food security. According to Bessaoud et al. (2019), Algeria's agricultural land is estimated at 3.5% of the total surface area, representing 8.5 million

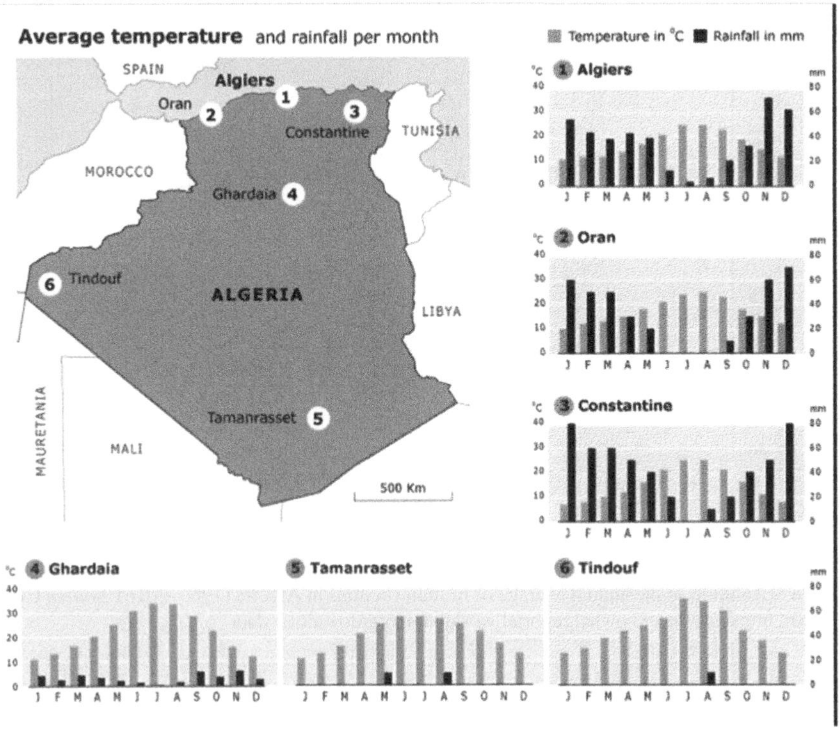

Fig. 4 Geography and Climate in Algeria. *Source* https://fanack.com/algeria/geography/

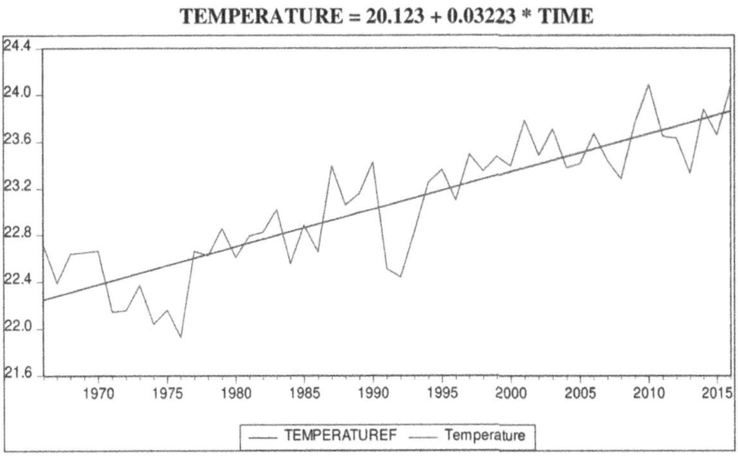

Fig. 5 Evolution of the annual average of temperature (in °C) in Algeria (1966–2016). *Source* CCKP (2020), https://climateknowledgeportal.worldbank.org/download-data

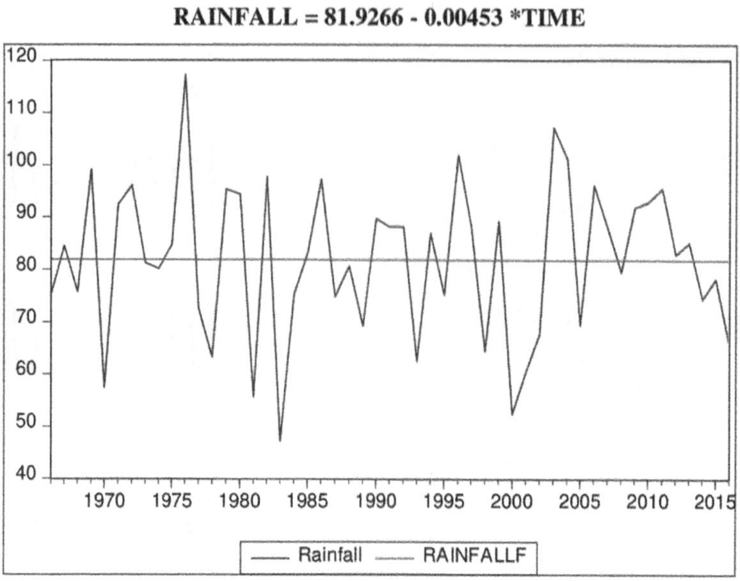

Fig. 6 Evolution of the annual average of rainfall (in mm) in Algeria (1966–2016). *Source* CCKP (2020), https://climateknowledgeportal.worldbank.org/download-data

Table 5 Impacts of climate change on food production

Temperature raise (°C)	Impacts of climate change on food production
1	Modest increases in cereal yields in temperate regions
2	Sharp declines in crop yield in tropical regions (5–10% in Africa)
3	150–550 additional millions at risk of hunger (if carbon fertilisation weak) Agricultural yields in higher latitudes likely to peak
4	Agricultural yields decline by 15–35% in Africa, and entire regions out of production (e.g.parts of Australia)
5	Continued increase in ocean acidity seriously disrupting marine ecosystems and possibly fish stocks
More than 5	Currently, very hard to capture with current models as temperatures would be so far outside human experience

Source Stern (2006)

hectares, and just only 12% is irrigated. However, the land devoted to cereal cultivation was around 35% during the period 2000–2011, but the production volume was strongly affected by the variability in precipitations. For instance, Fig. 7 shows the evolution of cereal yield (in kg per hectare) and the annual average of rainfalls (in mm) over the period 1966–2016. This figure reveals both curves follow similar trajectories where the variability in participation leads to a similar variability in cereal production, which confirms the positive relationship between the two factors. In sum,

Table 6 Main impacts of climate change on water and agriculture in Algeria

Sector	Main impacts of climate change in Algeria
Water	• Scarcity of water resources • Degradation of water quality • Hydraulic infrastructure deterioration • Intrusion of marine waters into coastal freshwater aquifers
Agriculture	• Desertification • Increase of erosion • Decrease in productivity due to disruption of plant cycle • Scarcity of productive factors, such as water and land • Degradation of food security

Source NCP-Algeria (2019), http://www.meer.gov.dz/a/?wpdmpro=plan-national-climat

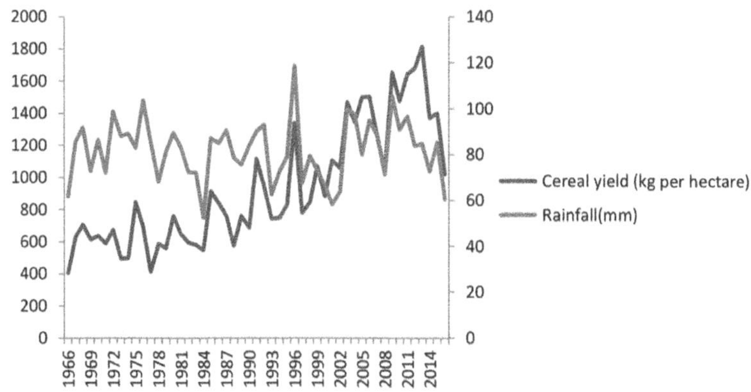

Fig. 7 Evolution of annual average rainfall and cereal yield. *Source* WDI (2020) and CCKP (2020)

climate change seriously affects food security which, in turn, hinders sustainable development.

6 Results and Discussion

Table 7 shows the results from the simultaneous equations model using 3SLS. Column (A) reports the estimates of Eq. (1) in which cereal yield is a function of rainfall, temperature, CO_2 emissions and trend. The results show that the sign of the estimated coefficient associated to precipitation variable is positive ($\beta_1 = 0.42$) and significant at 5%, which suggests a positive relationship with cereal yield variable. This means that if an annual average of rainfall during the growing season decreases, or increases, by one standard deviation, annual cereal yield would rise,

Table 7 Estimates of simultaneous equations model using 3SLS

	A	B	C
	CY	T	CO2
C	48.54 (35.73)	1.30*** (0.31)	-13.67*** (1.59)
P	0.42** (0.21)		
T	−14.13 (11.21)		
CO2	−0.76*** (0.31)	0.024*** (0.007)	
Trend	0.05*** (0.02)		
T(−1)		0.57*** (0.10)	
GDPC			1.78*** (0.19)
R^2	0.66	0.48	0.63
Obs	50	50	50

Notes Standard errors in brackets, *** and ** indicate significance at the 1%, 5% level respectively

or lower, by 6.72% (0.42*0.16*100). This result is consistent with that obtained in Lobell and Burke (2008) who found a strong positive relationship between rainfall and variability of agricultural production. It was shown that one standard deviation in a change in growing season precipitation leads to a change in agricultural production by 10%.

Regarding the variables of environmental degradation and climate change, the resulting temperature has a negative sign and is not statistically significant. This can be explained by the degree of annual average temperature in Algeria that does not yet reach the threshold which leads to negatively affect cereal yield. However, CO_2 emissions have a significant and negative effect on cereal yield. An increase by 1% in CO_2 emissions per capita results in a decrease in cereal yield by 0.76%. This result is consistent with Mulatu et al. (2016) who studied the effects of CO_2 emissions on agricultural productivity and household welfare in Ethiopia using a recursive dynamic Computable General Equilibrium (CGE) model. The simulation results, on the period 2010–2030, confirm that CO_2 emissions lower significantly agricultural productivity. Furthermore, the coefficient of the trend term is positive and significant at 1%, implying the cereal yield grows over time most likely due to the use of fertilizers and new seeds. Indeed, Bessaoud et al. (2019) found that total factor productivity in Algeria experienced an upward trend due to the use of mechanization, expansion of the surface of irrigated land, fertilizer and the continued improvement of seeds, and the adoption in 2005 of the national program of agricultural development

(PNDA). Therefore, agricultural productivity moved from 1.6% per year over the period 1990–2000 to 6.6% per year during the period of 2008–2013.

Column (B) shows the estimates of Eq. (2). The average annual temperature is positively correlated to the temperature of the previous year. Moreover, temperature is positively influenced by CO_2 emissions. If the later increase by 1%, the temperature raises also by 0.024%. More importantly CO_2 emissions lead to a continuous increase in temperature until it reaches the threshold that negatively impacts the cereal yield. The estimates of Eq. (3), in column (C), reveals that GDP per capita has a positive relationship with CO_2 emissions, which suggests that an increase in GDP per capita increases CO_2 emissions. This result is in line with that obtained by Bouznit and Pablo-Romero (2016) who found the economic growth in Algeria is the main factor fostering CO_2 emissions over the period 1970–2010 and leading to aggravate the warming climate. The results further show that GDP per capita also indirectly and negatively affects cereal yield through its effect on CO_2 emissions. Therefore, if GDP per capita grows by 1%, CO_2emissions increase by 1.78%, which in turn decreases cereal yield by 1.28% [(1.72)*(−0.76)]. According to Grossman and Krueger (1991), the amount of emitted CO_2 and the environmental degradation are highly dependent on the structure of economy. Indeed, the Algerian hydrocarbon sector contributes between 30 and 50% to GDP; 97% of total exports, 70% of tax revenue, and more than 98.5% of energy use are generated from the use of fossil fuels. Therefore, Algeria is classified as one of the important CO_2 emitters among developing countries, and thus highly vulnerable to climate change (Bouznit et al. 2020).Yet, GHG emissions can be notably reduced by promoting renewable energy and emery efficiency. As such, Algeria committed to promote technologies and adopt suitable measures to help rationalize domestic consumption and minimize energy losses which are estimated at 20% (Belaid and Abderrahmani 2013; Bouznit et al. 2020). Therefore, the expected quantities of natural gas which will be saved by 2030 are estimated at more than 300 billion cubic feet, which represents 8 times the domestic consumption in 2014 (SKTM 2018).

7 Conclusion

Recognizing the need to balance its international commitments and its sustainable development, Algeria adopted a set of measures to fight warming climate. The mitigation measures aim to reduce GHG emissions by 7% by 2030 through the development of renewable energy in order to reach the mix energy. The expected objective is to generate 22,000 MW from renewable energy resources by 2030, which represents 27% of the total energy production. Reaching this goal will save 300 billion m^3 of natural gas and reduce 348 $MtCO_2$ equivalents of CO_2 emissions. Therefore, greenhouse gas emissions should reach the turning point by 2024.The adaptation measures taken by Algeria include actions and projects to be implemented over the period 2020–2035 to protect the economic sectors from the negative effects of climate change. In this context, 64 adaptation actions have been identified; 36 actions to be implemented in the short term and 27 actions in the medium term.

The descriptive data confirm that the agricultural production is strongly linked to the variability of precipitations. The results from our simultaneous equations model, over the period 1966–2016, show that the agricultural yield was positively affected by rainfall and negatively impacted by CO_2 emissions. The latter are positively related to temperatures, which leads over time to aggravate the warming climate phenomenon. On the other hand, the results also show that GDP per capita indirectly lowers cereal yield through its effect on CO_2 emissions. Therefore, promoting renewable energy to reduce CO_2 emissions is the best route to fight climate change and reach food security in Algeria. In addition; some recommendations are offered to adapt the agricultural sector to climate change:

- Adopting an agricultural policy which considers the balance between food security and ecological sustainability
- Algeria must reach sustainable agriculture by putting in place mega projects
- Enhancing and encouraging local agricultural production
- Encouraging research and development in the agricultural sector
- Developing the agro-food industries
- Developing and generalizing the drip irrigation systems to increase the agricultural yield and save water
- Building new dams to increase the capacity of rainwater storage

References

Bessaoud O, Pellissier J-P, Rolland J-P, Khechimi W (2019) Rapport de synthèse sur l'agriculture en Algérie. [Rapport de recherche] CIHEAM-IAMM, p 82. ffhal-02137632f. Available at: https://hal.archives-ouvertes.fr/hal-02137632/document

Belaid F, Abderrahmani F (2013) Electricity consumption and economic growth in Algeria: a multivariate causality analysis in the presence of structural change. Energy Policy 55:286–295

Bouznit M, Pablo-Romero M, Sánchez-Braza A (2020) Measures to promote renewable energy for electricity generation in Algeria. Sustainability 12: 1468. https://doi.org/10.3390/su12041468

Bouznit M, Pablo-Romero M (2016) CO_2 emission and economic growth in Algeria. Energy Policy 96:93–104

Brown T, Gambhir A, Florin N, Fennell P (2012) Reducing CO_2 emissions from heavy industry: a review of technologies and considerations for policy makers, Grantham Institute for Climate Change paper N° 7

CEREFE (2020) Transition Energétique en Algérie: Leçons, Etat des Lieux et Perspectives pour un Développement Accéléré des Energies Renouvelables. Available at: http://www.cerefe.gov.dz/wp-content/uploads/2020/12/Rapport_CEREFE_2020_FINAL-30-11-2020.pdf

CCKP (2020) Climate change knowledge portal—World Bank. Available at: https://climateknowledgeportal.worldbank.org/download-data

Chabane M (2012) Comment concilier changement climatique et développement agricole en Algérie? Territoire en Mouvement, pp 73–91. https://doi.org/10.4000/tem.1754

Chandio AA, Jiang Y, Rehman A, Rauf A (2020) Short and long-run impacts of climate change on agriculture: an empirical evidence from China. Int J Climate Change Strategies Manage 12(2):201–221

Chen S, Binlei Gong B (2021) Response and adaptation of agriculture to climate change: evidence from China. J Dev Econ 148: 102557

De-Bernis DG (1971) Le plan quadriennal de l'Algérie 1970–1973. In: annuaire de l'Afrique de Nord, Centre national de la recherche scientifique, Centre de recherche et d'étude sur les sociétés méditerranéennes (CRESM) (éds), Paris, Editions du CNRS, vol P, pp 195–230 (1225p), contient bibl. Index, Sigles

Dell M, Jones Benjamin F, Olken Benjamin A (2008) Climate change and economic growth: evidence from the last half century. Working Paper 14132. National Bureau of Economic Research. Cambridge, MA 02138. Available at: https://www.nber.org/system/files/working_papers/w14132/w14132.pdf

DePaula G (2020) The distributional effect of climate change on agriculture: evidence from a Ricardian quantile analysis of Brazilian census data. J Environ Econ Manage 104: 102378

EC (2020) Available at: https://ec.europa.eu/clima/policies/international/negotiations/paris_en

Fanack (2020) Available at: https://fanack.com/algeria/geography/. Visited on 18 December 2020

Fankhauser S, Tol Richard SJ (2005) On climate change and economic growth. Resource and Energy Economics 27:1–17

Ghezloun A, Saidanc A, Oucher N, Merabet H (2015) Actual case of energy strategy in Algeria and Tunisia. Energy Proc 74:1561–1570

Grossman GM, Krueger AB (1991) Environmental impacts of a North American Free Trade Agreement. National Bureau of Economic Research, NBER, Cambridge, MA

INDC-Algeria (2015) Intended Nationally Determined Contribution INDC-Algeria. Available at: http://ww4.unfccc.int/ndcregistry/PublishedDocuments/Algeria%20First/Algeria%20-%20INDC%20

Lobell David B, Burke Marshall B (2008) Why are agricultural impacts of climate change so uncertain? The importance of temperature relative to precipitation. Environ Res Lett 3: 034007

Krishna KK, Rupa KK, Ashrit RG, Deshpande NR, Hansen JW (2004) Climate impacts on Indian agriculture. Int J Climatol 24(11):1375–1393

Mendelsohn R (2014) The impact of Climate Change on Agriculture in Asia. J Integr Agric 13(4):660–665

Mulatu DW, Eshete ZS, Gatiso TG (2016) The impact of CO_2 emissions on Agricultural Productivity and Household Welfare in Ethiopia: a computable general equilibrium analysis. Environment for Development Initiative. Available at: http://www.jstor.com/stable/resrep15044

Munir K, Riaz N (2020) Asymmetric impact of energy consumption on environmental degradation: evidence from Australia, China, and USA. Environ Sci Pollut Res 27:11749–11759

Nachmany M, Fankhauser S, Davidová J, Kingsmill N, Landesman T, Roppongi H, Schleifer P, Setzer J, Sharman A, Singleton CS et al (2015) The 2015 Global Climate Legislation Study—A Review of Climate Change Legislation in 99 Countries (summary for policy-makers), The Grantham Research Institute on Climate Change and the Environment, The London School of Economics and Political Science, London, UK. Available at: https://www.researchgate.net/publication/284149889

NCP-Algeria (2019) National Climate Plan of Algeria. Available at: http://www.meer.gov.dz/a/?wpdmpro=plan-national-clima

Nouaceur Z, Laignel B, Turki I (2013) Changements climatiques au Maghreb: vers des conditions plus humides et plus chaudes sur le littoral algérien? Physio-Géo: Géographie physique et environnement, vol 7, pp 307–323. https://doi.org/10.4000/physio-geo.3686

Olivier JGJ, Janssens-Maenhout G, Muntean M, Peters JAHW (2015) Trends in global CO_2 emissions: report, European Comission, Joint Research Centre: Ispra, Italy

Ollivier M (1985) L'économie algérienne vingt ans après 1966: L'indépedance nationale en question, Edictions du CNRS, Annuaire de l'Afrique du Nord, Tome XXIV

Palloix C (1980) Industrialisation et financement lors des deux plans quadriennaux (1970 1977). Tiers Monde, Tome 21(83):531–555
Pathak Tapan B, Maskey Mahesh L, Dahlberg Jeffery A, Faith K, Bali Khaled M, Zaccaria D (2018) Climate Change trends and impacts on California Agriculture: a detailed review. Agronomy 8(3): 25; doi: 10.3390
Porter JR, Semenov MA (2005) Crop responses to climatic variation. Philos Trans R Soc Lond B Biol Sci 360(1463):2021–2035
Rehman MU, Rashid M (2017) Energy consumption to environmental degradation, the growth appetite in SAARC nations. Renew Energy 111:284–294
Rezai A, Taylor L, Foley D (2018) Economic growth, income distribution, and climate change. Ecol Econ 146:164–172
Sahnoune F, Imessad K (2017) Analysis and impact of the measure to mitigate climate change in Algeria. Energy Proc 136:494–500
Sahnoune F, Belhamel M, Zelmat M, Kerbachi R (2013) Climate change in Algeria: vulnerability and strategy of mitigation and adaptation. Energy Proc 36:1286–1294
SKTM (2018) Projets réalisés dans le cadre du programme National des Energies Renouvelables. Available at: https://www.interieur.gov.dz/images/Projets-raliss-dans-le-cadre-du-prorgamme-national-des-nergies-renouvelables-.pdf
Stern N (2006) The economics of Climate Change. Cambridge University Press, Online publication date: March 2014. https://doi.org/10.1017/CBO9780511817434
UNFCCC (2020) Available at: https://unfccc.int/topics/adaptation-and-resilience/the-big-picture/what-do-adaptation-to-climate-change-and-climate-resilience-mean#:~:text=Adaptation%20refers%20to%20adjustments%20in,opportunities%20associated%20with%20climate%20change
U.S.-EIA (2019) US-Energy Information Administration. Available at: https://www.eia.gov/
Walthall CL, Hatfield J, Backlund P, Lengnick L, Marshall E, Walsh M, Adkins S, Aillery M, Ainsworth EA, Ammann C, Anderson CJ, Bartomeus I, Baumgard LH, Booker F, Bradley B, Blumenthal DM, Bunce J, Burkey K, Dabney SM, Delgado JA, Dukes J, Funk A, Garrett K, Glenn M, Grantz DA, Goodrich D, Hu S, Izaurralde RC, Jones RAC,Kim S-H, Leaky ADB, Lewers K, Mader TL, McClung A, Morgan J, Muth DJ, Nearing M, Oosterhuis DM, Ort D, Parmesan C, Pettigrew WT, Polley W, Rader R, Rice C, Rivington M, Rosskopf E, Salas WA, Sollenberger LE, Srygley R, Stöckle C, Takle ES, Timlin D, White JW, Winfree R, Wright-Morton L, Ziska LH (2012) Climate change and agriculture in the United States: effects and adaptation. USDA Technical Bulletin 1935. Washington, DC. 186 pages, US
WEC (2019) World Energy Council. World Energy Issues Monitors; World Energy Council: London, UK, Available at: https://www.worldenergy.org/assets/downloads/1.-World-Energy-Issues-Monitor-2019-Interactive-Full-Report.pdf
WDI (2020) World Development Indicators database. Available at: https://databank.worldbank.org/source/world-development-indicators
Wooldridge JM (2010) Econometric analysis of cross section and panel data. MIT press, US

Ready for Climate Change? An Assessment of Measures Adopted by 45 Mediterranean Coastal Cities to Face Climate Change

Xira Ruiz-Campillo, Olga Gil, and Cristina García Fernández

Abstract The IPCC has warned that countries such as France, Spain, and Italy will need to strengthen their coastal defences against an increased frequency of storms and of higher sea level. This information should lead to urgent city action, especially from those in coastal areas such as in the Mediterranean. Are cities ready for the impact of climate change? This work examines practices adopted by 45 coastal Mediterranean cities in eight countries (Croatia, Cyprus, France, Greece, Italy, Portugal, Spain, and Turkey) to address climate change, shedding light on initiatives being implemented and identifying areas lacking decisive action. The first source of data for this research was collected through information on the EU Covenant of Mayors website. Results show deficiencies in policy formulation and actions to respond to climate change through adaptation measures, while showing that cities are better at reduction in emissions and energy consumption. We also found that municipalities in the sample are more active in adapting to climate change when they have the support of programs at the national or European level.

X. Ruiz-Campillo (✉)
Department of International Relations and Global History, Faculty of Political and Social Sciences, Complutense University of Madrid, Madrid, Spain
e-mail: xiraruiz@ucm.es

O. Gil
Complutense Institute of Administrative Sciences (ICCA), Complutense University of Madrid, Madrid, Spain
e-mail: olgagil@ucm.es

C. García Fernández
Department of Applied Economics, Public Economics and Political Economy, Faculty of Political and Social Sciences, Complutense University of Madrid, Madrid, Spain
e-mail: cgarciaf@ucm.es

W. Leal Filho and E. Manolas (eds.), *Climate Change in the Mediterranean and Middle Eastern Region*, Climate Change Management,
https://doi.org/10.1007/978-3-030-78566-6_13

1 Introduction

In recent years, researchers have become increasingly involved in the study of cities as relevant actors in climate governance. Interest in this field of study has increased to the point that understanding how they can be governed to improve actions in resistance to climate change has turned into a global sustainability priority (Evans 2019; Castán Broto 2017). This is especially true with the global climate agreements for reducing greenhouse emissions, such as the Kyoto Protocol and the Paris Agreement. An enormous amount of research has been conducted to study the functioning of transnational networks and cooperation initiatives on climate change between subnational governments, regions, and other non-state actors (Khan 2013; Andonova et al. 2009; Bulkeley and Castán Broto 2013), such as the European Union's Covenant of Mayors (CoM), the C40 Cities Climate Leadership Group (C40), and the International Council for Local Environmental Initiatives (ICLEI). Literature on the experiments with cities has focused on an overwhelming number of issues including the transition to sustainability within cities (Smedby and Neij 2013; Nevens et al. 2013; Van den Heilingenberg et al. 2018), urban greening (Dorst et al. 2019), empirical studies on smart city initiatives (Gil 2021; Raven et al. 2019), and the categorization and characteristics of those experiments (Bernstein et al. 2009; Andonova et al. 2009; Castán Broto and Bulkeley 2013).

It is well known that cities account for between 60 and 80% of energy consumption and generate as much as 70% of human-induced greenhouse gas (GHG) emissions, mainly through the consumption of fossil fuels (UN Habitat 2019). With the increase in population, it is foreseen that two-thirds of the world's population could live in cities by 2050 (UNDESA 2014). As a result, it is a growing sentiment that much of the power to reduce climate emissions lies now with cities, not national governments (Gordon and Johnson 2017). Moreover, according to a recent study, the global impacts of sea-level rise and coastal flooding this century will likely be far greater than indicated by the most pessimistic past analyses, putting three times more people at risk of coastal flooding by 2050 than previously thought. A recent report found that over ten percent of the world's population—820 million people–live in coastal zones prone to sea level rise, and 86% of them live in urban or quasi-urban areas (Coalition for Urban Transitions 2019: 57).

In its publications, the Intergovernmental Panel on Climate Change (IPCC) warns that the frequency and intensity of droughts has increased in the Mediterranean and it will continue to increase, putting additional stress on land, increasing the number of people affected by wildfires and desertification, and exacerbating risks to livelihoods, vulnerability, human and ecosystem health, infrastructure, and food systems (IPCC 2020). In its 2014 report, the IPCC gives extensive information about its perspectives on the Mediterranean region. The frequency and intensity of droughts has likely increased in the Mediterranean, and the region is projected to be much drier and hotter in the warm seasons (Giorgi and Lionello 2008), making them the most significant trends in the Iberian Peninsula and Southern France. It is also projected that in the region there will be more frequent, longer, and more intense heat waves and warm spells. Although the total burned area has decreased since 1985 in the Mediterranean region, megafires triggered by extreme climate events have caused record maxima of burnt areas in some Mediterranean countries during the last decades (Ayanz et al. 2013), and future wildfire risk is projected to increase in Southern Europe (Lindner et al. 2010; Carvalho et al. 2011). The IPCC report also warns on the adaptive capacity of Southern European countries to climate change being lower in this region than in Nordic countries. Within Europe, the populations and infrastructure of coastal regions are likely to be more affected by sea level rise and previous research points out that the Mediterranean region will be impacted by multiple consequences of climate change, such as the increase of alien species, habitat losses, and degradation– both on land and in the sea—due to increasing heat waves and droughts (Ulbrich et al. 2012). All these data should lead to more urgent action coming from cities, especially those in coastal areas in the Mediterranean.

The IPCC warns that countries such as France, Spain, and Italy will need to strengthen their coastal defences since they will experience an increased frequency of storms and of higher sea levels. Many dwellings and buildings will suffer from occasional flooding and therefore will have to be adapted. Even so, "managed retreat" is likely to become a response in some cases (IPCC 2014). Moreover, food production and security are likely to be impacted with higher demand for food products and a decline in crops, fish and livestock yields due to climatic and other stress factors (Cramer et al. 2018). Despite the vulnerability of Mediterranean cities to this challenge, to date, scant attention has been paid to their level of preparedness against this global threat.

In the European Union (EU) there is an increasing awareness of the urgent need to address climate change at both national and local levels (Pietrapertosa et al. 2018, 2019). The analysis of the responses of European cities to climate change shows the influence that national urban climate change strategies and the participation in transnational municipal networks can have in the cities' adoption of mitigation and adaptation plans (Reckien et al. 2015, 2018; Heidrich et al. 2016; Heikkinen et al. 2020; Kumar 2019). It has also been established that the participation in the EU Covenant of Mayors has contributed to the diffusion of international best practices, to helping cities share their knowledge (Reckien et al. 2018; Domonerok et al. 2020), and that the support from other subnational governments such as regions or provinces can step up action at the local level (Melica et al. 2018). Despite the existence of previous analysis on the initiatives and policies adopted at the local level (Reckien et al. 2018; Pietrapertosa et al. 2019; Ruiz Campillo 2019; Heikkinen et al. 2020), there is less evidence of the impact of local participation. Are cities ready for the impact of climate change? How are they preparing for these projections in the coming years? It is undeniable that states have the obligation to promote national legislation to face climate change, but cities also have a responsibility in addressing this threat. Cities can have better mechanisms to cooperate between each other, a more direct communication with citizens, and greater pragmatism than higher levels of government (Hoornweg et al. 2011). Many cities in the Mediterranean, but specially in Spain and Italy, joined the EU Covenant of Mayors with the aim of making a firm commitment to reduce their CO_2 emissions and increase their adaptation to climate change.

Actions to face climate change both through mitigation and adaptation measures are being implemented at all levels. In 2013, the European Commission adopted an EU strategy on adaptation to climate change aimed at making Europe more climate resilient. At the EU level, the need for a joint approach has been stressed for urban adaptation to climate change in Europe (EEA 2012). At the Mediterranean level, there are several programmes (e.g., the MED Programme 2014–2020, the UNEP Mediterranean Action Plan, the UNEP-MAP Mediterranean Strategy for Sustainable Development 2016–2025, the Co-Evolve or the MPA-ADAPT) that aim to strengthen resilience to climate change, promote the co-evolution of human activities and natural systems in touristic coastal areas, protect and promote the Mediterranean natural and cultural resources, and foster low carbon strategies and energy efficiency, amongst many other initiatives.

Our analysis will focus on the progress cities have made in the coastal Mediterranean region to address and adapt to climate change. The work will examine the initiatives adopted by 45 cities that form part of the EU Covenant of Mayors to assess what types of action cities are undertaking and if they are centred in specific areas (e.g., reducing GHG emissions, increasing the share of renewable energy, or increasing the amount of green areas in the municipalities). This will provide information regarding those measures most popular amongst municipalities and those measures that are not being tackled.

For methodological purposes, we have restricted our research to the information cities published on the CoM website, which limits the scope of our work. Thus, it could be possible to find more cities active in mitigation and adaptation initiatives than those found in our results. It could also be possible to find cities with more advanced policies—but that are not updated on the website. However, our findings can be relevant to identify what type of initiatives are already underway and which ones are still missing within the framework of the EU Covenant of Mayors.

The chapter proceeds as follows: the first section comprises the framework of analysis, followed by the methodology adopted for this research. Then, the 45 cases of study are examined in alphabetical order, followed by the results found. The chapter concludes with an overview of our findings and potential future research areas.

2 The EU Covenant of Mayors as the Framework of Analysis

The EU Covenant of Mayors can be understood as an experiment that tries to mobilize cities in the fight against climate change so that they can contribute to—not substitute (Bansard et al. 2017)—the efforts of the European Union to comply with its international climate compromises (such as the Kyoto Protocol or the Paris Agreement) and to increase and shape the coordination of local sustainable energy and climate policies.

The EU Covenant of Mayors was launched in 2008 and in 2015, evolved into the EU Covenant of Mayors for Climate and Energy, bringing together local governments on a voluntary basis with the aim to implement not only EU climate but also energy objectives. As of November 2020, it gathers over 10.371 local and regional authorities across 57 countries, drawing on a worldwide multi-stakeholder effort, and with dedicated offices that offer technical and methodological support to signatories. The EU Covenant of Mayors identifies on its website some factors that contribute to its success: bottom-up governance, where cities of all sizes are mobilized to address climate change, cooperation and knowledge sharing at two levels (EU-cities and amongst cities), and a flexible and context-driven approach that allows cities to adopt the best measures possible for their municipalities. Most of their signatories are from the European Union but also from countries such as Azerbaijan, Georgia, Lebanon, and Morocco, making it one of the most populous initiatives to address climate change at the local level (if compared to initiatives such as ICLEI, Climate Alliance, or C40, with far fewer participants). The number of members of the CoM exemplifies that this European initiative has been very successful in mobilizing local authorities, cities, and towns of all sizes to commit to the fight against climate change (Melica et al. 2018; Domonerok et al. 2020).

Cities that join the CoM commit to a set of targets. The first one is to have been reached in 2020 and commits cities to reduce their emissions by 20%. A second target is the 2030 climate and energy framework, which commits cities to cut their GHG by 40% and to adopt a joint approach to tackling mitigation and adaptation to climate change. Cities inform on the initiatives they will implement through a Sustainable Energy and Climate Action Plan (SECAP), which includes information on the main sources of CO2 emissions (from sectors including municipal buildings, transport, waste, and the production of local energy) and includes the tools, policies, and budgets to meet the 2020 or 2030 goals they adhere to when joining the CoM. Within two years of the approval of the SECAP, cities must report on their progress and identify good practices of successful and innovative policies and tools implemented at the local level.

3 Methodology

As stated above, the aim of this chapter is to examine the main initiatives Mediterranean coastal cities and municipalities participating in the EU Covenant of Mayors are adopting to mitigate and adapt to climate change. This will lead us to conclude how and where cities are being more active in addressing climate change as well as which areas may need to be promoted by the cities that are part of the CoM. We do this work by analysing the information signatories must submit in their monitoring reports, following qualitative (policy measures) and quantitative (number of measures) approaches. Monitoring reports contain milestones or key actions– those actions that are considered important by the municipalities—that will serve as the basis for our analysis.

Data for this research was collected mainly through the EU Covenant of Mayors website, and in some cases through the specific city plan submitted to the CoM. The authors have examined information from 45 coastal cities from eight Mediterranean countries (see Annex 1): France and Turkey are represented with only one city each; Cyprus is represented with two cities; Portugal is included in the research with four cities[1]; Croatia and Greece have six and five cities respectively in the monitoring process that have been included in our research, while authors have had to limit the number of cities examined for Italy and Spain (limited to fourteen and twelve each) to avoid an unbalanced analysis. Although there are cities' signatories to the CoM in southern Mediterranean countries such as Morocco, Algeria, Tunisia, and Israel, for methodological purposes, only cities that are in the monitoring phase (i.e., they have submitted the monitoring report of their SECAPs) have been selected for the assessment carried out in this chapter. This does not mean that these cities are not implementing policies at the local level to address climate change, but if they are, authorities have not made that information public through the CoM website. On the same line, the analysis has been generally restricted in scope to the information published by cities in the CoM website, which many not include all the initiatives cities are undertaking to address climate change. Despite these limitations, we believe we have created an interesting manuscript, whose discussion and conclusion can contribute to filling the void in the literature. This chapter has an explanatory approach and is not meant to be a means to an end, but rather an aid for researchers to continue narrowing the gap between mitigation actions and adaptation to climate change at the municipal level.

The research follows a qualitative comparative analysis connected with public policy and examines initiatives taken by local government signatories of the EU Covenant of Mayors for Climate and Energy. We opted for this research design as it allows us to explore the actual needs and goals of policy and to connect with oriented research (Thomann 2020; Rihoux et al. 2011, 2013). Following Thomann and Maggetti (2017), we use a qualitative comparative analysis to compile a formalized data analysis based on data-set observations we have built for each city.

This qualitative comparative analysis will proceed as described by Thomann (2018), Ragin (2008), Rihoux and Ragin (2009), and Schneider and Wagemann (2012): the selection, conceptualization, and operationalization of cases, the outcome, (dependent variable, namely adaptation to climate change of 45 Mediterranean cities), and conditions (independent variables) follow the protocols of (qualitative) comparative research design, as well as measurement validity (Adcock and Collier 2001; Radaelli and Wagemann 2018; Toshkov 2016).

Since the forty-five cities examined are already in the monitoring process of the CoM, we have been able to access their reports on the actions they acknowledge to have implemented at the local level. To make the comparative analysis of city actions, we have structured them into three areas according to the aims of the CoM: (1) resilience to climate change, (2) reduction of emissions, and (3) energy consumption.

[1] Although Portugal is not strictly speaking a Mediterranean country, its proximity, its climate, social, and cultural characteristics have led us to include some of its cities.

We have also followed Rihoux and Ragin (2009) to elaborate a qualitative and comparative table analysis with logical minimization. This has allowed us to combine an analysis of complexity–preserving cases as configurations of attributes—with a systematic cross-case comparison to detect regularities as suggested by Thomann (2020), Berg-Schlosser et al. (2009), and Engeli et al. (2014). The results of this comparison are seen in the table in Annex II, comparing energy consumption, reduction of emissions, measures conductive to reducing emissions and energy consumption, and measures tackling resilience to climate change. The table also gives information on cities participating in European projects such as Life and other research and interregional projects.

4 Actions at the Local Level in the Mediterranean

4.1 Croatia

The actions adopted by the municipalities examined in Croatia are difficult to assess since many of them have identified as key action initiatives included in their SECAP, and not those that have been implemented. For instance, the municipality of Labin includes amongst its key actions the organization of workshops and lectures for tourist operators or adaptation of the drainage system in the Labin area to be implemented between 2020 and 2024. The same happens with the municipalities of Poreč-Parenzo, Rovinj-Rovigno or Pula-Pola. However, some of the actions recalled as key actions have already started and include important measures to act in the event of natural disasters in Rovinj-Rovigno, such as the improvement of the security monitoring system, the creation of protocols for connecting civil services, strengthening the capacity of the local fire community, and the implementation of a plan for health measures.

Some key actions that other Croatian municipalities have included are the installation of biomass boilers, the installation of efficient lighting (Rijeka), the co-financing of thermal solar energy systems, and the improvement of insulation in homes (Zadar and Rijeka) or in public buildings in the municipalities of Krk and Rijeka.

4.2 Cyprus

Both Larnaca and Paralimni have centered their efforts mainly in the promotion of the bicycle as sustainable transport. Larnaca includes the obligation to designate a 5% of parking places for bicycles in areas larger than 1200 m^2 and the inclusion of bicycle parking places in apartment buildings. The city of Paralimni has also focused on cyclist mobility with events to celebrate the European Mobility Week, the installation of recharge points for electric vehicles, and the adoption of the WASP

Tool European program to raise awareness among citizens and reduce the generation of waste.

4.3 France

France appears in this research with a sole coastal Mediterranean city in the monitoring process within the EU Covenant of Mayors: Nice (Metropole). This signatory has focused nearly all its efforts on installing more efficient lighting and reducing energy-consuming public lighting. This has been done through different means: the reduction by 20 minutes per day of public lighting, ensuring that their contracted power in all municipalities is adjusted to the amount they need (by comparing the electricity subscribed and the electricity consumed), and with projects that subsidize the public lighting in towns with fewer than 2000 inhabitants (13 towns) to reduce their electricity consumption by 50% and reduce their light pollution.

4.4 Greece

In the case of Greece, measures adopted by municipalities focus mainly on three areas: the improvement of the isolation of municipal buildings, the change to LED lighting, and the promotion of renewable energy. This last initiative has been implemented in the cities of Alexandroupolis and Thessaloniki. Most of the Greek municipalities examined have mentioned the use of LED technology for increasing the efficiency of public lighting as a key action. As additional key actions, they have also mentioned improving the efficiency of municipal buildings and facilities through the replacement of windows and frames, the insulation of roofs and walls (for example in Thessaloniki, Kavala, and Volos), and replacing the swimming pool covers to facilitate maintenance. Other examples of actions can be found in Alexandroupolis, which opened an office offering information on energy savings and the use of bioclimatic solutions such as sprinklers, water curtains, and tree planting in Chrimatistirio Square and the old commercial center of Thessaloniki.

4.5 Italy

In the case of Italy, measures adopted by the municipalities in terms of policy instruments focus on areas such as improving the insulation in municipal buildings including sports facilities and sometimes schools, which is pursued by Caglari, Genova, Messina, Napoli, La Spezia, and Trieste. Sometimes complemented with measures such as energy conservation awareness campaigns (in homes and/or aimed at primary school students), switching to LED lighting is yet another popular initiative that appears in the findings. Mixed measures in this orientation are adopted by Caglari, Genova, Messina, Napoli, La Spezia, and Venice. Energy renovation of facades and installation of solar thermal systems (with varying policies such as light replacements, renewable energy and gas, new monitoring systems, cogeneration, and green purchasing) are found in efforts pursued by Cagliari, Genova, Livorno, Messina, Napoli, and Trieste. Promotion of efficient transport is found in Cagliari and Livorno, targeting cycles and pedestrian/cycle paths, as well as the replacement of municipal vehicles with low-emissions vehicles. These policies are also found in Messina, Napoli, Trieste, La Spezia, and Venice in the form of an electric tramway.

The city of Livorno focuses on providing information to stakeholders including trade associations, citizens, and small and medium enterprises as well as developing the stakeholders' active participation in carrying out the expected actions. There is also evidence of work with stakeholders such as in Messina, involving the local government, working groups, stakeholders from civil society, the transport authority, the Association of Architects, and the Association of Engineers, amongst others.

The adaptation of regulatory frameworks, which is a key feature needed to tackle and implement measures for climate change, is not common among the Italian cases. We find adaptation of regulatory and of legislative frameworks only in Genova, a focus on planning land use solely in Napoli, and the exception of an urban regeneration plan in the city of Bitonto. We also found some plans to monitor air quality in specific zones, such as the port area for the city of Messina. Additional policy instruments include training meetings targeted at municipal technicians and local professionals, and digital agendas reaching the metropolitan areas—Messina being an interesting case. Finally, there are scant policies targeting recycling and waste reduction, such as those found in Napoli.

4.6 Portugal

As per the Portuguese cities examined, we have identified three main lines of action: the improvement of energy efficiency in schools and residential buildings (Almada), the development of a potential solar chart (Lisbon), and an innovative project for water supply in Porto.

Almada has created a huge set of tools for use in schools by students, teachers, and workers. Prizes were created as an initiative for citizens and households that achieve better results in terms of energy saving, in the form of the Energy House (a model of a real house with information on energy efficiency) and renewable and sustainable transport, amongst others. A smart tele-management system for Almada's public lighting allows the remote management of public lighting intensity by adjusting the luminous flux of all the lights throughout the day. The project has also included the conversion of all traffic lights to LED technology, which resulted in a reduction of energy consumption by 80%.

Through the Lisboa E-Nova project, Lisbon has developed a potential solar chart. This innovative tool is available online via the Google Maps application and covers every building in Lisbon. It allows for identification of preferable areas for investing in solar technologies, and provides an efficiency awareness tool for local authorities, investors, businesses, and citizens. Other projects refer to external wall insulation, the Risk Exposure Map in the framework of the RESCCUE/Horizon 2020 Project, the construction of a drainage infrastructure to reduce the risk of urban floods, and the reinforcement of green infrastructure connectivity, carried out by completing nine green corridors associated with active mobility solutions.

Porto has undertaken an interesting and innovative project titled Porto Gravitico. A new concept for the water supply system to Porto has been implemented to utilize gravity at the existing points of delivery, which (in normal conditions) eliminates the six pumping stations integrated into the original system. Through the creation of new loops, this solution allows alternative routes to each water reservoir.

4.7 Spain

Like most of the cities analysed in this chapter, cities in Spain such as Marbella, Altea, El Campello, and Tarragona have also focused on measures intended to save energy through improved lighting, LEDs, solar panels, and energy audits of public buildings. Optimizing waste collection has also been a target on the SECAPs examined and can also be found in Spanish cities. Improvement of energy efficiency and solar panels (e.g., Algeciras, Valencia) is another proposal for residential buildings and others under renovation. An urban mobility plan and the replacement of the municipal vehicle fleet, cycling paths and sidewalks, fiscal incentives, and green commerce programs are a few more examples of initiatives developed in most of the cities (e.g., Málaga, Mojácar, Línea de la Concepción). Málaga has also concentrated its efforts on an innovation program based on the extraction of biogas that has been generated because of degradation processes of organic matter in the mass of landfill waste.

Some cities like Algeciras and Valencia have opted to reduce GHG consumption in some buildings and facilities. Algeciras' energy efficiency project in the facilities in the Port Authority Bay as well as Valencia's bioclimatic architecture, energy efficiency, and integrated remote management of buildings are good examples of these efforts.

Cities are also increasing awareness among citizens through the development of climate change interpretation centres and exhibitions for the public (Valencia). Additionally, they are launching a series of campaigns and training courses through which to raise awareness on the correct use of our energy sources, recycling, and sustainable development (Mojácar, Algeciras, Marbella).

In addition to most of the improvements mentioned above, Barcelona has created an online platform where citizens who register can consult their energy consumption and receive information on how much energy they consume and how. Also, the development of a solar thermal and photovoltaic map of buildings in Barcelona informs citizens on how much energy they can produce on their roofs. Among the cities studied, Barcelona has possibly developed more adaptation measures to climate change than any other. The smaller city of Salobreña has also made impressive adaptation measures such as the building of breakwaters and the improvement of the municipal civil protection system which is intended to address emergencies derived from climate change.

4.8 Turkey

In Turkey, the only coastal city in the monitoring phase of the CoM is Nilüfer. According to its SECAP, Nilüfer has introduced initiatives expected to be implemented by 2030 which focus on the creation of green corridors, pedestrian and bicycle transport, controlled urban expansion, and decreased surface flow.

5 Results and Discussion: Improving the Reduction of Emissions but Lagging in Adaptation Measures

As exemplified above, the EU Covenant of Mayors has three main axes to augment actions at the local level: resilience to climate change, reduction of emissions, and energy consumption. An exam of the cities' action reports to the CoM show that all cities have adopted initiatives to reduce their emissions and their energy consumption. However, there are fewer cities implementing actions linked to increasing resilience to the impact of climate change.

Like previous studies examining the initiatives adopted by cities to face climate change in the CoM framework (Pietrapertosa et al. 2019; Ruiz Campillo 2019), our findings confirm that measures adopted by the municipalities in terms of policy instruments tend to focus on the reduction of emissions and energy consumption through several initiatives. These initiatives include improving the isolation of municipal buildings, substitution of LED lighting, energy conservation, energy renovation of facades, and the installation of solar thermal systems. Actions also include the promotion of efficient transport, cycles and pedestrian paths, the replacement of municipal vehicles with methane and electric vehicles, and (to a limited extent) providing information to and engaging with stakeholders. Cities have disseminated information on the impacts of common bad habits on climate change and how to address them through awareness campaigns and information points that can help citizens reduce their emissions (these are included in the column labelled "others" of the table in Annex II).

On the downside, in the cases studied we find a general lack of the adaptation of regulatory frameworks, of plans to monitor air quality, of specific training programs, and the increase of human resources. There is an absence of policies targeting coastal flooding, droughts, and heating island effects, which we would have expected there to be more of given that all cities examined are on the coast and therefore more vulnerable to the impacts of climate change.

Another unfortunate finding of ours is the lack of plans directed towards drought action, conservation of biodiversity, natural disaster risk management, public health, and the improvement of irrigation methods in agricultural production and food security. These themes are addressed at the national level but are either not re-addressed at the city level or are not included in SECAPs sent to the EU Covenant of Mayors website.

As seen above, most cities examined have undertaken initiatives that can be included in the axes of emission reduction and energy consumption. The fact that in some countries, cities do not have competences for adaptation measures (Heidrich et al. 2016) and that the EU Covenant of Mayors only just included adaptation as part of the actions to undertake in 2014 may help explain why municipalities have focused much more on those two axes rather than on resilience to climate change. This should lead the CoM to intensify its efforts in adaptation to climate change.

Indeed, actions conducive to adapt to climate change are not at all common among cities. Very few have actually implemented initiatives that enable authorities and citizens to respond to natural disasters as a consequence of climate change. In total, only nine cities in four countries (Croatia, Italy, Portugal, and Spain) seem to have understood that climate change requires not only mitigation but also adaptation measures. This is consistent with previous research that also points to the need of stronger adaptation policies in cities (Heikkinen et al. 2020; Pietrapertosa et al. 2019).

Among the cities studied, Barcelona is possibly the one which has developed the most adaptation measures to climate change. Barcelona has recently unified all the separate plans involving climate action into a single strategy—the Climate Plan 2018–2030—which has been created cooperatively with its citizens. It integrates adaptation and resilience, climate justice and mitigation, and promotes citizen action (García and Peek 2020). The Portuguese city of Almada is also working on adaptation. The National Programme for Action on Adaptation to Climate Change (P-3AC) is a draft plan which is on the verge of being released. The city has also implemented several measures related to dune restoration.

The case of Croatia is striking. Among the cities examined, three of them have been found to be very active in adapting to climate change (Pula-Pola, Rovign-Rovigno and Porec-Parenzo). The explanation can be found in the participation of these three cities in the Life-Sec Adapt program of the European Union, in which Italian and Spanish cities also participate. As other studies suggest, programs funded by the EU can contribute to increased action regarding mitigation and adaptation planning at the national and local level (Heidrich et al. 2016; Pietrapertosa et al. 2018, 2019). The membership of cities in climate and European networks seems then to be particularly relevant to the adoption of these plans at the local level, as previous studies show (Reckien et al. 2015, 2018). This reveals the effectiveness of various support lines for the mobilization of municipalities, most of which could not adapt to climate change without the support of programs at the national or European level. The mobilization of cities, therefore, does not occur from the bottom up. On the contrary, the data in this work shows that it is driven from the top down and when it exists, cities seem more likely to design adaptation plans.

So far, many cities' climate actions have already proven that a 20% reduction of GHG by 2020 is a reality in their territories. In fact, recent estimates by the Joint Research Centre indicate that a 25% reduction had already been achieved by 2017 (Bertoldi et al. 2020). Although this reduction is beyond the initial 20% commitment most cities had endorsed when joining the CoM and seems purely positive, the reality is that most of these goals have been reached by increasing energy efficiency at the municipal level and lacking focus on policies tackling resilience to climate change. This resilience is what is most needed to prepare for a better response in the event of floods and natural disasters in coastal locations. Overall, coastal cities in the Mediterranean are not responding to climate change with their geolocation in mind. In fact, most of the initiatives examined are no different to those of other inland cities (Ruiz Campillo 2018, 2019; Domonerok et al. 2020).

With the evidence gathered, we have also seen that policy transfer and best practices are taking place within the framework of the CoM, which is contributing to the sharing of initiatives and experiences among participants. Our results are congruent with previous literature on the EU Covenant of Mayors. Most of the cities examined have shared their initiatives, pointing to a successful established framework and reporting system that cities who form part of the CoM use, as also found by Domonerok et al. (2019). However, we consider that cities could benefit from a holistic approach to both mitigation strategies and adaptation to climate change, since integrated urban systems will make the governance of traditionally separate but interdependent urban services less complex and more efficient.

Overall, we would have expected a greater response from coastal cities given their vulnerability. A report published by C40 on the pathway cities need to follow in order to meet the Paris Agreement's goals confirmed that although cities have adopted plans and strategies to fight climate change, the volume of action must increase dramatically if a difference is to be made (C40 2016). Addressing climate change must go beyond changing light bulbs, and it is urgent that local actors receive both the financial support and the guidance from above levels, whether that be on a regional, national, or European level. Cities have proven that they can reduce their GHG emissions by 2020 as they originally committed to do (Bertoldi et al. 2020). However, new impetus should be given to the EU Covenant of Mayors, mobilizing municipal actions focused on adaptation to climate change.

6 Conclusions

This work has examined the measures that 45 coastal Mediterranean cities have adopted to mitigate and adapt to climate change. Our research has provided us with information on policies, both those most popular amongst municipalities and those we found absent. Findings have been presented in a comparative manner in Annex II and they have been discussed in the previous sections. Although this work is not intended to be a comprehensive and exhaustive overview of all mechanisms that cities included in this study have adopted to face climate change, it sheds light over the main initiatives that cities promote and examines whether cities have adopted adaptation and mitigation plans.

Generally, cities have been found better at reduction in emissions and energy consumption and are more active in adapting to climate change when they have the support of programs at the national or European level. On the contrary, the work finds deficiencies in policy formulation and actions to respond to climate change through adaptation measures such as in the event of natural disasters.

Our findings can be summarized in the following points: First, we have been able to produce empirically well-grounded and country-sensitive evidence about policy instruments in comparative perspective, thus contributing to the identification of public policy gaps in the field of local government and climate change governance. Secondly, the work has provided guidance to understand policy adopted in the field of energy consumption and reduction of emissions. This guidance is useful to work in future recommendations for technicians and policy makers in charge of adopting decisions in areas that need to generate more knowledge from the scientific community. Thirdly, when we asked whether cities are ready for the impact of climate change, we found a lack of policy formulation and actions to adapt to climate change, while cities are better at reduction in both emissions and energy consumption. Fourthly, even though national and regional regulatory frameworks are necessary to provide stability to measures adopted by cities, especially those of smaller size and with fewer capacities, regulatory frameworks are normally not addressed in the 45 cases under study. This can be the result of the late inclusion of adaptation measures in the EU Covenant of Mayors, but it can also be due to the fact that in some countries, mitigation and adaptation policies lie outside the administrative boundary of the city (Heidrich et al. 2016). And lastly, we found that municipalities in the sample studied are much more active in adapting to climate change when they have the support of programs at the national or European level. This is also to say that finance has been key to guaranteeing the substantive and differentiated advancement of plans in cities with positive results, and particularly with funds coming from the European Union.

As for prospects, we find two interesting fields for improvement. On the one hand, there is citizen participation. We have found that cities innovating in policy making—such as including vulnerable groups in the drafting of policies—are almost non-existent, and this could be an interesting field for improvement to identify risks, provide information on solutions, and to achieve a qualitative change in future policies addressing climate change. The second field of improvement is related to better preparing cities against climate hazards. As García and Peek (2020) suggest, cities need to adapt on time and there is currently a need to connect adaptation with smart growth.

Qualitative comparative analysis allows us to observe differences in quality and degree. However, we need further research of case-studies to, for example, be able to address efforts in monitorization and partially or fully successful/unsuccessful implementation in the cases chosen, as recommended by Dusa (2018) and Thiem (2014). In future studies, there is a need for further research in the form of in-depth case studies or on how networks can support context-based actions and more fundamental changes (Heikkinen et al. 2020). That would allow a better understanding of the role of international transnational networks, and the possible local differences in cultural, political, and administrative traditions, as suggested by Engeli et al. (2014: 89), Patton (1997), and Thomann (2018). Following Engeli et al. (2014: 88), we also suggest further research to work on how to identify conjectural patterns that might help to formulate some useful policy recommendations, such as appropriate 'policy mixes' to achieve future given policy goals.

Annex 1 Cities included in this research.

Italy (14 cities)
Agropoli
Bitonto
Cagliari
Cavallino Treporti
Genova
La Spezia
Licata
Livorno
Marsala
Messina
Napoli
Pietra Ligure
Trieste
Venice
Greece (5 cities)
Alexandroupolis
Kalamaria
Kavala
Tesalonica
Volos
Spain (12 cities)
Algeciras
Barcelona
Colera
Línea de la Concepción
Málaga
Marbella
Mojácar
Palamós

(continued)

(continued)

Italy (14 cities)
Roses
Salobreña
Tarragona
Valencia
Croatia (6 cities)
Krk
Porec-Parenzo
Pula-Pola
Rijeka
Rovinj-Rovigno
Zadar
Portugal (4 cities)
Almada
Faro
Lisbon
Porto
Cyprus (2 cities)
Larnaca
Paralimni
Turkey (1 city)
Nilüfer
FRANCE (1 city)
Nice

Annex II Table with the number of key actions examined.

Country/City	Energy consumption	Reduction of emmissions		Measures conducive to both reduction emmissions and energy consumption		Others	Resilience to climate change	Participating in life projects	Participating in research projects	Participating in interregional projects
	Public lighting	Electricity production, municipal + residential	Sustainable transport	Municipal buildings	Residential buildings + awareness campaigns	Others	Adaptation measures			
Croatia										
Krk				2	1					
Poreč-Parenzo					2		15	X		
Pula-Pola				1	4	1	11	X		
Rijeka				3						
Rovign-Rovigno					2	1	5	X		
Zadar					2	1			X	
Cyprus										
Larnaca			3							
Paralimni			2			1				
France										
Nice	3									X
Greece										
Alexandroupolis	1	1		1						
Kalamaria	2			1						
Kavala				3						
Thessalonica		1		2						X

(continued)

(continued)

Country/City	Energy consumption	Reduction of emmissions		Measures conducive to both reduction emmissions and energy consumption		Others	Resilience to climate change	Participating in life projects	Participating in research projects	Participating in interregional projects
	Public lighting	Electricity production, municipal + residential	Sustainable transport	Municipal buildings	Residential buildings + awareness campaigns	Others	Adaptation measures			
Volos				3					X	
Italy										
Agropoli		3								
Bitonto	1		1	1						
Cagliari		1	1	1		1				
Cavallino Treporti		3								
Genova	1			2						X
La Spezia	1	2	1	1	1					X
Licata		2				1				
Livorno	1		1	1						X
Marsala	1		2							
Messina	1	1	2	1	1	3				
Napoli	1	1	3	1		2			X	
Pietra Ligure		2	1	1						
Trieste	1	1	1	1		1				
Venice	1		1							X

(continued)

(continued)

Country/City	Energy consumption	Reduction of emmissions		Measures conducive to both reduction emmissions and energy consumption		Others	Resilience to climate change	Participating in life projects	Participating in research projects	Participating in interregional projects
	Public lighting	Electricity production, municipal + residential	Sustainable transport	Municipal buildings	Residential buildings + awareness campaigns	Others	Adaptation measures			
Portugal										
Almada	3		1	1			1		X	
Faro			2				1			
Lisbon		1		1		1	2		X	X
Porto						1	1			
Spain										
Algeciras	2	2	9	5	6	12				
Barcelona		3		1		2	4	X	X	X
Colera			1	2						
Línea de la Concepción		1	1			1				
Málaga	1	2	1						X	
Marbella		1	2							
Mojácar		3								
Palamós	1	2				1				
Roses	6	1	5	16	2	4				
Salobreña			1		2	1	3			

(continued)

(continued)

Country/City	Energy consumption	Reduction of emmissions		Measures conducive to both reduction emmissions and energy consumption		Others	Resilience to climate change	Participating in life projects	Participating in research projects	Participating in interregional projects
	Public lighting	Electricity production, municipal + residential	Sustainable transport	Municipal buildings	Residential buildings + awareness campaigns	Others	Adaptation measures			
Tarragona	3									
Valencia			1	1		1			X	
Turkey										
Nilüfer			1			3				
Total number of key actions	31	34	44	53	23	39	43	4	8	8

Source Data of this annex has been retrieved from the websites of the EU Covenant of Mayors(www.covenantofmayors.eu) and Climate Adapt (https://climate-adapt.eea.europa.eu/).

References

Adcock R, Collier D (2001) Measurement validity: a shared standard for qualitative and quantitative research. Am Political Sci Rev 95(3):529–546

Andonova L, Betsill M, Bulkeley H (2009) Transnational climate governance. Glob Environ Politics 9(2):52–73

Bansard JS, Pattberg PH, Widerberg O (2017) Cities to the rescue? Assessing the performance of transnational municipal networks in global climate governance. Int Environ Agreements 17:229–246

Berg-Schlosser D, De Meur G, Rihoux B, Ragin C (2009) Qualitative Comparative Analysis (QCA) as an approach. In: Rihoux B, Ragin CC (eds) Configurational comparative methods. Qualitative Comparative Analysis (QCA) and Related Techniques. Sage Publishing, London.

Bernstein S, Clapp J, Hoffmann M (2009) Reframing global environmental governance: results of a CIGI/CIS collaboration. The Centre for International Governance Innovation. Working Paper. Environmental Governance

Bertoldi P, Rivas S, Kona A, Hernández Y, Barbosa P, Palermo V, Baldi M, Lo Vullo E, Muntean M (2020) Covenant of Mayors: 2019 Assessment. JRC Science for Policy Report

Bulkeley H, Castán Broto V (2013) Governing by experiment? Global cities and the governing of climate change. Trans Inst Br Geogr 38:361–375

Carvalho A, Monteiro A, Flannigan M, Solman S, Miranda AI, Borrego C (2011) Forest fires in a changing climate and their impacts on air quality. Atmos Environ 45(31):5545–5553

Castán Broto V (2017) Urban governance and the politics of climate change. World Dev 93:1–15

Castán Broto V, Bulkeley H (2013) A survey from urban climate change experiments in 100 cities. Glob Environ Chang 23:91–102

Coalition for Urban Transitions (2019) Climate Emergency. Urban Opportunity. How national governments can secure economic prosperity and avert climate catastrophe by transforming cities. WRI, C40 Cities and the New Climate Economy. Available at: https://urbantransitions.global/wp-content/uploads/2019/09/Climate-Emergency-Urban-Opportunity-report.pdf

Cramer W, Guiot J, Fader M, Garrabou J, Gattuso JG, Iglesias A, Lange MA, Lionello P, Llasat MC, Paz S, Peñuelas J, Snoussi M, Toreti A, Tsimplis MN, Xoplaki E (2018) Climate change and interconnected risks to sustainable development in the Mediterranean. Nat Clim Chang 8:972–980

Domonerok E, Acconcia G, Bendlin L, Ruiz Campillo X (2020) Experiments in EU climate governance: the unfulfilled potential of the Covenant of Mayors. Glob Environ Politics 20(4):1–22

Dorst Hade, Van der Jagt A, Raven R, Runhaar H (2019) Urban greening through nature-based solutions—key characteristics of an emerging concept. Sustainable Cities and Society 49

Engeli I, Rihoux B, Allison, CR (2014) Intermediate-N comparison: configurational comparative methods. In: Engeli I, Rothmayr Alison C (eds) Comparative policy studies: conceptual and methodological challenge. Palgrave MacMillan, London

European Environment Agency (2012) Urban adaptation to climate change in Europe, EEA Report 2/2012

Evans J (2019) Governing cities for sustainability: a research agenda and invitation. Front Sustain Cities 1, Article 2

García C, Peek D (2020) Smart and sustainable? Positioning adaptation to Climate Change in the European Smart City. Smart Cities 3:511–526

Gil O (2021) The Green dimension in 11 Smart City Plans: is there an environmental ethic embedded in long term strategic commitments? Elsa Estevez et al (eds) Smart Cities and Smart Governance. Springer Nature. Public Administration, Information Tech, vol 37

Giorgi F, Lionello P (2008) Climate change projections for the Mediterranean region. Glob Planet Change 63(2–3):90–104

Gordon D, Johnson C (2017) The orchestration of global urban climate governance: conducting power in the post-Paris climate regime. Environ Politics 26(4):694–714

Heidrich O, Reckien D, Olazabal M, Foley A, Salvia M, De Gregorio HS, Orru H, Flacke J, Geneletti D, Pietrapertosa F, Hamman JJP, Tiwary A, Feliu E, Dawson RJ (2016) National climate policies across Europe and their impacts on cities strategies. J Environ Manage 168:36–45

Heikkinen M, Karimo A, Klein J, Juhola S, Ylä-Anttila T (2020) Transnational municipal networks and climate change adaptation: a study of 377 cities. J Cleaner Prod 257

Hoornweg D, Sugar L, Trejos Gómez C (2011) Cities and Greenhouse Gas emissions: moving forward. Environ Urban 23(1):207–227

IPCC (2014) Climate Change 2014. Impacts, adaptation, and vulnerability. Part B: regional aspects. Working Group II Contribution to the Fifth Assessment Report of the Intergovernmental Panel on Climate Change. Cambridge University Press, Cambridge. Available at: https://www.ipcc.ch/site/assets/uploads/2018/02/WGIIAR5-PartB_FINAL.pdf

IPCC (2020) Climate Change and Land. Summary for Policymakers. Available at: https://www.ipcc.ch/site/assets/uploads/sites/4/2020/02/SPM_Updated-Jan20.pdf

Khan J (2013) What role for network governance in urban low carbon transitions? J Clean Prod 50:133–139

Kumar N (2019) Think global, act local: European cities on the front lines of climate action. Kennedy School Rev

Lindner M, Maroschek M, Netherer S, Kremer A, Barbati A, Garcia-Gonzalo J, Seidl R, Delzon S, Corona P, Kolström M, Lexer MJ, Marchetti M (2010) Climate change impacts, adaptive capacity, and vulnerability of European forest ecosystems. For Ecol Manage 259(4):698–709

Melica G, Bertoldi P, Kona A, Iancu A, Rivas S, Zancanella P (2018) Multilevel governance of sustainable energy policies: The role of regions and provinces to support the participation of small local authorities in the Covenant of Mayors. Sustain Urban Areas 39:729–739

Nevens F, Frantzeskaki N, Gorissen L, Loorbach D (2013) Urban transitions labs: co-creating transformative action for sustainable cities. J Clean Prod 50:111–122

Patton MQ (1997) Utilization-focused evaluation. Sage, Thousand Oaks

Pietrapertosa F, Khokhlov V, Salvia M, Cosmi C (2018) Climate change adaptation policies and plans: a survey in 11 South East European Countries. Renew Sustain Energy Rev 81:3041–3050

Pietrapertosa F, Salvia M, De Gregorio HS, D'Alonzo V, Church J, Geneletti D, Musco F, Reckien D (2019) Urban climate change mitigation and adaptation planning: are Italian cities ready? Cities 91:93–105

Radaelli CM, Wagemann C (2018) What did I leave out? Omitted variables in regression and qualitative comparative analysis. Eur Political Sci 18:275–290

Ragin CC (2008) Redesigning social inquiry: fuzzy sets and beyond. University of Chicago Press, Chicago

Raven R, Sengers F, Spaeth P, Xie L, Cheshmehzangi A, De Jong M (2019) Urban experimentation and institutional arrangements. Eur Plan Stud 27(2):258–281

Reckien D, Flacke J, Olazabal M, Heidrich O (2015) The influence of drivers and barriers on urban adaptation and mitigation plans—an empirical analysis of European cities. PloS ONE 10(8)

Reckien D, Salvia M, Heidrich O, Church JM, Pietrapertosa F, de Gregorio-Hurtado S, Orru H (2018) How are cities planning to respond to climate change? Assessment of local climate plans from 885 cities in the EU-28. J Clean Prod 191:207–219

Rihoux B, Alamos P, Bol D, Marx A, Rezsohazy I (2013) From Niche to mainstream method? A comprehensive mapping of QCA applications in journal articles from 1984 to 2011. Polit Res Q 66(1):175–184

Rihoux B, Ragin CC (2009) Configurational comparative methods. Qualitative Comparative Analysis (QCA) and related techniques. Sage, Thousand Oaks

Rihoux B, Rezsöhazy I, Bol D (2011) Qualitative comparative analysis (QCA) in public policy analysis: an extensive review. German Policy Stud 7(3):9–82

Ruiz Campillo X (2018) La movilización de las ciudades españolas en la lucha contra el cambio climático. Tiempo De Paz 18:110–117

Ruiz Campillo X (2019) When fighting climate change leads to better cities: a study of actions implemented by 100 cities in Spain. In: Leal Filho W, Nagy G, Borga M, Chávez P, Magnuszewsk

A (eds) Climate Change, natural hazards and adaptation option: handling the impacts of a changing climate. Springer, Berlin
San-Miguel-Ayanz J, Moreno JM, Camia A (2013) Analysis of large fires in European Mediterranean landscapes: lessons learned and perspectives. For Ecol Manage 294:11–22
Schneider CQ, Wagemann C (2012) Set-theoretic methods for the social sciences. A guide to qualitative comparative analysis. Cambridge University Press, New York
Smedby N, Neij L (2013) J Clean Prod 50:148–158
Thiem A (2014) Unifying configurational comparative methods: generalized-set qualitative comparative analysis. Sociol Methods Res 43(2):313–337
Thomann E (2018) Customized implementation of European Union food safety policy: united in diversity? International Series on Public Policy, Palgrave Macmillan
Thomann E, Maggetti M (2017) Designing research with Qualitative Comparative Analysis (QCA): approaches, challenges, tools. Sociol Methods Res 49(2):356–386
Thomann E. (2020). Qualitative Comparative Analysis for comparative policy analysis. In: Handbook of research methods and applications in comparative policy analysis. Edward Elgar Publishing
Toshkov D (2016) Research design in political science. Palgrave Macmillan, London
Ulbrich U, Lionello P, Belušic D, Jacobeit J, Knippertz P, Kuglitsch FG, Leckebusch GC, Luterbacher J, Maugeri M, Maheras P, Nissen KM, Pavan V, Pinto JG, Saaroni H, Seubert S, Toreti A, Xoplaki E, Ziv B (2012) Climate of the Mediterranean: synoptic patterns, temperature, precipitation, winds, and their extremes. In: Lionello P (eds) The Climate of the Mediterranean Region: from the past to the future. Elsevier, London
UN Department of Economic and Social Affairs (UNDESA) (2014) World Urbanization Prospects, 2014 revision, p 1
UN Habitat (2019) The Strategic Plan 2020–2023. A better quality of life for all in an urbanizing world. UN Habitat. Available at: https://bit.ly/32BDKm3
Van den Heiligenberg H, Heimerikis G, Hekkert M, Raven R, Jifke S (2018) Contrasting regional habitats for urban sustainability experimentation in Europe. Sustainability 10(5)

Climate Change or Climate Crisis? Investigating the Views of Forestry Students on the Causes, Consequences and Tools for the Mitigation of Climate Change

Evangelia Karasmanaki, Georgios Tsantopoulos, and Evangelos Manolas

Abstract Scientists express different views on climate change and often disagree about the extent to which human activity affects the climate in the long run. While much has been published on citizens as well as primary and secondary school students, research on the attitudes of undergraduate students whose study field is related to environmental management is still limited. To fill this gap, this paper examines the views and attitudes towards the causes, consequences and mitigation measures of climate change among undergraduate students of the Department of Forestry and Management of the Environment and Natural Resources, Orestiada, Greece. Results show that the respondents perceived the pollution caused by industries and the lack of effective policies on environmental management as the most important causes of climate change while ice melting and extreme temperatures were regarded as the most concerning effects. In terms of measures, to mitigate the problem, the transition to a renewable energy-based energy system along with changing citizens' environmental behaviour were rated as the most effective measures. The results presented in this study could be useful to national and international climate policies in the future.

1 Introduction

As early as in 1938, Guy Stewart Callendar, an English steam engineer and inventor, claimed that the Earth's climate is changing and established the carbon dioxide theory of anthropogenic climate change (Callendar 1938). Unfortunately, Callendar

---

E. Karasmanaki · G. Tsantopoulos · E. Manolas (✉)
Department of Forestry and Management of the Environment and Natural Resources, School of Agricultural and Forestry Sciences, Democritus University of Thrace, Orestiada, Greece
e-mail: emanolas@fmenr.duth.gr

E. Karasmanaki
e-mail: evagkara2@fmenr.duth.gr

G. Tsantopoulos
e-mail: tsantopo@fmenr.duth.gr

W. Leal Filho and E. Manolas (eds.), *Climate Change in the Mediterranean and Middle Eastern Region*, Climate Change Management,
https://doi.org/10.1007/978-3-030-78566-6_14

did not manage to convince his peers (Graßl 2007) and perhaps a lot would have been avoided if the scientific community had paid more attention to his claims back then. Some decades later, people started to observe that springs arrived earlier and that the water in lakes and rivers was warmer than before. Eventually, hundreds of peer-reviewed scientific papers and reports from the Intergovernmental Panel on Climate Change (IPCC) provided evidence showing that the levels of carbon dioxide in the atmosphere have risen and global temperature has also increased (Oreskes 2004). In other words, there is clear evidence that the global climate has changed compared to the pre-industrial period and this change has already been exerting an influence on all living organisms and ecosystems as well as humans (IPPC 2018).

Climate change will have far-reaching biophysical impacts that are likely to pose a fundamental threat to places, species and people's livelihoods. Governments responded to this threat with world summits which reached agreements that set the policy framework for establishing policies and measures which would address the risk of climate change effectively. The most recent of these agreements is the Paris Agreement which was signed by 195 countries in 2015 and agreed "to keep the increase in global mean surface temperature to well below 2 °C, and to limit the increase to 1.5 °C, since this would significantly reduce the risks and impacts of climate change" (European Commission 2020b). To meet this objective, global emissions must decrease, and carbon neutrality should be achieved by 2050. The European Union ratified the Paris Agreement and set its own ambitious targets. In particular, it set the objective to decrease greenhouse gas emissions across EU by 80–95% by the year 2050 compared to the levels recorded in 1990 and each EU member state is bound to set and fulfil national emission reduction objectives. In the meantime, the EU has set the target to decrease greenhouse gas emissions by at least 40% compared to the 1990 levels by the year 2030. In addition, the EU has adopted legislation to promote the use of renewable energy sources, and to enhance the energy efficiency of household appliances. The EU also seeks to promote carbon capture and storage technologies to engage and store carbon dioxide which is released from power stations and other large-scale facilities (European Commission 2020b).

In her foreword in the book "Climate Change Denial: Heads in the Sand", Naomi Oreskes states that people denied the scientific evidence proving the reality of anthropogenic climate change and described themselves as sceptics. These "climate change sceptics" dismissed or debated the degree to which climate change is caused by human activity, its impacts on nature and human society, as well as the adaptation to global warming (McCright 2007). Undoubtedly, this debate was harmful and inhibited efforts to address climate change in the previous years. Climate change denial can perhaps explain why for many years the public responded to this severe issue rather moderately (Rosa 2001). At the same time, the indifference of the public to climate change has been recognized as a notable concern by environmental sociologists (Rosa 2001; Zerva et al. 2018). However, citizen activism, in the form of protests and other public expressions of concern about the climate, is the most effective way to achieve emission reductions, because governments are reluctant to enact carbon restrictions without public pressure while individual energy saving on its own cannot achieve a high level of emission reductions (Ockwell et al. 2009).

In this context, scientists and individuals whose expertise is closely related to the environment are entrusted with the task of communicating the science of climate change to the rest of us. To communicate it properly and to persuade the public about the reality of climate change, however, they need to be aware of climate change themselves; that is, they need to be able to identify what causes climate change, how it impacts life on Earth and how it could be mitigated.

As climate change is a complex problem the role of universities in climate change education is of great importance. Successfully combating climate change requires capable policymakers and an informed public. Educated people are also more capable in making responsible decisions at local, national and international levels (Leal Filho 2010). A collection of essays edited by Leal Filho (2010) reviews the approaches and methods to inform, communicate and educate university students and the public on climate change as these are implemented by universities around the world. These essays also introduce initiatives, projects and communication strategies used by universities in raising awareness on climate change related matters. Finally, the book disseminates some important on-going initiatives. Leal Filho et al. (2010) offer various examples of educational approaches and methods to handle climate changes in education. The criteria for choosing these examples and methods were making knowledge personal by contextualising it but, at the same time, revealing the global, regional and local impacts. Also, Manolas and Leal Filho (2011) discuss the use of cooperative learning in dispelling student misconceptions on climate change including putting forward some cooperative learning techniques teachers can use to handle the problem. McCright et al. (2013) offer three illustrative examples of interdisciplinary climate change-related STEM (science, technology, engineering and mathematics) education projects. Kirk et al. (2014) discuss motivations, strategies, successes and support concerning teaching climate change in higher education. Monroe et al (2019) conducted a literature review of research on effective climate change education strategies.

This study aims to investigate the views and attitudes towards the causes, consequences and mitigation measures of climate change among undergraduate students of the Department of Forestry and Management of the Environment and Natural Resources, in Greece—a Mediterranean country which is bound to experience droughts and wildfires due to climate change (European Commission 2020a). These students serve as an excellent sample for the purposes of this study because as foresters they will be required to communicate climate science to the public and they will take decisions which affect the environment directly. In addition, the students attend a high number of courses on climate, energy and policy as well as courses aiming at raising environmental awareness. Some of these courses are "Climate Change", "Climate Change and Forest Ecosystems", "Climate Change Scenarios—Adaptation and Vulnerability", "Renewable Energy Sources", "Energy and Environment", "Investments and Development of Forest Resources", "Environmental Education", "Environmental Policy" and "Environmental Communication".

With regard to the study program, it is expected that the students will exhibit adequate knowledge about the causes, effects and mitigation measures of climate change. Based on the findings, the curriculum of the department could be improved and strategies aiming at raising awareness about climate change could be designed. In addition, this study can provide the basis for extending the research to students of other environmental departments in Greece and elsewhere.

2 Theoretical Background

The current form of human species has lived on this planet for around 195,000 years which is little compared to the total of 3.56 billion years of life on Earth. For many centuries, humans lived in small groups and survived by hunting animals and gathering plants. It could be said that this was the only time in history that humans lived in absolute harmony with nature. Later, the Agricultural Revolution and the Industrial Revolution marked a major turning point in history and changed nearly every aspect of life. According to economists, the most critical impact of the Industrial Revolution was that the standard of living for a great part of the population in the western world increased steadily (Lucas 2002). Nevertheless, this prosperity did not come without a price. In comparison to the pre-industrial levels in 1850–1900, an increase—estimated at around 0.87 °C—in combined surface air and sea surface temperatures has been observed and consequently land and ocean regions are experiencing greater warming (IPPC 2018). These temperature increases are an indication that *the climate of Earth is changing.*

Although both anthropogenic and natural drivers have been scientifically recognized as causes of climate change, the human drivers are exceeding the natural ones (Washington 2013). As fossil fuels have been combusted to produce energy for industrial and electricity purposes for more than 150 years, the emissions related to energy services are responsible for most global emissions which trigger climate change (Karasmanaki et al. 2020; IPPC 1990). In addition, the elimination of forests, the ever-increasing livestock farming, the decline of natural carbon sinks such as mangrove swamps, the excessive use of fertilisers containing nitrogen, the use of fluorinated gases, changes in the hydrological regime of vast regions due to the conversion of areas into grasslands, as well as deforestation in the tropics have also been acknowledged as anthropogenic drivers of climate change (European Commission 2020a; Washington 2013; IPPC 2011, 1990). Moreover, the moderate public responses to the issue especially in the previous years have delayed the establishment of drastic measures and in a sense, this delayed response contributed to the issue (Rosa 2001). However, natural causes have also contributed to climate change. Most importantly, volcanic eruptions consist an important natural cause of climate change (Robock 2000) while changes in the climate have also been linked to the natural variability in stratospheric ozone as well as to shifts due to natural fluctuations in solar irradiance and carbon levels in the atmosphere (Washington 2013).

Scientists have predicted a multitude of impacts that would result from global climate change. Impacts can be placed into four basic categories: ecological, health, social, and economic. As far as ecology is concerned, the extant literature has suggested that almost all species would be affected adversely, and ecosystems would experience abrupt changes (Barnosky et al. 2011; IPCC 2014). At the same time, there are indications that climate change has already influenced the distribution and population of living organisms (Kerr et al. 2015; Parmesan and Yohe 2003) while it is expected to cause species extinction and loss of biodiversity (Johnson et al. 2017; Barnosky et al. 2011; Schneider and Lane 2006; IPPC 2001). It is also possible that higher surface temperatures will lead to sea level rise which will submerge coastal areas and islands. Based on scientists' estimations, climate change will have effects which will be felt by most people as there will be more frequent and more intense weather events including heat waves, hurricanes, storms, cyclones, floods and droughts (IPPC 2018; IPPC 2001).

In terms of human health, the temperature increase will accelerate the spread of disease (IPPC 2001) and millions of people will suffer hunger, malaria and water shortage (Parry et al. 2001). However, human societies will also be affected as there will be millions of refugees who will be displaced due to land loss, environmental degradation and natural disasters (Conisbee and Simms 2003; Assadourian 2013). While low-lying areas will be inundated by floods, other regions will face water scarcity. Water will become a vital issue, and some analysts foresee "water wars" as water supplies get scarcer (Shiva 2016). What is more, the effects of climate change will not be distributed evenly resulting in climate change inequities. Schneider and Lane (2006) divided the inequities into three categories: inter-country, intergenerational and inter-species equity. Beside health and society, economy will be influenced negatively and there is already considerable uncertainty surrounding the economic consequences of this phenomenon with investors already expressing interest in products which address climate risks (Engle et al. 2020; Molchanova et al. 2020).

Since emissions force the climate to change, the most effective step to mitigate climate change would be to curb the emissions from energy systems. The modern lifestyle, however, requires abundant energy with industries relying on energy to operate and homes needing electricity to provide a comfortable environment. For this reason, transition to a low carbon energy system in which renewable energies will account for a high share, is widely recognized as the most optimal path to follow in order to tackle climate change. That is because renewable energy sources generate low emissions and even when estimating the "life cycle" emission rates (the emissions produced during manufacturing, installation, operation and decommissioning of the facility), the emissions of renewable energy in comparison to fossil fuels are still notably lower (Karasmanaki and Tsantopoulos 2019; IPPC 2011). Another measure to lower carbon dioxide concentrations in the atmosphere would be to plant millions of hectares of new forests. Trees absorb carbon through their leaves as part of photosynthesis and consequently the new forests would capture huge amounts of carbon dioxide which would otherwise contribute to global warming (Turner et al. 2009; Trabucco et al. 2008). Addressing climate change should also involve adaptation strategies; that is, preparing for the management of inevitable consequences. In specific, there should be more focus on the planning for managing extreme climate and weather events (Rolnick et al. 2019).

As already stated, the public response to climate change was somewhat delayed and moderate in the previous decades probably because citizens had received little education and contradicting information related to climate change (Zerva et al. 2019a). The key to solving environmental issues is to recognize that social change is feasible when citizens unite and act together to bring about change (Zerva et al. 2019b). Remarkably, only 5–10% of the population of a community, region or country suffices to bring about quick social change. From this perspective, public mobilization could bring change and lead to the solution of climate change. However, this requires an increased level of awareness because individuals who have awareness about the environment and its problems are more likely to take action. To that end, education has a critical role to play in raising awareness among students about climate change and connecting the issue to social dimensions such as water scarcity, food shortage and energy (Rees 2003). Such a type of education would not only enable students understand why they should act for the climate, but it could also teach how individual habits affect the climate. In particular, individual lifestyle choices such as consuming less meat, saving energy and water or walking instead of driving, are only some habits which can lower our environmental footprint (Solomon and LaRocque 2019).

Acknowledging that current students will play an important role as future citizens in the solution of environmental issues, many social studies have focused on students' views on climate change. There is research indicating that most US college students (Wachholz et al. 2014) as well as German and Chinese university students majoring in environmental science (von Storch et al. 2019) perceived that climate change is largely induced by humans. Meanwhile, a considerable share of engineering students in different colleges and universities in the US were neutral or disagreed with the contribution of humans in climate change (Shealy et al. 2017).

The extant literature has also examined how dangerous students regard climate change to be. On a general note, students seemed to be appreciably aware of the risks stemming from climate change. In particular, strong majorities of environmental students in China and Germany were certain that climate change poses severe threats to humanity (von Storch et al. 2019). Risks were acknowledged also by Chinese students majoring in preventive and medical medicine who perceived that humans would be negatively affected in terms of health (Liao et al. 2019). In another study, it was indicated that many school students believed that many people would die due to floods, extreme temperatures, water and food shortage (Shepardson et al. 2011, 2009). The impact of climate change on global food production and human health was also recognized as a risk by Greek school students (Liarakou et al. 2011), as well as by Tanzanian university students who perceived that climate change would influence adversely the spread of disease (Carr et al. 2015). In addition, it appears that students are aware that climate change is responsible for shifts in weather and seasons as well as plants and animals. Indicatively, Greek school students noticed that the average global temperature is increasing over time and perceived that extreme weather events will occur more frequently and intensely (Liarakou et al. 2011).

Similarly, the overwhelming majority of their US counterparts observed that the weather was likely to get warmer with hotter days, shorter winter seasons and longer summers, less rainfalls, higher evaporation as well as rises in humidity levels (Shepardson et al. 2009). The occurrence of extreme weather events like droughts and floods were mentioned as adverse impacts of climate change by junior and senior university students in Tanzania (Carr et al. 2015). In relation to plant and animal species, US school students thought that multiple animals and plants would die or become extinct (Shepardson et al. 2009) while Tanzanian university students knew little about the effect of climate change on biodiversity (Carr et al. 2015).

Environmental issues often create emotions among individuals and climate change was no exception. Interestingly, most students in Oman and U.S. reported a high level of concern about climate change (Ambusaidi et al. 2012; Wachholz et al. 2014) while business students were less worried about the issue (Artz et al. 2012) and the concern among Chinese students was rather neutral (Deng et al. 2017).

To conclude, Earth's climate is changing, and this change is accompanied by catastrophic consequences for humans. As human activities have a discernible impact on warming trends, there should be more focus on analysing individuals' views about climate change and its causes, effects and solutions. The environmental students whose views are analysed in this study will become experts in the years to come and will possibly affect the environment with their decision-making. For this reason, the continuous investigation of students' attitudes towards climate change is imperative to make timely changes in curricula and to help these students become key actors in future efforts to tackle climate change.

3 Methodology

The findings presented in this paper are part of a broader research, which was approved by the Research Ethics Committee of the Democritus University of Thrace and whose aim was to investigate the perceptions and attitudes towards climate change among undergraduate students of the Department of Forestry and Management of the Environment and Natural Resources at the Democritus University of Thrace, which is located in the town of Orestiada in Greece (Fig. 1).

To achieve the aim of the study, the authors designed a questionnaire which was six pages long and consisted of 29 items with all of them being closed-ended questions. Given that the students would answer it during classroom time, the closed-ended type was considered as the most appropriate type as it requires less time and minimum effort to complete. It is also worthwhile to note that all items drew on the findings of previous relevant studies and were specifically designed to capture students' attitudes with regard to various climate change-related topics. In specific, the focus was on studies exploring the awareness, opinions, perceptions and attitudes of school and undergraduate students towards climate change.

The items of the questionnaire had to be formed in such a way that they would not only deliver the necessary information, but also be reliable and comprehensible.

Fig. 1 The location of the Democritus University of Thrace (Orestiada, Greece). *Source* Google Maps

In order to ensure that the questionnaire would give accurate and coherent findings, it was deemed necessary to carry out a pilot study on a limited scale. Based on the results of the pilot study certain minor changes were made and the final questionnaire was ready to be administered.

The questionnaires were administered by professors of each class at the beginning of regular class periods. In total, 187 undergraduate students completed the questionnaire from November 2019 to December 2019. The chosen sampling method was multistage sampling and the collected data were analysed using the Statistical Package for the Social Sciences (SPSS). To scrutinize the collected data and draw meaningful conclusions, descriptive statistics was applied to all variables and then different statistical tests were conducted including the non-parametric Friedman test. The Friedman test was applied in order to compare the values of three or more correlated groups of variables. The distribution of the Friedman test is X^2 distribution with degrees of freedom (df) df $= k - 1$, where k is the number of groups or samples. This test classifies the values of variables for every subject separately and calculates the mean rank of classification values for each variable.

Finally, the findings of this study have to be seen in the light of some limitations. First, the curriculum of the forestry department may have affected students' views on climate change. This means that the findings presented in this study do not apply to and do not represent the students majoring in other forestry departments which provide different curricula. Instead, the findings represent only the undergraduate students who study at the Department of Forestry and Management of the Environment and Natural Resources at the Democritus University of Thrace. Second, it is possible that students may have been affected by the climatic conditions of the town where the department is located since this is where they spend most of their time. As Orestiada is located at the northernmost part of Greece where average temperatures are lower than temperatures observed in continental areas, it is possible that the students may have been affected by the temperatures in Orestiada in evaluating the effects of climate change.

4 Results

This section focuses on the presentation of the demographic characteristics of students as well as on the presentation of views but also on the ranking of evaluations with regard to the causes of climate change.

4.1 Demographic Characteristics

The questionnaire included a separate section gathering information about the students' gender, year of study as well as parental occupational and educational level. As it can be seen in Table 1, the sample consisted of 187 students of which 74 were male (39.6%) and 113 (60.4%) were female.

Next, Table 2 presents the respondents' year of study. It is worth-noting that the duration of study in the Department of Forestry is five years. Participation in the third and fourth years of study is lower, while the first and fifth years enjoy higher percentages.

The students' family background with regard to parental occupation and education level were also examined. As it is shown in Table 3, a notable share of students' fathers is employed in the public (23%) and private (21.4%) sectors, whereas only 5.3% of the fathers were unemployed. As for the occupation of the mothers of respondents, substantial percentages were employed in public and private sectors (19.8 and 21.4%,

Table 1 Frequency and percentages of students' gender

Gender	Frequency	Percentage
Male	74	39.6
Female	113	60.4
Total	187	100.0

Table 2 Frequency and percentages regarding students' year of study

Year of study	Frequency	Percentage
1	49	26.2
2	32	17.1
3	23	12.3
4	26	13.9
5	57	30.4
Total	187	100.0

Table 3 Frequency and percentages regarding students' fathers' occupational level

	Fathers' occupation		Mothers' occupation	
	Frequency	Percentage	Frequency	Percentage
Employed in the public sector	43	23.0	37	19.8
Employed in the private sector	40	21.4	40	21.4
Free-lancer	44	23.5	29	15.5
Household	0	0.00	47	25.1
Farmer	23	12.3	13	7.0
Unemployed	10	5.3	8	4.3
Pensioner	27	14.4	13	7.0
Total	187	100.0	187	100.0

Table 4 Frequency and percentages of the students' parental educational level

	Educational level of fathers		Educational level of mothers	
	Frequency	Percentage	Frequency	Percentage
Compulsory education	53	28.3	28	15.0
Secondary education	72	38.5	76	40.6
Higher education	62	33.2	83	44.4
Total	187	100.0	187	100.0

respectively) while a considerable share of mothers (25.1%) was not involved in paid work but was involved with household duties. Finally, only very few mothers were unemployed (4.3%).

With regard to the education level of students' parents, Table 4 shows that considerable shares of fathers (33.2%) and mothers (44.4%) were university graduates, while the proportion of mothers having completed compulsory education was significantly lower.

4.2 *Causes of Climate Change*

With regard to the causes of climate change, 52.9% of the respondents regards the pollution caused by businesses and industries very important and 34.8% extremely important while 45.5% of the respondents regards the lack of effective policies as extremely important. Another cause worth mentioning is deforestation which is regarded as very important by 48.1% of the respondents and extremely important

Table 5 Percentage ratios regarding the causes of climate change

	Not at all	Slightly	Moderately	Much	Very much
Increase of greenhouse gases	1.1	7.0	26.7	50.8	14.4
Pollution caused by businesses and industries	2.1	2.1	8.0	52.9	34.8
Aerosols	2.1	12.3	38.5	37.4	9.6
Use of fossil fuels for energy production	1.1	5.3	20.3	50.8	22.5
Use of fossil fuels for transportation	0.0	7.0	19.8	52.4	20.9
Volcanic explosions	7.0	41.7	27.8	18.2	5.3
Deforestation	1.6	3.2	21.4	48.1	25.7
Overpopulation	2.7	9.6	25.7	40.1	21.9
Lack of effective policies	1.1	5.3	16.0	32.1	45.5

Table 6 Ranking of the causes of climate change

	Mean rank
Increase of greenhouse gases	4.76
Pollution caused by businesses and industries	6.19
Aerosols	4.06
Use of fossil fuels for energy production	5.38
Use of fossil fuels for transportation	5.33
Volcanic explosions	2.61
Deforestation	5.57
Overpopulation	4.85
Lack of effective policies	6.24
$N = 187$ Chi-Square $= 333.587$ df $= 8$ $p < 0.001$	

by 25.7% of the respondents. On the other hand, 41.7% regards volcanic explosions as a rather insignificant cause while 38.5% of the respondents regards aerosols as a fairly significant cause (Table 5).

Next in order to rank the causes of climate change Friedman's test was applied. The application of the above test showed the main cause is lack of effective policies with mean rank 6.24 with the one coming next being pollution caused by businesses and industries with mean rank 6.19 (Chi-Square $= 333.587$ df $= 8$ $p < 0.001$). It is worth noting the last cause is volcanic explosions with mean rank 2.61 (Table 6).

4.3 Impacts of Climate Change

Table 7 presents the impacts of climate change. 48.1% regards extreme temperatures

Table 7 Percentage ratios regarding the impacts of climate change

	Not at all	Slightly	Moderately	Much	Very much
Melting of the ice	0.0	1.6	15.0	38.5	44.9
Floods	0.5	7.5	28.9	39.0	24.1
Drought	0.5	10.2	25.7	39.6	24.1
Fires	0.0	5.9	18.2	40.6	35.3
Loss of biodiversity	0.5	4.3	21.9	37.4	35.8
Destruction of food chain	1.1	4.3	22.5	34.8	37.4
Desertification and steppe formation	1.6	11.8	34.8	32.6	19.3
Extreme temperatures		2.7	10.7	38.5	48.1
Spreading of pathogens	1.1	10.7	25.1	30.5	32.6

as an extremely important impact of climate change and 38.5% a very important impact. 38.5% regards the melting of the ice a very important impact. 10.7% think that climate change has little impact on the spreading of pathogens and 10.2% that climate change has little impact on drought.

In order to rank the impacts of climate change Friedman's test was applied (Table 8). The application of the above test showed that the most important impact of climate change is extreme temperatures with mean rank 6.07 with the one coming next being the melting of the ice with mean rank 5.89 (Chi-square = 164.195 df 9 = 8 Asymp. $p < 0.001$). It is worth-noting that in the last ranking positions we find drought with mean rank 2.61 and desertification and steppe formation with mean rank 3.83.

Table 8 Ranking of the impacts of climate change

	Mean rank
Melting of the ice	5.89
Floods	4.46
Drought	4.49
Fires	5.26
Loss of biodiversity	5.19
Destruction of food chain	5.17
Desertification and steppe formation	3.83
Extreme temperatures	6.07
Spreading of pathogens	4.64
$N = 187$ Chi-Square = 164.195 df = 8 $p < 0.001$	

4.4 Measures for Confronting Climate Change

With regard to confronting climate change the adoption of legislation which will contain stricter sanctions on industries and businesses which are characterized by excessive emission of pollutants, the transition to an energy system which will be based on renewable energy sources and the reduction of carbon emissions as a result of change in the environmental behavior of citizens are regarded as extremely important measures by 44.9, 39.6 and 37.4% respectively. On the contrary 15.5 and 10.2% think that the abolition of subsidies which support fossil fuels and the frequent and systematic realization of programs of environmental communication and education aimed at creating climate awareness are of little importance (Table 9).

Next in order to rank the measures for confronting climate change Friedman's test was applied. The application showed that the most important measure is the transition to a system of energy which will rely on renewable energy sources with mean rank 5.70 with the one coming next being the adoption of legislation which will contain stricter sanctions on industries and businesses which are characterized by excessive emission of pollutants as a result of change in the environmental behavior of citizens with mean rank 5.60 (Chi-Square = 108.262. df = $p < 0.001$). Last with regard to

Table 9 Percentage ratios with regard to measures for confronting climate change

	Not at all	Slightly	Moderately	Much	Very much
Reforestation of tropical forests	0.5	7.5	23.5	34.8	33.7
Planting of trees in urban centers	1.1	9.6	21.9	38.0	29.4
Reduction of carbon emissions as a result of changes in the environmental behavior of citizens	0.0	2.7	17.1	42.8	37.4
Abolition of subsidies which support fossil fuels	1.1	15.5	31.0	36.9	15.5
Transition to an energy system which will rely on renewable energy sources	0.5	3.2	15.5	41.2	39.6
Adoption of stricter models of energy saving to all household devices but also energy measures for buildings and vehicles	0.0	5.3	29.9	34.8	29.9
Adoption of sustainable practices in agriculture and livestock farming		6.4	24.6	39.6	29.4
Adoption of legislation which will contain stricter sanctions on industries and businesses which are characterized by excessive emission of pollutants	1.1	4.3	18.7	31.0	44.9
Frequent and systematic realization of programs of environmental communication and education aimed at creating climate awareness	0.5	10.2	19.3	35.8	34.2

Table 10 Ranking of measures for confronting climate change

	Mean rank
Reforestation of tropical forests	4.93
Planting of trees in urban centers	4.82
Reduction of carbon emissions as a result of changes in the environmental behavior of citizens	5.60
Abolition of subsidies which support fossil fuels	3.68
Transition to an energy system which will rely on renewable energy sources	5.70
Adoption of stricter models of energy saving to all household devices but also energy measures for buildings and vehicles	4.85
Adoption of sustainable practices in agriculture and livestock farming	4.88
Adoption of legislation which will contain stricter sanctions on industries and businesses which are characterized by excessive emission of pollutants	5.61
Frequent and systematic realization of programs of environmental communication and education aimed at creating climate awareness	4.94
$N = 187$ Chi-Square = 108.262 df = 8 $p < 0.001$	

ranking is the abolition of subsidies which support fossil fuels with mean rank 3.68 (Table 10).

5 Discussion

This research examined the views of environmental students on the causes and effects of climate change and, most importantly, it revealed the solutions that the students recommend in order to tackle this grave issue. As one would expect, the students had an adequate understanding of the major processes of climate change, but some knowledge gaps were also identified.

In terms of the causes of climate change, the students ranked the lack of effective policies, pollution caused by industries, deforestation and the combustion of fossil fuels for energy production as the most important causes of the issue. Since students ranked human activities higher than natural drivers, it can be stated that the respondents were familiar with scientific evidence showing that although climate change is ascribed to both human and natural drivers, human activity is more accountable for climate change (Washington 2013). That being said, the use of fossil fuels has been recognized as a major cause of global warming which, in turn, triggers climate change (European Commission 2020a) but our respondents ranked it behind pollution induced by industries and deforestation. Moreover, volcanic eruptions received the lowest ranking even though they consist a significant natural cause of climate change (Robock 2000). The lower ratings the students gave to the combustion of fossil fuels and volcanic eruptions as drivers of climate change suggest that the respondents were perhaps confused about the degree to which fossil fuels and volcanoes have contributed to climate change. In addition, even though the students knew that human activities have affected the climate more than natural causes, it seems that they could not evaluate the contribution of each human driver to the issue.

To our surprise, the respondents ranked the lack of effective policies as the most important cause of climate change. Given that students have correctly ranked pollution from industrial activities and deforestation as important causes of the issue, it can be argued that the respondents were well aware that the lack of policies is not a direct nor the most important cause of climate change but with this ranking they possibly expressed their desire for more effective policies on climate change. In the past, scientific evidence on climate change has been debated acutely and this debate resulted in serious delays in establishing stricter emissions policies on polluting industries. Hence, it is possible that students were aware that this debate had a negative effect on mobilizing authorities to establish stricter measures.

The Mediterranean will be particularly vulnerable to climate change and will possibly experience pronounced warming, drought and wildfires (European Commission 2020a; Giorni and Lionello 2008). In our study, the students ranked extreme temperatures, ice melting, fires and biodiversity loss as the most important impacts of climate change whereas drought, desertification and steppe formation received the lowest rankings. It seems that although the students had a good grasp of the ecological impacts of climate change, tended to place less importance on drought which, however, is an impact that will be intense in the Mediterranean region. In addition, the spreading of pathogens was ranked quite low although it has been acknowledged as a severe impact of climate change. These rankings suggest that students may have some misconceptions about the impacts of climate change.

The European Union is often described as a global leader in environmental policy and has established a multitude of policies, directives and measures to address climate change through the promotion of environmentally friendly energy sources (European Commission 2020a). Remarkably, students' responses on mitigation measures resonate with the measures that the Union is already implementing. In specific, the students ranked transition to an energy system based on renewables as the most important mitigation measure followed by stricter legislation on polluting industries and emission reduction resulting from changes in citizens' environmental behavior. Therefore, it seems that the students are familiar with the current policies being implemented across EU and appreciate their effectiveness. Students' awareness about effective mitigation measures may be attributed to the curriculum; that is, the Department of Forestry and Management of the Environment and Natural Resources of the Democritus University of Thrace provides the highest number of renewable energy-related courses, but also courses on climate change, environmental policy and environmental communication.

6 Conclusions

The purpose of the present study has been to capture the views of forestry students on climate change. The investigation of their views on this severe issue is meaningful for two reasons. First, their level of knowledge and awareness about climate change is likely to affect their decisions as foresters on efforts intended to mitigate climate change in the future. Second, the examination of their views on topics related to climate change can reveal the effectiveness of the current curriculum of the department so that timely changes can be made.

With regard to the findings presented in this study and the discussion that followed, certain conclusions may be drawn. Most importantly, it may be inferred that the students attending the Department of Forestry and Management of the Environment and Natural Resources had an adequate level of knowledge about the causes and impacts of climate change and recommended mitigation measures which resonate with the key policies that the European Union has already been applying to tackle climate change.

Even though the respondents were found to be knowledgeable about the causes, effects and mitigation of climate change, some degree of confusion was detected. That is, the students did not identify drought as an important effect despite the fact that climate change has an influence on a variety of factors which are contributing to drought in areas such as the Mediterranean. Future foresters, however, have to be well prepared in order to address climate change in their careers and, therefore, possible misconceptions need to be addressed while they are still attending forestry studies. For this reason, material teaching in depth the effects of climate change to students should be added in the relevant undergraduate courses. Moreover, as climate change will not have only ecological impacts but will also affect human health, society and economy, it is highly advisable to reform the curriculum of the department and include the latest research findings on the multifaceted impacts of climate change so that the students can form a comprehensive understanding of the different ways in which climate change will affect the world.

Nevertheless, the students had a good grasp of the mitigation measures which can tackle climate change effectively and it could be inferred that the curriculum of the Department, which encompasses courses on climate science, energy and environmental policy, has succeeded in disseminating knowledge on ways to mitigate climate change. In specific, it was striking that the students recommended the use of renewables as the most important measure to mitigate climate change. The emphasis on renewable energy sources not only suggests that students value their potential but also shows that they are familiar with the EU policies promoting renewable energies as a tool to lower the emissions which trigger climate change.

The findings of this study can serve as the foundation to extend the research on awareness about climate change to undergraduate students from other forestry and environmental departments both in Greece and elsewhere. The findings of such studies could be used to improve the curricula and to enhance the engagement of students with climate change. In addition, given that there are rapid advances in the scientific understanding of climate change, it is recommended that similar studies exploring students' attitudes to climate change are carried out on a more frequent basis and in a systematic manner so that it is discernible whether students keep up with the new knowledge. Finally, the present findings could be used by policymakers, curriculum designers and educational institutions to develop new strategies in order to raise awareness about climate change. To conclude, climate change is a potential threat that would have catastrophic consequences for humanity, but education is perhaps the most important element of the global response to climate change especially for individuals who are likely to be key actors in the environmental sector in the years to come.

References

Ambusaidi A, Boyes E, Stanisstreet M, Taylor N (2012) Omani students' views about global warming: beliefs about actions and willingness to act. Int Res Geogr Environ Edu 21(1):21–39

Artz N, Chene D, Wachholz S (2012) Warming to the challenge: are business students learning about climate change?. In: Competition forum (vol 10, no 1). American Society for Competitiveness, p 191

Assadourian E (2013) The rise and fall of consumer cultures. 21st Century, 2 (12)

Barnosky AD, Matzke N, Tomiya S, Wogan GO, Swartz B, Quental TB, Marshall C, McGuire JL, Lindsey EL, Maguire KC, Mersey B (2011) Has the Earth's sixth mass extinction already arrived? Nature 471(7336):51–57

Callendar GS (1938) The artificial production of carbon dioxide and its influence on temperature. Q J R Meteorol Soc 64(275):223–240

Carr P, Buggy CJ, McGlynn G (2015). Climate change awareness amongst secondary level students' in a Dar es Salaam University College of Education (DUCE) affiliated school in urban Tanzania. In: United Nations sustainable development network 3rd annual international conference on sustainable development practice, Columbia University, New York, USA, 23–24 Sept 2015

Conisbee M, Simms A (2003) Environmental refugees: the case for recognition. New Economics Foundation, London, pp 17–18

Deng Y, Wang M, Yousefpour R (2017) How do people's perceptions and climatic disaster experiences influence their daily behaviors regarding adaptation to climate change?—a case study among young generations. Sci Total Environ 581:840–847

Engle RF, Giglio S, Kelly B, Lee H, Stroebel J (2020) Hedging climate change news. Rev Financ Stud 33(3):1184–1216

European Commission. (2020a). Causes of climate change. Available at: https://ec.europa.eu/clima/change/causes_en

European Commission. (2020b). Climate change mitigation. Available at: https://www.eea.europa.eu/themes/climate/intro

Giorgi F, Lionello P (2008) Climate change projections for the mediterranean region. Glob Planet Change 63(2–3):90–104

Graßl H (2007) Guy stewart callendar: A pioneer of anthropogenic climate change theory. GAIA-Ecol Perspect Sci Soc 16(3):222–224

IPCC, 2001: Climate Change 2001: the scientific basis. contribution of working Group I to the Third Assessment Report of the Intergovernmental Panel on Climate Change. In: Houghton JT, Ding Y, Griggs DJ, Noguer M, van der Linden PJ, Dai X, Maskell K, Johnson CA (eds) Cambridge University Press, Cambridge, United Kingdom and New York, NY, USA, pp 881

IPCC (2011) Summary for policymakers. In: Edenhofer O, Pichs-Madruga R, Sokona Y, Seyboth K, Matschoss P, Kadner S, Zwickel T, Eickemeier P, Hansen G, Schlömer S, von Stechow C (eds)

IPCC special report on renewable energy sources and climate change mitigation, Cambridge University Press, Cambridge, United Kingdom and New York, NY, USA
IPCC (2014) Climate Change 2014: synthesis report. Contribution of working groups I, II and III to the fifth assessment report of the intergovernmental panel on climate change (Core Writing Team, R.K. Pachauri and L.A. Meyer (eds)). IPCC, Geneva, Switzerland, pp 151
IPCC (2018) Global Warming of 1.5 °C. In: Masson-Delmotte V, Zhai P, Pörtner H-O, Roberts D, Skea J, Shukla PR, Pirani A, Moufouma-Okia W, Péan C, Pidcock R, Connors S, Matthews JBR, Chen Y, Zhou X, Gomis MI, Lonnoy E, Maycock T, Tignor M, Waterfield T (eds) An IPCC Special Report on the impacts of global warming of 1.5 °C above pre-industrial levels and related global greenhouse gas emission pathways, in the context of strengthening the global response to the threat of climate change, sustainable development, and efforts to eradicate poverty. Intergovernmental Panel on Climate Change
IPPC (1990) First Assessment Report, vol III. WG III Formulation of Response Option Strategies.
Johnson CN, Balmford A, Brook BW, Buettel JC, Galetti M, Guangchun L, Wilmshurst JM (2017) Biodiversity losses and conservation responses in the anthropocene. Science 356(6335):270–275
Karasmanaki E, Tsantopoulos G (2019) Exploring future scientists' awareness about and attitudes towards renewable energy sources. Energy Policy 131:111–119
Karasmanaki E, Ioannou K, Katsaounis K, Tsantopoulos G (2020) The attitude of the local community towards investments in lignite before transitioning to the post-lignite era: the case of Western Macedonia, Greece. Res Policy 68:101781
Kerr JT, Pindar A, Galpern P, Packer L, Potts SG, Roberts SM, Rasmont P, Schweiger O, Colla SR, Richardson L, Wegner DL, Wagner DL, Gall LF, Sikes DS, Pantoja A (2015) Climate change impacts on bumblebees converge across continents. Science 349(6244):177–180
Kirk KB, Gold AU, Ledley TS, Sullivan SB, Manduca CA, Mogk DW, Wiese K (2014) Undergraduate climate education: motivations, strategies, successes, and support. J Geosci Educ 62:538–549
Leal Filho W (2010) Climate change at universities. Results of a world survey. In: Leal Filho W (Ed) Universities and climate change: introducing climate change to university, Springer, Berlin, pp 1–19
Leal Filho W, Pace P, Manolas E (2010) The contribution of education towards meeting the challenges of climate change. J Balt Sci Educ 9(2):142–155
Liao W, Yang L, Zhong S, Hess JJ, Wang Q, Bao J, Huang C (2019) Preparing the next generation of health professionals to tackle climate change: are China's medical students ready? Environ Res 168:270–277
Liarakou G, Athanasiadis I, Gavrilakis C (2011) What Greek secondary school students believe about climate change? Int J Environ Sci Educ 6(1):79–98
Lucas RE (2002) Lectures on economic growth. Harvard University Press
Manolas E, Leal Filho W (2011) The use of cooperative learning in dispelling student misconceptions on climate change. J Balt Sci Educ 10(3):168–182
McCright AM (2007) Dealing with climate change contrarians. In: Creating a climate for change: communicating climate change and facilitating social change, pp 200–212
McCright AM, O'Shea BW, Sweede RD, Urquhart GR, Zeleke A (2013) Promoting interdisciplinarity through climate change education. Nat Clim Chang 3:713–716
Molchanova TK, Yashalova NN, Ruban DA (2020) Environmental concerns of Russian businesses: top company missions and climate change agenda. Climate 8(4):56
Monroe MC, Richard R, Plate RR, Oxarart A, Bowers A, Willandia A, Chaves WA (2019) Identifying effective climate change education strategies: a systematic review of the research. Environ Educ Res 25(6):791–812
Ockwell D, Whitmarsh L, O'Neill S (2009) Reorienting climate change communication for effective mitigation: forcing people to be green or fostering grass-roots engagement? Sci Commun 30:305–327
Oreskes N (2004) The scientific consensus on climate change. Science 306(5702):1686–1686

Parmesan C, Yohe G (2003) A globally coherent fingerprint of climate change impacts across natural systems. Nature 421(6918):37–42
Parry ML, Arnell NW, McMichael T, Nicholls RJ, Martens P, Kovats RS, Livermore MTJ, Rosenzweig C, Iglesias A, Fischer G (2001) Millions at risk: defining critical climate threats and targets. Glob Environ Chang 11(3):1–3
Rees WE (2003) Impeding sustainability. Plan High Educ 31(3):88
Robock A (2000) Volcanic eruptions and climate. Rev Geophys 38(2):191–219
Rolnick D, Donti PL, Kaack LH, Kochanski K, Lacoste A, Sankaran K, Ross AS, Milojevic-Dupont N, Jaques N, Waldman-Brown A, Luccioni A (2019) Tackling climate change with machine learning. arXiv preprint arXiv:1906.05433
Rosa EA (2001) Global climate change: background and sociological contributions. Soc Nat Resour 14(6):491–499
Schneider SH, Lane J (2006) An overview of 'dangerous' climate change. In: Avoiding dangerous climate change. In: Joachim Schellnhuber H, Cramer W, Nakicenovic N, Wigley T, Yohe G (Eds) Avoiding dangerous climate change, Cambridge University Press, pp 7–22
Shealy T, Valdes-Vasquez R, Klotz L, Potvin G, Godwin A, Cribbs J, Hazari Z (2017) Half of students interested in civil engineering do not believe in anthropogenic climate change. J Prof Issues Eng Educ Pract 143(3):D4016003
Shepardson DP, Niyogi D, Choi S, Charusombat U (2009) Seventh grade students' conceptions of global warming and climate change. Environ Educ Res 15(5):549–570
Shepardson DP, Niyogi D, Choi S, Charusombat U (2011) Students' conceptions about the greenhouse effect, global warming, and climate change. Clim Change 104(3–4):481–507
Shiva V (2016) Water wars: privatization, pollution, and profit. North Atlantic Books
Solomon CG, LaRocque RC (2019) Climate change—a health emergency. N Engl J Med 380(3):209–211
Trabucco A, Zomer RJ, Bossio DA, van Straaten O, Verchot LV (2008) Climate change mitigation through afforestation/reforestation: a global analysis of hydrologic impacts with four case studies. Agr Ecosyst Environ 126(1–2):81–97
Turner WR, Oppenheimer M, Wilcove DS (2009) A force to flight global warming. Nature 462:278–279
Von Storch H, Xue-En CHEN, Birgit PE, Dennis BRAY, Ullmann A (2019) Attitudes of young scholars in Qingdao and Hamburg about climate change and climate policy–the role of culture for the explanation of differences. Adv Clim Chang Res 10(3):158–164
Wachholz S, Artz N, Chene D (2014) Warming to the idea: university students' knowledge and attitudes about climate change. Int J Sustain High Educ 15:128–141
Washington H (2013) Climate change denial: heads in the sand. Routledge. https://doi.org/10.4324/9781849776738
Zerva A, Tsantopoulos G, Grigoroudis E, Arabatzis G (2018) Perceived citizens' satisfaction with climate change stakeholders using a multicriteria decision analysis approach. Environ Sci Policy 82:60–70
Zerva A, Manolas E, Skanavis C, Tsantopoulos G (2019) Effectiveness of communication strategies in confronting climate change: the views of the citizens of Greece. Addressing the challenges in communicating climate change across various audiences. Springer, Cham, pp 499–516
Zerva A, Tsantopoulos G, Manolas E, Tampakis S (2019) The views of citizens on the issue of participation in confronting climate change: the case of Greece. Addressing the challenges in communicating climate change across various audiences. Springer, Cham, pp 457–480

Development of a Water Security Index Incorporating Future Challenges

Deniz Marangoz and Irem Daloglu

Abstract Climate change and population increase are worsening the problem of water security, one of the major challenges faced by society. Especially in agricultural regions, water security is crucial for achieving food security and improving livelihoods. The provision and sustainable management of water are key indicators for sustainable development. This chapter presents a Water Security Index constructed using the Pressure-State-Response (PSR) framework. Variables affecting water security are associated with 15 indicators distributed over four dimensions: "Resources", which provides a measure of existing water quality and quantity; "Access and Use", which considers the impact of population and access to existing resources; "Capacity", which examines the factors influencing the efficient use of water; and "Sustainability", which quantifies the impact of land use and management. The index is tested by applying it to a predominantly agricultural study area comprised of those parts of four neighboring provinces (Aydın, Denizli, Muğla and Uşak) in Turkey's Aegean Region that account for around 80% of the Büyük Menderes Basin. Next, scenarios are constructed to explore the impacts of climate change (RCP 4.5 and RCP 8.5) and population dynamics on regional water security. Under both scenarios, the region is expected to become heavily water stressed in the near future, with changing population dynamics exacerbating the problems associated with climate change. Because this index is generic in nature, it can be applied to other agricultural areas to assess water security at the provincial level. With specific information regarding resource use, access, capacity and sustainability of existing water resources, policymakers can use this index to construct guidelines for sustainable water management under future climatic and demographic conditions.

D. Marangoz · I. Daloglu (✉)
Institute of Environmental Sciences, Boğaziçi University, Istanbul, Turkey
e-mail: irem.daloglu@boun.edu.tr

I. Daloglu
Center for Climate Change and Policy Studies (iklimBU), Boğaziçi University, Istanbul, Turkey

W. Leal Filho and E. Manolas (eds.), *Climate Change in the Mediterranean and Middle Eastern Region*, Climate Change Management,
https://doi.org/10.1007/978-3-030-78566-6_15

1 Introduction

Achieving water security is one of the major challenges to have emerged in environmental governance and resource management. Climate change, population increase and rapid urbanization exacerbate this challenge by increasing the demand for water and degrading the quality of existing resources. The challenge posed by water insecurity is not unique to developing nations, but currently affects nearly 80% of the world's population (Vörösmarty et al. 2010).

The definition of water security and its methods of assessment vary widely across disciplines, and while the literature reflects sectoral as well as geographical variations, the definitions share basic common elements (Cook and Bakker 2012). Earlier definitions focused mainly on human needs; however, in a current definition popularized by the Global Water Partnership (GWP), water security not only entails ensuring that every individual at every level, from the household to the global, has access to sufficient safe water at an affordable cost to lead a clean, healthy and productive life, it also includes ensuring that the natural environment is protected and enhanced (GWP 2000). The United Nation's comprehensive and well-accepted definition of water security emphasizes the notion that water security means that sustainable access to a decent quality and quantity of water necessary for humans, economies and the ecosystem is provided in an environment of peace and political stability (UN 2013).

The dynamic nature of water security requires adaptive management strategies that account for various multiple possibilities, such as a decrease in water supply due to climate change and an increase in water demand due to population growth. Indicators and indices that consolidate complex, interrelated dimensions into single numbers can be useful tools for the quantification of water security, which, as Babel and Shinde (2018) point out, is a prerequisite for the development and implementation of adaptive, efficient water-management practices. Assessing water security makes it possible to identify current problems, set targets and evaluate strategies implemented to achieve improvements (Beek and Arriens 2014).

In one of the earliest attempts to evaluate water scarcity through a comprehensive index, Falkenmark et al. (1989) developed the Falkenmark Water Stress Index to describe the annual amount of renewable fresh water available per capita. Later, Raskin et al. (1997) presented the Water Resources Vulnerability Index, which focuses on water withdrawal by calculating the ratio of total annual withdrawals to available water resources. However, Srinivasan et al. (2017) criticize both these approaches for measuring "physical water scarcity" only, without including capacity constraints such as inadequacies in the water infrastructure. The Water Poverty Index developed by Sullivan (2002) goes a step beyond the aforementioned indexes by adding an economic component, and the "Watershed Sustainability Index", introduced by Chaves and Alipaz (2006) is even more integrative, since it is based on hydrological and environmental water-policy issues and actions. On a more specific level, Forouzani and Karami (2010) developed an Agricultural Water Poverty Index (AWPI) that they used to evaluate temporal changes in agricultural water security in Iran.

The need for an extensive and comprehensive framing of water security is evident. As Mason and Calow (2012) point out, uncertainties with respect to future socio-economic development and climate change play a significant role in risk-based approaches for achieving water security. Whether the aim is to increase water security in a region, city, or basin, future uncertainties must be managed, with decisions on investments and interventions directed accordingly. The water security index examined in this chapter was constructed to provide a local perspective on water security, incorporating future climate-change and population-growth scenarios to aid in the process of decision-making under uncertainty in an agricultural region.

The index is applied to four neighbor provinces in the Aegean region of Turkey, which constitute more than 80% of the Büyük Menderes River Basin, a dominantly agricultural basin. It is structured according to the Pressure-State-Response framework introduced by the OECD (1993) to aid policymakers and other stakeholders in conducting environmental assessments that reflect the interconnectedness between human activities and pressures on the environment. The PSR framework is a useful tool for the systematic organization of data from a variety of sources that has been employed by various studies in the water resources-management literature (Chaves and Alipaz 2006; Chung and Lee 2009; Perez-Foguet and Gine 2011; Wang et al. 2019), as well as more general sustainability and environmental evaluation studies (Hazbavi et al. 2020; Lu et al. 2016; Wu et al. 2020).

2 Materials and Methods

2.1 Conceptualizing an Index for Water Security

Water security indexes have been used to assess water security at the community (Shrestha et al. 2018; Thapa et al. 2018), city (Assefa et al. 2018; Jensen and Wu 2018; Ray and Shaw 2019), basin (Chavez and Alipaz 2006; Babel and Shinde 2018; Hazvabi et al. 2019; Jiang et al. 2018; Perez-Foguet and Garriga 2011; Men and Liu 2018) and regional (Lawrence et al. 2002) levels. Quantifying water security at different scales can be done through an aggregation of indicators that provide a valid and reliable way of measuring achievement, assessing performance and reflecting changes connected to an intervention (Mason and Calow 2012).

Various methodologies and criteria have been used in the selection of indicators by different water-assessment studies (Juwana et al. 2012; Schwemlien et al. 2016). Since the physical availability of water is not the only factor to affect water security, the inclusion of indicators representing water-management issues is critical. Based on their systematic review of the various tools used in previous studies, Plummer et al. (2012) highlight the importance of including indicators for social factors such as governance, adaptation, and institutions.

On a pragmatic level, data availability and quality play a strong role in indicator selection. The OECD (2003) lists policy relevance and utility for users, analytical soundness, and measurability among their criteria for indicator selection. Similarly, according to Dunn and Bakker (2011), indicators should be easy to access, easy to understand, timely, relevant, credible, transparent and accurate. By representing the physical, socio-economic and institutional aspects of water security and the inherent complexity of the challenges involved in terms of decision-making, indicators are powerful tools that can provide valuable information for making spatial comparisons and assessing future conditions (Ray and Shaw 2019).

Guided by the UN Sustainable Development Goal SDG6, namely ensuring availability and sustainable management of water and sanitation for all, the comprehensive Water Security Index presented here integrates dimensions pertaining to sustainability, societal well-being and the environment (OECD 2003; Cook and Bakker 2012; Plummer et al. 2012). Because the assessment of water security requires easily measurable variables, the variables included in this index are quantitative by nature. In developing the index, the PSR approach was utilized to evaluate present conditions and expectations for the future under different scenarios of climate change and population growth. The PSR framework (Fig. 1) reflects the concept of causality, which is illustrated in the cause-effect relationships of variables for "Pressure", "State" and "Response": *pressures* stemming from human activities alter the *state* of environmental resources (both qualitatively and quantitatively), triggering a *response* by society in terms of environmental, economic and sectoral policies (OECD 1993).

The index contains three levels of information for water security assessment: dimensions, indicators and variables (Table 1). A total of 15 indicators are distributed across the four dimensions of the index—"Resource", "Access and Use", "Capacity" and "Sustainability". "Pressure", "State" and "Response" variables that reflect causal relationships are provided for each of the indicators, except in a few cases where either the variable is irrelevant or the relevant data is unavailable.

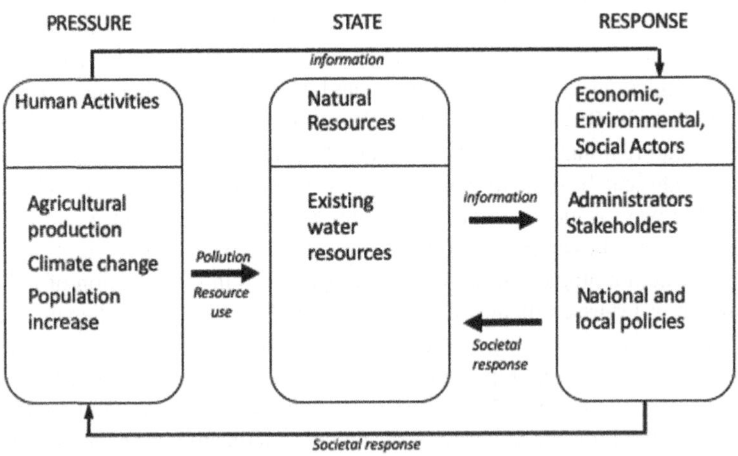

Fig. 1 Pressure-state-response framework. Adapted from OECD (1993)

Table 1 Water security index: dimensions, indicators, and corresponding PSR variables

Dimensions	Indicators	State variable	Pressure variable	Response variable
Resource	Water Budget (WB)	Rate of change in water stored	Annual variation in water budget	Improvement in study period
	Aridity Index (AI)	Degree of dryness	Annual variation in aridity index	Improvement in study period
	Water quality	Surface water contamination	Annual variation in surface nitrate levels	% increase in all WWTP capacities
	Reservoir storage	Mean occupancy rates (%)	Annual variation in occupancy rates	Change in occupancy rates during study period
Access and use	Water use per capita	Average water use per person	Annual variation in water use per person	Improvement in water use
	% Population in sewage network	Long term average % of access to sanitation	Annual variation in improved sanitation accessibility	Improvement in water infrastructure
	% Population in drinking water treatment service	Long term average % of access to drinking water	Annual variation in improved drinking water accessibility	Improvement in water infrastructure
	Population density		Number of people per km^2 (%)	
Capacity	Efficient agricultural practice (%)	Share of efficient agricultural practices in total irrigation area	Cost ratio of efficient agricultural practice	Frequency of policies supporting efficient water use
	Household income	Average household income	% of households failing to meet basic needs	
	Annual water investments	Total expenditures for water-related investments	% of water-related projects as a share of total investments	Evolution in investment incentives
	Income from agricultural activities	Mean agricultural GDP per capita	Agricultural GDP per capita / GDP per capita (%)	Evolution in agricultural subsidies
	Years of education	Mean years of education	Annual variation in years of education	Improvement in education levels
Sustainability	% of study area under natural vegetation	% of study area under natural vegetation	% change in green areas	

(continued)

Table 1 (continued)

Dimensions	Indicators	State variable	Pressure variable	Response variable
	WW treatment plant discharge per capita	Average WWTP discharge per capita	Annual variation in WWTP discharge	Improvement in municipal WWTP capacities

The "Resource" dimension represents the quality and quantity of existing water resources, and its four indicators—"Water Budget", "Aridity Index", "Water Quality" and "Reservoir Storage"—were selected especially for their relevance to agricultural settings. The "Water Budget" indicator measures the physical quantity of water within the study area (Jorgensen et al. 2012; Uhlenbrook and Savenije 2006) using the formula "Water Budget (WB) = Precipitation (P)—Evapotranspiration (E)—Streamflow (Q)". The "Aridity Index" (AI) reflects the dryness of the regional climate and strongly influences agricultural production. It is calculated according to Thornthwaite (Karlsson and Pomade 2013; Ferrer and Candela 2015) as the ratio of annual precipitation (P) to potential evapotranspiration (PET) (Barrow 1992). The "Water Budget" indicator, together with the "Reservoir Storage" indicator, provide an assessment of water availability, as reservoirs in the study area are used mainly for irrigation purposes.

The "Access and Use" dimension closely corresponds to SDG6, as it measures the access to water resources in conjunction with the pressure of population dynamics. It is comprised of four indicators: "Water Use Per Capita", "Percentage of Population in Sewage Network", "Percentage of Population in Drinking Water Treatment Service", and "Population Density". The "Water Use Per Capita" indicator measures daily water use (liter/person/day) on a provincial basis. As availability of water resources does not necessarily imply access to them, infrastructure coverage and access to clean water by end-users are represented by two indicators, "Percentage of Population in Sewage Network" and "Percentage of Population in Drinking Water Treatment Service". Lastly, the "Population Density" indicator is included due to the important role that density plays in determining water availability and access.

The "Capacity" dimension demonstrates the capacity of society to take steps in favor of sustainable use of water resources. It consists of five indicators: "Percentage of Efficient Agricultural Practice", "Annual Household Income", "Annual Water Investments", "Annual Income from Agricultural Activities" and "Years of Education". Of these, "Years of Education" and "Annual Household Income" play an influential role in increasing the capacity of society to invest in and stimulate the adoption of environmentally friendly practices. "Annual Income from Agricultural Activities" demonstrates the dependence of these communities on the agricultural profits, thereby reflecting their vulnerability to fluctuations in yields, which are subject to water availability. Water consumption due to agricultural production is highly dependent upon the technology used for irrigation purposes, and the indicator "Percentage of Efficient Agricultural Practice" reflects the proportion of water-efficient irrigation. Finally, the "Annual Water Investments" indicator demonstrates the effect of national expenditure on infrastructure by revealing the total expenditures on water-related investments.

The final dimension of the Water Security Index, "Sustainability", is represented by two indicators, "Percentage of Study Area Under Natural Vegetation" and "Treatment Plant Wastewater Discharges Per Capita". Whereas natural vegetation cover affects resource sustainability through its important role in the hydrologic cycle, the efficiency of water treatment plants contributes to the health of existing resources.

3 Results and Discussion

3.1 Calculating the WSI

Data was obtained for the four provinces that comprise the majority of the Büyük Menderes Basin and aggregated based on the area of each province that falls within this basin (95.11% of Aydin, 70.32% of Denizli, 67.87% of Usak, 19.71% of Mugla). For each indicator, reference values for each PSR variable are normalized between 0–100, and the variable is scored between 0 and 1 accordingly (Table 2). Depending on the indicator's relationship with water security, scores indicate either a positive or negative relationship with the reference value. When a higher score indicates a desirable condition—for example, a high rather than a low long-term average of reservoir occupancy rate for the "State" variable of the "Reservoir Storage" indicator—a positive relationship is said to exist. Conversely, when a higher score indicates an unfavorable condition—for example, a high rather than a low fluctuation in average reservoir storage volume for the "Pressure" variable of the same indicator—a negative relationship is said to exist.

Each indicator is given a score by calculating the arithmetic mean of its pressure, state and response scores (PSR values), with each variable given equal weight. (For indicators with only two relevant variables, means are calculated accordingly, and for indicators with only one variable, the indicator score is the same as the variable score.) Similarly, each dimension is scored by calculating the arithmetic mean of its indicator scores, with each indicator given equal weight, and an overall WSI score is

Table 2 Scoring according to sub-ranges of normalized values

Sub-ranges of normalized values	Scores (positive relationship)	Scores (negative relationship)
0–20	0	1
20–40	0 0.25	0 0.75
40–60	0 0.5	0 0.5
60–80	0 0.75	0 0.25
80–100	1	0

Table 3 Interpretation of the water security index

Score	Meaning	Description for the overall index
(−0.6–0)	Extremely poor	Extremely severe water-related problems and extreme water insecurity, management and governance are highly inadequate
(0–0.2)	Very poor	Severe water-related problems and high levels of water insecurity, management and governance are inefficient
(0.2–0.4)	Poor	Some water-related problems and water insecurity, management and governance need improvement
(0.4–0.6)	Medium	Scattered water-related problems and mixed water security, management is fair, but needs improvement to yield better results
(0.6–0.8)	Good	Limited water-related problems and adequate water security, management instruments are yielding good results
(0.8–1)	Excellent	No water-related problems and high levels of water security, management and governance are highly efficient

obtained by calculating the arithmetic mean of the dimensions, with each dimension given equal weight.

Indicator, dimension, and overall WSI scores range between −0.6 and 1, with the scores corresponding to one of six descriptive categories ranging from "Excellent" to "Extremely Poor" (Table 3). Scores in the negative range only occur in calculations for future scenarios, as they represent a decrease below "very poor" baseline conditions.

3.2 *Applying the WSI Index*

The Water Security Index described above was applied in a study area comprised of four neighboring provinces (Aydın, Denizli, Muğla and Uşak) located in the Aegean coast of Turkey that together constitute more than 80% of the area of the Büyük Menderes River Basin (Fig. 2).

With its diverse regional climate suitable for growing a variety of crops, the study area is one of the most productive agricultural regions in Turkey, with 79% of regional water resources used for agricultural activities (Büke et al. 2013). More than 70% of the total annual precipitation falls in the high-flow season between November and May (Durdu 2010). During summer months, evaporation may exceed precipitation, causing runoff to cease. Furthermore, extensive cultivation of water-demanding crops (e.g. cotton, fig and some grains) requires a level of irrigation that has led to overexploitation of water resources, including groundwater. When this overexploitation coincides with long dry periods, the water balance is substantially disrupted. Furthermore, farmers in the study area typically practice inefficient surface irrigation for the cultivation of their water-demanding crops, and although the Ministry of Agriculture and Forestry has been granting investments in efficient irrigation systems (such as drip irrigation) since 2006, farmers are reluctant to adopt these water-saving

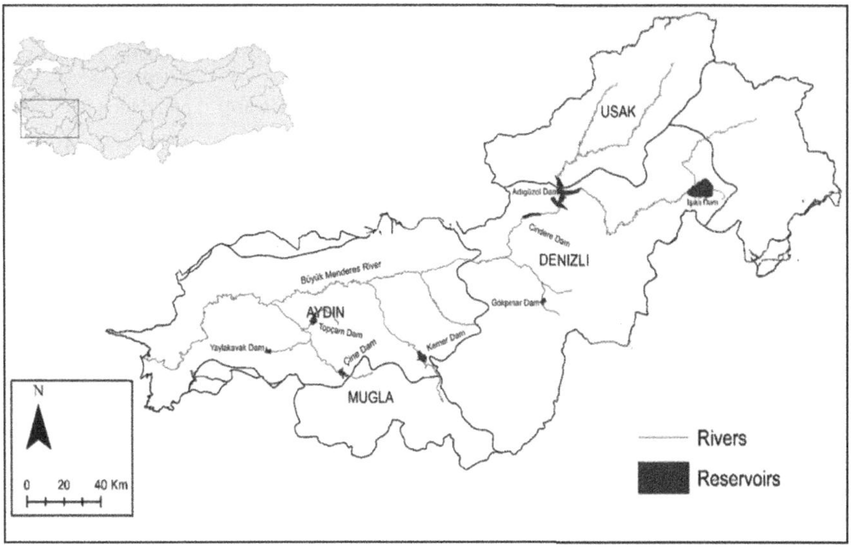

Fig. 2 Boundaries of the Büyük Menderes Basin (blue), showing the borders of the provinces Aydın, Denizli, Muğla and Uşak

technologies because high levels of pollution reduce their efficiency (Büke et al. 2013).

"Baseline" WSI scores for the provinces of Aydın, Denizli, Muğla and Uşak were calculated from the mean scores for the years 2000–2018, derived by the process described above. A "Baseline" WSI score for the entire Büyük Menderes Basin study area was then calculated from the four provincial scores, and it is presented, together with the scores for all four dimensions and all 15 indicators, in Table 4.

As the table shows, the highest baseline score for any indicator in the WSI was the "*good*" score recorded for the "Population Density" indicator of the "Access and Use" dimension, whereas the lowest baseline scores were the "*very poor*" scores recorded for the "Water Budget" indicator of the "Resource" dimension and the "Annual Water Investments" indicator of the "Capacity" dimension. With the exception of the indicator for "Efficient Agricultural Practice" of the "Capacity" dimension, which was scored as "*poor*", the baseline scores for all remaining indicators as well as the overall WSI were "*medium*", indicating that the region has water-related problems and has yet to achieve water security, and that management is fair, but needs improvement to yield better results.

Table 4 Water security index scores for each indicator under future scenarios

Dimensions	Indicators	Baseline		Scenario 1 (RCP 4.5 + population increase)		Scenario 2 (RCP 8.5 + population increase)	
		Score	Average	Score	Average	Score	Average
Resource	Water Budget (WB)	0.013	0.376	−0.318	0.291	−0.507	0.155
	Aridity Index (AI)	0.534		0.525		0.339	
	Water quality	0.456		0.456		0.456	
	Reservoir storage	0.500		0.500		0.333	
Access and Use	Water use per capita	0.500	0.563	0.313	0.484	0.313	0.484
	% population in sewage network	0.500		0.500		0.500	
	% population in drinking water treatment service	0.500		0.500		0.500	
	Population density	0.688		0.625		0.625	
Capacity	Efficient agricultural practice (%)	0.250	0.407	0.250	0.407	0.250	0.407
	Household income	0.530		0.530		0.530	
	Annual water investments	0.194		0.194		0.194	
	Income from agricultural activities	0.583		0.583		0.583	
	Years of education	0.477		0.477		0.477	
Sustainability	% of study area under natural vegetation	0.519	0.468	0.519	0.468	0.519	0,468
	WW treatment plant discharge per capita	0.417		0.417		0.417	

(continued)

Table 4 (continued)

Dimensions	Indicators	Baseline		Scenario 1 (RCP 4.5 + population increase)		Scenario 2 (RCP 8.5 + population increase)	
		Score	Average	Score	Average	Score	Average
	Water security index overall score		0.442		0.413		0.379

3.3 *Representing Future Scenarios in the Water Security Index*

The availability and management of water resources are subject to spatial and temporal variations that inherently result in uncertainty and variability. Climate change adds a large layer of uncertainty over the complex challenge of water security in terms of hydrological hazards, namely either a lack or excess of water, while population dynamics also play an important role in determining the availability of water for activities such as agricultural production that rely heavily on water. Not only will climate change and population increases exacerbate the existing risks and challenges to water security, it is also likely that under these pressures, existing poor water management and governance will become even more inefficient (Ray and Shaw 2019). Addressing variability and risk are key themes with regard to water security assessments (Mason and Calow 2012), and while the literature offers a wealth of methodologies on this subject, few rely on the use of indices to examine scenarios that incorporate climate change and population growth.

Here, to examine the effects of climate change on water security in the study area, the WSI is recalculated according to two different scenarios based on two different emission levels for the period 2018–2100. Scenario 1 is calculated for Representative Concentration Pathway (RCP) 4.5, which describes a situation in which emissions peak in 2040 and then decline, whereas Scenario 2 is calculated for RCP 8.5, which describes a situation in which emissions continuously increase through the end of the century. For both scenarios, the variables associated with "Water Budget" and "Aridity Index" were recalculated based on relevant hydroclimatic data on temperature, precipitation, and evapotranspiration obtained from the Centre National de Recherches Météorologiques General Circulation Model (GCM) CNRM-CM5.1. According to these climate projections, the provinces of Denizli and Uşak will experience increases in precipitation, while in Muğla and Aydin, precipitation is expected to decrease (Fig. 3), and temperature is expected to decrease in all four provinces, but at varying rates (Fig. 4). In addition to these changes, the variables associated with the "Reservoir Storage" indicator for the "Resource" dimension of the WSI under Scenarios 1 and 2 were recalculated using data obtained from Erkol (2020), which modeled changes in reservoir storage volumes in the Büyük Menderes Basin using the Water Evaluation and Planning System (WEAP).

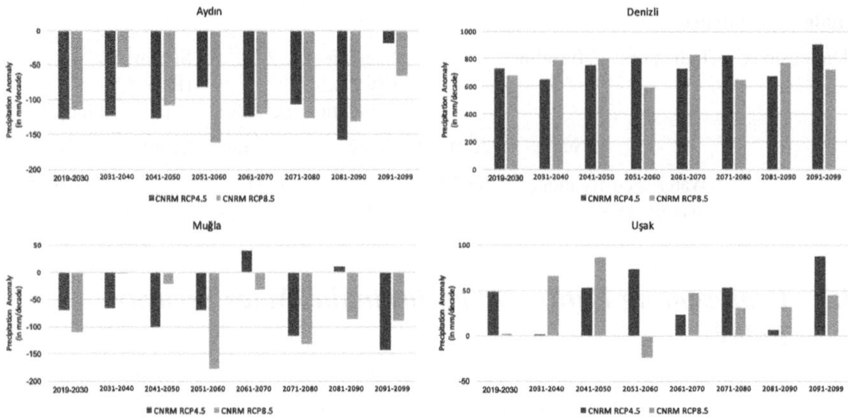

Fig. 3 Decadal precipitation anomaly (2018–2100) for Aydın, Denizli, Muğla and Uşak

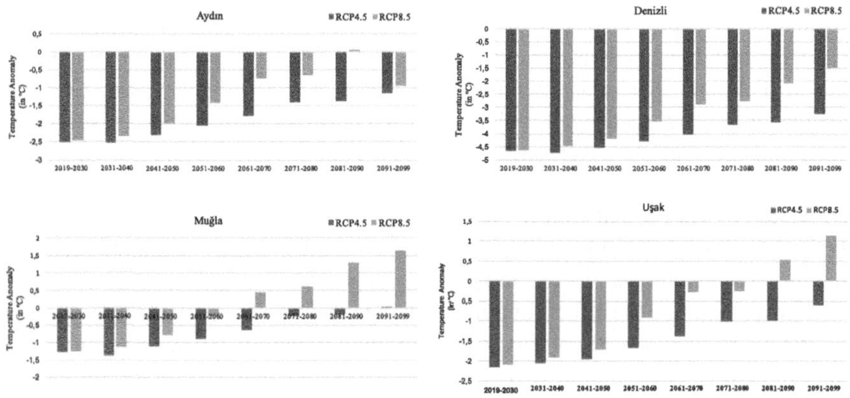

Fig. 4 Decadal temperature anomaly (2018–2100) for Aydın, Denizli, Muğla and Uşak

To examine the effects of population growth on water security under both RCP 4.5 and RCP 8.5 climate conditions, data on the projected rates of population increase in the study area were obtained from the Turkish Statistical Institute (TUIK) and used to recalculate the variables associated with the "Water Use Per Capita" and "Population Density" indicators for the "Access and Use" dimension of the WSI in both Scenarios 1 and 2. As Fig. 5 shows, the data indicates that by the year 2080, the provinces in the study area will experience population increases ranging between 45–80% over the average of the 2000–2018 period used to describe the baseline.

As Table 4 shows, under Scenario 1 (RCP 4.5 + population increase), while the overall WSI score is still "*medium*", the study area is expected to experience water insecurity. Although climate change in this scenario does not affect the scores for the indicators "Aridity Index" or "Reservoir Storage", the score for the "Water Budget" indicator decreases from "*very poor*" to "*extremely poor*", causing a significant drop

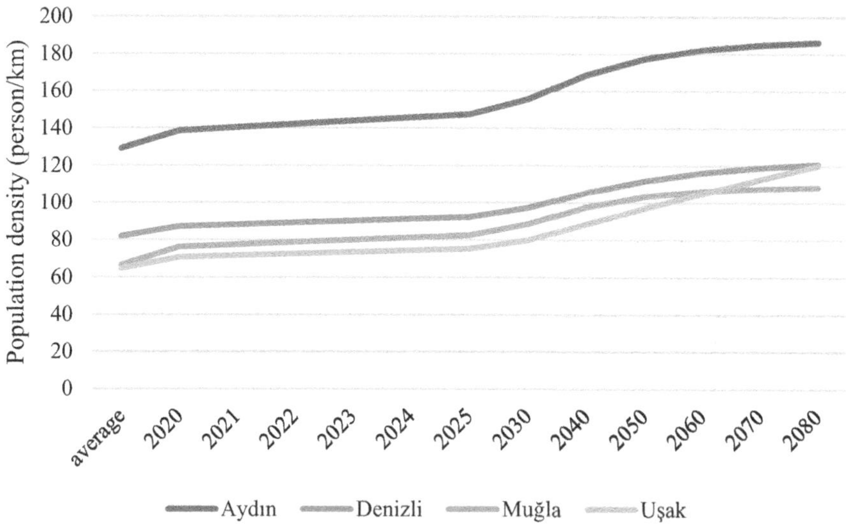

Fig. 5 Future population increase projection for Aydın, Denizli, Muğla and Uşak. source: TUIK website (http://data.tuik.gov.tr)

in the "Resource" dimension score as compared to baseline (from 0.376 to 0.291). Moreover, although the impact of population increase on the "Population Density" indicator is minimal in this scenario, a decrease in "Water Availability Per Person" due to population growth is reflected in a drop in the score of the corresponding indicator from "*medium*" to "*poor*".

Under Scenario 2 (RCP 8.5 + population increase), the water security of the study area is significantly challenged, with the overall WSI score dropping to "*poor*" from a baseline score of "*medium*" as a result of the drop in the "Resource" dimension score from "*poor*" to "*very poor*". The drop in this dimension score, in turn, can be explained by drops in the "Water Budget" indicator score from "*very poor*" to "*extremely poor*" and in drops in the scores of both the "Aridity Index" and "Reservoir Storage" indicators from "*medium*" to "*poor*". As with the RCP 4.5 scenario, the "Water Availability Per Person" indicator score under the RCP 8.5 scenario drops from "*medium*" to "*poor*" due to an increase in population, while the effect of population growth on the "Population Density" indicator is only minor.

Overall, the water security of the study area is significantly challenged under Scenario 2 (RCP 8.5 + population increase), indicating high levels of water insecurity. Some information as to the resilience of the study area in the face of problems related to water insecurity can be found in the indicators related to the "Capacity" dimension of the index. Considering the "*poor*" score for "Annual Investments for Water Infrastructure" and the "*medium*" score for "Water-efficient Agricultural Practices" at baseline, it can be surmised that problems related to existing agricultural management practices and governance of resources could impede any attempts at adapting to future challenges.

3.4 Data Challenges

While integrating plausible future scenarios into this index to assess the possible impacts of population increase and climate change, various data limitations were encountered with regard to the representation of these two factors.

For example, this index was constructed using downscaled GCM-CNRM-CM5.1 outputs; however, the use of a single downscaled climate model rather than a combination of different GCM outputs introduces spatial variability and uncertainty into hydroclimatic variable estimations (Jaramillo and Nazemi 2018). According to Eekhout et al. (2018), the redistribution of water, which directly impacts on water security, is driven by an increase in extreme precipitation and a decrease in precipitation frequency, whereas changes in annual precipitation volume have relatively less impact. Unfortunately, because hydroclimatic parameters obtained from downscaled CNRM-GCM models are annual averages, they lack the temporal resolution needed to reveal changes in the frequency of precipitation events or changes in precipitation volume by event. Considering that the index presented here is designed to provide decisionmakers with basic insights that will assist them in formulating guidelines for sustainable water management under future climatic and demographic conditions, increasing the temporal resolution of the hydroclimatic parameters falls outside the index focus; nevertheless, the possibility of underestimation should be kept in mind when interpreting the overall index scores.

It is also important to remember that while the complex relationships between climate and ecosystems suggests that climate change is likely to affect many WSI indicators, in this index, the impact of climate change is represented in its hydrological sense only, mainly through the indicators that fall within the "Resource" dimension of the index (i.e. water budget, aridity index, and reservoir storage). The reason for this is simply that an assessment of all the convoluted, dynamic impacts of climate change on social-ecological systems requires complex and very detailed modelling techniques, whereas this index is designed to be easily implemented by policymakers using the most readily available data.

Regarding the representation of population increases in future scenarios, the index incorporates projections for future population density and water-use per capita in its calculations for Scenarios 1 and 2. However, a growing population might also affect the indicators "Percentage of Population in Sewage Network" and "Percentage of Population in Drinking Water Treatment Service". Future water-security levels could also be significantly affected by the amount of future investment in sewage and drinking water networks, and in this regard, the already low annual investments in water infrastructure, as reflected by the indicator "Annual Water Investments", suggest that the index may overestimate future water security.

4 Conclusion

Water security plays a crucial role in achieving food security and enhancing livelihoods in agricultural regions. Changing climatic conditions and growing populations threaten water security. Climate change directly affects frequency and intensity of extreme hydroclimatic events as well as available water quantity and quality, all of which affect both ecosystems and livelihoods. Population growth exerts additional stress on both water availability and the infrastructure that provides access to existing resources and sanitation by increasing demand. Given the potential of climate change and population increase to exacerbate risks to water security, it was essential for climate- and population-related variables to be integrated into the Water Security Index of the four neighboring provinces that comprise a region central to the agricultural production and sustainable water security of Turkey.

The WSI scores for the study area under baseline conditions demonstrate the existence of water-related problems in the region as well as a fair level of management that could be improved to yield better results. An increase in water insecurity in the study area is observed when future climate-change and population-growth scenarios are integrated into the Water Security Index. Under Scenario 1 (RCP 4.5 + population increase), the WSI score is reduced considerably to borderline "*medium*" conditions, whereas in Scenario 2 (RCP 8.5 + population increase), the WSI score drops significantly to "*poor*" conditions. The drop in the index score highlights the opportunity for development in water resource management as reservoir storage volumes decrease and pressures on water budget and aridity index exacerbate. This analysis highlights two main challenges faced in the study area: (1) current low levels of agricultural investment decreasing the adaptive capacity of stakeholders to future challenges; (2) existing low technological capacity (denoted by adoption of "Water-efficient Agricultural Practices") leading to the possibility of technological lock-ins and ultimately requiring higher investment efforts and costs. Results of the water security index for four cities in the Büyük Menderes Basin illustrates the necessity of managing future uncertainties and directing decisions on investments and interventions accordingly and can help policy and decision-makers to target limited resources effectively.

The Water Security Index proposed in this chapter aims to integrate future projections into water security assessments in a manner that reflects the cause-and-effect relationships between environmental, social and climate-related pressures and responses, all which are central to the long-term sustainability of the region. The index is designed to be replicable and easily used with available data, and the equal-weighting approach also simplifies implementation while expanding the applicability for stakeholders and policymakers undertaking water-security assessments. Even data-scarce regions should find that the data required by the four dimensions and 15 indicators comprising the index are all relatively accessible. While these variables may not span the entire number of potential water security-assessment indicators, a higher number of variables might challenge the applicability of the index. In its present form, by highlighting problematic dimensions and indicators that need attention, the water security index can support policymakers in the formulation of guidelines for sustainable water management under future climatic and demographic changes based on specific information about usage, accessibility, capacity and sustainability of existing water resources.

References

Assefa Y, Babel M, Sušnik J, Shinde V (2018) Development of a generic domestic water security index, and its application in Addis Ababa. Water 11(1):37

Babel M, Shinde V (2018) A framework for water security assessment at basin scale. APN Sci Bull 8(1):27–32

Barrow CJ (1992) World atlas of desertification. In Middleton N, Thomas DSG (eds). Edward Arnold London 1992

Beek EV, Arriens WL (2014) Water security: putting the concept into practice (No 20) Global Water Partnership

Büke A, Gültekin M, Aksoy A, Dıvrak BB, Göcek Ç, Berke MÖ, Çeşmeci H (2013) Büyük Menderes havza atlası. World Wildlife Fund-Turkey, Istanbul, pp 6–167. Available at: https://wwftr.awsassets.panda.org/downloads/atlas_web_download.pdf

Chaves HML, Alipaz S (2006) An integrated indicator based on basin hydrology, environment, life, and policy: the watershed sustainability index. Water Resour Manage 21(5):883–895

Chung ES, Lee KS (2009) Identification of spatial ranking of hydrological vulnerability using multi-criteria decision making techniques: case study of Korea. Water Resour Manage 23(12):2395–2416

Cook C, Bakker K (2012) Water security: debating an emerging paradigm. Glob Environ Chang 22(1):94–102

Dunn G, Bakker K (2011) Fresh water-related indicators in Canada: an inventory and analysis. Can Water Res J 36(2):135–148

Durdu OF (2010) Effects of climate change on water resources of the Büyük Menderes river basin, western Turkey. Turk J Agric for 34(4):319–332

Eekhout JPC, Hunink JE, Terink W, de Vente J (2018) Why increased extreme precipitation under climate change negatively affects water security. Hydrol Earth Syst Sci 22(11):5935–5946

Erkol I (2020) Modeling the impacts of climate change on water supply and demand balance of Büyük Menderes Basin, Master's thesis, Boğaziçi University, Istanbul

Forouzani M, Karami E (2010) Agricultural water poverty index and sustainability. Agron Sustain Dev 31(2):415–431

Falkenmark M, Lundquist J, Widstrand C (1989) Macro-scale water scarcity requires micro-scale approaches: aspects of vulnerability in semi-arid development. Nat Resour Forum 13(4):258–267

GWP (Global Water Partnership) (2000) Towards water security: a framework for action. © GWP, Stockholm, Sweden. pp 11–111. Available at: https://www.gwp.org/globalassets/global/toolbox/references/towards-water-security.-a-framework-for-action.-mobilising-political-will-to-act-gwp-2000.pdf

Hazbavi Z, Sadeghi SH, Gholamalifard M, Davudirad AA (2020) Watershed health assessment using the Pressure–State–Response (PSR) framework. Land Degrad Dev 31(1):3–19

Jaramillo P, Nazemi A (2018) Assessing urban water security under changing climate: challenges and ways forward. Sustain Urban Areas 41:907–918

Jensen O, Wu H (2018) Urban water security indicators: development and pilot. Environ Sci Policy 83:33–45

Jiang G, He L, Jing J (2018) Water security evaluation in yellow river basin. In: AIP Conference of Proceedings 1944, 020059-1-020059-6, AIP Publishing LLC

Jorgensen S, Tundisi JG, Tundisi TM (2012) Handbook of inland aquatic ecosystem management. CRC Press

Juwana I, Muttil N, Perera BJC (2012) Indicator-based water sustainability assessment—a review. Sci Total Environ 438:357–371
Karlsson E, Pomade L (2013) Methods of estimating potential and actual evaporation. Department of Water Resources Engineering, Salt Lake City
Lawrence PR, Meigh J, Sullivan C (2002) The water poverty index: an international comparison. Department of Economics, Keele University, Straffordshire, UK
Lu Y, Wang X, Xie Y, Li K, Xu Y (2016) Integrating future land use scenarios to evaluate the spatio-temporal dynamics of landscape ecological security. Sustainability 8(12):1242
Mason N, Calow R (2012) Water security: from abstract concept to meaningful metrics. ODI working paper no. 357. Overseas Development Institute (ODI), London. Available at: https://www.odi.org/sites/odi.org.uk/files/odi-assets/publications-opinion-files/7866.pdf
Men B, Liu H (2018) Water resource system vulnerability assessment of the Heihe River Basin based on Pressure-State-Response (PSR) model under the changing environment. Water Supply 18(6):1956–1967
OECD (1993) OECD core set of indicators for environmental performance reviews: a synthesis report by the group on the state of the environment. In: OECD environmental monographs, vol 83. Organisation for Economic Co-operation and Development (OECD), Paris, France
OECD (2003) OECD environmental indicators development, measurement and use. Organisation for Economic Co-operation and Development (OECD), Paris, France
Pascual Ferrer J, Candela Lledó L (2015) Water balance on the Central Rift Valley. In: GDEE (ed) Case studies for developing globally responsible engineers. Global Dimension in Engineering Education, Barcelona. Available at: https://upcommons.upc.edu/bitstream/handle/2117/89152/CS_10.pdf
Pérez-Foguet A, Giné Garriga R (2011) Analyzing water poverty in basins. Water Resour Manage 25(14):3595–3612
Plummer R, de Loë R, Armitage D (2012) A systematic review of water vulnerability assessment tools. Water Resour Manage 26(15):4327–4346
Raskin P, Gleick PH, Kirshen P, Pontius RG, Strzepek K (1997) Comprehensive assessment of the freshwater resources of the world. Stockholm Environmental Institute, Sweden
Ray B, Shaw R (eds) (2019) Developing water security index for urban areas. In: Urban drought. Disaster risk reduction (methods, approaches and practices). Springer, Singapore, pp 53–68
Schwemlein S, Cronk R, Bartram J (2016) Indicators for monitoring water, sanitation, and hygiene: a systematic review of indicator selection methods. Int J Environ Res Public Health 13(3):333
Shrestha S, Aihara Y, Bhattarai AP, Bista N, Kondo N, Futaba K, Shindo J (2018) A novel water security index and well-being at micro level in urban areas of developing countries. SSM Popul Health 6:276–285
Srinivasan V, Konar M, Sivapalan M (2017) A dynamic framework for water security. Water Secur 1:12–20
Sullivan C (2002) Calculating a water poverty index. World Dev 30(7):1195–1210
Thapa B, Ishidaira H, Pandey V, Bhandari T, Shakya N (2018) Evaluation of water security in Kathmandu Valley before and after water transfer from another basin. Water 10(2):224
Uhlenbrook S, Savenije HHG (2006) Hydrology of catchments, river basins and deltas. UNESCO-IHE Institute for Water Education, Netherlands
United Nations (2013) Water security and the global water agenda: a UN-Water analytical brief. United Nations University Institute for Water Environment and Health, Canada
Vörösmarty CJ, McIntyre PB, Gessner MO, Dudgeon D, Prusevich A, Green PS, Glidden S, Bunn E, Sullivan CA, Reidy Liermann C, Davies PM (2010) Global threats to human water security and river biodiversity. Nature 467(7315):555–561
Wang Q, Li S, Li R (2019) Evaluating water resource sustainability in Beijing, China: combining PSR model and matter-element extension method. J Clean Prod 206:171–179
Wu J, Wang X, Zhong B, Yang A, Jue K, Wu J, Zhang L, Liu Q (2020) Ecological environment assessment for Greater Mekong subregion based on pressure-state-response framework by remote sensing. Ecological Indicators. 117:106521

An Approach to Adapting Urban Drainage Design to Climate Change: Case of Northern Morocco

Mohammed Moujahid, Laila Stour, and Ali Agoumi

Abstract Intense rainfall can lead to urban flooding because of the inability of Urban Drainage (UD) systems to evacuate runoff water. In addition, the performance of a UD system is limited by the rainfall event that exceeds the capacity of UD structure. However, according to climate projections, extreme rainfall events will be more intense in future, and so the Climate Change (CC) will impact directly the performance of UD structures by increasing the number of floods in urban areas. This study aims to propose an approach to adapting the design criteria of UD to CC in Northern Morocco through the incorporation of two parameters that characterize: (1) the variation in the probability of occurrence of the rainfall events, and (2) the UD operation under extreme events based on the management feedback of the urban catchment.

1 Introduction

The management of the urban water cycle is based on the performance of the urban drainage (UD) infrastructures, which characterize the sustainability of the system to evacuate the rainwater in urban catchment. Moreover, several studies have shown that the UD system will be affected by climate change (CC), which should lead to significant changes on the probability of occurrence of the rainfall events (Teegavarapu 2012; Sarhadi and Soulis 2017; Moujahid et al. 2018; García-Marín et al. 2019). Indeed, CC causes floods in urban areas due to the incapacity of the UD system to maintain its performance at the same level of service, for which infrastructures have been primarily designed. In addition, the design of UD infrastructures has been constantly established on the hypothesis of the "*climate system stationarity*", which

M. Moujahid (✉) · L. Stour
Faculty of Sciences and Technics of Mohammedia, Department of Process and Environment Engineering, Hassan II University, Casablanca, Morocco

A. Agoumi
Department of Hydraulic, Environment and Climate, Hassania School of Public Works, Casablanca, Morocco

W. Leal Filho and E. Manolas (eds.), *Climate Change in the Mediterranean and Middle Eastern Region*, Climate Change Management,
https://doi.org/10.1007/978-3-030-78566-6_16

means that future climate could be considered as a representation of the current climate. However, with the CC, it is not the case (Grum et al. 2006; Willems et al. 2012; Willems 2013; Dale et al. 2018).

Several studies have shown that rainfall events will continue to be intensified with CC. For example, Acquaotta et al. (2018) have shown that the maximum annual 3-hourly and 6-hourly rainfall rates are increasing in the Mediterranean area particularly in recent years (2004–2016). Also, Wang et al. (2019) have found an increasing trend on runoff level in urban areas under the RCP 2.6 and RCP 4.5 scenarios for the period of 2015–2099.

This study aims to propose an approach to adapting the criteria used currently by UD system designers, through incorporating the impact of CC into the design method. The main idea of the approach is taking into consideration the CC as a component of the design criteria, by characterizing the variability of the probability of occurrence of the rainfall event usually used in UD design.

After a description of the methodology (Sect. 2.1), Sects. 2.2 and 2.3 presents its application on northern Morocco as a study area. In Sect. 3, an analysis and discussion of results and their utilities for the hydrologic UD design will be presented. The last section will be devoted to summarize conclusions.

2 Methodology

2.1 Concept and Principles

In context of CC, the designer of UD structures should consider the variability of the probability of occurrence of the rainfall event (Moujahid et al. 2015). Indeed, the adaptation of the UD system to CC can be reached by characterizing the variation of the probabilities of non-exceedance between two principal periods (Fig. 1). The first one is what it could be called "*Design Period*" during which the designer predict the

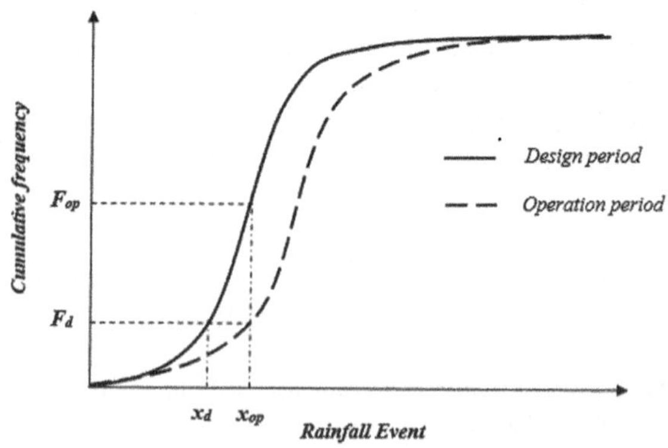

Fig. 1 Concept of the variability of the cumulative frequency under climate change

capacity of the UD to evacuate a rainfall event X in the future (Te Chow 2010; Meylan et al. 2011). This prediction is based on defining and implementing a *frequency model*, which is a mathematical description of the statistical behavior of a rainfall event through its probability distribution function, or more commonly, *cumulative frequency* F(x) defined as (Meylan et al. 2011):

$$F(x) = P[X < x] \tag{1}$$

Otherwise, several studies have shown an intensification of the rainfall extreme events due to CC (Tabari and Willems 2018; Armal et al. 2018; Myhre et al. 2019). For the UD management system, this intensification can be interpreted as a variability in the probability of occurrence of the rainfall events between the climate, during which, UD have been designed (*Design Period*) and the climate when the UD will be operated (*Operation Period*). Indeed, during the Design Period the designer uses a probability of occurrence F_d and assumes that it will remain unchanged during the *Operation Period* until the end of the structure lifetime. However, under the CC impact, this hypothesis should be reviewed taking into consideration that (Fig. 1):

- For a given probability of occurrence F_d, if the UD structure is designed to evacuate rainfall event $X_d(F_d)$, then it will be forced, in the operation period, to evacuate an event $X_{op}(F_d)$ greater than $X_d(F_d)$.

$$X_d(F_d) \leq X_{op}(F_d) \tag{2}$$

- The presumed UD dysfunction under CC can be surpassed by using the probability of occurrence F_{op} of the expected event X_{op} to design the UD structure instead of F_d.

$$X_{op}(F_d) = X_d(F_{op}) \Leftrightarrow F_d \leq F_{op} \tag{3}$$

The approach proposed to adapt UD system to CC is based on the assessment of the variability of the probability of occurrence of the rainfall events used to design UD structures. For this, a parameter φ will be incorporated to the design criteria and linked to the intensification of the rainfall event due to CC. this parameter will be defined as:

$$\varphi = \frac{F_{op}}{F_d} - 1 \tag{4}$$

where:

- F_{op} the probability of occurrence of the design rainfall event X_d during the operation period.
- F_d the probability of occurrence used practically to design the design UD structure.

Instead of the cumulative frequency, the concept of return period T defined as:

$$T = \frac{1}{1 - F} \tag{5}$$

And so, Eq. 4 may be written as:

$$T_{op} = \frac{T_d}{1 - \varphi(T_d + 1)} \tag{6}$$

with:

- T_d the probability of occurrence used theoretically to design UD structure. Generally, for UD structures, it will be equal to 2, 5, 10, 25 and 50 years.
- T_{op} the return period that the designer should use to adapting UD structure to CC impact during operation period.

Finally, the parameter φ should be assessed through a frequency analysis of the rainfall data available in the urban catchment. The methodology proposed is based on the feedback of the UD system operation in the urban catchment where the structure designed will be installed. For this, the rainfall data in the project area must be analyzed taking into consideration the history of rainfall events that caused a UD dysfunction during operation period. This analysis will allow identifying the period when the UD system service level have been exceed during operations.

2.2 *Study Area and Data Analysis*

The study area is located in northwestern Morocco where the climate is influenced by the Mediterranean Sea (Fig. 2). In general, precipitation exceeds 700 mm/year and the torrential rainfall is the main cause of the violent floods that affect the

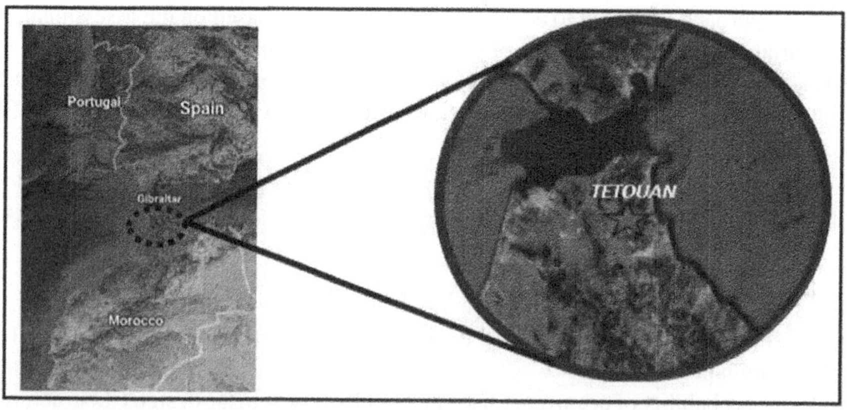

Fig. 2 Geographical location of the study area. Adapted from Google Map, 2020

region (example of rainfall: 14 December 2009, 06 March 2010, 29 August 2013, 22 November 2016, 22–23 March 2019).

In this study, Annual Maximum Precipitation (AMP) data is used. The data were recorded in the meteorological station of Tetouan city (TET) in northern Morocco and were available, by the National Meteorological Office of Morocco (DMN), in series of Annual Maximum Precipitation (AMP) over various durations that are 1, 2, 6 and 12 h.

The available data has been compared to historical rainfall events that caused a major dysfunction for the UD structures installed in Northen Morocco. Indeed, more than 160 rainfall events have been analyzed in order to detect when the intensity of a given rainfall event was greater than that used during the design period.

Because of the limitation of the data, the analysis was limited to the structures designed for a return period 2, 5 and 10 years, which are UD networks. Primary, this analysis was based on the identification of rainfall events X_i that were caused a dysfunction in UD system from the history of the operation services in urban catchment of Tetouan city. In this context, from 160 rainfall events from 1970 to 2015, only 31 events have caused a major dysfunction of the networks of the UD system. Then, the return period T_i of each event X_i have been determined using the Intensity–Duration–Frequency (IDF) curves of the region. Finally, all events with T_i > 10 years have been considered as exceptional events that the level of UD operation has been conditioned by.

The results of the analysis (Fig. 3) shown that two periods can be considered in the frequency analysis of the events. Indeed, during 1970–1989 when only 4 events were surpassed the capacity of the UD networks installed in the catchment of the Tetouan city, while 27 events have been recorded between 1990 and 2015 which have caused major dysfunction in the structures designed for T = 10 years. Therefore, in the next, the AMP of two analysis periods, that are 1970–1989 and 1990–2015, will be used to assessing φ parameter through a frequency analysis.

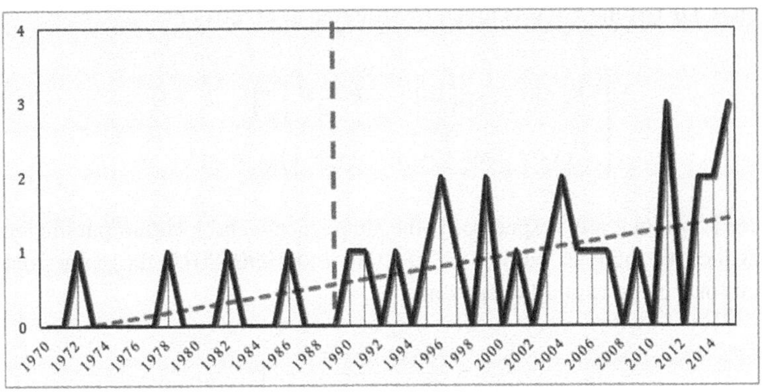

Fig. 3 The number of rainfall events exceeding the UD networks capacity over the period 1970–2015

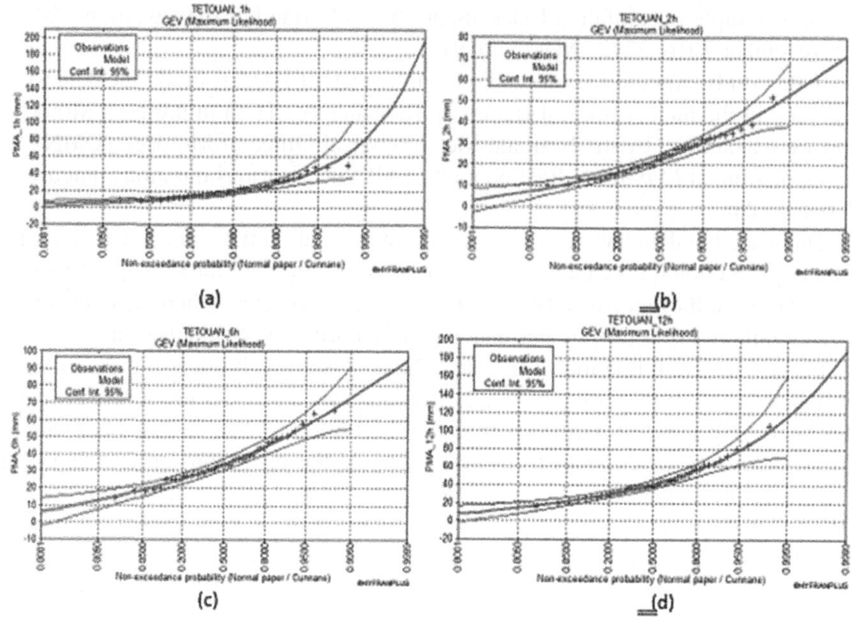

Fig. 4 The GEV distribution fitted to the sample of AMPs of the station TET: **a** 1 h, **b** 2 h, **c** 6 h and **d** 12 h

The frequency analysis has been done using the General Extreme Values (GEV) distribution, because of the worldwide use of this distribution to modeling the rainfall maxima (Zheng et al. 2016; Sugi et al. 2017; Switzman et al. 2017; Nguyen and Nguyen 2019). Otherwise, the Chi- Square (χ^2) test (Preacher 2001) also shows that the GEV distribution can be selected with a significance level of 5% (Fig. 4).

Following the parameterization done by Jenkinson, the cumulative distribution function (CDF) of the GEV is given by (Meylan et al. 2011):

$$\begin{cases} F(x) = \exp\left(-\left[1 - k\frac{(x-\xi)}{\alpha}\right]^{\frac{1}{k}}\right); \quad k \neq 0 \\ F(x) = \exp\left(-\left[1 - \frac{(x-\xi)}{\alpha}\right]\right); \quad k = 0 \end{cases} \tag{7}$$

where k, α and ξ are, respectively, the shape, scale and location parameters.

Over the two analysis periods 1970–1989 and 1990–2015, the φ parameter can be given for a given rainfall event x as:

$$\varphi = \frac{F_{1990-2015}(x)}{F_{1970-1989}(x)} - 1 \tag{8}$$

With $F_{1990-2015}$ and $F_{1970-1989}$ are, respectively, the probability of occurrence of the rainfall event X during the period 1990–2015 and 1970–1989.

2.3 Peak Discharge Assessment

To estimate the peak discharge of a specific return period T, rational method is used. It is defined as (Freire Diogo and Antunes do Carmo, 2019):

$$Q(T,\ t_c) = C \cdot i(T,\ t_c) \cdot A \tag{9}$$

where Q is the peak discharge required for sizing UD structure, i(F,t_c) is the intensity of rainfall event of chosen return period T and for a duration equaling the time of concentration t_c of the catchment, C runoff coefficient which usually empirically determined and A the area of the catchment.

Generally, designers use Montana model to describe the intensity i (Meylan et al. 2011):

$$i(F, t_c) = a(F) \cdot t_c^{-b(F)} \tag{10}$$

where a(F) and b(F) are the two parameters to be estimated for a given region, based on a frequency analysis of the rainfall data.

The frequency analysis allows to assessing the intensity of a given rainfall event X associated to return period F and duration t_c, by the flowing relationship (Meylan et al. 2011):

$$i(F,\ t_C) = \frac{x(F,\ t_C)}{t_C} \tag{11}$$

The GEV distribution function have been translated to the following inverse cumulative or quantile functions (Meylan et al. 2011):

$$\begin{cases} x(F) = \xi + \frac{\alpha}{k}\left(1 - [-\ln(F)]^k\right); & k \neq 0 \\ x(F) = \xi - \alpha(\ln[-\ln(F)]); & k = 0 \end{cases} \tag{12}$$

From Eqs. 10–11, Montana model can be written as (Meylan et al. 2011):

$$x(F,\ t_c) = a(F) \cdot t_c^{1-b(F)} \tag{13}$$

For a given probability of occurrence F, Montana parameter can be assessed using linear interpolation through following relationship:

$$\ln(x) = \ln(a) + (1 - b) \cdot \ln(t_c) \tag{14}$$

The time of concentration t_c have been estimated with the most commonly used formula by designer in Northern Morocco, that is Kirpich formula, which defined as (Longo and Rooney 2018):

$$t_c = 0.01947 L^{0.77} S^{-0.385} \tag{15}$$

where t_c is the time of concentration in min, L the maximum length of the principal pipe in UD network, S the slope of the network.

Finally, following Eqs. 9–15, UD structures can be designed using the peak discharge adapted to CC in Northern Morocco, defined as:

$$Q_p = \theta \cdot C \cdot A \cdot L^{\beta} \cdot S^{\gamma} \tag{16}$$

With θ, β and γ are function of Montana parameters a_φ and b_φ associated to φ parameter:

$$\theta = \frac{0.000325^{-b_\varphi} \times a_\varphi}{360} \tag{17}$$

$$\beta = -0.55 b_\varphi \tag{18}$$

$$\gamma = 0.385 b_\varphi \tag{19}$$

3 Results and Discussion

To analyze the AMP variability, the GEV was used distribution of the rainfall data over the analysis periods 1970–1989 and 1990–2015. The estimation of the parameters of the distribution was evaluated by the method of moments which gave the following results (Table 1):

Using the approach described in Sect. 2, the results shows changes in the probability of occurrence of rainfall events over the analysis periods (Fig. 5).

Thus, for all durations (1, 2, 6 and 12 h), the conclusion is, for a specific return period T over the analysis periods 1970–1989 and 1990–2015:

$$X_{1970-1989}(T) \leq X_{1990-2015}(T) \tag{20}$$

Table 1 Values of GEV parameters for the study area over 1970–1989 and 1990–2015

h	1970–1989			1990–2015		
	k	α	ξ	k	α	ξ
1	0.0546	4.53	14.39	0.1616	5.28	16.81
2	0.0750	7.08	19.96	0.1178	8.17	23.50
6	0.1167	8.91	28.46	0.2037	10.93	32.31
12	0.0915	13.02	36.05	0.0536	14.03	42.98

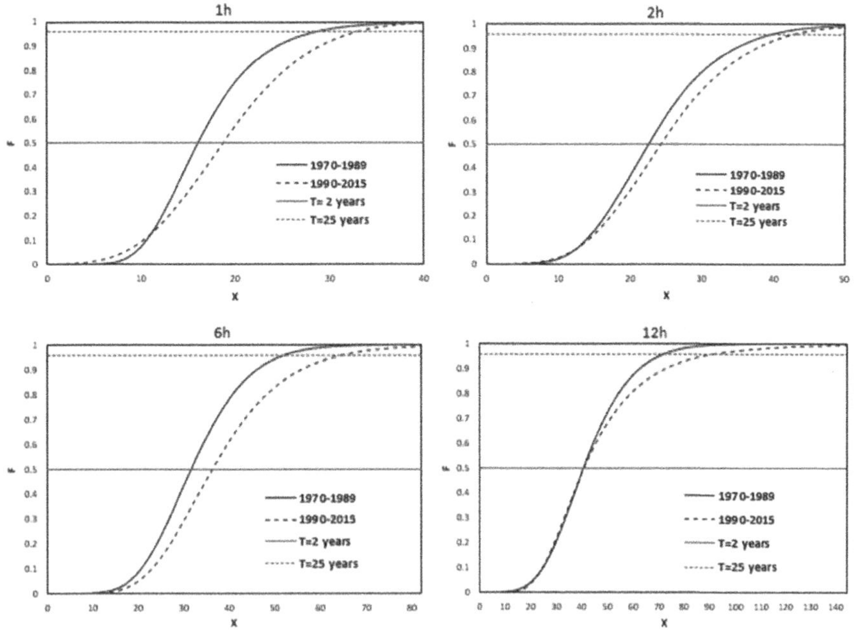

Fig. 5 Changes in the probability of occurrence of AMP over analysis periods for all durations (1, 2, 6 and 12 h)

As observed, the results (Table 2) allow to conclude on the intensification of the AMP between the analysis periods. Indeed, Table 2 shows an increase in all rainfall events, especially those with 1–2 h which affect directly UD system, the increase is ranged from 7 to 20% depending on the return period. Based on this conclusion, the φ parameter was evaluated for $T = 2, 5, 10, 25$ using Eq. 8 (Table 3).

As shown in Table 3, the UD networks designed for $T = 2$, 5, 10 and 25 years during 1970–1989, have been confronted to events with $T = 3$, 12, 27 and 68 years

Table 2 AMP of various return period for different durations over analysis periods

Period	Duration h	T = 2 years	T = 5 years	T = 10 years	T = 25 years	T = 50 years	T = 100 years
1970–1989	1	16.0	21.2	24.5	28.8	31.9	35.0
	2	22.5	30.0	34.6	40.1	43.9	47.5
	6	31.7	40.7	46.1	52.2	56.4	60.2
	12	40.7	54.3	62.5	72.2	78.8	84.9
1990–2015	1	18.7	25.4	29.1	33.3	36.0	38.3
	2	24.2	32.5	37.5	43.1	46.9	50.3
	6	36.3	48.2	55.7	64.9	71.5	77.9
	12	40.9	59.0	72.6	91.9	107.9	125.4

Table 3 Values of the φ parameter in the study area over the analysis periods

$F_{1970–1989}$	$F_{1990–2015}$	φ	$T_{1970–1989}$ (years)	$T_{1990–2015}$ (years)
0.5	0.682	**0.36**	2	3
0.8	0.916	**0.15**	5	12
0.9	0.963	**0.07**	10	27
0.96	0.985	**0.03**	25	68

during the operation period 1990–2015. Indeed, the results confirm also the high number of the events that caused dysfunction for UD networks during 1990–2015 (Fig. 3).

Therefore, through the frequency analysis (Sect. 2, Tables 2 and 3), the IDF curves was updated (Fig. 6) and allowed assessing Montana parameters $a_{\varphi}(T)$ and $b_{\varphi}(T)$ (Table 4) for the study area using Eq. 14.

Using Eqs. 16–19, the peak discharge adapted to CC in urban catchment of Tetouan city is given by:

$$Q_p(T = 2\ years) = 12.2\ C \cdot A \cdot L^{-0.372} \cdot S^{0.261} \tag{21}$$

$$Q_p(T = 5\ years) = 13.9\ C \cdot A \cdot L^{-0.361} \cdot S^{0.253} \tag{22}$$

$$Q_p(T = 10\ years) = 13\ C \cdot A \cdot L^{-0.348} \cdot S^{0.244} \tag{23}$$

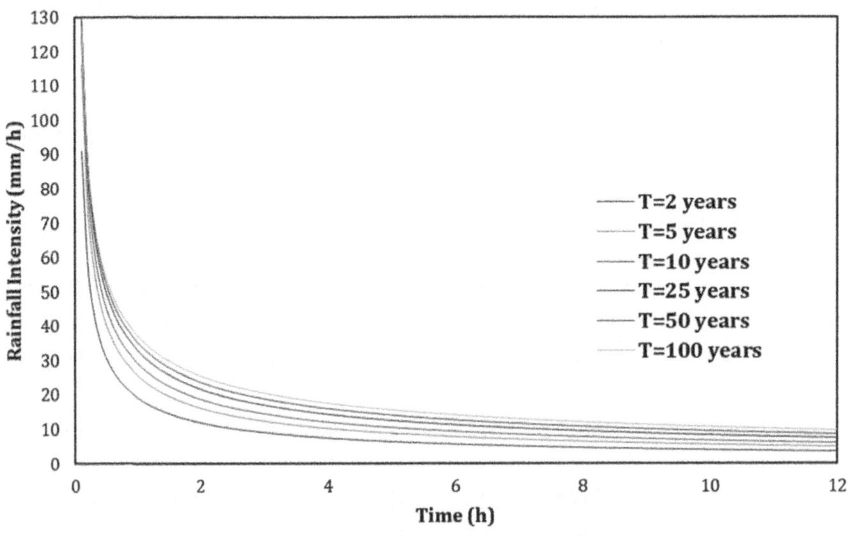

Fig. 6 IDF curves in the study area over analysis periods

Table 4 Assessment of montana parameters $a_\varphi(T)$ and $b_\varphi(T)$ in study area

F	T	a_φ	b_φ
0.50	2	19.18	0.677
0.80	5	25.56	0.657
0.90	10	29.06	0.633
0.96	25	32.77	0.597
0.98	50	35.07	0.567
0.99	100	37.02	0.536

$$Q_p(T = 25\ years) = 11\ C \cdot A \cdot L^{-0.312} \cdot S^{0.218} \quad (24)$$

The last step of analysis was devoted to provide an evaluation of UD system performance under the hypothesis of CC. To achieve this aim, a watershed in the urban area of Tetouan city was adopted as study area (Fig. 7). With an area around 250 ha, the analyzed watershed is highly urbanized (65–80% impervious areas mainly buildings, roads and squares) and it is known by recurrent floods. The study area is drained by a UD system designed between 1995 and 2011 with approximately 35 km of pipes ranging between 300 and 2000 mm. The concentration time of watershed is around 1 h and the network of the UD system was designed for T = 5 and 10 years return periods.

Equations 21–23 were applied to eighteen (18) sub catchments of the study area, the results have shown that the design peak discharge have increased by up to 51%. In other words, without the adoption of a CC adaptation procedure, the conveyance capacities of UD system have been under designed by up to 62% (Fig. 8). Indeed, using conservative design practice, UD pipes diameters still have to be increased by 25–62% to provide the same degree of protection (Fig. 8).

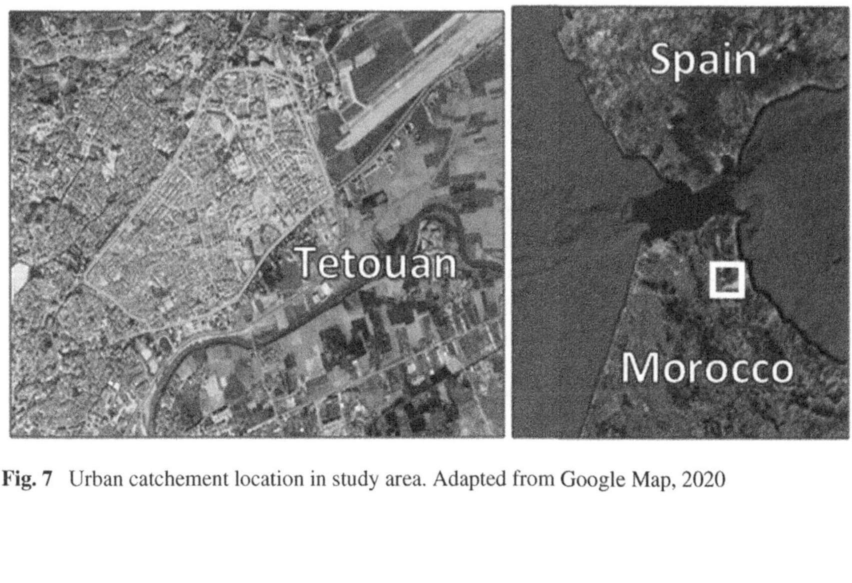

Fig. 7 Urban catchement location in study area. Adapted from Google Map, 2020

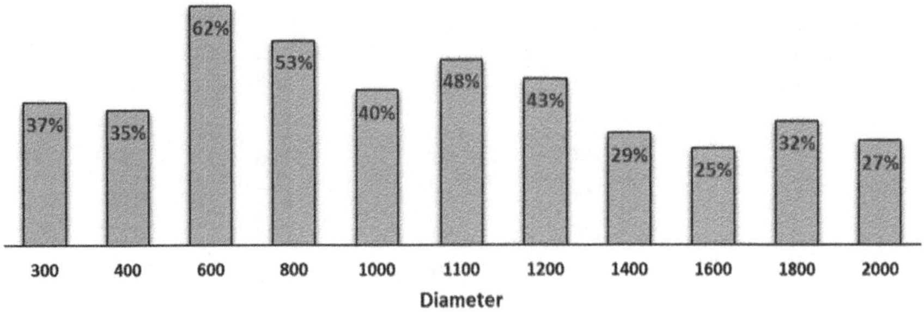

Fig. 8 Increase of pipes diameter on the study area under climate change

Due to unavailability of a small timescale rainfall data (5–10 min), the findings of this study have to be seen in light of some limitations. Indeed, there are two major limitations that could be addressed in future research:

(1) The study focused on the rainfall variability between two periods (1970–1989 and 1990–2015) and assumed that results could be extrapolated to future climate, therefore, they could be used to design UD structures. Actually, this hypothesis is acceptable only for structures designed for return period less than 10 years (e.g. collectors, pipes). That's why the case study has been focused on the conveyance capacity of UD system.
(2) The periods (1970–1989 and 1990–2015) used in frequency analysis (Fig. 3) have been chosen based on the operation services feedback in urban catchment of Tetouan city. Indeed, if rainfall data were available for the majority of extreme events recorded in study area, it would be possible to analyze rainfall patterns in the watershed, and so, avoid any subjectivity of the information emanating from operation services.

Finally, the limitations presented do not influence the reliability of the analysis procedure proposed in this work.

4 Conclusions

An approach to adapting UD structures to CC is proposed in this work. With this purpose, extreme rainfall data series in urban watershed located in northern Morocco were used. Indeed, the properties of rainfall events affecting directly the UD system were evaluated by applying frequency analysis method over two periods 1970–1989 and 1990–2015. In this context, the GEV probability distribution function was used, and the quantiles were obtained. The results have shown an increase in all rainfall events, especially those with 1–2 h, the increase is ranged from 7 to 20% depending on the return period.

The rainfall quantiles obtained through the approach were fitted by the Montana IDF curve. Then, using the factor φ, the rational method was followed to assess the peak discharge adapted to CC in urban catchment of Tetouan city.

The approach was applied in northern Morocco and the results have shown that the design peak discharge has increased by up to 51% under CC conditions. Considering the obligation to maintain service level and avoid excessive flood in urban area, designer has to analyze and update periodically the IDF relationship to reflect the possible changing of future climate conditions.

This study is a part of researches carried out in northern Morocco to adapt the UD system design to the Climate Change (CC). The next step will be focused on the governance of the UD management system and operating procedures in the context of Climate Change.

References

Acquaotta F, Faccini F, Fratianni S, Paliaga G, Sacchini A (2018) Rainfall intensity in the genoa metropolitan area: secular variations and consequences. Weather 73:356–362. https://doi.org/10.1002/wea.3208

Armal S, Devineni N, Khanbilvardi R (2018) Trends in extreme rainfall frequency in the contiguous United States: attribution to climate change and climate variability modes. J Clim 31(1):369–385. https://doi.org/10.1002/2016GL072201

Dale M, Hosking A, Gill E, Kendon E, Fowler HJ, Blenkinsop S, Chan S (2018) Understanding how changing rainfall may impact on urban drainage systems: lessons from projects in the UK and USA. Water Pract Technol 13(3):654–661. https://doi.org/10.2166/wpt.2018.069

Freire Diogo A, Antunes do Carmo, J (2019) Peak flows and stormwater networks design—current and future management of urban surface watersheds. Water 11(4):759. https://doi.org/10.3390/w11040759

García-Marín AP, Morbidelli R, Saltalippi C, Cifrodelli M, Estévez J, Flammini A (2019) On the choice of the optimal frequency analysis of annual extreme rainfall by multifractal approach. J Hydrol 575:1267–1279. https://doi.org/10.1016/j.jhydrol.2019.06.013

Grum M, Jorgensen AT, Johansen RM, Linde JJ (2006) The effect of climate change on urban drainage: an evaluation based on regional climate model simulations. Water Sci Technol 54(6–7):9–15. https://doi.org/10.2166/wst.2006.592

Longo E, Rooney JP (2018) Development of a general protocol to enhance the hydrological analysis techniques for urban catchments in Ireland. Geosciences 8(7):252. https://doi.org/10.3390/geosciences8070252

Meylan P, Favre AC, Musy A (2011) Predictive hydrology: a frequency analysis approach. CRC Press

Moujahid M, Stour L, Agoumi A (2015) Urban drainage and climate change: limits of the modeling. Houille Blanche—Revue Internationale De L'eau 1(1):71–75. https://doi.org/10.1051/lhb/2015009

Moujahid M, Stour L, Agoumi A, Saidi A (2018) Regional approach for the analysis of annual maximum daily precipitation in northern Morocco. Weather Clim Extremes 21:43–51. https://doi.org/10.1016/j.wace.2018.05.005

Myhre G, Alterskjær K, Stjern CW, Hodnebrog Ø, Marelle L, Samset BH, Sillmann J, Schaller N, Fisher E, Shulz M, Stohl A (2019) Frequency of extreme precipitation increases extensively with event rareness under global warming. Sci Rep 9(1):1–10. https://doi.org/10.1038/s41598-019-52277-4

Nguyen TH, Nguyen VTV (2019) Decision-support tool for constructing robust rainfall IDF relations in consideration of model uncertainty. J Hydrol Eng 24(7):06019004. https://doi.org/10.1061/(ASCE)HE.1943-5584.0001802

Preacher KJ (2001) Calculation for the Chi-square test: an interactive calculation tool for chi-square tests of goodness of fit and independence. Ohio State University. Available at: http://quantrm2.psy.ohio-state.edu/kris/chisq/chisq.htm

Sarhadi A, Soulis ED (2017) Time-varying extreme rainfall intensity-duration-frequency curves in a changing climate. Geophys Res Lett 16 44(5): 2454–2463. https://doi.org/10.1002/2016GL072201

Sugi M, Imada Y, Nakaegawa T, Kamiguchi K (2017) Estimating probability of extreme rainfall over Japan using extended regional frequency analysis. Hydrol Res Lett 11(1):19–23. https://doi.org/10.3178/hrl.11.19

Switzman H, Razavi T, Traore S, Coulibaly P, Burn DH, Henderson J, Fausto E, Ness R (2017) Variability of future extreme rainfall statistics: comparison of multiple IDF projections. J Hydrol Eng 22(10):04017046. https://doi.org/10.1061/(ASCE)HE.1943-5584.0001561

Tabari H, Willems, P (2018) Anomalous extreme rainfall variability over Europe—interaction between climate variability and climate change. In: International Conference on Urban Drainage Modelling. Springer, Cham, pp 375–379. https://doi.org/10.1007/978-3-319-99867-1_63

Te Chow V (2010) Applied hydrology. Tata McGraw-Hill Education.

Teegavarapu RS (2012) Floods in a changing climate: extreme precipitation. Cambridge University Press

Wang J, Xu Y, Wang Y, Wang Y, Yuan J, Wang Q (2019) Non-stationarity analysis of extreme water level in response to climate change and urbanization in the Taihu Basin, China. Stoch Environ Res Risk Assess 33:891–904. https://doi.org/10.1007/s00477-019-01666-2

Willems P (2013) Revision of urban drainage design rules after assessment of climate change impacts on precipitation extremes at Uccle, Belgium. J Hydrol 496:166–177. https://doi.org/10.1016/j.jhydrol.2013.05.037

Willems P, Arnbjerg-Nielsen K, Olsson J, Nguyen VTV (2012) Climate change impact assessment on urban rainfall extremes and urban drainage: methods and shortcomings. Atmos Res 103:106–118. https://doi.org/10.1080/15730620802600924

Zheng Y, Xue M, Li B, Chen J, Tao Z (2016) Spatial characteristics of extreme rainfall over China with hourly through 24-hour accumulation periods based on national-level hourly rain gauge data. Adv Atmos Sci 33(11):1218–1232. https://doi.org/10.1007/s00376-016-6128-5

Mainstreaming Climate Change Adaptation into Water Development Plan Case of Morocco

Laila Oualkacha, Laila Stour, and Ali Agoumi

Abstract Mainstreaming climate change adaptation into development plans requires sustained improvement of the political, economic, social, technical, environmental, and legal aspects. International organizations have developed processes to support developing countries in mainstreaming climate risks into water development plans. However, the implementation of this process remains a significant challenge, while many adaptation projects are generally managed outside the existing processes of planning and development. Decision-makers and critical players in the ground agree that there is a deficiency in "how to operationalize the mainstreaming process in the water sector?" This article aims to answer the above question by exploring a structured approach that decision-makers can use to operationalize mainstreaming climate change adaptation into the water development plan at the local level. This approach considers a completed strategic analysis of political, economic, social, technical, environmental, and legal aspects. The results of this approach have generated the proposed theoretical tool designed to reflect the dynamism of climate change phenomena through its seven iterative steps, which permit to identify most vulnerable units, climate change impacts, prioritized adaptation actions, mechanisms to access climate funds, processes for implementation, monitoring and evaluation. The proposed approach has been tested in two Moroccan river basins: the Loukkos and the Tensift RBs. Similar river basins of Algeria and Tunisia will also be tested using the same processes.

L. Oualkacha (✉) · L. Stour
Laboratory of Process and Environment Engineering, Faculty of Sciences and Techniques of Mohammedia, Hassan II University of Casablanca, B.P. 146, Mohammedia, Morocco

A. Agoumi
Laboratory of Civil Engineering, Hydraulics, Environment and Climate, Hassania School of Public Works, Casablanca, Morocco

W. Leal Filho and E. Manolas (eds.), *Climate Change in the Mediterranean and Middle Eastern Region*, Climate Change Management,
https://doi.org/10.1007/978-3-030-78566-6_17

1 Introduction

Water is a crucial resource for economic growth and development in Morocco. It contributes widely to the socio-economic balance and the Gross Domestic Product. However, water is endangered by Climate Change (CC), and Morocco seems to be one of the most vulnerable countries globally concerning water scarcity, according to the fifth report of the Intergovernmental Panel on Climate Change (IPCC 2014).

Moroccan climate data depict a noticeable warming trend over the past 40 years and a net increase in the frequency of droughts and floods within the country (Moujahid et al. 2015). Future climate trends in Morocco include rising temperatures of 1–1.5 °C by 2050 and a decrease in average precipitation by 10–20% across the country by 2100 (ME 2016).

In this challenging context, Morocco has adopted some plans and strategies, such as National Strategy against CC and National Plan against CC. Despite all undertaken efforts, implementation of mainstreaming CC adaptation into development plans remains a significant challenge at national and local levels. The national strategies have highlighted some of these adaptation measures; however, these measures are usually managed outside the existing planning process, which is mainly the case of the water sector.

Mainstreaming CC adaptation into development plans' "mainstreaming process" is defined by the United Nations Development Program (UNDP) as an iterative process to integrate CC aspects and adaptation into policy decisions and budgeting at national and local levels (UNDP 2011). The mainstreaming process requires coordination among multiple stakeholders and institutions, making the journey from a plan on paper to action on the ground slow (Mogelgaard et al. 2018).

Despite growing political attention and tools to support mainstreaming, an implementation gap persists (Mogelgaard et al. 2018). Furthermore, international organizations (UNDP, the United States Agency for Integrated Development, and the Deutsche Gesellschaft fur international cooperation (GIZ)) have developed processes to support developing countries (DC) to operationalize mainstreaming process. However, these processes provide only general guidance, and efforts have been slow to translate mainstreaming into concrete actions and programs at national and local levels (Mogelgaard et al. 2018).

Implementation of the mainstreaming process depends on the synergy between sectors at national and local levels (Eisenack et al. 2014). It also requires sustainable strategic coordination, explaining the challenges and the complexity of implementing this process (Waters et al. 2014). Thus, operationalizing the mainstreaming process remains a significant challenge (Mogelgaard et al. 2018), and the key question is "how to operationalize the mainstreaming process"?

Considering the gaps emphasized previously regarding mainstreaming process implementation, this article aims to present an approach that permits to operationalize "mainstreaming process" at the local level within the water sector: 'the Tool for Mainstreaming CC (TMCC). The TMCC is developed using qualitative and quantitative approaches. It is composed of seven iterative steps, harmonized in a coherent manner, which generates a practical and comprehensive approach that supports decision-makers to operationalize the mainstreaming process and enhance the River Basin (RB) resilience.

The TMCC's findings help identify the most vulnerable units, CC biophysical and socio-economic impacts, and prioritized adaptation actions, funding and implementation processes. These results permit to improve water resources (WR) management and planning, including climate risks. The TMCC has been tested in two Moroccan RBs: the Loukkos and the Tensift RBs. Similar RBs of Algeria and Tunisia will also be tested using the same processes.

2 Methodology and Tool

2.1 Combination of Qualitative and Quantitative Approaches

Mainstreaming process is a complicated procedure to address through a combination of qualitative and quantitative approaches (Benjamin et al. 2017). This combination of methods has been described as the best approach to address CC adaptation issues (Harrison et al. 2013). Thus, selecting this method to conduct the research.

The qualitative approach is used to identify the most vulnerable units, CC impacts and actions for adaptation, while the quantitative approach is used to develop scores and evaluation criteria.

2.2 Adopted Strategy

The strategy adopted to design the TMCC is interactive; indeed, the TMCC provides guidelines to decision-makers through a collection of questions that allow to create resonance regarding CC mainstreaming process and identify barriers that inhibit the implementation of this process.

The answers to each question include strategic options developed considering several dimensions: efficiency, equity, feasibility and acceptability by the stakeholders within the water sector. Based on the suggested answers, decision-makers can choose the most suitable option that fits the context of their organization.

2.3 Design of the TMCC's Steps

Figure 1 presents the TMCC's steps: preparation, vulnerability analysis, adaptation actions, integration of the relevant actions into the Integrated WR Development Plan (IWRDP), funding, implementation and monitoring processes. The 7 iterative steps

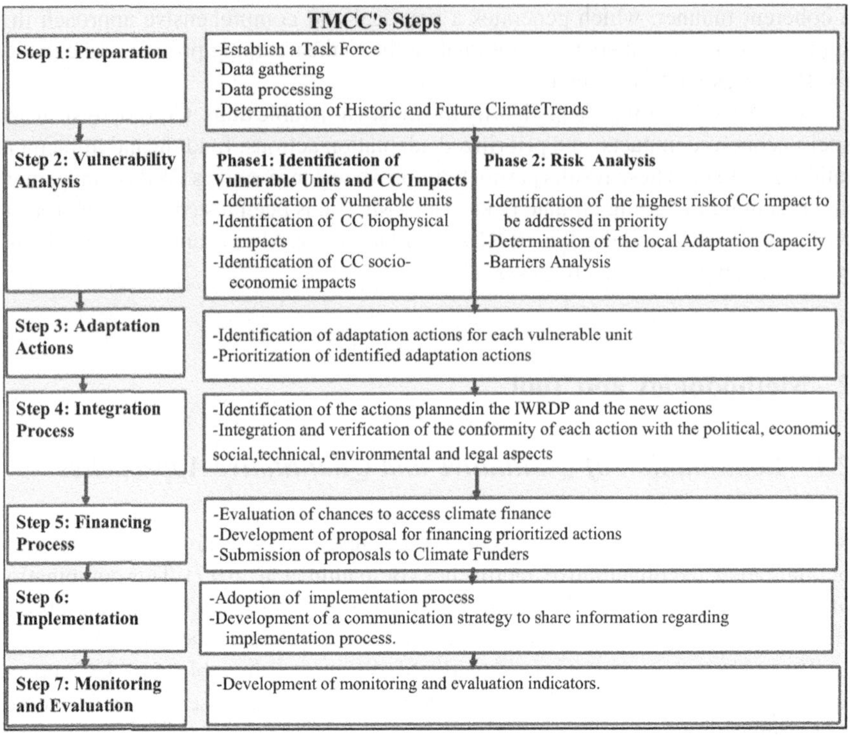

Fig. 1 The TMCC's steps design

of the TMCC are harmonized in a coherent manner, where each step is an essential part in the chain of actions within the methodology; the output of one step is the input for the following step.

Step 1 of the TMCC: Preparation

The TMCC suggests a participatory approach that engages different stockholders. Therefore during the first step, the decision-maker can establish a Task Force (TF) to monitor and coordinate mainstreaming process implementation. Step 1 involves building CC knowledge within the RB. It consists of gathering and analyzing CC data and determining historical and future climate trends. The climate data analysis requires the combination of CC physical concepts (temperature and rainfall) that can affect all aspects of planning and inhibit CC mainstreaming process implementation (Nouri and Costa 2017).

The combinations of CC physical factors have permitted to identify four situations: (i) Situation S1: decreased rainfall and increased temperature. (ii) Situation S2: decreased rainfall and decreased temperature. (iii) Situation S3: increased rainfall and increased temperature and (iv) Situation S4: increased rainfall and decreased temperature.

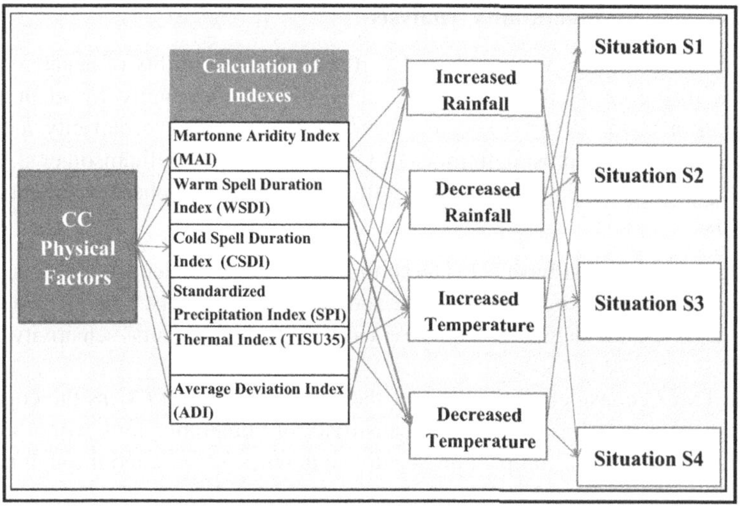

Fig. 2 The design of the TMCC's step 1

To determine how CC physical concept can influence historical and future climate trends, the TMCC suggests the calculation of six indexes: (a) Martonne Aridity Index; (b) Warm Spell Duration Index; (c) Cold Spell Duration Index; (d) Standardized Precipitation Index; (e) Thermal Index; (f) Average Deviation Index.

Figure 2 presents the design of TMCC's Step 1. It shows that the six proposed indexes' calculation allows determining historical and future climate trends (S1, S2, S3 or S4) within the RB.

Step 2 of the TMCC: Vulnerability Analysis

Based on the results of step 1, the TMCC suggests an approach to analyze the RB's vulnerability. Step 2 includes two phases: (i) Phase 1: identification of CC vulnerable units and impacts and (ii) Phase 2: risk analysis.

Step 2–Phase1: Identification of Vulnerable Units and CC Impacts

An analysis of policy, economic, social, technical, environmental and legal (PESTEL) aspects has been undertaken to identify the most vulnerable units and CC impacts within the water sector. The results of this analysis support in identifying six (6) potential vulnerable units: (a) water supply; (b) water demand; (c) water quality; (d) territory, people and economic activities; (e) coastal zone and (f) water governance. To guide the decision-makers, the TMCC offers an inventory of potential CC biophysical and socio-economic impacts for each vulnerable unit. Indeed 55 CC impacts have been identified, including 25 biophysical impacts and 30 socio-economic impacts.

The results of step 2–phase1 allow the user to identify vulnerable units and CC expected biophysical and socio-economic impacts.

Step 2–Phase 2: Vulnerability Analysis

Local communities are vulnerable to CC if there is a probability of negative consequences that exceeds the intrinsic capacity of the community to adapt to CC (Williamson et al. 2014). Exposure, sensitivity and adaptive capacity are three common elements in most definitions of CC vulnerability (Williamson et al. 2014). Therefore, this second phase of step 2 will include three sub-phases related to these elements.

Step 2, Phase 2, Sub-Phase 2.1: The sub-phase 2.1 aims to identify the most significant risk of CC effects that should be addressed as a priority. Therefore, the first task is to assess the degree of exposure and the vulnerable units' sensitivity within the RB.

The TMCC considers the impact of the adverse effects of CC as the combined effect of the degree of exposure and sensitivity of vulnerable units within the RB. The importance of CC's impact on negative outcomes is evaluated through ranking scores from 1 to 5 (1 = very low and 5 = very high). A score of 1 means that the importance of CC's negative effects on vulnerable units is not very important. A score of 5 means that the significance of negative effects is expected to be very high.

The characterization of vulnerability includes the CC future uncertainty (Gleeson and Coll 2011). Therefore, the CC impacts are further evaluated through a risk analysis based on their probability of occurrence in the future. The TMCC has adopted five probability criteria rated from 1 to 5 for the 2050 horizon (1 = very low and 5 = very high). A probability of 1 signifies that the occurrence of the negative effect is not very likely. A probability of 5 means that this occurrence is considered to be very high.

After evaluating the importance of CC negative effects and their probability of occurrence in the future, climate risk is calculated by the following equation:

$$R = P * I$$

R is the climate risk;
P is the probability of future occurrence of CC negative effects;
I refers to the importance of negative effects of CC impacts.

The results of this sub-phase allow the user to determine the list of CC impacts to be treated as a priority.

Step 2-Phase 2-Subphase 2.2: the aim of this sub-phase is to estimate the RB's adaptive capacity. The TMCC considers CC adaptive capacity (AC) as "the ability of the RB to moderate CC potential damage and take advantage of existing opportunities to operationalize the mainstreaming process."

An analysis has been undertaken to identify key indicators to assess the AC at the RB level. 8 indicators are selected and adopted by the TMCC to estimate the AC. The selected indicators list is not exhaustive; it presents significant markers that provide information about the RB's AC. The estimation of these indicators permits to identify 5 levels of AC.

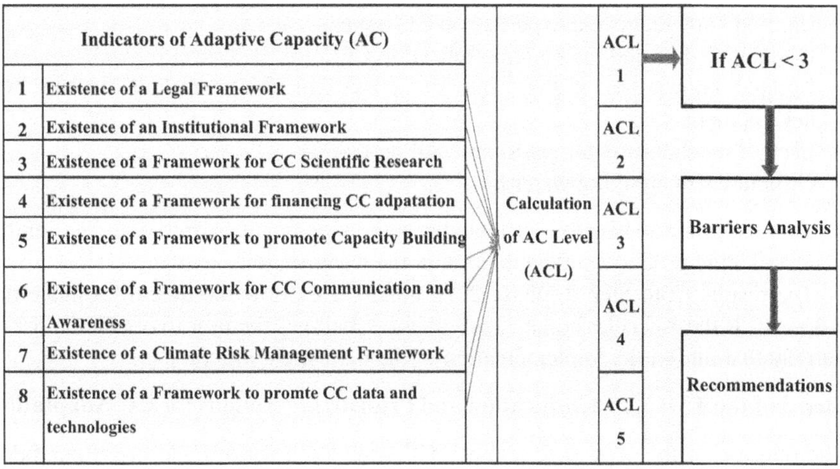

Fig. 3 Design of the Step 2, Phase 2, Sub Phase 2.2 of the TMCC

Figure 3 presents the design of the Step 2, phase 2-sub-phase2.2 of the TMCC. It represents the 8 selected indicators and the 5 levels of AC. Level 1 of AC indicates that tools, institutions, and mechanisms which create mainstreaming process implementation opportunities are not available. Level 2 of AC indicates that these tools, institutions and mechanisms are available. Level 3 of AC suggests that the local government is engaged in implementing these tools and mechanisms (the local government has recently administered these tools into their agenda and other essential issues in the government agenda). Level 4 of AC indicates that the implementation of these tools and mechanisms are limited. Level 5 of AC implies that the implementation plan of these tools and mechanisms are advanced.

Figure 3 shows that the TMCC has adopted level 3 of AC as a reference level required to operationalize mainstreaming process implementation. Indeed, if AC is less than 3, the TMCC suggests conducting a barriers analysis. This analysis is related to the 8 selected areas of opportunities; furthermore, the TMCC presents recommendations to overcome barriers and make progress in the mainstreaming process implementation.

Step 2-Phase 2-Sub-phase 2.3: this sub-phase aims to analyze barriers that inhibit mainstreaming process implementation. PESTEL analysis has allowed identifying 12 barriers:

- Limited government prioritization of mainstreaming process at the local level
- Poor coordination between National and Local Government
- Difficulty to access climate funds
- Limited financing mechanisms to strengthen the mainstreaming process
- Limited involvement of local community
- Limited ways to build on local traditional knowledge and practices
- Limited belief and understanding of CC's impact at local level

- Limited knowledge and expertise on CC issues
- Inadequate data and information related to CC
- Limited knowledge of the value of ecosystems that contribute to the WR development
- Lack of mechanisms for environmental protection
- Difficulties in applying water law.

This list is not exhaustive; it includes some significant barriers that can inhibit mainstreaming process implementation in the water sector.

The results from Step 2, phase 2, sub-phase 2.2 allow the user to estimate the RB's AC. If the AC is less than 3, sub-phase 2.3 allows the user to obtain the list of barriers that inhibit the implementation of the mainstreaming process.

Step 3 of the TMCC: Identification and Prioritizing Actions for CC Adaptation

The purpose of this step is to identify CC adaptation actions and prioritize them, as it has been acknowledged that we cannot implement all of the identified actions at the same time. To minimize CC impacts on vulnerable units, the TMCC offers a list of 60 adaptation actions and adopted six strategic axes related to the six most vulnerable units identified above.

The TMCC allows users to classify adaptation actions into six groups of Actions (GA1; GA2; GA3; GA4; GA5; GA6). GA1 includes actions with high priority. The actions of the GA6 and GA5 are not necessarily optional or urgent; often, it is their essential cost that downgrades them; therefore, there is a need to search for significant additional investment to achieve these actions.

The Climate Proofing for Development (CPDev) approach, developed by the GIZ (GIZ 2011), has been used to prioritize adaptation actions at the RB level. The TMCC has adopted this approach based on the following aspects: (i) CPDev is an approach that allows to mainstreaming adaptation actions into existing development plans (GIZ 2011); (ii) CPDev is an approach that uses criteria adapted to the water sector context (Stour and Agoumi 2013); (iii) CPDev is based on principles that considered the socio-economic and political aspects (GIZ 2011).

According to the CPDev approach, Table 1 presents the criteria adopted by the TMCC to prioritize adaptation actions and the rating scores. Rating scores are identified on a scale of 1–5, 1, meaning the lowest intensity, and 5 meaning the high intensity to reduce CC vulnerability while applying the identified actions.

The results of the TMCC's Step 3 allow users to have a list of prioritized adaptation actions to be urgently implemented.

Step 4 of the TMCC: Integration of Prioritized Actions into the IWRDP

Step 4 of the TMCC allows the user to integrate a prioritized list of adaptation actions into the IWRDP. At this stage, the TMCC presents guidance for the separation of new actions, and those already emphasized in the IWRDP. Furthermore, the TMCC offers an opportunity to achieve a conformity test with the PESTEL context. According to the results of this test, congruent actions with the PESTEL context are directly

Table 1 Scoring criteria

Key criteria	Designation	Qualification (Q)	Scores
C1	Cost of Identified Action	**Q1**: Very High cost: national government or international organizations will support the cost **Q2**: High cost but the region can support the cost **Q3**: Medium cost tolerable by the city **Q4**: Low cost supportable by the commune **Q5**: Very low cost supportable by a group of people	1 2 3 4 5
C2	Intensity of vulnerability reduction ensured through the implementation of the identified action	Huge spread of intensity Very high intensity High intensity Medium intensity Low intensity	5 4 3 2 1
C3	Scope of the adaptation action	At familial level At the village level At the commune level At the city level At the region level	1 2 3 4 5
C4	Actions with/without regret	Actions with regret Actions without regret	1 2
C5	Action implementation	The action is difficult to implement The action is easy to implement	1 2

integrated into the IWRDP. However, the non-congruent actions must be modified and adjusted to the PESTEL context, facilitating its integration into the IWRDP.

The results of this Step allows user to dispose of: (i) the list of the adaptation actions emphasized in the IWRDP to be integrated directly in the IWRDP; (ii) the list of actions emphasized in the IWRDP to be adjusted; (iii) the list of new actions to be integrated directly in the IWRDP; (vi) the list of new actions to be adjusted or modified to facilitate their integration into the IWRDP. Therefore, these results permit to update the current IWRDP, which provides a practical basis plan for the next 5 years.

Step 5 of the TMCC: Financing Prioritized Adaptation Actions

The prioritized actions must be implemented, which requires funding resources. If the decision-maker has these resources, it could go directly to implementation. Otherwise, the TMCC offers an approach to guide the decision-maker to access climate funds.

The prioritized actions can be grouped on Program of Actions (PA). The TMCC has adopted an approach to evaluate the PA's chances to access climate funds. This approach is based on the evaluation criteria developed by the European Union within

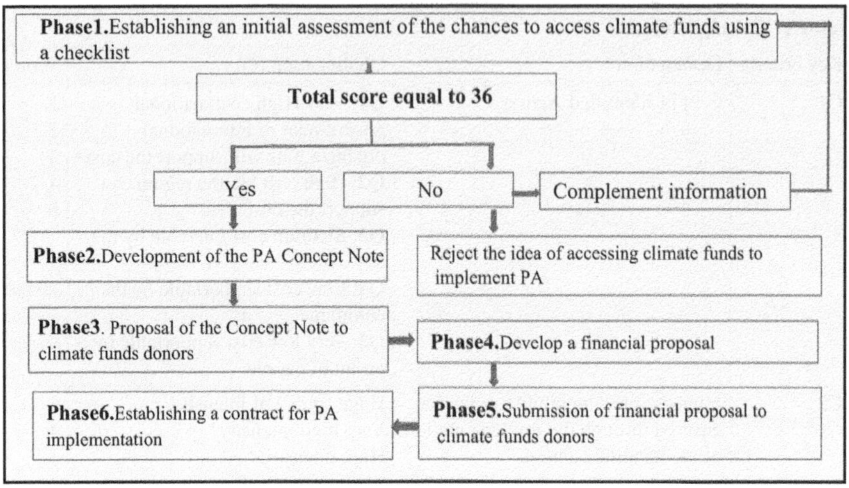

Fig. 4 Design of the step 5 of the TMCC (Stour 2017)

the ClimaSouth project (ClimaSouth 2016). Indeed, a checklist has been established based on the five keys developed within the ClimaSouth project: (a) description of action; (b) expected results; (c) compliance with national priorities; (d) clarity of the implementation plan; and (e) estimation of the budget and co-financing indicative (ClimaSouth 2016).

The five keys outlined above include 12 aspects required by climate fund donors. These aspects provide specific information and issues to consider in order to access climate funds. In this regards, three scores have been identified for each aspect to assess the availability of information required by climate donors (CD) (i): No information is available; (ii): Information partially available, with a chance to be provided later; (iii): Information is fully available.

Figure 4 presents the procedure adopted by the TMCC to guide decision-makers to access climate funds.

The results of the TMCC's Step 5 allow the user to determine the chances of accessing climate funds for action or PA implementation.

Step 6 of the TMCCC: Implementation Process

To facilitate the implementation process, the TMCC suggests to develop an implementation strategy with the following considerations: (a) identify key stakeholders and their responsibilities; (b) specify the implementation delays; (c) develop methodological guides to facilitate implementation; (d) ensure synergies between sectors: water-energy-food security; (e) strengthen coordination with RB neighbors communities; (f) mobilize civil society; (g) strengthen capacity building of the TF representatives; (h) develop communication strategies and awareness (websites to share information and results); (i) create some Champion patterns for implementation.

The results of Step 6 provide users with further guidelines regarding the implementation process. At the same time, the established TF will monitor the integration of the prioritized adaptation actions into the newly revised water plan during the five coming years considering CC science innovation and local community evolution.

Step 7 of the TMCC: Monitoring and Evaluation

The TMCC has adopted the monitoring and evaluation (ME) method generated from a local approach. This approach enables identifying quantitative and qualitative indicators that emphasize effectiveness and the role of mainstreaming adaptation action into existing plans, policies, and budgets at the local community level (Bours et al. 2013). The TMCC suggests some ME indicators that characterize barriers inhibiting the PA process implementation. These indicators are harmonized with the barriers outlined above. Table 2 presents the ME indicators adopted by the TMCC.

The results of the TMCC's Step 7 allow the user to have a guiding mechanism for monitoring the implementation.

3 Results

The TMCC has been tested in the two RBs of Loukkos (LRB) and Tensift (TRB). The data used to generate results is collected from the Loukkos RB agency (LRBA) and the Tensift RB agency (TRBA). The available data are collected during 1960–2016, especially in the Tangier station (LRBA) and Marrakech station (TRBA).

3.1 Presentation of the Loukkos and Tensift River Basins

In Morocco, each river basin is managed by an agency according to the 10–95 water law. The agency is responsible for the advancement of the water development plan and procedures for water management. It is also responsible for data collection, data analysis, and data compilation for water management at the RB level.

In 2012, the LRBA and the TRBA developed their IWRDP that presents WR status and identifies potential solutions to ensure water security within the RB. However, these plans did not include CC impacts and considerations. Therefore, the LRBA and the TRBA are currently instructed by the national authority to update their IWRDP to anticipate and mainstream CC issues into the new water development plan.

The area of actions of the LRBA presented in Fig. 5 is located in the north of Morocco. Studies available for the LRB announced trends of decrease in annual average rainfall of −10 to −20% during 2016–2035 with a pronounced decreasing trend (−30%) during 2081–2100 (M.E 2016). Furthermore, these studies announced: (i) trends of increase in annual average temperatures of 4–5 °C at the 2100 horizon (M.E 2016); (ii) an accelerated Mediterranean Sea level rise that may exceed 90 cm during the end of this century (M.E 2016).

Table 2 Indicators for monitoring-evaluation process

Aspects	ME Indicators	Key elements of characterization	Score
Policy	Local government priority	PA implementation is not in the local government agenda	1
		local government has just put PA in his agenda (other issues are important in the government agenda)	2
		PA is a priority in the local government agenda	3
Institutional	Availability of processes for coordination between national and local government	Coordination processes not available	1
		Coordination processes available, but not easily implemented or inadequate	2
		Coordination processes available and easily implemented	3
Financial	Availability of mechanisms to support local actors to implement PA	Public mechanisms not available	1
		Public mechanisms available but not easily understood or inadequate	2
		Public mechanisms available and easily implemented	3
Social	Availability of mechanisms to inform vulnerable groups	Mechanisms not available	1
		Mechanisms available, but not easily understood or inadequate	2
		Mechanisms available and easily implemented	3
Technical	Availability of climate scenarios and process to strengthen capacity building	Scenarios and capacity building process not available	1
		Scenarios and capacity building processes available but not easily understood or inadequate	2
		Scenarios and capacity building processes available and easy implemented	3
Environmental	Availability of environmental protection mechanism to strengthen PA implementation	Mechanisms not available	1
		Mechanisms available, but not easily understood or inadequate	2
		Mechanisms available and easily implemented	3
Legal	Availability of legal mechanism to strengthen PA implementation	Mechanisms not available	1
		Mechanisms available, but not easily understood or inadequate	2

(continued)

Table 2 (continued)

Aspects	ME Indicators	Key elements of characterization	Score
		Mechanisms available and easily implemented	3

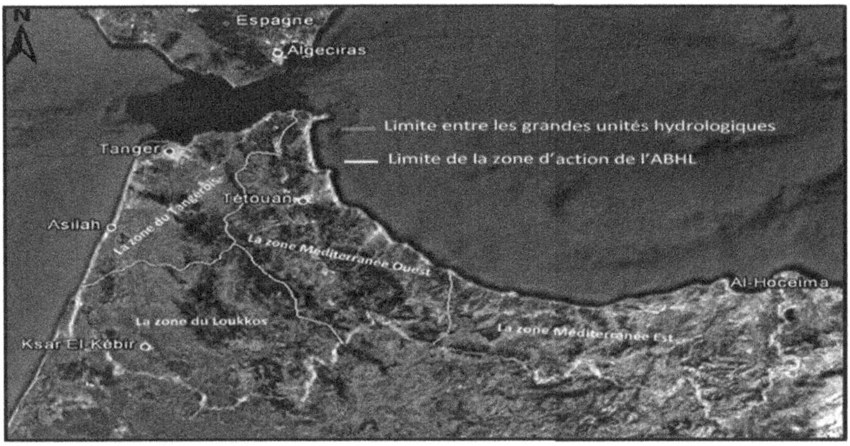

Fig. 5 The LRBA area of action

The area of action of the TRBA presented in Fig. 6 is located in the Midwest of the country. Climate projections available for the TRB show that according to the optimistic scenario (RCP 2.6), annual average temperatures will increase by 0.5–1 °C during 2016–2035 and by 1.0–2.5 °C during the end of this century (M.E 2016). However, according to the pessimist scenario (RCP 8.5) results, the annual average

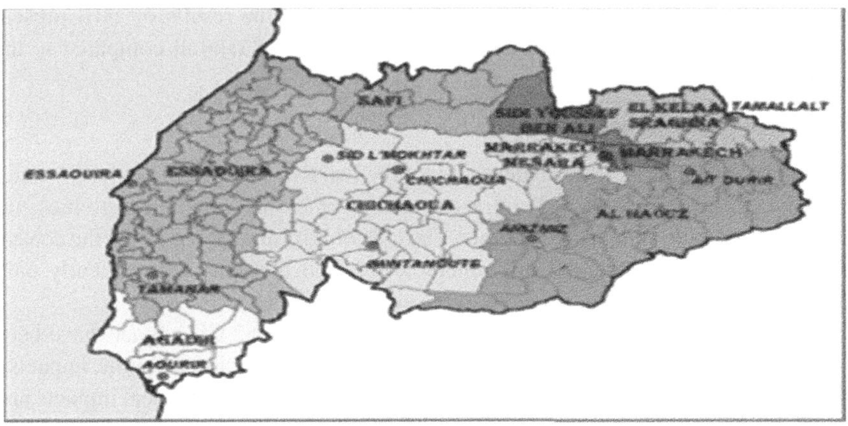

Fig. 6 The TRBA area of action

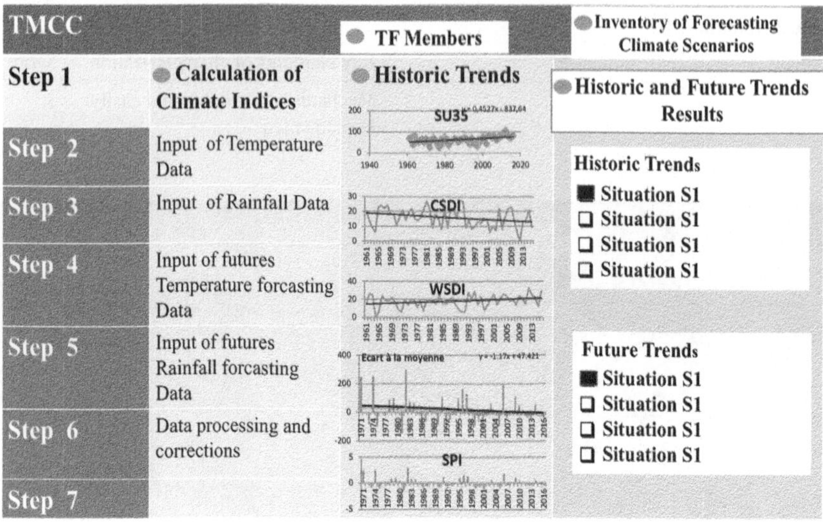

Fig. 7 Results of the step 1: case of the Loukkos River Basin

rainfall will decrease by −10 to −20% on the 2065 and by −40% on the 2100 horizon (M.E 2016). Furthermore, the results from the pessimist scenario show that the annual average temperature will increase from +1.5 °C to +4 °C on the 2065 horizon and +5 °C to +7 °C by the end of this century (M.E 2016).

Results of Step 1 of the TMCC

The results of Step 1 are quite similar for the two RBs. Figure 7 presents the results from Step 1 of the LRB. It shows that situation S1 (increased temperature and decreased rainfall) characterizes the LRB's historical and future trends. However, there is a need to note that: (i) the results of TRB present pronounced historical and future warming trends compared to the LRB; (ii) the results of TRB present pronounced historical and future trends of a decrease of rainfall compared to the LRB.

Results of Sep 2 of the TMCC

Six vulnerable units have been identified for the two RBs: (a) water supply; (b) water demand; (c) water quality; (d) infrastructure, people and economic activities; (e) water governance; and (f) coastal zone. The lack of information regarding the coastal zone of the TRB has excluded this vulnerable unit from the study. Consequently, only five vulnerable units are considered in the case of the TRB.

The results of Step 2-phase 1 of the TMCC show that 52 CC impacts have been inventoried in the LRB (22 biophysical impacts and 30 socio-economic impacts), while 44 CC impacts have been inventoried in the TRB (18 biophysical impacts and 26 socio-economic impacts).

TMCC									
Steps		AC Indicators	Score of Adaptive Capacity adopted by the TMCC						Estimation of AC Level
Step 1	1	Existence of a Legal Framework promoting mainstreaming process implementation	1	2	3	4	5	6	3
Step 2 P 1	2	Existence of an Institutional Framework promoting mainstreaming process implementatio	1	2	3	4	5	6	3
Step 2- P2- SP 2.1	3	Existence of a Framework enhancing Scientific Researches related to CC	1	2	3	4	5	6	1
Step 2- P2- SP 2.2 Step 2- P2-SP 2.3	4	Existence of a Funding Framework promoting mainstreaming process implementatio	1	2	3	4	5	6	1
Step 3	5	Existence of a Framework improving Capacity Building	1	2	3	4	5	6	2
Step 4	6	Existence of a Communication Framework	1	2	3	4	5	6	1
Step 5	7	Existence of a Framework for CC Risk	1	2	3	4	5	6	3
Step 6 Step 7	8	Existence of a Framework for promoting climate data and technology development	1	2	3	4	5	6	3
Level of Adaptive Capacity within the Tensift RB (Average)									2

Fig. 8 Results of Step 2, Phase 2, Sub-phase 2.2 of the TMCC: Case of the Tensift RB

The results of Step 2-phase 2-sub-phase 2.1 of the TMCC are quite similar for the two RBs. The following CC impacts are classified as impacts with a high level of risk (catastrophic risk): (a) loss of human lives; (b) water scarcity; (c) increased cost for water treatment; (d) increased water demand and (e) limited institutional tool. Furthermore, the increased sea level and the loss of livelihoods are also impacts classified with very high risk in the case of the LRB where the coastal area has been studied.

The results of Step 2-phase 2-sub-phase 2.2 of the TMCC are similar for the two RBs. Figure 8 presents the results of the sub-phase 2.2 of the TMCC for the TRB. It shows that the average level of AC of this RB is estimated at 2. Therefore, according to the TMCC procedures outlined above, this level of AC when less than 3 requires analyzing barriers that inhibit mainstreaming process implementation.

The results of the sub-phase 2.3 are also similar for the two RBs. These results show that all barriers (12 barriers) adopted by the TMCC are identified as major barriers that inhibit the mainstreaming process implementation within the two RBs

Results of Sep 3 of the TMCC

The prioritizing process is applied according to the TMCC procedures outlined above. The results of Step 3 show that the cost of identified action (C1) and the intensity of vulnerability reduction ensured through the action implementation (C2) are identified as weighing criteria playing an essential role in the prioritizing process within the two RBs. The results also show that most proposed actions concern water governance (18 proposed adaptation actions).

The classification of adaptation actions on GA is applied according to the TMCC procedures outlined above. The findings show that priority is given to water governance. Indeed 100% of the GA1 actions (GA1 presents the highest priority action group) are required for improving water governance within the two RBs. This result does not mean that the actions of the other strategic axes are less important; however, it seems that the success of mainstreaming process implementation required strong action on water governance that can support overcoming challenges of the other strategic axes as well as political, institutional, legal, human, and financial challenges.

Results of Step 4 of the TMCC

According to the TMCC procedures outlined above, the results from step 4 of the TMCC show that the conformity test has allowed identifying actions already outlined in the IWRDP. These need to be reinforced to make it easier to mainstream (22 actions within the LRB and 23 actions within the TRB). The conformity test has also permitted to identify the new actions congruent with PESTEL aspects (30 actions within the LRB and 21 actions within the TRB). These results permit to have a practical and completed new plan for the next 5 years.

Results of Step 5 of the TMCC

The LRBA and the TRBA are preparing a PA, including prioritized actions related to 'water governance'. The PA will be submitted shortly to the climate funds for approval. The chances of funding the PA have been estimated to 24 for the two RBs, which explains the lack of information required by climate donors. Consequently, the LRBA and the TRBA are currently focusing their efforts on providing complementary information to achieve the financing process described in Fig. 4.

Results of Step 6 of the TMCC

The LRBA and the TRBA are preparing the implementation strategy.

Results of Step 7 of the TMCC

The LRBA and the TRBA establish a ME system that will monitor the progress of the mainstreaming process implementation.

4 Limitations

The mainstreaming process should be designed within an integrated territorial perspective considering synergies between various development sectors. Some identified adaptation actions have considered this perspective; however, the ability to design the territorial approach in its overall vision is a limitation of the TMCC's structure.

The TMCC provides a list of barriers to the mainstreaming process based on the PESTEL analysis; it does not claim to record all possible mainstreaming process barriers. Furthermore, the methodology used did not trait barriers interconnections analysis, which is a limitation of this work.

The TMCC has adopted an extensive list of clear indicators to monitor the mainstreaming process based on the PESTEL analysis. It does not claim to record all possible indicators that can monitor the implementation progress, which is a limitation of this work.

A small sample size of respondents, from national and local levels, was interviewed to discuss different steps of the TMCC, which is a limitation of this work. The constraints of time and resources have prevented conducting a large sample size of respondents.

5 Conclusion

Climate change is one of the greatest challenges for the present and future generations; it has threatening consequences in the water sector at the core of economic growth. The need to mainstream CC into the water development plan is growing and increasingly recognized by the political community. However, despite growing attention to support mainstreaming, there is a need to close the implementation gaps.

The international community has undertaken considerable efforts to support countries to implement mainstreaming processes; the undertaken efforts have enriched the literature but did not provide a clear answer on "how to implement in a practical and operational way mainstreaming process on the ground?".

The challenges of closing the implementation gaps are decreasing as there are movements in considering the inter-linked PESTEL aspects ensemble and assessing barriers and synergies, explaining the complexity of implementing the mainstreaming process.

The TMCC has considered the analysis of PESTEL aspects ensemble and barriers showing support for decision-makers to operationalize the mainstreaming process on the ground. It contributes to enrich the literature through the main finding: a systematic methodology that decision-makers can use to update the water development plan considering CC and promoting climate resilience within the RB.

The application of the TMCC in the Loukkos and Tensift RB has allowed updating the water development plan considering the main results: future climate trends are moving towards an increased temperature and a decreased rainfall, which requires specific actions to overcome this challenge.

Loss of human lives, water scarcity, increased water demand, and inadequacy of the institutional tools are the impacts classified with very high risk to be anticipated. The increased sea level and the loss of livelihoods are also impacts classified with very high risk in the case of the LRB where the coastal area has been studied.

This working paper encourages decision-makers to improve water governance based on its role and power leverage to better and operationalize the mainstreaming process. Barriers related to governance, political and institutional aspects should most certainly be addressed in the early stages of the mainstreaming process implementation.

Barriers interconnections analysis is not expected in this work; it is a complex process that needs specialized focus. Accordingly, the barrier interconnections analysis can be further improved depending on the needs of future researches.

The monitoring and evaluation indicators help planners follow the progress of mainstreaming process and generate more robust and resilient program actions. However, the identified indicators can be further updated, including vulnerability and climate risk indicators depending on future research needs.

References

Benjamin P, Warner BP, Kuzdas C, Yglesias C, Childers D (2017) Mixed-methods approaches to understanding climate change risks to agrarian-based livelihoods

Bours D, McGinn C, Pringle P (2013) Monitoring and evaluation for climate change adaptation: a synthesis of tools, tools and approaches. SEA change community of practice and UKCIP. Available at: https://www.ukcip.org.uk/wp-content/PDFs/SEA-change-UKCIP-MandE-review.pdf

Intergovernmental Panel on Climate Change (IPCC) (2014) Impacts, adaptation, and vulnerability. Part A: global and sectoral aspects. In: Contribution of working group II to the 5th assessment report of the intergovernmental panel on climate change. Cambridge University Press. Available at: https://www.ipcc.ch/site/assets/uploads/2018/02/WGIIAR5-FrontMatterA_FINAL.pdf

ClimaSouth (2016) improve access to finance for climate action. Detailed guide for practitioners. Guide developed by the European Union within the ClimaSouth project. Available at: http://www.climasouth.eu/docs/E_Handbook_N8_EN.pdf

Deutsche Gesellschaftfür Internationale Zusammenarbeit (GIZ) (2011) Climate proofing for development, climate change adapting and risk reduction. Available at: https://www.adaptationcommunity.net/?wpfb_dl=34

Eisenack K, Moser SC, Hoffmann E, Klein RJT, Oberlack C, Pechan A, Rotter R, Termeer CJAM (2014) Explaining and overcoming barriers to climate change adaptation. Published by nature climate change NCLIMATE 2350. Available at: https://research.wur.nl/en/publications/explaining-and-overcoming-barriers-to-climate-change-adaptation

Gleeson J, coll (2011) Practitioner's guide to climate change adaptation in Ontario's ecosystems. Available at: http://www.climateontario.ca/doc/Tools/A%20Practitioners%20Guide%20to%20ClimateChange%20Adaptation%20in%20Ontario's%20Ecosystems%20Ver%201%202011.pdf

Harrison PA, Holman IP, Cojocaru G, Kasper K, Kontogianni A, Metzger MJ, Gramberger M (2013) Combining qualitative and quantitative understanding for exploring cross-sectoral climate change impacts, adaptation and vulnerability in Europe. Reg Environ Change 13(4):761–780. https://doi.org/10.1007/s10113-012-0361-y

Ministry of Environment (M.E) (2016) the 3rd Moroccan national communication to the United Nations convention on climate change. Available at: https://unfccc.int/sites/default/files/resource/Marnc3.pdf

Mogelgaard K, Dinshaw A, Ginoya N, Gutiérrez M, Preethan P, Waslander J (2018) From planning to action: mainstreaming climate change adaptation into development. Working paper. Washington, DC, World Resources Institute. Available at: https://www.wri.org/publication/climate-planning-to-action.

Moujahid M, Stour L, Agoumi A (2015) Urban drainage and climate change: limits of modeling, international water review "La Houille Blanche", N1, pp 71–75. https://doi.org/10.1051/lhb/2015009

Nouri AS, Costa JP (2017) Place making and climate change adaptation: new qualitative and quantitative considerations for the "Place Diagram. J Urbanism Inter Res Place Mak Urban Sustain 10(3):356–382. https://doi.org/10.1080/17549175.2017.129509
Stour L (2017) Companies and climate in Morocco: guide published by the general confederation of enterprises of morocco within the climate business initiative Morocco, p 98
United Nations Development Program (UNDP-UNEP) (2011) Mainstreaming climate change adaptation into development planning: a guide for practitioners. available online at www.unpei.org
Waters E, Barnett J, Puleston A (2014) Contrasting perspectives on barriers to adaptation in Australian climate change policy. Clim Change 124:691–702. https://doi.org/10.1007/s10584-014-1138-8
Williamson TB, Price DT, Beverly JL, Bothwell PM, Parkins JR, Patriquin MN, Pearce CV, Stedman RC, Volney WJA (2014) A framework for assessing vulnerability of forest-based communities to climate change. Nat. Resour. Can., Can. For. Serv., North. For. Cent., Edmonton, Alberta. Inf. Rep. NOR-X-414

Consumption Patterns and Public Attitudes Toward Organic Foods: The Role of Climate Change Awareness

Elena Raptou and Evangelos Manolas

Abstract Consumer demand for environmentally-friendly food options is considered the crucial step in the transitions to sustainable production and climate change mitigation and adaptation. The present study sought to investigate attitudes toward climate change and the organic food production system and also explore consumption patterns for organic foods after adjusting for various socio-demographic characteristics and food choice indicators. Individual-level data for 807 adults were selected via a formal questionnaire in supermarkets and food stores in Northern Greece. Factor and cluster analysis elicited consumer behavior toward climate change and organic foods. The methodological approach also employed a univariate ordered probit model to estimate consumption behavior for organic foods and distinguish purchase patterns among non-consumers, occasional and regular consumers. Two segments were identified with differences in the level of climate change awareness (climate change concerns, climate change skepticism, and activity involvement) and the level of support toward organic agro-food networks (price fairness of organics, organic foods benefits, healthy food-choice habits and suspicion toward the organic food control system). Furthermore, estimations of the univariate ordered probit model and the calculated marginal effects showed that consumers who acknowledged the benefits of the organic foods production system had a higher likelihood by 4.4% to purchase organics on a regular basis. Similarly, consumption of organic foods was positively associated with health consciousness ($\beta = 0.150$, $p < 0.10$) and climate change concerns ($\beta = 0.277$, $p < 0.01$). On the other hand, non-consumers were more likely to raise considerable doubts over the reliability of food control in organics produce by 10.3% and express skepticism on climate change ($\beta = -0.306$, $p < 0.01$).

E. Raptou
Department of Agricultural Development, School of Agricultural and Forestry Sciences, Democritus University of Thrace, 193 Pantazidou Street, 68200 Orestiada, Greece
e-mail: elenra@agro.duth.gr

E. Manolas (✉)
Department of Forestry and Management of the Environment and Natural Resources, School of Agricultural and Forestry Sciences, Democritus University of Thrace, 193 Pantazidou Street, 68200 Orestiada, Greece
e-mail: emanolas@fmenr.duth.gr

W. Leal Filho and E. Manolas (eds.), *Climate Change in the Mediterranean and Middle Eastern Region*, Climate Change Management,
https://doi.org/10.1007/978-3-030-78566-6_18

Overall, this study suggested that established attitudes toward climate change and environmentally-friendly food products seem to be the key factor in purchase and consumption behavior. In order to increase green products acceptance, policy makers and stakeholders should better focus on the emotional-evoked attributes of organic foods rather their sensory characteristics and link sustainable foods with consumer altruism and social responsibility.

1 Introduction

In contemporary society, the global food system, including agricultural production and mainstream food markets, has a catalytic effect on environmental degradation and climate change. Recent evidence has noted that agricultural systems account for 24% of total global emissions, whereas the global food industry may contribute up to 35% of global greenhouse gas emissions (Vermeulen et al. 2012; Niles et al. 2018). In the transitions to sustainable production and climate change mitigation and adaptation, consumer demand for more environmentally-friendly food options seems to be crucial for reducing greenhouse gas emissions and the usage of synthetic pesticides/fertilizers, and has gained prominence in food market (Grunert et al. 2014; Allievi et al. 2015; Pirsich and Weinrich 2019).

The demand for organic agro-food products and the organic market have substantially increased in most developed countries, especially in Europe and North America (Monier-Dilhan and Bergès 2016), transforming a niche market into a well-established industry in the last two decades (Aygen 2012; Oroian et al. 2017). In the European Union, the total organic area raised from 5.0 million in 2002 to 11.0 million hectares in 2015, reaching 13.4 million hectares in 2018 and indicating an upward trend over the coming years (Oroian et al. 2017; Eurostat 2020). The organic food industry in Europe constitutes a profitable market and retail sales increased up to 29.8 billion Euros in 2015 (Wunsch 2020). The organic market in Greece was estimated to be approximately 60 million Euros in 2010 and although it presented a slow growth until 2010, economic recession restrained further growth and development (Willer et al. 2014).

European consumers present high levels of awareness for organic claims and also express willingness to pay a premium for these claims (Wunsch 2020). Understanding the main drivers that make consumers purchase organic agro-food products despite their higher prices is a topic of special interest for all the actors involved in the organic market (Monier-Dilhan and Bergès 2016). Several researches have noted that the upward trend in organic food production has followed consumers' needs for food safety and quality, environment and animal welfare (Smith and Paladino 2010; Oroian et al. 2017; Kim et al. 2018). In addition, consumers' mistrust in foods produced by novel technologies (e.g. genetic modification, food irradiation) may direct them to natural foods like organics (Van Huy et al. 2019).

A considerable body of literature suggests that there is a variety of factors motivating organic foods consumption. According to Hughner et al. (2007), there are nine motives that influence individual's purchase decisions on organic foods, namely healthiness, taste, environmental concerns, food safety, animal welfare, support of the local economy, wholesomeness, reminiscence of the past and fashionableness. The authors concluded that despite significant differences in individuals' perceptions toward organics, health and nutrition concerns were ranked first as the primary motives for organic food choice, implying that egoistic motives are better predictors of organic products purchase compared with altruistic motives (Magnusson et al. 2003; Hughner et al. 2007). In recent years consumers are well-informed about food choices and have become more aware of the relationship between food and health (Oroian et al. 2017). Recent researches have underlined the strong relationship between health consciousness and organic food choice, indicating that health attributes have become of major importance during the decision making process (Aygen 2012; Meixner et al. 2014; Higuchi and Avadi 2015; Lee and Yun 2015; Kapuge 2016; Oroian et al. 2017; Petrescu et al 2017).

Perceptions that specific eating behaviors, such as organic consumption, may contribute to environmental protection and especially climate change mitigation can be adopted by environmentally conscious consumers (Vittersø and Tangeland 2015). Ravn Heerwagen et al. (2014) found that individuals linked food patterns, such as organic food consumption and meat consumption reduction, with efforts to mitigate climate change. Environmental concerns also have a major importance in determining organic food purchase, perceived as an environment-friendly behavior (Smith and Paladino 2010). There is evidence that consumers with high awareness of environmental issues show a stronger preference to organic foods (Al Mamun et al. 2018), whereas environmental concerns stimulate the purchase of organic agro-food products (Pagiaslis and Krontalis 2014; Yadav and Pathak 2016; Ghali 2020).

Thus, the dynamics of the organic food networks depend on various factors that may affect the characteristics of the market segments, which these networks intend to approach. This study's objective is to explore attitudes toward climate change and the organic food production system and investigate consumption behavior for organic foods after adjusting for various socio-demographic characteristics and food choice indicators. Participants stated their opinion on climate change and organic products and were classified according to the level of climate change awareness and environmental consciousness in consumer segments with similar attitudes, values and characteristics. Consumption patterns are also explored for three distinct groups, namely non-consumers, regular and occasional consumers in order to define organic consumers' profile and determine the factors that influence organic product choices.

2 Materials and Methods

2.1 Data Selection

To get insight into consumption behavior for organic agro-food products, an empirical research was designed and carried out in Northern Greece, in the geographic division of Macedonia. The survey also focused on attitudes toward the organic food produce system and climate change awareness of adult consumers and sought to investigate differentiations in perceptions and consumption patterns across potential market segments. The research procedure lasted for eight months and was completed in October 2018 after incorporating both qualitative and quantitative research techniques. More specifically, in the qualitative research implementation, data were obtained through a semi-structured group interview process, in which 40 individuals of various sociodemographic profiles were enrolled in six focus groups discussions. The qualitative research analysis shed light into consumers' perceptions on climate change, environmental protection and local organic food networks and how those perceptions affected eating patterns and food choice.

In the quantitative research procedure, individual-level data were collected via a formal questionnaire constructed according to the main findings of the preceding qualitative research design. Participation was voluntary and the average time for the questionnaire completion was 12 min. The sample size was estimated according to the total population size in the selected area for a 3% margin of error using a 95% confidence interval (Brooks et al. 2017), whereas the total sample size was classified in 36 regional units according to a simple percentage. Participants were approached by the survey experimenters who were instructed to randomly select respondents in food venues and introduce them to the research topics. The food venues (selected as the sampling points) were super markets and medium-sized food stores that sold both organic and conventional agro-food products. Once informed on the present study goals, potential respondents were asked to complete the questionnaire on a desk set outside the food store and return it to the survey experimenter. In order to minimize social desirability biases, each completed questionnaire was accumulated in a box located in the entrance of the food store during the subsequent data selection process (Pino et al. 2012). The response rate for this study was approximately 76% resulting to the selection of 807 completed and valid questionnaires.

2.2 Measures

Participants were asked to report the consumption/purchase patterns of organic agro-food products. Responses were rated on a unipolar seven-point frequency scale ranging from "hardly" to "every day" (i.e. hardly—sometimes—less than once a week—once a week—twice a week—more than twice a week but not every day - every day) (Armitage and Conner 1999; Scalvedi and Saba 2018). Then, a three-point ordinal indicator was constructed after aggregating the seven cut-off points of the frequency scale in order to define specific consumer groups, namely non-consumers, occasional consumers (organic consumption less than once a week, aggregating sometimes and less than once a week) and regular consumers (at least once a week, including more frequent consumption) (O'Doherty Jensen et al. 2011; Scalvedi and Saba 2018).

In the context of consumer ethics, participants were asked if they prioritize fair trade issues with the following item: "Are you concerned about fair trade certifications on the food products you purchase?". A brief explanation was included to explain fair trade in consumption choice (i.e. fair price and fair working environment for producers and suppliers, equitable trading agreements). Fair trade products may be perceived as organic substitutes since increasingly feature organic certification (O'Doherty Jensen et al. 2011). Furthermore, there is a strong complementarity between organics and fair trade foods because they may be both linked with behaviors promoting sustainable development (D'Alessio et al. 2007; Monier-Dilhan and Bergès 2016). Thus, a dichotomous indicator was constructed to describe fair trade concerns, taking the value 1 for respondents who pay attention to fair trade certifications on food packaging and zero otherwise.

Food origin as a crucial component of food choice was also considered in the questionnaire design. Recent literature has shown that environmental protection cues, including food origin, comprise driving forces for purchase intentions and food choice (Petrescu et al. 2020). Participants were asked to declare if they evaluate food origin during the consumption process and a dichotomous indicator was employed reflecting the impact of food origin in consumer purchase behavior (the value of 1 for consumers seeking information on food origin and the value of zero otherwise).

Preference to locally produced foods was also depicted with a dichotomous variable to distinguish between consumers opting for foods from their home regions (the variable takes the value of 1) and consumers selecting conventional foods (the variable takes the value of 0). There is a significant body of research showing that consumers' mistrust to the food system transparency has increased their interest in local products because they are seen as healthier, safer, and more environmentally-and-climate friendly than imported foods (Hempel and Hamm 2016). Furthermore, Wägeli and Hamm (2016) noted that consumer motives for local food purchases may overlap with the motives for organics as the basic production principles in both food systems tend to be similar.

Furthermore, home food preparation skills were measured with a dichotomous indicator taking the value 1 for individuals reporting lack of time for preparing their meals at home and zero otherwise. Respondents who explained that they have poor cooking skills were also included in the consumer segment with poor home food preparation skills. Previous research has indicated that home food preparation is positively related with lower consumption of ultra-processed foods, higher food involvement and a subsequent increase in consumption of organic products (Chen 2007; Slater and Mudryj 2016; Lam and Adams 2017; Martins et al. 2020).

To explore climate change awareness, participants were asked to score a 13-item Likert type rating scale from “totally disagree” to “totally agree” (totally disagree, disagree, neither disagree nor agree, agree, totally agree). The construction of the 13-item scale was based on the main findings of the preceding qualitative research and the multi-item variables adopted in previous research (Wells et al. 2011). The 13-item scale adopted to express consumers’ perceptions toward climate change is presented in Appendix 1. The outcomes of the primary qualitative approach were also incorporated in the construction of a 15-item variable rated on a five-point Likert scale (totally disagree, disagree, neither disagree nor agree, agree, totally agree) employed to investigate consumers’ attitudes toward various aspects of the organic food production system (Appendix 2).

The standard covariates pertained to socio-demographic characteristics, such as gender (binary variable, 1 for women, 0 for men), age (continuous variable), educational attainment (binary variable, 1 for higher educational level including BSc and postgraduate education, 0 for lower educational level including primary and secondary education), marital status (three dichotomous indicators: married including registered partnership, divorced/widowed, single) and household income (three dichotomous indicators: lower income up to 1000 Euros, middle income 1001–2000 Euros and higher income over 2000 Euros).

2.3 *Methodology*

In order to summarize the data and capture an impression of the underlying data structure, our sample was first analyzed with univariate methods. The frequency distributions of sociodemographic characteristics, food choice/preparation indicators, and consumers’ attitudes toward climate change and the organic food production system were first explored to provide a primary overview of the data.

2.3.1 Factor Analysis

Then, an exploratory factor analysis was conducted to reduce the number of variables into a smaller set, and also define the underlying dimensions among the variables and the latent factors (Taherdoost et al. 2014). Factor analysis could also help further support the construct validity of the self-reported scales (Taherdoost et al. 2014; Gupta and Falk 2017). Although the scales on climate change and consumers’ attitudes toward the organic food production system were based on a primary qualitative research design, exploratory factor analysis (EFA) was employed to group the different variables (items) into a manageable number of factors that may affect consumers’ choices and test the construct validity of the two adopted scales.

For the present study, two EFA applications were considered. First, the 13 items of the climate change awareness scale were factor-analyzed by employing principal component analysis (PCA) with the Varimax rotation method in order to aggregate variables that load highly on a specific factor. This procedure was also adopted to analyze the 15 items of the second scale expressing individuals' attitudes toward the organic food production system. The quality of the factor analysis was evaluated by the Kaiser-Meyer-Olkin (KMO) criterion and the Bartlett test for sphericity with subsequent reliability analysis. In particular, the KMO criterion of sampling adequacy and the Bartlett's test of Sphericity were applied to test the fitness of the data (Oroian et al. 2017). KMO estimations over the value of 0.60 signify data adequacy (Kaiser 1974; Ding and He 2004; Oroian et al. 2017). Furthermore, all extracting factors whose eigenvalues were higher than 1 were retained, whereas the rest extracting factors with eigenvalues less than 1 were discarded without losing much of the original variability. To achieve the optimal factor solution, all variables with estimated factor loadings over 0.40 on more than one factor were removed from the analysis (Beer and Theuvsen 2019). Finally, reliability analysis (Cronbach's α) was conducted to evaluate whether the variables representing each factor measure its concept with accuracy. All factors with Cronbach's α coefficient above 0.60 were accepted in this study (Hair et al. 1998).

2.3.2 Cluster Analysis (Discriminant Analysis)

Based on the factors identified, a cluster analysis was conducted to assign consumers into different homogenous segments, describe differences in characteristics and ascertain common features according to their attitudes toward climate change and organic foods. The clusters were defined by both hierarchical and non-hierarchical (k-means) clustering techniques and verified by discriminant analysis.

First, a hierarchical cluster analysis was employed to consolidate cases into homogenous clusters by combining them together one at a time in a series of sequential steps (Blei and Lafferty 2009). Yim and Ramdeen (2015) provide a thorough presentation of hierarchical clustering application and interpretation. Cases in each cluster are assumed to be similar to each other and dissimilar to cases in the other clusters. In the present study, the squared Euclidean distance between each pair of cases was adopted to measure dissimilarity between groups, where shorter distances indicated greater similarity (Malhotra 2006; Carvalho et al. 2015). The Ward's hierarchical clustering method was used to determine the optimal number of clusters according to the climate change awareness and the aspects of organic foods industry factors. Thus, the dendrogram of the merger process and the elbow-criterion were used to identify the optimal number of clusters (Beer and Theuvsen 2019). To enhance the validity of hierarchical clustering, a k-means algorithm was employed, in which it was set a priori the number of clusters resulted from the Ward's hierarchical clustering method. Finally, the cluster structure was further validated by discriminant analysis seeking to estimate a discriminant function for examining the accuracy by which the study participants were allocated in the clusters (Carvalho et al. 2015; Beer and Theuvsen 2019; Pirsich and Weinrich 2019).

2.3.3 The Ordered Probit Model

Finally, an ordered probit model was adopted to estimate purchase decisions on locally produced organic foods and distinguish among never-consumers, occasional and regular consumers' eating habits and food choices and attitudes toward the organic foods production system and climate change (Greene 2012).

According to classical economic theory, rational individuals are assumed to make choices among alternative combinations of goods/services that maximize their utility. The level of satisfaction that individuals derive from the choice of the *i*th alternative may be presented as:

$$U_i = \mathbf{X_i}\boldsymbol{\gamma} + e_i \tag{1}$$

where **X** is a vector of independent variables representing values for each of factors of attributes of the *i*th alternative, $\boldsymbol{\gamma}$ is a vector of coefficients to be estimated and e is the error term reflecting unobserved attributes of the alternatives, random choice behavior or measurement error. Empirically, an individual's satisfaction level (U_i) for each of the alternatives is not directly observed. In lieu of U_i, what may be gleaned from an opinion survey is the value of a polychotomous categorical variable Y on the respondent's alleged disposition, where:

$$\begin{aligned} Y &= 0 \text{ for non-consumers of organic foods} \\ &= 1 \text{ for occasional organic food consumers} \\ &= 2 \text{ for regular organic food consumers} \end{aligned} \tag{2}$$

Based on the polychotomous probit model developed by McKelvey and Zavoina (1975), the present study employed the ordered probit model, as suggested by Greene and Hensher (2010), built around the aforementioned latent regression. It is also assumed that the e_i values are normally distributed and the observed variable (Y_i, i = 0, 1,2) is related to the true unobserved satisfaction level (U_i), hence a one-to-one correspondence could exist between them. For instance, it is specified that:

$$\begin{aligned} Y_i &= 0 \text{ if } \ U_i \le \mu_1 \\ &= 1 \text{ if } \ \mu_1 < U_i \le \mu_2 \\ &= 2 \text{ if } \ \mu_2 < U_i, \end{aligned} \tag{3}$$

where this is a form of censoring and μ_1, μ_2 represent unknown parameters. More specifically, their values define the boundaries of the intervals for the unobserved satisfaction that correspond to the observed individual's ranking. Finally, the parameters to be estimated are the μ_i and μ_2 values and the γ vector, assuming that μ_1 is normalized to zero. Therefore, in Eq. 3, μ_2 can be substituted with μ for reasons of simplification, hence:

$$Y_i = 0 \text{ if } \ U_i \le 0$$

$$
\begin{aligned}
&= 1 \text{ if } \quad 0 < U_i \leq \mu \\
&= 2 \text{ if } \quad \mu < U_i.
\end{aligned} \tag{4}
$$

With the normal distribution assumption [e ~ (0,1)], the probability that a particular state of disposition i is chosen may be derived as:

$$
\begin{aligned}
&\Pr(Y = 0) = 1 - \Phi(\mathbf{X}_i\boldsymbol{\gamma}) \\
&\Pr(Y = 1) = \Phi(\mu - \mathbf{X}_i\boldsymbol{\gamma}) - \Phi(-\mathbf{X}_i\boldsymbol{\gamma}) \\
&\Pr(Y = 2) = 1 - \Phi(\mu - \mathbf{X}_i\boldsymbol{\gamma}),
\end{aligned} \tag{5}
$$

where Φ is the cumulative distribution function assumed to be normal for the probit specification.

Given a cross section data set of sample size n generated by simple random sampling, $\boldsymbol{\gamma}$ and μ estimates are obtained by the method of maximum likelihood according to the following function:

$$
\mathrm{LnL} = \sum_{t=1}^{n}\sum_{i=0}^{2} k_{ti} \ln P(Y_i = i), \tag{6}
$$

where

$$
\begin{aligned}
k_{ti} &= 1 \text{ if } Y_i = i, \text{ for } i = 0, 1, 2 \text{ and n: sample size} \\
&= 0 \text{ otherwise.}
\end{aligned} \tag{7}
$$

According to econometric theory, the estimates of the ordered probit model give ambiguous explanations for the intermediate categories (Kennedy 1999; Greene 2012). Note that if a change in a X_i value increases the value of an index, the probability of having the last rating (regular consumers) increases, the probability of having the first rating (non-consumers) decreases, but the probability of having the intermediate category (occasional consumers) could move in each direction. Thus, the calculation of marginal effects is necessary when the ordered probit model is employed. For the three probabilities, the marginal effects of changes in the regressors are:

$$
\begin{aligned}
&\frac{\partial \mathrm{Prob}[Y = 0]}{\partial X} = -\varphi(\mathbf{X}_i\boldsymbol{\gamma})\boldsymbol{\gamma} \\
&\frac{\partial \mathrm{Prob}[Y = 1]}{\partial X} = [\varphi(-\mathbf{X}_i\boldsymbol{\gamma}) - \varphi(\mu - \mathbf{X}_i\boldsymbol{\gamma})]\boldsymbol{\gamma} \\
&\frac{\partial \mathrm{Prob}[Y = 2]}{\partial X} = \varphi(\mu - \mathbf{X}_i\boldsymbol{\gamma})\boldsymbol{\gamma},
\end{aligned} \tag{8}
$$

where φ is the density function (Greene 2012).

For the purpose of this study, the vector of explanatory variables encompasses the sociodemographic indicators (i.e. gender, age, educational level, marital status, personal disposable income), the food choice/preparation indicators (i.e. home food preparation, fair trade concerns, information on food origin, preference to locally produced foods) and the components resulted from the exploratory factor analyses. To create dichotomous measures for the econometric framework, the detached items of each factor (component) were added and the sums were dichotomized according to the factor mean.

3 Results

3.1 Descriptive Analysis

Our sample characteristics are analytically provided in Table 1. As expected, the prevalence of female consumers was 55.30%, exceeding that of men (44.7%). Although potential respondents of both genders were approached, women were more likely to accept participation in the present research after being introduced to the research topics. This may have to do with the gendered patterning of food provisioning, in which females still continue to carry major responsibility (Mehta et al. 2020). Thus, females are more likely to show a greater responsiveness to food research topics and may feel more confident than males to respond to questions related with food consumption. Furthermore, the majority of participants were either singles (51.30%) or married (41%) with substantial participation rates in higher education (52.80%). The mean sample age was 35.75 years and nearly half reported a middle-sized household income (50.10%).

With regard to food choice, approximately one quarter of consumers (24.8%) declared the importance of food origin and its influential role on food selection and also showed a clear preference to locally produced foods (27.40%). Respondents' ethical intentions to purchase fair trade foods, ensuring a fair price and fair working conditions for producers and suppliers were weak as only 9.80% expressed their interest in fair trade certification foods. The lack of time for home food preparation has been reported by a significant proportion of the sample participants (27.5%). Furthermore, F-tests (Analysis of variance-ANOVA) explored statistically significant differences in food selection indicators (i.e. lack of time/skills for home food preparation, fair trade concerns, information on food origin, preference to locally produced food) among different consumption levels of organics. In all tests, the *p*-values reached statistical significance showing that consumption patterns of organics may be associated with specific food/diet choices (Table 1).

Participants' views on the organic food production system are presented in Table 2. The great majority of consumers were convinced that organics are healthier than conventional foods (73.9%) and with a superior quality (56.2%). Furthermore, organic production has been put to critical advantage since the majority of respondents seem to strongly support its lower climate impact (85%) and environmental pollution (81.1) and the better animal welfare in organic livestock systems (47%). A

Table 1 Sample characteristics (N = 807)

Variables	Non-consumers	Occasional consumers	Regular consumers	Total	F-test	*P*-value
Total sample	383	305	119	807		
	47.60%	37.90%	14.50%	100.00%		
Female	209	174	63	446	0.355	0.701
	26.00%	21.60%	7.70%	55.30%		
Male	174	132	55	361		
	21.60%	16.30%	6.80%	44.70%		
Age	34.91a	36.37a	35.65a	35.57a	0.848	0.429
	15.461b	14.368b	12.379b	14.636b		
Lower education	209	125	47	381	**8.063**	**0.000**
	25.90%	15.50%	5.80%	47.20%		
Higher education	174	181	71	426		
	21.60%	22.40%	8.80%	52.80%		
Single	210	150	54	414		
	26.00%	18.60%	6.70%	51.30%		
Married	143	132	56	331		
	17.70%	16.40%	6.90%	41.00%	1.278	0.279
Divorced/widowed	28	23	8	59		
	3.50%	2.90%	1.00%	7.30%		
Lower income	121	68	17	206		
	15.00%	8.40%	2.10%	25.50%		
Middle income	181	162	61	404		
	22.40%	20.10%	7.60%	50.10%	**7.509**	**0.001**
Higher income	73	68	33	174		
	9.00%	8.40%	4.10%	21.60%		
Lack of time for home food preparation	126	73	23	222		
	15.60%	9.00%	2.90%	27.50%	**5.775**	**0.003**
Fair trade concerns	32	26	21	79		
	4.00%	3.20%	2.60%	9.80%	**5.063**	**0.007**
Information on food origin	85	67	48	200		
	10.50%	8.30%	5.90%	24.80%	**9.557**	**0.000**
Preference to locally produced foods	87	88	46	221		
	10.80%	10.90%	5.70%	27.40%	**6.310**	**0.002**

Statistically significant variables are marked in bold, a: mean, b: standard deviation

Table 2 Consumer attitudes toward the organic food production system (%)

Attitudes	Strongly disagree	Disagree	Neither disagree nor agree	Agree	Strongly agree
Organics are more healthy than conventional foods	2.1	5.2	18.8	45.3	28.6
When I buy locally produced organic foods, I support preserving local jobs	2	6.2	17.5	37.3	36.9
Organic foods have better quality than conventional food products	5.2	13.5	25.1	37	19.2
Organic production system has a lower climate impact than conventional agriculture	2.7	3	9.2	32.3	52.7
I have serious concerns about the authenticity of organic foods	41.3	25.5	18.3	10.2	4.7
I believe that the control system for organic foods is inadequate	43.7	28.4	13.8	9.5	4.7
Animals engaged in organic production have better animal welfare than in conventional livestock production system	4.4	17.7	31	26.9	20.1
Even if the price of organic foods was higher, I'd definitely consume them	4.6	8.6	17.8	35.2	33.7
Organic foods have a reasonable price	4.1	6.8	21.1	36.8	31.1
Food quality is very important to me when I make my food choice	1.9	4.5	12.3	43	38.4
I believe that organic food producers are only interested in making money and not producing better quality food	20	33.1	32.3	10.7	3.9
Organic food production pollutes the local environment less than the conventional food production system	1.7	5.4	11.8	30.5	50.6
I believe I follow a healthy eating lifestyle	13.4	22.3	36.6	21.5	6.2
For me healthy eating habits are the key to good health	6.1	14.8	25.1	40.4	13.6
I try to make informed food decisions	6.2	14.9	25.9	39.9	13.1

N = 807

significant proportion of respondents believed that organic production may enhance local economy and rural development since it may help preserve local jobs (74.2%). Concerning organic foods pricing, consumers were willing to be charged a price premium for organic products declaring that even in case of higher prices they would still prefer organics to conventional foods (68.9%). Our study participants seemed to be confident about the authenticity and the control system of organics (66.8% and 72.1%, respectively) and showed trust to organic producers' efforts to deliver better quality food to the public (55.1%). Regarding eating patterns, although the large majority of consumers underlined their need to make informed food decisions (53%) and purchase quality food items (81.4%), only the 27.7% believe that they follow a proper diet.

Participants' perceptions on climate change are analytically presented in Table 3. A considerable proportion of consumers seemed to favor personal responsibility and noted that they should take action on climate change (42.8%) and should also try to reduce emissions at individual level (57.8%). However, they were rather suspicious

Table 3 Consumer awareness toward climate change (%)

Attitudes	Strongly disagree	Disagree	Neither disagree nor agree	Agree	Strongly agree
I don't see why I'd take action on climate change	17.6	25.2	30.4	18.1	8.7
I don't see why I'd try to reduce emissions at individual level	21.8	36	21.8	13.5	6.8
I'd prefer to buy products from companies that are working to reduce climate change	6.4	14.7	40.6	28.7	9.4
The economic growth of developing countries is the greatest threat to the world's climate	22.9	25	23	19.7	9.3
It's too late to do anything about climate change	13	27.3	30.3	22.6	6.7
If the government enforced more strict environmental policies it should prevent climate change	11.8	20.9	18.5	32.8	16
The media is exaggerating the potential impact of climate change	16.7	20.7	23.3	24	15.3
Climate change has occurred many times in the past and it's part of the natural shifting of the environment	24.7	32	22.4	16.4	4.5
I'm concerned about the consequences of climate change	3.3	8.3	32.1	41.5	14.7
I want more information from businesses on how they address climate change impact	7.7	13.9	40.2	31.1	7.1
Consumers can help reduce the impact of climate change	8.4	18.6	30.7	27.8	14.4
I trust companies to take action when it comes to climate change	16.9	28.7	25.1	18.5	10.9
Attempts to prevent climate change should be coordinated at an international level	6.5	14.9	38.6	28.3	11.8

N = 807

of companies' actions to help decrease climate change impacts (45.6%) and they believed that the economic growth of developing countries has damaged world's climate (47.9%). They also supported that efforts to prevent climate change should be coordinated at international level (40.1%) and strict environmental policies could help prevent climate breakdown (48.8%). However, the role of mass media was rather ambivalent in disseminating climate change impact issues. Nearly half of the sample reported that individuals' actions could reduce the consequences of climate change (42.2%), and expressed concerns about the climatic change aftermath (56.2%).

3.2 Factors Analysis Results

Explorative factor analysis (EFA) through principal component analysis (PCA) with varimax rotation was employed to evaluate the dimensionality of the most important factors describing attitudes toward the organic food production system and climate change awareness. The tests that examined the quality of EFA met the common requirements. In particular, the Kaiser–Meyer–Olkin (KMO) measure of sampling adequacy was 0.761, showing that the data used to delineate consumers' attitudes toward the organic food production system were suitable for the PCA (Pirsich and Weinrich 2019). The Bartlett's test of Sphericity was highly significant (Chi-square = 3357.716, $p < 0.01$). The EFA of the 15 variables provided a four-component solution and the total variance explained was 58.074%. To assess the reliability of the four components, the Cronbach's α coefficients were evaluated indicating that the internal consistency of each factor is satisfactory since reliability coefficients ranged from 0.648 to 0.807 (> 0.60) (Backhaus et al. 2016; Beer and Theuvsen 2019).

Table 4 shows the EFA and reliability analysis output on consumers' attitudes toward the organic food production system. The first factor labeled as "price fairness of organic foods" included four variables explaining 17.857% of the total variance and had a reliability coefficient of 0.782. This factor loaded attributes related to organic food pricing, consumers' price responsiveness and willingness to pay-and-accept higher prices in organic foods. The second factor labeled "organic foods benefits" comprised six variables, which explained 14.527% of the total variance with a reliability coefficient of 0.668. This factor involved attributes that focus on the organic food production and its impact on environment, health and society. The third factor was comprised of three variables and was labeled as "health consciousness". This factor had a Cronbach α reliability coefficient equal to 0.648 and explained 13.326% of the total variance, whereas involved attributes related to consumers' lifestyle and nutrition perceptions. Finally, the fourth factor entitled as "suspicion toward the organic food system control" loaded two items regarding respondents' beliefs on the authenticity of organics and the adequacy of their quality control. This factor explained the 12.364% of total variance and had a reliability coefficient of 0.807 (Table 4).

With respect to the application of EFA in the climate change awareness 13-item scale, the KMO measure of sampling adequacy was 0.801, implying data suitability

Table 4 Factor analysis (PCA) and reliability analysis output on consumers' attitudes toward the organic food production system

Eigenvalue	Explained variance %	Factors	Factor loading	M	S.D.	Cronbach α
		Factor 1: Price fairness of organic foods (m = 3.837, S.D. = 0.810)				
		Organic foods have a reasonable price	0.859	4.00	1.068	
4.023	17.857	Even if the price of organic foods was higher, I'd definitely consume them	0.828	4.00	1.119	0.782
		Food quality is very important to me when I make my food choice	0.727	4.00	0.918	
		I believe that organic food producers are only interested in producing better quality food and not for making money	0.587	4.00	1.047	
		Factor 2: Organic foods benefits (m = 3.897, S.D. = 0.621)				
		Organic production system has a lower climate impact than conventional agriculture	0.753	5.00	0.949	
		Organic food production pollutes the local environment less than the conventional food production system	0.568	5.00	0.972	
2.068	14.527	Animals engaged in organic production have better animal welfare than in conventional livestock production system	0.531	3.00	1.123	0.668
		When I buy locally produced organic foods, I support preserving local jobs	0.488	4.00	0.987	
		Organic foods have better quality than conventional food products	0.476	4.00	1.104	
		Organics are more healthy than conventional foods	0.450	4.00	0.932	
		Factor 3: Health consciousness (m = 3.213, S.D. = 0.835)				
		I try to make informed food decisions	0.842	4.00	1.083	
1.376	13.326	For me healthy eating habits are the key to good health	0.838	4.00	1.084	0.648
		I believe I follow a heathy eating lifestyle	0.460	3.00	1.097	

(continued)

Table 4 (continued)

Eigenvalue	Explained variance %	Factors	Factor loading	M	S.D.	Cronbach α
		Factor 4: Suspicion toward the organic food control system (m = 2.074, S.D. = 1.083)				
1.245	12.364	I have serious concerns about the authenticity of organic foods	0.853	2.00	1.191	0.807
		I believe that the organic foods control system is inadequate	0.808	2.00	2.174	

M median, *SD* standard deviation

for the PCA (Pirsich and Weinrich 2019). Furthermore, the Bartlett's test of Sphericity was highly significant (Chi-square = 2490,696, $p < 0.01$). The three components derived from EFA explained the 53.164% of the total variance. The calculated Cronbach's α reliability coefficients ranged from 0.626 to 0.794, showing that the internal consistency of each factor was satisfactory (Backhaus et al. 2016; Beer and Theuvsen 2019).

The results of the EFA and reliability analysis are analytically presented on Table 5. The first factor entitled as "climate change concerns" included six variables that explained 23.293% of the total variance and had a reliability coefficient of 0.794. This factor loaded attributes related to participants' perceptions on climate change impact, climate change prevention and business environmental responsibility. The second factor, "climate change skepticism", incorporated five variables, explaining 17.929% of the total variance with a reliability coefficient of 0.667. This factor involved attributes regarding consumers' doubts on climate as a physical phenomenon and doubts about the efficacy of organized actions to fight climate change. The third factor labeled as "climate change activity involvement" explained the 17.942% of the total variance and had a reliability coefficient of 0.626. This factor loaded two items concerning individual's willingness to take action on climate change (Table 5).

3.3 Cluster Analysis Results

A two-step cluster analysis was conducted on the basis of the seven identified factors in the preceding EFA to reveal consumers' segments according to attitudes toward the organic food production system and climate change awareness. First, Ward's hierarchical clustering method was employed to determine the optimal number of clusters, followed by k-means algorithm for stabilization purposes (Oroian et al. 2017).

Cluster analysis resulted to the identification of two consumer segments. T-test for equality of means showed that statistically significant differences existed between the two clusters regarding consumers' attitudes toward organic foods and climate change

Table 5 Factor analysis (PCA) and reliability analysis output on consumers' awareness toward climate change

Eigenvalue	Explained variance %	Factors	Factor loading	M	S.D.	Cronbach α
		Factor 1: Climate change concerns (m = 3.192, S.D. = 0.755)				
		I trust companies to take action when it comes to climate change	0.771	3.00	1.239	0.794
		Attempts to prevent climate change should be coordinated at an international level	0.748	3.00	1.051	
3.713	23.293	I'd prefer to buy products from companies that are working to reduce climate change	0.694	3.00	1.015	
		I want more information from businesses on how they address climate change impact	0.680	3.00	1.008	
		Consumers can help reduce the impact of climate change	0.665	3.00	1.156	
		I'm concerned about the consequences of climate change	0.567	4.00	0.954	
		Factor 2: Climate change skepticism (m = 2.834, S.D = 0.833)				
		It's too late to do anything about climate change	0.740	3.00	1.123	0.667
1.715	17.929	Climate change has occurred many times in the past and it's part of the natural shifting of the environment	0.698	2.00	1.162	
		The media is exaggerating the potential impact of climate change	0.650	3.00	1.507	
		The economic growth of developing countries is the greatest threat to the world's climate	0.616	3.00	1.277	
		If the government enforced more strict environmental policies it should prevent climate change	0.481	3.00	1.268	
		Factor 3: Climate change activity involvement (m = 2.613, S.D. = 1.008)				
1.484	17.942	I should take action on climate change	0.819	2.00	1.169	0.626
		It's very important to help reduce emissions at an individual level	0.778	3.00	1.194	

M mean, *M* median, *SD* standard deviation

perceptions. The clusters were labeled according to the factors that were considered to be of high importance for each of them. Table 6 presents cluster analysis results for consumers' attitudes toward organic agro-food products and climate change. To provide a complete profile of consumer segments, cross-tabulation and Pearson's χ^2 statistics were adopted to depict sociodemographic characteristics and decisions on food choice/preparation of participants in each cluster (Table 7).

Cluster 1 (N = 334, 41.39%), labeled as "skeptical to climate change and environmental issues", represented the 41.39% of participants. This segment had the highest mean score in "climate change skepticism" (3.245 vs 2.250, t-test = 12.767, $p <$

Table 6 Cluster analysis results for consumers' attitudes toward climate change and organic agro-food products

Factors	Cluster 1 (41.39%)		Cluster 2 (58.61%)		t-test for equality of means	
	Mean	S.D.	Mean	S.D.	t-test	*P*-value
Price fairness of organic foods	3.419	0.765	4.134	0.702	−13.687	0.000
Organic foods benefits	3.545	0.601	4.147	0.503	−15.384	0.000
Health consciousness	2.929	0.873	3.415	0.744	−8.474	0.000
Suspicion toward the organic food control system	3.000	0.930	1.423	0.602	29.162	0.000
Climate change concerns	2.899	0.711	3.396	0.716	−9.708	0.000
Climate change skepticism	3.245	0.782	2.550	0.744	12.767	0.000
Climate change activity involvement	2.911	1.026	2.409	0.939	7.183	0.000

Table 7 Cluster Profile—Sociodemographic characteristics

	Variables	Cluster 1 (41.39%)	Cluster 2 (58.61%)	Pearson chi-square	*P*-value
Gender	Female	56.3%	54.4%	0.306	0.580
	Male	43.7%	45.6%		
Marital status	Singles	69.9%	38.7%		
	Married	25.3%	52.1%	75.772	0.000
	Divorced/widowed	4.8%	9.2%		
Educational attainment	Higher education	54.5%	52.0%	0.489	0.484
	Lower education	45.5%	48.0%		
Household income	Lower income	39.4%	17.2%		
	Middle income	45.7%	55.7%	52.211	0.000
	Higher income	14.9%	27.2%		
				t-test	*P*-value
Age[a]	Age (continuous variable)	30.06 (13.427)	39.44 (14.247)	−0.404	0.000

[a]means and standard deviations in parentheses

0.01), the highest mean score in "suspicion toward the organic food control system" (3.000 vs 1.423, t-test = 29.162, $p < 0.01$) and the lowest mean score in "climate change concerns" (2.899 vs 3.396, t-test = −9.708, $p < 0.01$) compared with Cluster 2 (Table 6). Consumers included in this cluster are more skeptical and less aware of climate change issues and seem to be more suspicious toward the organic food control system. The first cluster grouped younger consumers (mean age: 30.1 years), mostly singles (69.9%) with lower and middle-sized income (39.4% and 45.7%, respectively) (Table 7).

Cluster 2 (N = 473, 58.61%), labeled as "environmentally conscious", included the 58.61% of respondents characterized by positive attitudes toward "price fairness of organic foods" (4.134), "organic food benefits" (4.147), "climate change concerns" (3.396) and "health consciousness" (3.415) (Table 6). Therefore, participants are more willing to consume organic foods and they are also convinced of their quality attributes. They are more environmentally responsible and aware of climate change issues. This cluster grouped the highest percentage of married adults (52.1%) and exhibited the highest proportion in middle and high-sized income (55.7% and 27.2%, respectively) (Table 7).

Finally, discriminant analysis was adopted as a validity criterion to confirm the classification. The analysis showed that the exactness of classification was 97.8%.

3.4 *Ordered Probit Model Results*

An ordered probit regression was adopted to model the consumption frequency of organic foods and distinguish attitudes toward the organic food production system and climate change among never-consumers, occasional and regular consumers. The Likelihood Ratio test indicated that the explanatory variables employed in the estimation process are appropriate (LR test = 133.66, $p < 0.01$). The ordered probit model results are analytically presented in Table 8.

With respect to attitudes toward the organic food production system, several regressors were found to have statistical significance in the analysis. In particular, individuals' persuasiveness on organic foods benefits was positively associated with the level of organic foods consumption ($\gamma = 0.217$, $p < 0.05$). Furthermore, the estimated marginal effects revealed that beliefs about organic production benefits increase the probability for regular consumption of organics by 4.4%, while decrease the probability for non-consumption by 8.6%. Health consciousness was also positively linked with consumption of organics ($\gamma = 0.150$, $p < 0.10$), with participants oriented to healthy nutrition habits to have a higher likelihood to consume organic foods either on occasional (3%) or regular basis (2.9%). Individuals declaring mistrust to the organic food control system were less likely to frequently purchase organics ($\gamma = -0.258$, $p < 0.01$). Thus, doubts about the authenticity of organics and uncertainty for the adequacy of the control system decreased the probability for organic consumption by 10.3% (Table 8)

Table 8 Ordered probit model estimates for organic foods consumption intentions (N = 807)

Variables	Ordered probit model estimates	Marginal effects		
		Non consumers	Occasional consumers	Regular consumers
Sociodemographic characteristics[1]				
Gender	−0.092	0.037	−0.019	−0.018
	(0.087)	(0.035)	(0.018)	(0.017)
Age	**−0.008***	**0.003***	**−0.002***	**−0.002***
	(0.004)	**(0.002)**	**(0.001)**	**(0.001)**
Higher educational attainment	**0.346*****	**−0.137*****	**0.069*****	**0.068*****
	(0.092)	**(0.036)**	**(0.019)**	**(0.018)**
Single	−0.070	0.028	−0.014	−0.014
	(0.191)	(0.076)	(0.038)	(0.038)
Married	−0.054	0.021	−0.011	−0.011
	(0.166)	(0.066)	(0.034)	(0.033)
Lower income	**−0.406*****	**0.161*****	**−0.089*****	**−0.072*****
	(0.113)	**(0.044)**	**(0.027)**	**(0.018)**
Higher income	0.106	−0.042	0.020	0.022
	(0.108)	(0.043)	(0.020)	(0.023)
Attitudes toward organics				
Price fairness of organic foods	−0.128	0.051	−0.026	−0.025
	(0.096)	(0.038)	(0.020)	(0.018)
Organic products benefits	**0.217****	**−0.086****	**0.042****	**0.044****
	(0.096)	**(0.038)**	**(0.018)**	**(0.020)**
Health consciousness	**0.150***	**−0.060***	**0.030***	**0.029***
	(0.089)	**(0.035)**	**(0.018)**	**(0.017)**
Suspicion toward the organic food control system	**−0.258*****	**0.103*****	**−0.053*****	**−0.049*****
	(0.094)	**(0.037)**	**(0.021)**	**(0.017)**
Climate change awareness				
Climate change concerns	**0.277*****	**−0.110*****	**0.054*****	**0.056*****
	(0.088)	**(0.035)**	**(0.017)**	**(0.018)**
Climate change skepticism	**−0.306*****	**0.121*****	**−0.062*****	**−0.060*****
	(0.093)	**(0.037)**	**(0.020)**	**(0.018)**
Climate change activity involvement	0.058	−0.023	0.011	0.012
	(0.089)	(0.035)	(0.018)	(0.018)
Food choice/preparation				
Home food preparation (lack of time)	**−0.452*****	**0.179*****	**−0.099*****	**−0.079*****
	(0.100)	**(0.039)**	**(0.024)**	**(0.016)**

(continued)

Table 8 (continued)

Variables	Ordered probit model estimates	Marginal effects		
		Non consumers	Occasional consumers	Regular consumers
Fair trade concerns	**0.352****	**−0.137*****	**0.055*****	**0.082****
	(0.141)	**(0.053)**	**(0.016)**	**(0.038)**
Information on food origin	**0.246****	**−0.097****	**0.045*****	**0.052****
	(0.105)	**(0.041)**	**(0.017)**	**(0.024)**
Preference to locally produced foods	**0.318*****	**−0.125*****	**0.057*****	**0.068*****
	(0.102)	**(0.039)**	**(0.017)**	**(0.024)**
μ_1	−0.225			
	(0.291)			
μ_2	1.022			
	(0.292)			
Log likelihood	**−739.330**			

Standard errors are given in parentheses; statistically significant variables are marked in bold
[1]Marital status: divorced/widowed (omitted variable), Household income: middle income (omitted variable)
$^{*}p < 0.1$, $^{**}p < 0.05$, $^{***}p < 0.01$

Climate change awareness also played a crucial role in determining the level of organic foods consumption. Climate change skepticism was inversely related with organic foods consumption ($\gamma = -0.306, p < 0.01$) implying that unconvinced individuals about the magnitude and the impact of climate change are less likely to regularly purchase organic foods by 6% and more likely to consume conventional foods by 12.1%. In addition, individuals concerned with the consequences of climate change and the role of companies in reducing climatic fluctuations have a higher probability to purchase organics on a regular basis by 5.6% and a higher probability to restrict organics from their diet by 11%. As expected, fair trade concerns ($\gamma = 0.352$, $p < 0.05$), information on food origin ($\gamma = 0.246$, $p < 0.05$) and preference to locally produced food ($\gamma = 0.318, p < 0.01$) were also positively associated with the level of organic foods consumption. On the other hand, consumers' lack of time for home food preparation was inversely linked with organic consumption frequency and non-organic consumers were less likely to have sufficient time for cooking by 17.9% (Table 8).

With respect to sociodemographic characteristics, age was negatively associated with organic food consumption frequency ($\gamma = -0.008$, $p < 0.10$), noting that older individuals are less likely to consume organics. Higher educational attainment increased the likelihood of regular consumption by 6.8%, whereas lower income was inversely related with occasional and regular consumption of organic foods by 8.9% and 7.2%, respectively.

4 Discussion

This study explored in detail consumers' attitudes toward organic foods and climate change and examined the dimensionality of the main components that describe perceptions toward organic food networks and climate change awareness. Consumers' profiles were also defined in terms of environment-oriented behavior adoption, health consciousness and reactions to organic foods attributes. Finally, the consumption of organic foods was modeled to determine the factors affecting different consumption frequency patterns (non-consumption, occasional consumption, regular consumption), as well as the direction and the magnitude of their influence.

According to our findings, the great majority of consumers strongly evaluate the underlying values of organic foods based on the lower climate impact, environmental protection and support to local economies. Although a high percentage of respondents note that organics are healthier, premium foods of a better quality, attitudes toward the environment and climate change seem to be the primary driver for organic product consumption. Previous research shows mixed results regarding how consumers perceive organic foods. In particular, several studies argue that organic food consumption is first of all motivated by environmental considerations, whereas other studies support that attributes, such as taste, freshness or health are the main reasons behind organic food choices (Hamzaoui-Essoussi et al. 2013; Higuchi and Avadi 2015). Examining the driving and hindering factors that affect consumers' perceptions toward organics is of major importance in order to understand consumers' personality traits and their purchase motives. Petrescu et al. (2017) underlined that altruistic or ethical purchases are related with attributes, such as product and food safety, fair trade issues and environmental impact. According to Crane (2001), ethical goods are often named with the term "organics", which is widely linked with both environmental and social issues. On the other hand, non-altruistic and egocentric purchases are usually motivated by individual's health awareness or interest in social image. Concerning the latter, Hill and Lynchehaun (2002) have shown that there is a consumer share that perceives organic foods as fashionable because of the extended coverage of organics in the media and their premium pricing.

Factor analysis provided the underlying main components to categorize individuals' attitudes toward organic agro-foods. Climate change awareness resulted to a three-component solution, namely "climate change concerns", "climate change skepticism" and "climate change activity involvement", whereas perceptions toward the organic food production system were categorized in four components, namely "price fairness of organic foods", "organic foods benefits", "health consciousness" and "suspicion toward the organic food control system". Previous research has also suggested attitudinal choice factors, which encompass health and environmental concerns, taste preferences and food origin (Stolz et al. 2011). Lee and Yun (2015) also noted five components of organic food attributes (nutrition, naturalness, ecological welfare, sensory appeal, and price) to distinguish between hedonic and utilitarian attitudes toward organic foods.

According to our results, Greek consumers seem to highly evaluate organic food benefits (including climate and environmental considerations, animal welfare, preserving local jobs, superior quality, and healthiness) and believe that the organics pricing system is fair and justified. They are concerned about climate change and its consequences and although they seem health conscious and try to make informed food choices, their eating habits are somewhat imbalanced. Our sample presented high levels of trust to the organic food control system (authenticity, adequacy), whereas a minority of respondents expressed skepticism toward climate change. In addition, consumers expressing environmental and climate change considerations, also acknowledged the quality superiority and healthiness of organic foods.

Cluster analysis organized the observed data into taxonomies according to various combinations of the main factors. Although all factors describing attitudes toward organic food networks and climate change awareness implied statistically significant differences between clusters, variations in attitudes toward climate change and skepticism, mistrust to the organic food control system and nutrition behaviors were clearly detected between the two clusters. Key stakeholders, such as organic food producers and distributors, food scholars and policy instructors should take into account the main attitudinal differences in the cluster profiles and consider the appropriate strategies to enhance organic food markets and improve eating habits. Firms should adapt promotional strategies to the characteristics of each market segment in order to effectively communicate the superior characteristics of organic foods in terms of quality, healthiness and sustainability ethics.

Finally, an ordered probit model analyzed organics consumption behavior and revealed the main differences in attitudes and food choice indicators across three consumer segments (i.e. non-consumers, occasional and regular consumers). Our findings showed that attitudes toward organic foods and climate change awareness were highly correlated with individuals' consumption level. Thus, supporters of organic foods benefits were more likely to purchase organics on an occasional or regular basis, whereas the probability for systematic consumption was slightly higher. Positive attitudes toward the contribution of organic food production to environmental protection and local- regional development, as well as perceived healthiness and superiority in quality of organics had a major influence in purchasing behavior and consumption frequency. Furthermore, climate change concerns were also positively associated with organics consumption, whereas skeptical consumers had a significantly higher probability to avoid organic foods. Following previous research, this study further supported the influence of sustainability in organic food consumption (Pagiaslis and Krontalis 2014; Yadav and Pathak 2016; Oroian et al. 2017; Al Mamun et al. 2018; Ghali 2020). Environmental considerations comprise significant purchase motivators that stimulate ethical consumerism (Seyfang 2006), which in turn inspires consumers to purchase "green" products in order to satisfy their needs for ethical responsibility (Cho and Krasser 2011).

Individuals who pay attention to fair trade certifications had a higher probability of purchasing organic products either occasionally or systematically. O'Doherty Jensen et al. (2011) noted that fair trade products may overlap with organics and can be considered as substitutes since increasingly feature organic certification.

In addition, fair trade and organic labels are among the most widespread sustainability labels available in the market (Zander and Hamm 2010; Andorfer and Liebe 2012). Thus, potential complementarity between organics and fair trade products may exist because they both have symbolic values promoting sustainability besides their tangible values (D'Alessio et al. 2007; Monier-Dilhan and Bergès 2016). Defining values that are important to organic food consumers could help marketers understand the purchase decision process and position their product in the market accordingly.

Preference to locally produced foods was found to augment the likelihood of both occasional and regular organic consumption/purchase. In general, individuals use to highly evaluate local foods due to expected benefits associated with taste, food safety, animal welfare standards and lower environmental impact. Recent research has shown that there is some overlap between consumers who prefer locally produced foods and consumers who choose organic foods (Costanigro et al. 2014; Hempel and Hamm 2016). However, the nature of the relationship between organics consumption and locally produced food consumption is still unclear (whether organic and local foods are complementary or competitive) (Hempel and Hamm 2016). Further research should be conducted to explore potential differences in preferences for organics and locally produced foods besides their common expected values. Food marketers and retailers should take into consideration the material and immaterial qualities in both categories and explore potential interaction effects.

To further support previous literature, our findings showed that health consciousness is positively linked with organic food consumption (Aygen 2012; Meixner et al. 2014; Higuchi and Avadi 2015; Lee and Yun 2015; Kapuge 2016; Oroian et al. 2017; Petrescu et al. 2017). The growth of the organic market has been attributed to concerns about health and preferences to products that contain at least smaller amounts of additives, pesticides and hormones than conventional foods. Although the adoption of healthy eating habits has a positive influence on organic food consumption, the magnitude of the influence (marginal effect) is slightly smaller compared with the marginal effects assessed to evaluate the influence of organic products benefits and climate change concerns on organics consumption. It seems that sustainable-minded consumers have a higher likelihood to regularly (or occasionally) consume organics compared to health conscious individuals, implying that more altruistic motives may have a stronger impact in organic demand than more egocentric incentives. Therefore, the ordered probit estimations further support and stabilize the main findings of the primary descriptive analysis, as discussed above. Examining the driving factors affecting individuals purchase behavior will help scholars understand consumers' motivation process and design the appropriate communication strategies to promote organic foods consumption.

This study noted that mistrust to the organic food networks (authenticity, organic food control) was inversely associated with the level of organic food consumption. Recent evidence has suggested various obstacles in purchasing organic foods, such as lack of trust in organic product certification/labeling (Aertsens et al. 2009; Gerrard et al. 2013; Xie et al. 2015), premium pricing (Aygen 2012; Bryla 2016; Yadav and Pathak 2016), awareness (Biao et al. 2015; Oroian et al. 2017) and lack of availability (Aygen 2012; Oroian et al. 2017). However, as Chiciudean et al. (2019) explained, mistrust in food authenticity/control may stem from mistrust in the organic food itself. Lack of information and limited knowledge in technological issues and production system may hold back consumption/purchase intentions of organic foods and increase consumers' insecurity and vulnerability. Therefore, marketers should identify or design the appropriate communication strategies to disseminate information and knowledge in favor of organic foods.

4.1 Limitations

The main limitations of this study are summarized as follows. First, the cross sectional design does not allow inferring causal relationships. A longitudinal follow-up may provide additional information on the impact of climate change awareness and organic food networks on organic food consumption. Second, the study participants were recruited from Northern Greece, so there may be uncertainty about the relevance and the applicability of the findings to the rest of the country. Regional differences in culinary culture may have an effect on eating patterns and preferences, which can influence consumers purchase behavior toward organic foods.

5 Concluding Remarks

Using individual level data from Greece, the present study explored attitudes toward climate change and the organic food production system and also analyzed consumption patterns for organic foods after adjusting for various socio-demographic characteristics and food choice indicators. Our findings showed that the majority of consumers highly evaluate the underlying values of organic foods based upon sustainability ethics, such as lower climate impact, environmental protection, animal welfare and support to local/regional economies. Participants also presented high levels of trust to the authenticity and adequacy of the organic food control system, whereas a minority of the sample expressed skepticism toward climate change. Furthermore, individuals' consumption level and purchase behavior were significantly associated with organic foods benefits (including environmental concerns, healthiness and superior quality) and climate change concerns. Defining the symbolic values of organics, besides their tangible values will help marketers find out the purchase motives and develop the appropriate communication strategies to promote organic foods in the marketplace.

Appendix

Appendix 1 Climate change awareness

Climate Change Awareness[a]
I don't see why I'd take action on climate change
I don't see why I'd try to reduce emissions at individual level
I'd prefer to buy products from companies that are working to reduce climate change
The economic growth of developing countries is the greatest threat to the world's climate
It's too late to do anything about climate change
If the government enforced more strict environmental policies it should prevent climate change
The media is exaggerating the potential impact of climate change
Climate change has occurred many times in the past and it's part of the natural shifting of the environment
I'm concerned about the consequences of climate change
I want more information from businesses on how they address climate change impact
Consumers can help reduce the impact of climate change
I trust companies to take action when it comes to climate change
Attempts to prevent climate change should be coordinated at an international level

[a]all items were rated on a five-point Likert scale ranging from "totally disagree" to "totally agree"

Appendix 2 Attitudes toward the organic food production system

Attitudes toward the organic food production system[a]
Organics are more healthy than conventional foods
When I buy locally produced organic foods, I support preserving local jobs
Organic foods have better quality than conventional food products
Organic production system has a lower climate impact than conventional agriculture
I have serious concerns about the authenticity of organic foods
I believe that the organic foods control system is inadequate
Animals engaged in organic production have better animal welfare than in conventional livestock production system
Even if the price of organic foods was higher, I'd definitely consume them
Organic foods have a reasonable price
Food quality is very important to me when I make my food choice
I believe that organic food producers are only interested for making money and not producing better quality food
Organic food production pollutes the local environment less than the conventional food production system
I believe I follow a healthy eating lifestyle
For me healthy eating habits are the key to good health
I try to make informed food decisions

[a]all items were rated on a five-point Likert scale ranging from "totally disagree" to "totally agree"

References

Aertsens J, Verbeke W, Mondelaers K, Van Huylenbroeck G (2009) Personal determinants of organic food consumption: a review. Br Food J 111:1140–1167

Al Mamun A, Fazal S, Bin Ahmad G, Yaacob M, Mohamad MR (2018) Willingness to pay for environmentally friendly products among low-income households along coastal peninsular Malaysia. Sustainability 10:1316

Allievi F, Vinnari M, Luukkanen J (2015) Meat consumption and production—analysis of efficiency, sufficiency and consistency of global trends. J Clean Prod 92:142–151

Andorfer VA, Liebe U (2012) Research on fair trade consumption—a review. J Bus Ethics 106:415–435

Armitage CJ, Conner M (1999) Distinguishing perceptions of control from self-efficacy: predicting consumption of a low-fat diet using the theory of planned behavior. J Appl Soc Psychol 29:72–90

Aygen FG (2012) Attitudes and behavior of turkish consumers with respect to organic foods. Int J Bus Soc Sci 3:262–273

Backhaus K, Erichson B, Plinke W, Weiber R (2016) Multivariate Analysemethoden. Eine anwendungsorientierte Einführung. Springer, Berlin

Beer L, Theuvsen L (2019) Conventional German farmers' attitudes towards agricultural wood and their willingness to plant an alley cropping system as an ecological focus area: a cluster analysis. Biomass Bioenerg 125:63–69

Biao X, Wang L, Yang H, Wang Y, Zhang M (2015) Consumer perceptions and attitudes of organic food products in Eastern China. Br Food J 117:1105–1121

Blei D, Lafferty J (2009) Topic models. In: Srivastava A, Sahami M (eds) Text mining: classification, clustering, and applications. Taylor & Francis Group, Boca Raton, pp 71–94

Brooks MJ, Bear T, Hacker K, Ricci EM, Foulds A, Anderson H, Raible C, Miller E (2017) An alternative technique for youth risk surveillance outside of the school system. J Adolesc Health 60:72–78

Bryla P (2016) Organic food consumption in Poland: motives and barriers. Appetite 105:737–746

Carvalho NB, Rodrigues Minim VP, Nascimento M, Ribeiro Vidigal MCT, Marques Ferreira MA, Arruda Gonçalves AC, Minim LA (2015) A discriminant function for validation of the cluster analysis and behavioral prediction of the coffee market. Food Res Int 77:400–407

Chen M-F (2007) Consumer attitudes and purchase intentions in relation to organic foods in Taiwan: moderating effects of food-related personality traits. Food Qual Prefer 18:1008–1021

Chiciudean GO, Harun R, Ilea M, Chiciudean DI, Arion FH, Ilies G, Murean GC (2019) Organic food consumers and purchase intention: a case study in Romania. Agronomy 9:145

Cho S, Krasser AH (2011) What makes us care? The impact of cultural values, individual factors, and attention to media content on motivation for ethical consumerism. Int Soc Sci Rev 86:3–23

Costanigro M, Kroll S, Thilmany D, Bunning M (2014) Is it love for local/organic or hate for conventional? Asymmetric effects of information and taste on label preferences in an experimental auction. Food Qual Prefer 31:94–105

Crane A (2001) Unpacking the ethical product. J Bus Ethics 30:361–373

D'Alessio M., De Devitiis B, Maietta O (2007) A comparative analysis of the purchase motivations of fair trade products: the impact of social capital. J Perspectives Econ Political Soc Integ J Mental Change 13(1/2):39–87

Ding C, He X (2004) K-means clustering via principal component analysis. In: Proceedings of the 21st international conference on machine learning, Banff, AB, Canada, 4–8 July 2004, pp 29–37

Eurostat (2020) Organic farming statistics. Available at: https://ec.europa.eu/eurostat/statistics-explained/pdfscache/5461.pdf. Retrieved at 28 Nov 2020

Gerrard C, Janssen M, Smith L, Hamm U, Padel S (2013) UK consumer reactions to organic certification logos. Br Food J 115:727–742

Ghali Z (2020) Motives of willingness to buy organic food under the moderating role of consumer awareness. J Sci Res Rep 25:1–11

Greene WH, Hensher DA (2010) Modeling ordered choices: a primer and recent developments. Cambridge University Press, Cambridge
Greene WH (2012) Econometric analysis. Prentice-Hall, Boston
Grunert KG, Hieke S, Wills J (2014) Sustainability labels on food products. Consumer motivation, understanding and use. Food Policy 44:177–189
Gupta R, Falk TH (2017) Latent factor analysis for synthesized speech quality-of-experience assessment. Qual User Exp 2:2
Hair JF, Anderson RE, Tatham RL, Black WC (1998) Multivariate data analysis, 5th edn. Prentice Hall, Upper Saddle River
Hamzaoui-Essoussi L, Sirieix L, Zahaf M (2013) Trust orientations in the organic food distribution channels: a comparative study of the Canadian and French markets. J Retail Consum Serv 20:292–301
Hempel C, Hamm U (2016) Local and/or organic: a study on consumer preferences for organic food and food from different origins. Int J Consum Stud 40:732–741
Higuchi A, Avadi A (2015) Organic purchasing factors and consumer classification through their preferences in the metropolitan area of Lima. Peru Agronomía Colombiana 33:271–279
Hill H, Lynchehaun F (2002) Organic milk: attitudes and consumption patterns. Br Food J 104:526–542
Hughner RS, McDonagh P, Prothero A, Shultz CJ, Stanton J (2007) Who are organic food consumers? A compilation and review of why people purchase organic food. J Consum Behav 6:94–110
Kaiser HF (1974) Index of factorial simplicity. Psychometrika 39:31–36
Kapuge KDLR (2016) Determinants of organic food buying behavior: Special reference to organic food purchase intention of Sri Lankan customers. Procedia Food Sci 6:303–308
Kennedy P (1999) A guide to econometrics. Blackwell Publishers Ltd, United Kingdom
Kim G, Seok JH, Mark TB (2018) New market opportunities and consumer heterogeneity in the U.S. Organic food market. Sustainability 10:3166
Lam MCL, Adams J (2017) Association between home food preparation skills and behaviour, and consumption of ultra-processed foods: cross-sectional analysis of the UK National Diet and nutrition survey (2008–2009). Int J Behav Nutr Phys Act 14:68
Lee HJ, Yun ZS (2015) Consumers' perceptions of organic food attributes and cognitive and affective attitudes as determinants of their purchase intentions toward organic food. Food Qual Prefer 39:259–267
Magnusson MK, Arvola A, Hursti U, Aberg L, Sjoden P (2003) Choice of organic food is related to perceived consequences for human health and to environmentally friendly behaviour. Appetite 40:109–117
Malhotra NK (2006) Análise de Clusters. Pesquisa de Marketing: uma orientação aplicada (4th edn). Bookman, Porto Alegre, pp 570–590
Martins CA, Machado PP, da Costa Louzada ML, Levy RB, Monteiro CA (2020) Parents' cooking skills confidence reduce children's consumption of ultra-processed foods. Appetite 144:
McKelvey RD, Zavoina W (1975) A statistical model for the analysis of ordinal level dependent variables. J Math Sociol 4:103–120
Mehta K, Booth S, Coveney J, Strazdins L (2020) Feeding the Australian family: challenges for mothers, nutrition and equity. Health Promotion Int 35:771–778
Meixner O, Haas R, Perevoshchikova Y, Canavari M (2014) Consumer attitudes, knowledge, and behavior in the russian market for organic food. Int Food Syst Dynam 5:110–120
Monier-Dilhan S, Bergès F (2016) Consumers' motivations driving organic demand: between self-interest and sustainability. Agricult Resour Econ Rev 45:522–538
Niles M, Ahuja R, Barker T, Esquivel J, Gutterman S, Heller MC, Mango N, Portner D, Raimond R, Tirado C, Vermeulen S (2018) Climate change mitigation beyond agriculture: a review of food system opportunities and implications. Renew Agric Food Syst 33:297–308
O'Doherty Jensen K, Denver S, Zanoli R (2011) Actual and potential development of consumer demand on the organic food market in Europe. NJAS Wageningen J Life Sci 58:79–84

Oroian CF, Safirescu CO, Harun R, Chiciudean GO, Arion FH, Muresan IC, Bordeanu BM (2017) Consumers' attitudes towards organic products and sustainable development: a case study of Romania. Sustainability 9:1559
Pagiaslis A, Krontalis AK (2014) Green consumption behavior antecedents: environmental concern, knowledge, and beliefs. Psychol Mark 31:335–348
Petrescu DC, Petrescu-Mag RM, Burny P, Azadi H (2017) A new wave in Romania: organic food. Consumers' motivations, perceptions, and habits. Agroecol Sustain Food Syst 41:46–75
Petrescu DC, Vermeir I, Petrescu-Mag RM (2020) Consumer understanding of food quality, healthiness, and environmental impact: a cross-national perspective. Int J Environ Res Publ 17:169
Pino G, Peluso AM, Guido G (2012) Determinants of regular and occasional consumers' intentions to buy organic food. J Consum Aff 46:157–169
Pirsich W, Weinrich R (2019) The impact of sustainability aspects in the meat sector: a cluster analysis based on consumer attitudes and store format choice. J Int Food Agribus Mark 31:150–174
Ravn Heerwagen L, Mørch Andersen L, Christensen T, Sandøe P (2014) Can increased organic consumption mitigate climate changes? Br Food J 116:1314–1329
Scalvedi ML, Saba A (2018) Exploring local and organic food consumption in a holistic sustainability view. Br Food J 120:749–762
Seyfang G (2006) Ecological citizenship and sustainable consumption: examining local organic food networks. J Rural Stud 22:383–395
Slater JJ, Mudryj AN (2016) Self-perceived eating habits and food skills of Canadians. J Nutr Edu Behav 48:486–495
Smith S, Paladino A (2010) Eating clean and green? Investigating consumer motivations towards the purchase of organic food. Australas Mark J (AMJ) 18:93–104
Stolz H, Stolze M, Hamm U, Janssen M, Ruto E (2011) Consumer attitudes towards organic versus conventional food with specific quality attributes. NJAS-Wageningen J Life Sci 58:67–72
Taherdoost H, Sahibuddin S, Jalaliyoon N (2014) Exploratory factor analysis; concepts and theory. In: Advances in applied and pure mathematics, pp 375–382
Van Huy L, Chi M, Lobo A, Nguyen T-N, Phan HL (2019) Effective segmentation of organic food consumers in Vietnam using food-related lifestyles. Sustainability 11:1237
Vermeulen SJ, Campbell BM, Ingram JSI (2012) Climate change and food systems. Annu Rev Environ Resour 37:195–222
Vittersø G, Tangeland T (2015) The role of consumers in transitions towards sustainable food consumption. The case of organic food in Norway. J Clean Prod 92:91–99
Wägeli S, Hamm U (2016) Consumers' perception and expectations of local organic food supply chains. Org Agricult 6:215–224
Wells VK, Ponting CA, Peattie K (2011) Behaviour and climate change: consumer perceptions of responsibility. J Mark Manage 27(7–8):808–833
Willer H, Bteich MR, Meredith S (2014) Country reports: Greece in organic in Europe: prospects and developments FOAM EU GROUP. In: Stephen M, Helga W (eds) FIBL and CIHEAM-IAMB. FOAM EU Group, Brussels
Wunsch N-G (2020) Organic food market in Europe—statistics and facts. STATISTA. Available at: https://www.statista.com/topics/3446/organic-food-market-in-europe/. Retrieved at 28 Nov 2020
Xie B, Wang L, Yang H, Wang Y, Zhang M (2015) Consumer perceptions and attitudes of organic food products in Eastern China. British Food J 117:1105–1121
Yadav R, Pathak GS (2016) Intention to purchase organic food among young consumers: Evidences from a developing nation. Appetite 96:122–128
Yim O, Ramdeen KT (2015) Hierarchical cluster analysis: comparison of three linkage measures and application to psychological data. Quant Methods Psychol 11:8–21
Zander K, Hamm U (2010) Consumer preferences for additional ethical attributes of organic food. Food Qual Prefer 21:495–503

Climate Change in Lebanon and the Impact to Water Resources

Jalal Halwani and Bouchra Halwani

Abstract Lebanon is affected by climate change, the decrease in precipitation, the changes in the rainfall pattern, and the increase in evapotranspiration were exacerbated by water shortage due to urbanization and population growth. In the past decades water resources continue to decrease. We have studied this phenomenon and analyzed different indices of weather parameters and compared different records in graphical illustrations and numerical values. The study looked at the minimal (low) and maximal (high) temperatures as well as the precipitations. The data used was provided from four principal meteorological stations in Lebanon: Rafic Hariri International Airport in Beirut, Tripoli, Cedars, and Rayak from 1930 to 2019. Several observations regarding the variation of the temperature and precipitation were noted. The hottest temperature recorded in Beirut was (38 °C) reported in August 2015. Historically, the average temperature from July to September is normally around 26 °C; in 2012 the average temperature was around 28.5 °C, with an average daily temperature as high as 33 °C. An elevated rainfall quantity (25 mm) was recorded on June 13th, 2011 in Tripoli (North Lebanon), and in 2013 it was the worst precipitation year in past 15 years. We can conclude that the average temperature has increased between 1 to 3 °C, and globally, the monthly average precipitation was slightly decreased in the availability of water resources, between 6 and 12%. Lebanon was considered rich in water resources compared to Middle Eastern neighbouring countries; Lebanon is now experiencing a shortage in water resources close to hydrological drought.

1 Introduction

Global warming is the gradual heating of Earth's surface, oceans and atmosphere, is caused by human activity, primarily the burning of fossil fuels that pump carbon dioxide (CO_2), methane and other greenhouse gases into the atmosphere. Climate

J. Halwani (✉) · B. Halwani
Water and Environment Science Laboratory, Lebanese University, Tripoli PH3, Lebanon
e-mail: jhalwani@ul.edu.lb

W. Leal Filho and E. Manolas (eds.), *Climate Change in the Mediterranean and Middle Eastern Region*, Climate Change Management,
https://doi.org/10.1007/978-3-030-78566-6_19

change is a phenomenon where an increase in the average temperature of the oceans and the atmosphere occurs globally, and constitutes a significant challenge for which international environmental organizations must solve. The increased urbanisation and the population growth demands abusive usage of all-natural resources, and the consumption rate remains dependant on fossil fuels emitting greenhouse gases causing "climate change". Climate experts are categorical; the earth is warming at a sustained rate. The increases in greenhouse gases of human origin, with the consequences of health, ecological, and humanitarian crises are the beginnings. Despite this, the fight against climate change is a real opportunity for the transition to a low carbon society.

Globally, an average temperature rise of 0.8 °C has considerable consequences at the local level, both on ecological balances and on our societies (Allen 2018). Climate change and its effects now appear to be proven and linked, at least in part, to the additional anthropogenic greenhouse effect, which is reflected in the inclusion of this major issue on global political agendas. Because it is at the heart of development issues, it is not a minor environmental problem, nor a related issue, but it is becoming a geopolitical object with "global warming". Internationally, negotiations between states are progressing. After the COP21 and the Paris Agreement (UN 2015), hope is emerging to reduce GHG emissions and limit the warming to 2 °C or even 1.5 °C by 2100.

In the Mediterranean region, global warming is occurring at a greater pace than in other parts of the world (Lionello and Scarascia 2018) with marked changes in temperatures and precipitation, the temperature increase in temperature in the Mediterranean basin has already reached 1.5 °C above pre-industrial levels, for a global average of 1.1 °C, threatening agricultural and water resources (MedEcc 2020). If this phenomenon continues, it should reach 2.2 °C in 2040 and it could rise to 3.8 °C, in parts of the Mediterranean basin by the end of the century. Heat wave episodes could become more frequent and/or more intense, according to the MedEcc report. This rise in temperatures will be accompanied by a reduction in precipitation in the coming decades, which can reach up to 30% in some Mediterranean areas. An increase in episodes of torrential rains is also expected. The decrease in precipitation associated with the intensification of warming is contributing to strong trends towards a drying up of the climate and the Mediterranean population considered to be "water-poor" (that is to say with less than 1,000 m^3 per capita per year) could drop from 180 million (people) in 2013 to more than 250 million in the next 20 years. Crops could be affected by soil depletion, droughts, or heat waves, while fishing could be affected by overexploitation and the disappearance of species due to the warming of the sea. "The inevitable calamity" due to climate change threatens the Mediterranean at a faster rate than we thought.

Although the above mentioned are considered sceptical by many, the climate change influences all countries, including Lebanon, which has been witnessing in the past few years, warm during months normally is cold and rain, and rain during months normally hot (MoE/GEF/UNDP 2011). In fact, an elevated rainfall quantity (25 mm) was recorded on June 13th, 2011 in the meteorological station of Tripoli, North Lebanon, and in 2013, it was the worst year during the past 15 years in precipitation. The average temperature from July to September is normally around 26 °C; however, in 2012 the average temperature was 28.5 °C with an average daily temperature higher than 33 °C. The hottest temperature experienced in Lebanon, was 38.0 °C, recorded in Beirut on August 14th, 2015, and the coldest temperature recorded along the coastal area was reported in Tripoli (−0.6 °C) on January 10th, 2016.

The changes in weather patterns will reduce the total volume of available water and reduce the snow coverage, and increase the drought periods. During the last decades, water resources in Lebanon show an abrupt volumetric decrease with direct consequences on water availability for users in particular the agriculture sector (MoA 2003).

2 Aims and Objectives

Based on statistical observations, the parameters linked to climate change (that is an increase of low and high temperatures and the fluctuation of precipitations) were studied and analyzed using recorded data from four principal meteorological stations in Lebanon: Beirut, Tripoli, Cedars and Rayak. The data was collected from the department of meteorology at Rafic Hariri International Airport in Beirut. The results obtained were available water resources figures that are less than historical levels, evidence that Lebanon is experiencing a climate change impacting its water resources and the environment.

The main objective of this study is to alert the national authorities on the current climatic variations and its impact on water resources and to encourage officials to take this natural phenomenon into account in their strategic decisions in order to sustain water resources availability for future generations.

3 Materials and Methods

The study was completed applying statistical analysis based on authentic numeric records. The data was collected from the department of Meteorology at Rafic Hariri International Airport in Beirut, which monitors meteorological parameters including temperature, humidity, rainfall, wind speed and direction, sunshine hours and barometric pressure.

Prior to the civil war, Lebanon had more than eighty stations spread across the country. Many were damaged during the war and meteorological records were interrupted for several years. Several stations were rehabilitated in the 1990s (some with continuous and automatic data logging systems) and their operations have resumed. In addition, several research centers and universities operate auxiliary weather stations. A statistical monthly bulletin is published by the Central Administration of Statistics which provides meteorological records for Beirut, Tripoli and Rayak.

The data includes day-to-day low and high temperatures and the rainfall records. The data were recorded in four different meteorological stations in Lebanon located in Beirut, Tripoli, Cedars and Rayak. Beirut and Tripoli stations are coastal cities located respectively in the middle, Cedars is a northern mountainous region in Bcharre (Altitude 2000 m), and Rayak is located in Bekka Valley near Zahle east of Lebanon (Altitude 1020 m,) between the east and west mountains (Fig. 1).

It should be noted that the rainfall and temperature data contain many gaps due to the absence of numeric values in several years; these gaps are caused by the civil war technical problems at the stations or for other unknown reasons. In order to ensure a fine spatial coverage of all the years, it was beneficial to use statistical means on every station in order to detect any changes happening in all the areas in Lebanon.

The statistical methods have been carried out on a monthly, seasonal and annual basis, in order to evaluate the intensity of this "change", and to confirm the authenticity of the exaggeration disseminated at times in the media.

4 Climate of Lebanon

Despite its little area (10,452 km^2) Lebanon enjoys various climatic zones. In addition of coastal region, there are two mountain chains expanding perpendicularly to the atmospheric circulation and constituting the core of the country, the Bekaa Valley (Fig. 1). The climate of Lebanon is typically Mediterranean and characterized by hot, dry summers and cool, moist winters. Humidity can reach 90% and temperatures are commonly over 30 °C between July and August. During the winter, the annual average temperatures vary from 15 °C at the coast to 6 °C in the mountains, and 8 °C in the Bekaa Valley.

Most rainfall, which comes sometimes in dramatic bursts, occurs from October to May. Precipitation is irregular and varies considerably from the narrow and relatively moist coastal plain to the almost desert-like conditions in the Anti-Lebanon mountain chain. Rainfall patterns result from extra-tropical weather systems from Europe and the Atlantic Ocean that bring clouds and cold air. The typical rainfall distribution from these systems results in a south-to-north gradient. This leads to the northern, mountainous parts of the country receiving the most precipitation. Throughout the year, some convective clouds develop occasionally, even during the dry season, when subtropical maritime air masses converge in the region. In rare cases, these develop into heavy thunderstorms.

The dry season coincides with summer period, which starts in June till the end of September. During this period, no rain is recorded and a state of high pressure dominates the whole country, with a general tendency toward Northeast. In summer period, the occidental winds coming from Greece humidify the air on the coast and in the mountains and lower slightly the temperatures. The presence of natural districts

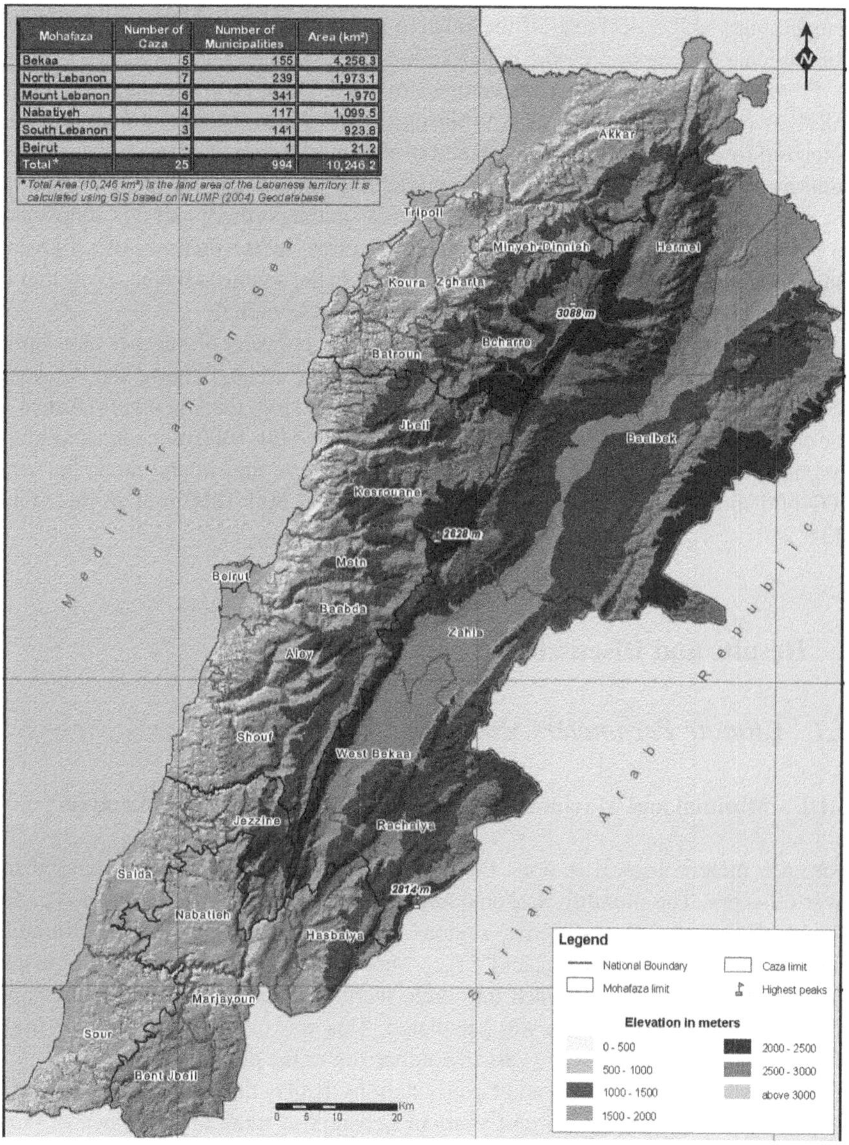

Fig. 1 Lebanon Map. *Source* MoE/UNDP/ECODIT, SOER (2011)

at the levels of Marjaoun in the South and Dahr el Baidar in the central Mount-Lebanon permit to the occidental winds to reach partly the southern and central parts of the Bekaa valley. Whereas, in the northern Bekaa Valley, the dry and hot winds originating from the Arab Peninsula contribute to an increase of the day/night temperatures and a decrease of the relative humidity of the air, that increases by consequence the vapor pressure deficit of the air.

Precipitation in Lebanon is unevenly distributed. Up to 90% of total precipitation falls between November and April. While the coastal and mountainous areas are characterized by abundant rainfall distributed over winter season, the Bekaa Valley has a semi-arid to continental climate with unpredictable rainfall and recurrent drought. In the central part the climate is semi-arid, whereas in the northern part it is almost arid to continental, since it is separated from the sea effect by the presence of a high and ridge mountain chain, which height near 3000 m. In the southern Bekaa, a sub-humid Mediterranean climate is dominant, with more reliable rainfall.

As a result, there is a season of stable rainfall between November and April, annual precipitation can reach up to 2000 mm in some areas (Bcharre and Faraya in the western mountain range) and barely makes the 200 mm mark in others (Aarsal in the eastern mountain range). Coastal areas experience 600–1000 mm annual rainfall; neighboring mountain areas between 1000–1400 mm; and inland areas between 200–600 mm in the North and Central regions and 600–1000 mm in the South (Fig. 2).

5 Results and Discussion

5.1 Climate Parameters Variation

5.1.1 Minimal and Maximal Temperature (TT_{min} and TT_{max}) Analysis

For each meteorological station, TT_{min} and TT_{max} were individually analysed for over 60 years: The monthly, seasonal and annual averages of TT_{min} and TT_{max} for Tripoli, Beirut, Rayak and Cedars regions were studied. In addition, the minimum and maximum values of TT_{min} and TT_{max} were also analysed. Moreover, for deeper analysis, the charts were divided into three sections, mainly according to the gaps, and the average for every era was calculated. This was important in order to study the average of the past 10–15 years in comparison to the past years.

Figures 3 and 4 for Tripoli and Cedars stations display the variation of TT_{max} and TT_{min} respectively during the past years in Lebanon. A trend line is drawn to show the variation slope.

5.1.2 Precipitations

For each meteorological station, the monthly, seasonal and annual RR for over 60 years, for Tripoli, Beirut, Rayak and Cedars regions were studied (Figs. 5 and 6).

In addition, the percentage of years where RR was below average was calculated in Table 1.

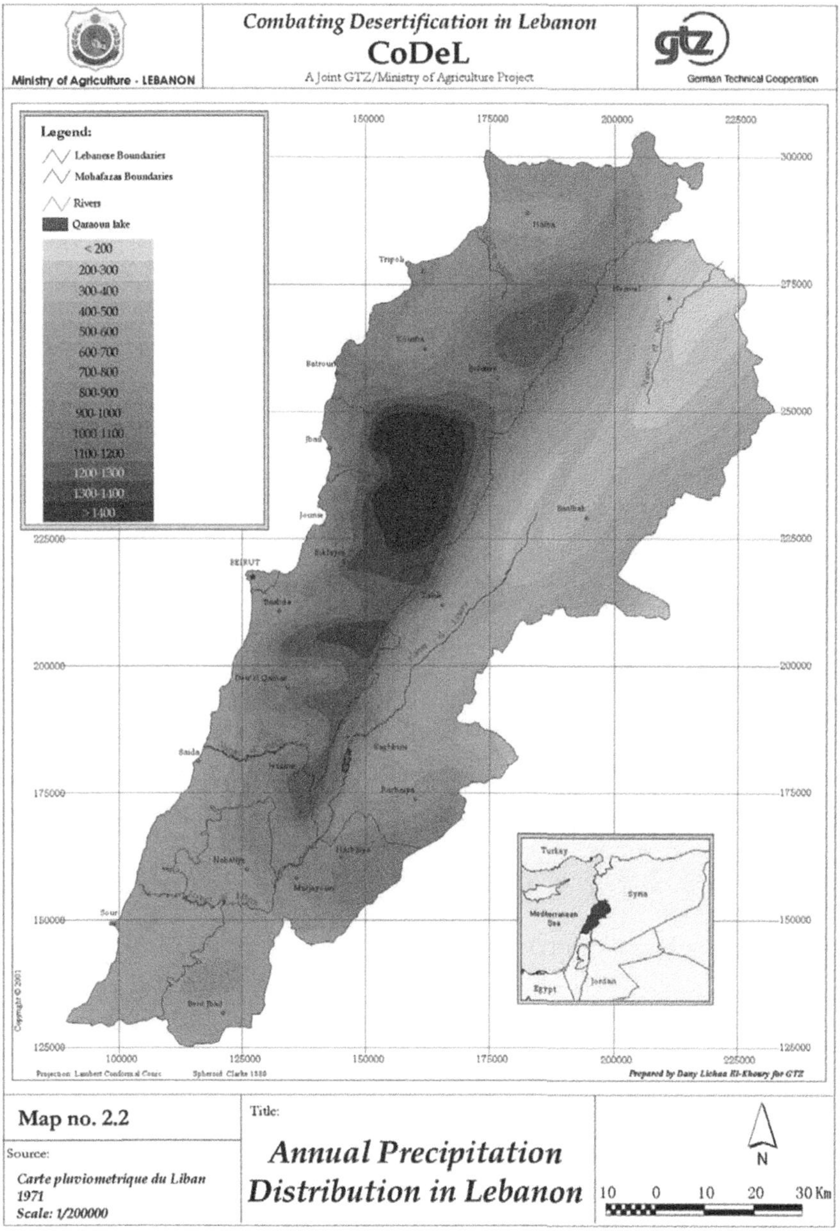

Fig. 2 Annual Precipitation distributions in Lebanon. *Source* Ministry of Agriculture, Lebanon, June 2003

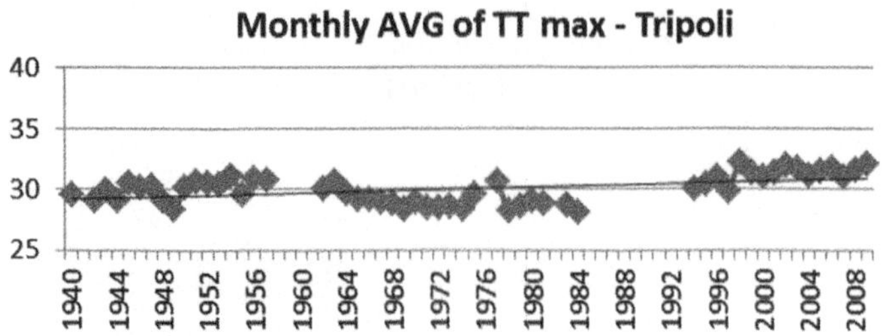

Fig. 3 Monthly average of TTmax in Tripoli from 1940 to 2008

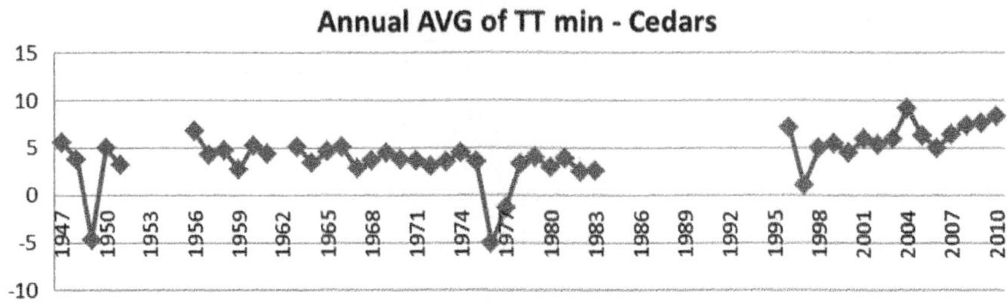

Fig. 4 Annual average of TTmin in Cedars from 1947 to 2010

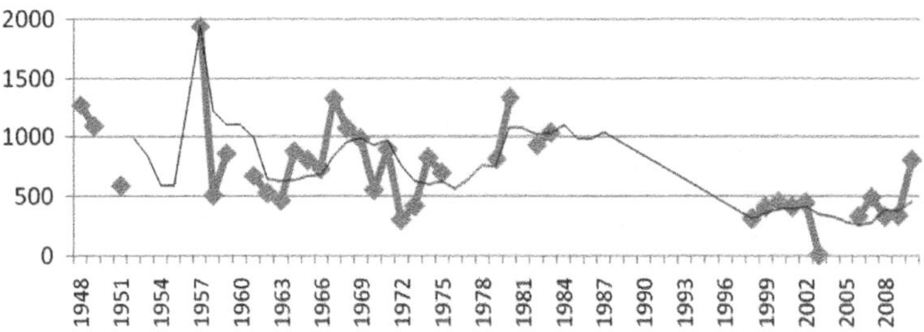

Fig. 5 Monthly average (February) of rainfall records in cedars from 1948 to 2008

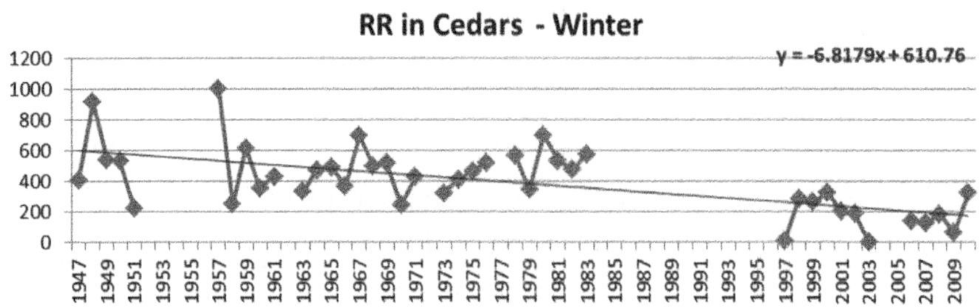

Fig. 6 Seasonal average (winter) of rainfall records in cedars from 1947 to 2010

Table 1 Percentage of years where the rainfall records are below average in all 4 stations

	1930–1950	1950–1970	1970–1990	1990–2010
Tripoli	42.86	64.29	66.67	57.89
Beyrouth	54.55	55.00	40.00	75.00
Rayak	31.25	45.00	69.23	70.00
Cedars	–	55.56	57.14	88.89

Observational precipitation records for the drier inland regions show annual values ranging from less than 30% to more than 200% of the mean value. Such variability is more moderate in the coastal regions (±60 to 80% of mean value). Figure 7 presents average rainfall in Beirut for two time periods, 1933–63 and 1961–90, and Table 2 shows recorded rainfall in Beirut, Tripoli and Zahle for the 1996–2000 years. In Fig. 8 we can see the cumulative precipitation rate in Lebanon between 1965 and 2015 for the main stations.

Recent weather monthly parameters distributions in Tripoli and Beirut for 2019 are given in Fig. 9. In 2019, Beirut gains an average of 825.5 mm of rainfall per year, or 68.8 mm per month. On average there are 80 days per year with more than 0.1 mm of rainfall. The driest weather is in July when an average of 0.3 mm of rainfall occurs. The wettest weather is in January when an average of 190.9 mm of rainfall occurs. Tripoli is very similar to Beirut.

In order to get an overall impression of annual precipitation variation in Beirut from 1876 to 2015, the data registered in American University of Beirut meteorological station has been reproduced in Fig. 10. The average precipitation from 1876

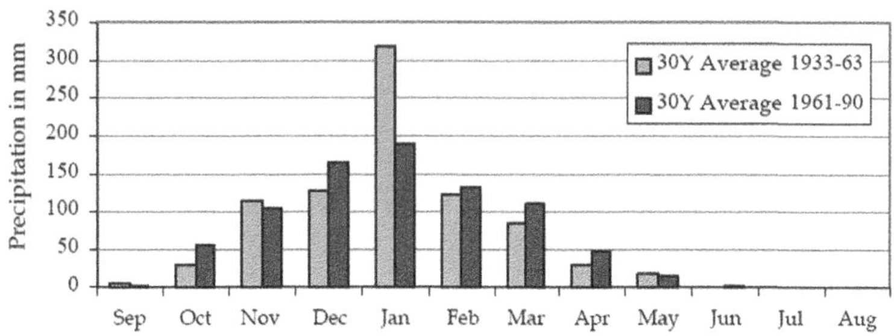

Fig. 7 Average rainfall measured at Beirut during two time periods

Table 2 Recorded rainfall for years 1996–2000

Station	Year					Average	
	1996	1997	1998	1999	2000	1996–2000	1961–1990
Beirut	827	733	658	487	834	708	825
Tripoli	844	639	694	378	872	685	NA
Rayak	798	686	533	311	614	588	NA

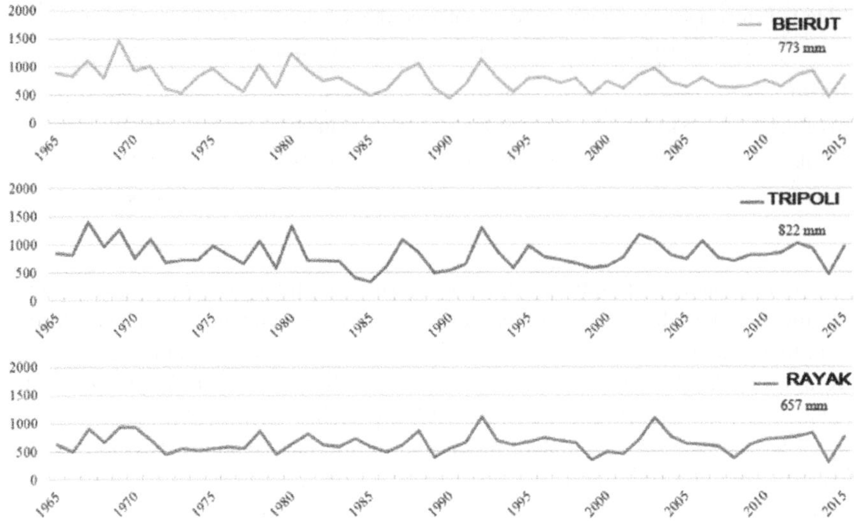

Fig. 8 Cumulative precipitation rates in Lebanon between 1965 and 2015

until 2015 was 900 mm; the record low rainfall was 335 mm in 1932 and the high record 1600 mm in 1969.

5.2 *Origin of Climate Change*

In Lebanon, energy is believed to be responsible for more than 50% of greenhouse gas emissions (MoE/GEF/UNDP 2011). They include carbon dioxide produced during consumption of solid, liquid, gas fuels, and gas flaring. Carbon dioxide emissions are particularly those stemming from the burning of fossil fuels and the manufacture of cement. In the year 2000, Lebanon's total GHG emissions recorded 18,507 Gg (18.5 Million tonnes (Mt)) of CO_2 equivalent (CO_2 eq.), recording an average of 2.77% per year increase from 1994 (15,901 Gg CO_2 eq.). The energy sector is the main source of GHG emissions, accounting for 74.86% of the national emission (53.45% energy production and 21.41% transport). This is followed by industrial processes and waste sectors which account for 9.62 and 9.40% respectively. Emissions from agriculture and land use change and forestry make up 5.76%, and 0.36% of total CO_2 eq. respectively. Carbon dioxide is the main emitted GHG with 84.13% of emissions in 2000, while CH_4 and N_2O constitute 10.19 and 5.68% respectively. The main contributors of CO_2 emissions are energy production and transport with 63 and 25% respectively whereas the waste sector constitutes the main source of CH_4 emissions (88%). The main contributor to N_2O emissions is the agriculture sector with 88%.

The average value of CO_2 emissions for Lebanon from 1960 to 2016 was 10,807 kt with a minimum of 2582 kt in 1960 and a maximum of 24,796 kt in 2016 (Fig. 11a).

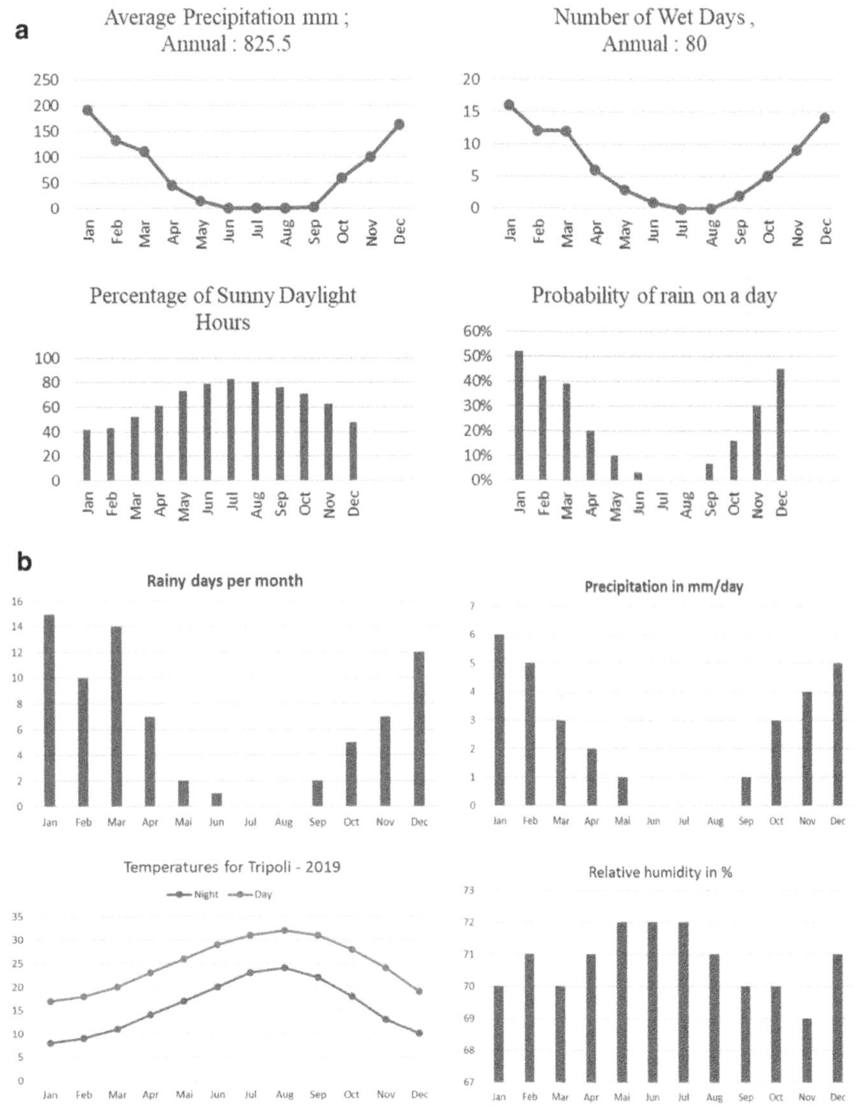

Fig. 9 **a** Values of weather parameters in Beirut for 2019 year. **b** Values of weather parameters in Tripoli for 2019 year

For comparison, the world average in 2016 based on 193 countries is 171158 kt (World Bank 2016a, b). The maximum CO_2 emissions per capita is 4.995 metric tons was marked in 1998, and the minimum during the last 30 years is 2.824 for 2007 (Fig. 11b).

The Paris Agreement on climate change was adopted by 196 Parties at COP 21 in Paris, on 12 December 2015 and entered into force on 4 November 2016

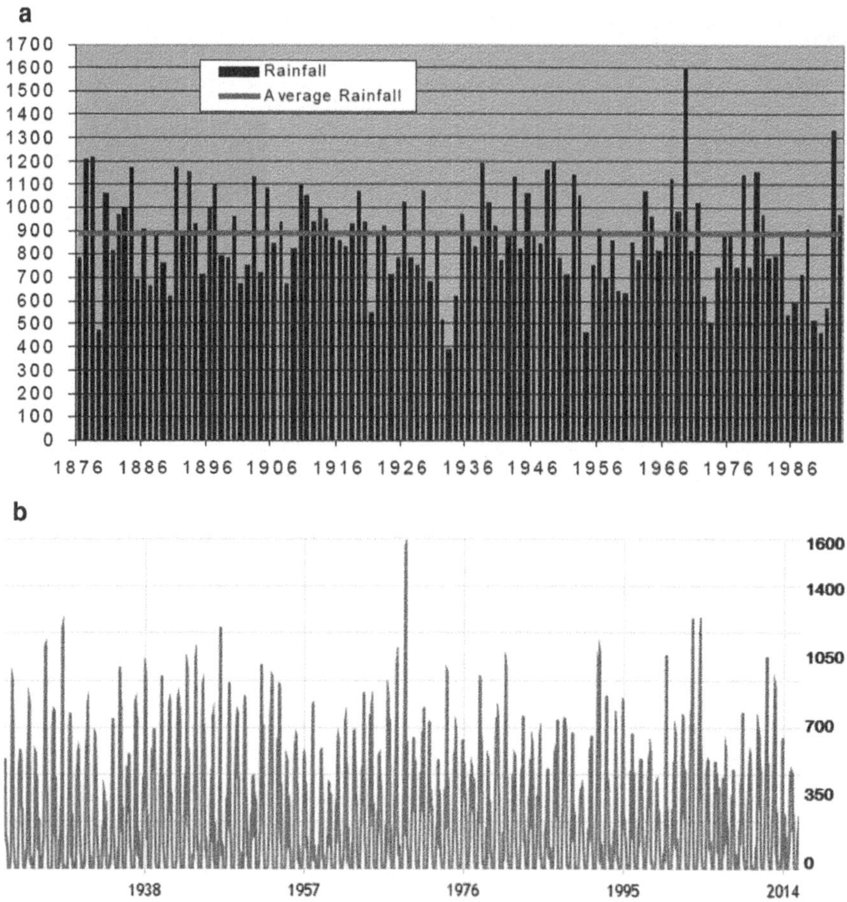

Fig. 10 **a** Values of annual precipitation (mm) in Beirut since 1876–1990 registered at AUB station. *Source* Hakim (1993). **b** Values of annual precipitation (mm) in Beirut since 1901–2015 registered at AUB station. *Source* World Bank (2016a, b)

(United Nations 2015). Its goal is to limit global warming to well below 2 °C, preferably to 1.5 °C, compared to pre-industrial levels. All countries of the southern Mediterranean which has participated in this international conference have ratified the Paris Agreement, except Lebanon despite agriculture and water resources are the most affected by the phenomenon of climate change.

Unfortunately, in its political agenda, Lebanon does not give high priority to the issue of climate change and the protection of the environment although the country is starting to feel the effects of climate change. In addition, ratifying the Paris Agreement would require Lebanon to apply the provisions of the treaty, which would influence public policies and protection mechanisms. As the authorities do not see this issue as a priority, the agreement is seen as a "heavyweight" in terms of legislation but also in terms of spending.

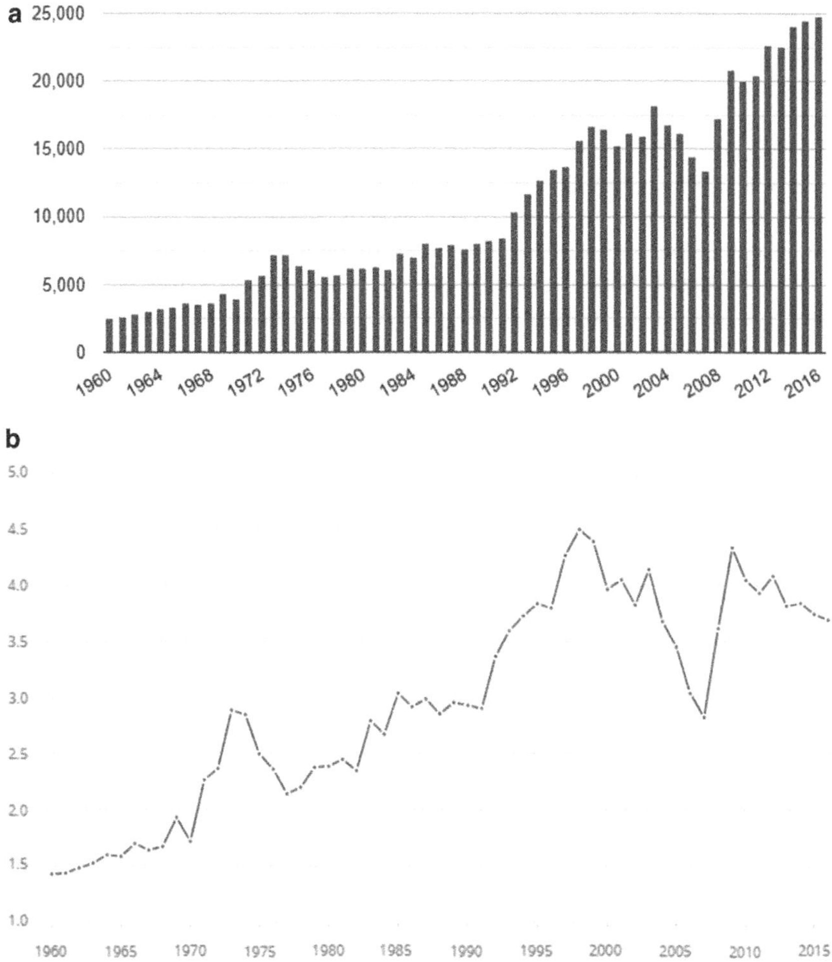

Fig. 11 **a** Carbon dioxide emissions in Lebanon in thousands of tonnes. *Source* https://www.theglobaleconomy.com/Lebanon/carbon_dioxide_emissions/. **b** CO_2 emissions (metric tons per capita) in Lebanon from 1960 to 2016. *Source* World Bank, Carbon Dioxide Information Analysis Center (2016a, 2016b) https://data.worldbank.org/indicator/EN.ATM.CO2E.PC?locations=LB

5.3 *Impact of Climate Change to Water Resources*

Water and climate change are inextricably linked, as the consequences of climate change manifest themselves first in the realm of water: through droughts, floods and storms (Tremblay et al. 2020). These natural disasters can destroy water supplies or leave contaminated water in their wake, endangering the lives of millions of people. Without clean water, people are at risk of contracting diseases like diarrhea, which kills around 525,000 children under five each year worldwide, and cholera, with 95,000 deaths per year (WHO 2019).

Lebanon currently faces major water challenges related to the sustainable management of water resources and the delivery of water services for domestic, agricultural, and industrial use (MoEW and UNDP 2014). Like several Mediterranean countries, Lebanon is influenced by climate change, which is viewed not only from fluctuation in the climatic parameters but also from its influence on the availability of water resources which could go to hydrologic drought (ESCWA-UN 2016). Indeed, climate change and climate variability can increase water challenges impacting the quantity and quality of available water resources, and generate secondary effects that influence environmental sustainability (World Bank 2018).

Contrary to current notions about water abundance in Lebanon, the country is facing serious problems of unavailability, both in quantity and quality (Shabane 2020). Even though this is due to several existing human-related aspects, i.e. management, ill-finance, non-integration and lack of relevant awareness on water use, the new dimension of climate change is unfortunately adding to the water problems in the country. Since the last four decades, the amount of available water in Lebanon is in a descending trend as a result of combined effect of climate change and human interference (Halwani 2009).

Databases in Lebanon are not quite consistent and accurate, implying difficulty in interpreting correctly trends of change. Besides, the most usable and dependent conventional water resources were highly affected by hydrologic droughts. That was viewed from the decrease in the number of springs and their discharge as well as the number of local reservoirs having the most excessive influence, as they were a major source of water before the 1960s. The human interference is reflected by direct pumping from rivers and an over exploitation of groundwater (Halwani 2008).

The climate exhibits its influence through the decrease in the amount of water from rainfall and snow cover. Actual records on precipitation reflect a decreasing pattern over the last century. This is in agreement with predicted trends of temperature and precipitation values obtained from the GCM models over Lebanon. The HadCM2/HHGGax model predicts an average of 1.6 °C increase in temperature, and an equivalent average of about 3% less precipitation (Khawli 2001). If the averages of temperatures increased between 1 to 3 °C, the monthly averages of precipitation were slightly change conducting to a decrease in the availability of water resources between 6 and 12% of the volume of water resources (MoE/GEF/UNDP 2011). Lebanon, which was considered as the richest country in the Middle East in water resources is faced now with a shortage of water and we are close to the hydrological drought.

Water quantity would be reduced as the total amount of precipitation would be less, evaporation rate would increase, therefore, springs water yield and rivers flow would decrease as well. Because population is increasing and social attitude is becoming more consumptive, demand for water would also increase. Reduced water amounts and increased exploitation of less available water would affect by further deterioration in water quality.

6 Conclusions

Climate change is a global phenomenon that goes beyond national borders. It affects all economic sectors, weather patterns, causes intense drought, warming of oceans that can directly harm animals, and wreak havoc on people livelihoods and communities. It is an issue that requires coordinated effort and solutions across international borders, working together, to lessen the impact of climate change.

It's difficult to estimate with sufficient evidence the changes occurring seasonally. However, our analysis showed fluctuations in every parameter regarding the temperature and precipitation in the most seasons at all weather stations.

An increase in temperature around 2 °C was observed compared with the average of the low and high temperatures in all weather stations during the last 60 years. This increase which has been occurring especially in the primary 10 years of the century remains constant, and the signs of climatic warming phenomenon was confirmed in Lebanon.

Although it appears that precipitation was not predicted to decrease significantly, the temperature increases of 2 °C could impact the water balance and reduce the available water resources. Indeed, we have noted that the rainfall monthly averages were slightly decreased in all regions, and we recorded in the mountainous region a clear decrease in the amount of precipitation which is a direct consequence of the climate change influenced by the Mediterranean Sea.

It is evidenced that the water resources in Lebanon are influenced by the climate change phenomenon, the elevation of the temperature and the fluctuation of the precipitation indicates an uncertain environmental future and its impact on the country water resources and agriculture (World Bank 2018). Currently Lebanon does not have a reliable hydrological data collection stations throughout the country, and desperately in need of modern observation network on its 18 rivers to better collect future hydrological data.

Additionally, Lebanon needs a new strategy to confront climate change that includes a continuous review of management plans for the various sectors related to Water, Energy, Agriculture and the Environment.

7 Recommendations

Lebanon is affected by climate change, evidence of the past decades of recorded events of heat waves, drought conditions, and wildfires that are becoming more frequent, and with the increase in temperature during the summer with a temperature as high as 38 °C, at sea level, and as high as 40 °C in Bekaa. In addition, Lebanon is experiencing moderate temperatures during the winter season; the snow is melting earlier in spring, and a decreasing rainfall between January and February months. Based on this trend, in the coming decades the impact of the climate change is likely to further decrease the supply of available water, increase the risk of wildfires, and threaten the coastal development and ecosystems.

Climate change is happening and there are plenty of things Lebanon can do to mitigate its impact on the environment. The government can take action implementing policies change to reduce greenhouse gases and emissions from different sectors of the economy, the inclusion and participation of NGOs and local citizens to assist in the effort to reduce greenhouse gases. The most effective strategies are to limit or reduce the use of fossil fuels such as fuel oil to generate electricity, industrial pollutions, agricultural, and other sectors. The reduction of inefficient method of public transportation, the use of control technology that capture emissions at the source, and replacing power generation with renewable and cleaner sources of energy that cut the CO_2 emission. Climate change requires change and it starts with environmental education in school, general public awareness.

As mentioned earlier, the energy sector in Lebanon is the highest contributor of CO_2 emission. Lebanon needs to modernize its existing power plants; the replacement of fuel oil with natural gas that include best available technology that reduces GHG, and future investment in renewable energy sources such as solar, wind power, bio energy, hydropower and onshore wind.

8 Limitations of the Work

This study was completed with the available climatic data at the time; this is due to multiple factors, such as government priorities and crisis interruptions in the country and the lack of resources. We acknowledge it might not present itself with an accurate picture, we believe continuous effort from government, and NGOs and the private sectors should make a difference in addressing climate change in Lebanon. Although Lebanon is a small country, Lebanon would require more than ten metrological stations distributed in different climatic zones, to better collect reliable data that can be analyzed and understood in future studies. During the civil war between 1975 and 1990, the existing metrological stations were damaged or not functional and comprehensive data was not available, lost or misplaced. Despite all these challenges a great effort was taken with the available climate data for this study. Our desire is to continue our work in future studies related to climate change impacting Lebanon and the Middle East; we hope that future data would be better captured and recorded that would show a better picture on the effect of climate change on Lebanon and its environment.

Acknowledgements We express our profound gratitude to Mr. Marc Wehaibe, the chief of the Meteorology department at the Rafic Hariri International Airport in Beirut, for sharing data.

References

Allen MR et al (2018) Framing and context. In: IPCC special report: global warming of 1.5 °C. The Intergovernmental Panel on Climate Change, Geneva. Available at: https://www.ipcc.ch/sr15/

ESCWA-UN (2016) Climate change adaptation in the environment sector using Integrated Water Resources Management (IWRM) and Ecosystem-Based Management Tools. In: 2030 Agenda for Sustainable Development Workshop, Economic and Social Commission for Western Asia, UN building, Beirut, Lebanon, 22–24 March 2016. Available at: https://www.unescwa.org/events/workshop-climate-change-adaptation-environment-sector-using-integrated-water-resources

Halwani J (2008) Assessment of the water situation in Lebanon. In: The 3rd international conference on water resources and arid environments, King Saud University, Riyadh, Saudi Arabia, 16–19 November 2008. Available at: https://www.icwrae-psipw.org/images/stories/2008/Water/3.pdf

Halwani J (2009) Climate change and water resources in Lebanon. In: Climate change: global risks, challenges and decisions, IOP Publishing, IOP conference series: earth and environmental science 6 (2009), Copenhagen, Denmark p 292011. https://doi.org/10.1088/1755-1307/6/29/292011

Hakim B (1993) Question de L'Eau au Liban, Ressource et Besoins. In : Conference on "Water as an element of cooperation and development in the Middle East", Ankara, 4/8 Oct 1993

Khawlie MR (2001) The impacts of climate change on water resources of Lebanon—Eastern Mediterranean. In: Fondazione Eni Enrico Mattei, Note di Lavoro Series Index, Milano, Italy. https://doi.org/10.2139/ssrn.278513

Lionello P, Scarascia L (2018) The relation between climate change in the mediterranean region and global warming. Reg Environ Change 18:1481–1493. https://doi.org/10.1007/s10113-018-1290-1

Med ECC (2020) Climate and environmental change in the mediterranean basin—current situation and risks for the future. First Mediterranean Assessment Report. In: Cramer W, Guiot J, Marini K (eds) Union for the Mediterranean, Plan Bleu, UNEP/MAP, Marseille, France, p 600. Available at: https://www.medecc.org/first-mediterranean-assessment-report-mar1/

MoEW, UNDP (2014) Assessment of groundwater resources of Lebanon, Ministry of Energy and Water Beirut. Available at: https://www.undp.org/content/dam/lebanon/docs/Energy%20and%20Environment/Publications/Assessment%20of%20Groundwater%20Resources%20of%20Lebanon.pdf

MoE/GEF/UNDP (2011) Lebanon's second national communication to the United Nations framework convention on climate change, Ministry of Environment, Beirut. Available at: http://climatechange.moe.gov.lb/viewfile.aspx?id=19

MoE/UNDP/ECODIT (2011) State and trends of the Lebanese environment, Ministry of Environment, Beirut. Available at: http://www.undp.org.lb/communication/publications/downloads/SOER_en.pdf

Ministry of Agriculture (2003) National Action Programme to combat desertification, Beirut. Available at: https://knowledge.unccd.int/sites/default/files/naps/lebanon-eng2003.pdf

Shaban A (2020) Water resources of Lebanon. Springer International Publishing. Series: World Water Resources, p 219. https://doi.org/10.1007/978-3-030-48717-1

Tramblay Y, Llasat MC, Randin C, Coppola E (2020) Climate change impacts on water resources in the Mediterranean. Reg Environ Change 20:83. https://doi.org/10.1007/s10113-020-01665-y

United Nations (2015) Paris Agreement. United Nations Framework Convention on Climate Change. UNFCCC, New York, USA. Available at: https://unfccc.int/sites/default/files/english_paris_agreement.pdf

World Bank (2016a) World Development Indicators. Washington DC, USA. Available at: https://datatopics.worldbank.org/world-development-indicators/themes/environment.html#carbon-dioxide-emissions

World Bank (2016b) Carbon dioxide information analysis center, environmental sciences division, Oak Ridge National Laboratory, Tennessee, United States. Available at: https://data.worldbank.org/indicator/EN.ATM.CO2E.PC?locations=LB

World Bank (2018) Droughts and agriculture in Lebanon: causes, consequences, and risk management, Washington DC, USA. Available at: https://openknowledge.worldbank.org/handle/10986/30595
World Health Organization (2019) Safer water, better health, WHO, Geneva. Available at: https://www.who.int/publications/i/item/9789241516891

Enhancing Urban Microclimates Towards Climate-Resilient Cities: The Potential of Courtyards

Carmen Galán-Marín, Carlos Rivera-Gómez, Jesus Lizana, Jorge Roa-Fernández, Eduardo Diz-Mellado, and Victoria López-Cabeza

Abstract The increasingly urgent phenomenon of global warming has a critical epicentre in the topic of urban thermal comfort, which is significantly influenced by the urban heat island effect. In this built environment, creative thinking is required to shape pleasant, healthy, and sustainable microclimates, and not only urban planners and designers but also researchers and software developers are all involved in the search for feasible solutions, tools and opportunities. This chapter evaluates the potential use of one of the most dominant urban configurations in historic Mediterranean cities, the courtyard, as a promising thermal tempering solution to mitigate the impact of climate-related events. The methodology to evaluate the potential of the courtyard microclimate as a climate-responsive strategy follows a top-down approach. Firstly, the relevance of courtyards at city-scale is evaluated in two historic city centres in Spain, Seville and Cordoba. Secondly, six representative courtyards are characterised and monitored to evaluate their thermal benefits. Thirdly, alternatives to improve courtyard performance are discussed and tested in two scenarios. Finally, an urban CFD software to support efficient courtyard design is evaluated in case studies. The results show that this building configuration is highly representative of both historic urban contexts, with approximately 80% of existing plots having inner courtyards. Moreover, the monitored data demonstrates the potential thermal benefits of courtyard microclimates, which can reduce outdoor peak temperature from 6.8 up to 14.3 °C during the hottest days. The analyses show that courtyards with a height/width relation (aspect ratio) above 3 perform better, especially with additional shading devices to reduce solar gains. Finally, the study demonstrated the need to develop more procedures to accurately simulate the specific microclimate of these deep, small-scale spaces as a climate-resilient strategy for buildings and cities, to efficiently mitigate the impact of extreme heat wave events.

C. Galán-Marín (✉) · C. Rivera-Gómez · J. Lizana · J. Roa-Fernández · E. Diz-Mellado · V. López-Cabeza
Instituto Universitario de Arquitectura y Ciencias de La Construcción, Universidad de Sevilla, Avda. Reina Mercedes 2, 41012 Seville, Spain
e-mail: cgalan@us.es

W. Leal Filho and E. Manolas (eds.), *Climate Change in the Mediterranean and Middle Eastern Region*, Climate Change Management,
https://doi.org/10.1007/978-3-030-78566-6_20

1 Introduction

Human activity is estimated to have already caused approximately 1.0 °C of global warming above pre-industrial levels, which is likely to reach 1.5 °C between 2030 and 2052 (IPCC 2018). Changes in many extreme weather and climate events in Europe have been observed since them, and in particular after 2000, including heat waves, which now occur more often and last longer (IPCC 2014; European Environment Agency 2019). Moreover, the projected frequency of heat waves is greater for southern and south-eastern Europe, especially in the Mediterranean areas (Cramer et al. 2018, 2019), where the effects will be exacerbated in large cities due to the urban heat island effect (European Environment Agency 2019), whose magnitude ranges from 0.4 to 11 °C (Garshasbi et al. 2020). Such heat waves will reduce work productivity, and increase morbidity (e.g.: dehydration, heat stroke and heat exhaustion) and mortality (IPCC 2014; WMO and WHO 2015). Urgent actions to strengthen resilience to climate-related hazards in vulnerable hot-dry cities are required, in line with Sustainable Development Goal 13 (SDG13) defined by the (United Nations 2015). Climate-resilient pathways should be addressed from building-scale to urban planning in order to mitigate climate risks and improve the comfort and well-being of citizens (Garshasbi et al. 2020).

In this context, the tempering and mitigation potential of specific urban technologies began to receive greater attention in the last decade, in the development of climate-resilient urban solutions, including the production of highly reflective materials, cool pavements, water-based technologies, urban greenery or green roof technologies, among others (Akbari et al. 2016). However, most of these solutions focused on outdoor public spaces (Akbari et al. 2016), with few studies evaluating the mitigation potential of semi-outdoor spaces, or the inner courtyards of building blocks arranged as buffer spaces (Yang et al. 2020). Inner courtyards represent an important, representative feature of Mediterranean cities, especially in the case of historic city-centre layouts.

Courtyards are an efficient strategy to improve the thermal and microclimatic conditions of urban spaces (Zamani et al. 2018). Moreover, they play an essential role in the buildings´ environmental performance because they create specific microclimates that differ markedly from regular outdoor climatic conditions (Naboni and Havinga 2019). The courtyard acts as storage for cool air during the night when the temperature drops to minimal values. The cool air is then distributed around the indoor space during the hotter periods of the day (Hassan et al. 2016). The thermal performance of the courtyard also depends on proportion, orientation, geometry, opening characteristics and materials (Zamani et al. 2018). Deep courtyards perform better thermally than shallow courtyards in hot, dry areas (Johansson 2006). Temperatures are lower in and around the courtyards due to a decrease in sun exposure compared to the surrounding areas (Naboni and Havinga 2019). Furthermore, this effect can be improved if the courtyard is surrounded by a balcony or an arched portico, or by the presence of vegetation, water and shading devices (Aldawoud 2008; Zamani et al. 2018). In addition, the shaded areas provided by the courtyard to the building have a considerable impact on indoor cooling energy demand due to the reduction of solar gains (Cantón et al. 2014).

Recent studies have demonstrated the benefits of courtyard blocks compared to other block typologies. Natanian et al. 2019 applied an automated parametric analysis to evaluate the performance of different urban building configurations. The results highlighted the courtyard archetype as the most favourable urban building design in hot climates. Chi et al. 2020 demonstrated the potential use of a courtyard as a passive cooling and heating strategy to reduce energy consumption by 23%. Almumar 2019 showed how positioning the courtyard within the land plot so that main façades receive sun in the winter and block out most of it in the summer contributed to the natural heating and cooling of the house. Asfour 2020 evaluated the optimal configuration of courtyards and atriums to improve daylighting performance while reducing energy demand. The results showed how the most efficient, balanced alternative was the courtyard or atrium configuration, with a window-to-wall ratio of 30%, together with shading devices. (Guedouh et al. 2019) demonstrated that courtyard depth, opening ratio and orientation remain crucial elements in optimizing the balance between courtyards' thermal and luminous environments. Rodríguez-Algeciras et al. 2018 highlighted the aspect ratio (AR) as the most important parameter influencing courtyard thermal conditions.

However, the area required and the associated urban land cost are the major reasons for rejecting the internal courtyard technique, despite its beneficial function as a thermal regulator (Hassan et al. 2016). Further studies are still required to evaluate and highlight the importance of this vernacular building configuration, and to provide design guidelines and decision-support tools to enable correct implementation of these strategies to enhance the climate-resilient capacity of buildings and cities that will experience extreme climate-related events. Currently, no urban modelling tools exist to evaluate the benefits of these urban microclimates. Moreover, current building energy simulation (BES) tools do not consider the tempering potential of these buffer spaces, opting for historical weather data that fail to take into account the urban microclimate effects. New procedures are needed to evaluate the urban climate implications in building performance (Shen et al. 2019) and the potential thermal interactions of buildings with specific urban microclimates (Samuelson et al. 2020).

This paper evaluates the peak-temperature mitigation potential of traditional courtyards in two historic Spanish cities in the Mediterranean area, Seville and Cordoba. The aim is to quantify the benefits of these building typologies and define decision-support criteria for future urban planning in order to mitigate the impact of heat waves in hot climates. The relevance and impact of these urban microclimates are evaluated from city-scale to specific courtyard configuration and simulation in order to emphasize the function of these traditional spaces in historic urban contexts, quantify their benefits and identify the optimal courtyard configuration.

The chapter is structured as follows. First, the methodology is described. Material and methods for each assessment scale are detailed, and urban case studies are characterised. Second, the results are presented and discussed in five sections: courtyard relevance at city-scale; performance of representative courtyards; testing of improvement techniques; validation of simulation tools; and limitations of the study. Finally, future research activities to satisfy existing needs in the implementation of these urban microclimates are reported.

2 Methodology

The methodology to evaluate the potential of courtyards as a climate-responsive strategy to mitigate the impact of extreme heat events in cities consisted of a top-down approach. The evaluation procedure was divided into four steps, from city-scale to specific courtyard monitoring, testing and simulation. An overview of the methodology is illustrated in Fig. 1.

The methodological steps were: the evaluation of the relevance of courtyards at city-scale in two historic, non-coastal cities in southern Spain; the characterisation and monitoring of six representative courtyards to evaluate their thermal benefits; the testing of alternatives to improve courtyard performance in two case studies; and, the validation of existing urban CFD tools for courtyard simulation in two scenarios. The materials and methods used in each methodological step are detailed in Sect. 2.1. Finally, the urban contexts used as a case study are defined and characterised in Sect. 2.2.

2.1 Materials and Methods

In the first step, the existence and relevance of courtyards in the selected urban cities were evaluated using the QGIS geographic information system (QGIS 2017) according to the methodology defined by Rojas-Fernández et al. (2017). Data for urban context characterisation was collected from a Spanish government database capable of generating shapefile vector files (Gobierno de España 2017). These files

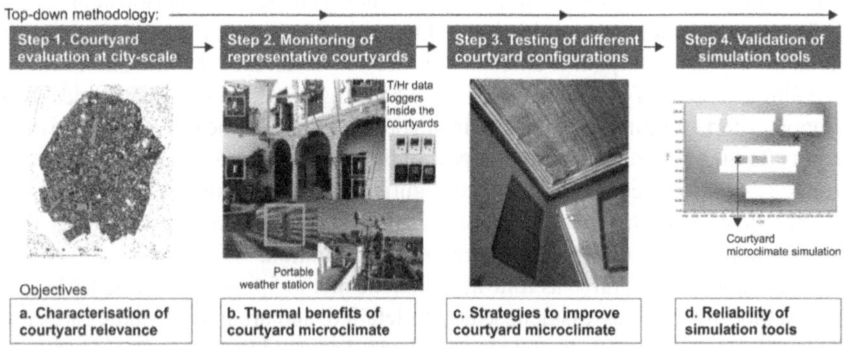

Fig. 1 Overview of the methodological top-down assessment

follow the infrastructure for spatial information provided by the European Union's INSPIRE directive (EU 2007). Then, the percentage of plots with courtyards in the urban context studied, the plot porosity index (PI) and net plot porosity index (NPI) were calculated. PI is the ratio between the void area and total plot area, considering all urban plots in the studied area according to Eq. (1). NPI provides the mean percentage of the courtyard surface using the mean ratio between void area and plot area, considering only the urban plots with courtyards according to Eq. (2). Finally, the number of courtyards per size and the most representative courtyard geometry were obtained through the mean ratio of height and width according to Eq. (3), namely the aspect ratio (AR). AR is an indicator that characterizes courtyard geometry, widely used in the literature, with another common parameter, the sky view factor (SVF), which shows the fraction of visible sky according to Eq. (4) (De Wolff 2008; Dirksen et al. 2019).

$$PI = \frac{\sum \text{Void area per plot}}{\sum \text{Total area of plot}} \tag{1}$$

where

PI Plot porosity index

$$\text{NPI} = \frac{\sum \text{Void area per plot}}{\sum \text{Total area of plot with courtyard}} \tag{2}$$

where

NPI Net plot porosity index

$$AR = \frac{H}{W} \tag{3}$$

where

AR Aspect ratio
H Courtyard height
W Courtyard width

$$SVF_{2D} = \cos\left(\arctan\left[\frac{H}{0.5W}\right]\right) \tag{4}$$

where

SVF Sky view factor

H Courtyard height
W Courtyard width.

In the second step, the performance of the most representative courtyards was evaluated during a one-week monitoring campaign, following the procedure reported by Rivera Gómez et al. 2019. A set of six courtyards was selected, all located inside buildings. Selection criteria were based on the most representative characteristics in terms of the courtyard area, geometry and surface albedo. The selected courtyards had AR values ranging between 0.8 and 7.1. The outdoor urban temperature in each urban context was recorded by a portable weather station (model PCEFWS) located on the roof of each building; and the courtyards' indoor temperature was monitored by data loggers protected from direct solar radiation (model TESTO 174 T) placed at different heights (1 and 2 m). The final courtyard temperature obtained was a mean value between both measurements. Different monitoring campaigns were carried out in the courtyards throughout the summer season with a minimum duration of one week each.

In the third step, the improvement potential of courtyard performance through the implementation of shading devices was tested in two case studies in Seville and Cordoba, following the criteria reported by (Lopez-Cabeza et al. 2020). Two courtyards were selected in Seville and Cordoba with an AR value of 0.72–0.96 and 1.57–1.46, respectively. Both courtyards have a similar wall finish, white mortar coating and some windows. The same shading device was installed in both cases, a black canvas mesh installed at the top of the courtyard, with some separation from the walls to allow ventilation. The fabric of the canvas cover was black polyethylene of 60–70% UV filter and 70 g/m^2 density. The equipment and procedure for data collection were similar to the previous campaign. Both courtyards were monitored with, and without, the shading devices during a full summer day, with peak temperatures at over 40 °C.

Finally, the effectiveness of powerful calculation software, based on computational fluid dynamics (CFD), to simulate the performance of these specific microclimates was evaluated in two courtyard scenarios. ENVI-met software was selected as the most suitable tool to evaluate the performance of outdoor spaces (Acero and Herranz-Pascual 2015; Ketterer and Matzarakis 2015). ENVI-met can simulate interactions between buildings, soil, vegetation and air. The effectiveness of this CFD tool in the prediction of the thermodynamic performance of courtyards was evaluated by comparing field data with simulated results in two case studies in Seville, following the procedure defined by (López-Cabeza et al. 2018). The selected courtyards had an AR value of 4.1 and 1.4. The equipment and procedure for data collection were similar to those previously defined in step 2. The case studies were modelled and calibrated through a manual iterative calibration process. Finally, different standardised statistical indices were calculated to ascertain the accuracy and reliability of the proposed tool for simulating the courtyard microclimate. The indices selected were Root Mean Square Error (RMSE), Systematic Root Mean Square Error (RMSEs), Unsystematic Root Mean Square Error (RMSEu) and the coefficient of determination

(R2), widely used in the literature (Coakley et al. 2014). Two temperature indicators were compared using these statistical indices in each case study: one inside the courtyard and another in the building surroundings.

2.2 *Study Areas*

Two historic urban contexts in Seville and Cordoba (Spain), illustrated in Fig. 2, were selected to evaluate the existence and performance of traditional Mediterranean courtyards. These two historic city centres are characterised by narrow streets of irregular shape, and building topologies based on traditional and vernacular architecture, influenced by the Roman, and later Muslim, historic periods. Their climates are similar, both are located in the valley of the Guadalquivir river, with relatively mild winters and very hot summers.

These scenarios were selected due to the marked presence of the traditional Mediterranean courtyard, and to the increase in heat wave events during last decade (AEMET 2019), which registered temperatures higher than those previously recorded for the 2003 heat wave that affected southern and central Europe, or for the 2010 heat wave that hit eastern Europe (European Environment Agency 2019). Moreover, the projected monthly mean temperature in the twenty-first century for these Andalusian regions is set to rise by up to 4.2 and 4.8 °C (Fig. 3), for Seville and Cordoba, respectively (Junta de Andalucía 2019), considering the A1B scenario of the IPCC's Fourth Assessment Report (IPCC 2007).

3 Results and Discussion

The results are presented as follows. First, the relevance of courtyards in the selected urban cities is shown. Second, the performance of the most representative courtyards

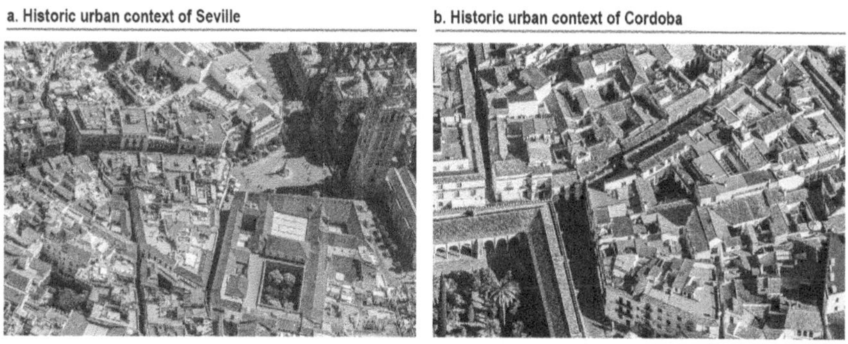

Fig. 2 **a** Historic urban context of Seville. **b** Historic city centre of Cordoba

is analysed. Third, strategies to enhance courtyard performance are evaluated. Fourth, courtyard performance is simulated using a CFD urban modelling tool, highlighting the needs and requirements for efficient courtyard installation for climate-resilient cities. Finally, limitations of the study are discussed.

3.1 The Relevance of Courtyards in the Two Mediterranean Cities

The results of the percentage of plots with courtyards and the total courtyard surface in the selected urban contexts are summarised in Table 1. The porosity index (PI) and net porosity index (NPI) were also calculated. PI provides the percentage of the void area considering all urban plots. NPI shows the percentage of the void area considering only plots with courtyards.

Courtyards are highly representative of both these historic city centres due to their Mediterranean tradition. Seville and Cordoba are characterised by a large number of plots with courtyards, exceeding 6000 and 3000 inner courtyards, respectively. The percentage of plots with a courtyard ranges between 77 and 82%. Moreover, the mean surface percentage of courtyards is higher in Cordoba (NPI: 19%) than in Seville (NPI: 16%).

Figure 4a illustrates the plot porosity index (PI) in both urban contexts, Fig. 4b is a zoomed picture of each urban context, and Fig. 4c shows the most typical size of courtyards in each historic city centre.

The historic urban context of Seville is characterised by a mean plot porosity of 12%, lower than the historic urban context of Cordoba, with 15%. The relevance of courtyards can easily be identified in the zoomed picture of each urban context (Fig. 4b), where each white inner square represents existing inner courtyards. The right bar graphs (Fig. 4c) also show that the most representative courtyard surfaces range between 5–10 m^2 and 10–15 m^2, whose mean AR values are 3.7 and 3.1 for Seville and Cordoba, respectively.

Table 1 Porosity and percentage of plots with courtyards in the selected urban contexts

	Seville	Cordoba
Total urban context size (m^2)	2,853,558	1,830,447
Total urban plots	8139	4625
Plots with courtyard	6263	3790
Percentage of plots with courtyards (%)	77%	82%
Total courtyard surface (m^2)	384,504	267,956
PI (%)	12%	15%
NPI (%)	16%	19%

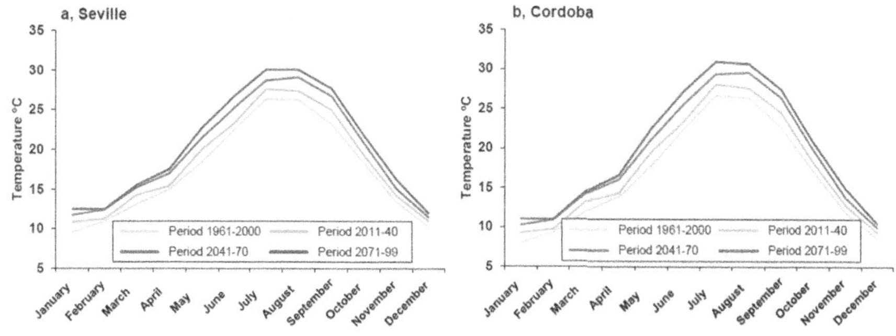

Fig. 3 Projected monthly mean temperature throughout the twenty-first century. Data collected from the (Junta de Andalucía 2019). **a** Region of Seville. **b** Region of Cordoba

3.2 The Benefits of Courtyard Microclimate as a Passive Cooling Solution

The performance of six representative courtyards was evaluated to assess their benefits as a passive cooling strategy. The schematic plans and sections of the selected case studies are illustrated in Fig. 5. Table 2 summarises the characteristics of the courtyards.

The selected case studies were monitored on different days, with a minimum of one week during summer, with maximum temperatures always higher than 40 °C. Figure 6 shows the evolution of maximum and minimum daily temperatures throughout the monitored period inside the courtyard compared to outside urban temperature oscillation.

The results show that maximum peak temperature reduction was obtained during hotter days. Seville courtyards reduced peak temperature by up to 14.3, 14.8 and 9.7 °C for CSA1, CSA2 and CSA3, respectively, during day 4, which had the maximum peak temperature of 42.8 °C. Cordoba courtyards reduced peak temperature by up to 7.3, 8.9 and 6.8 °C for CSB1, CSB2 and CSB3, respectively, during day 7.

The mean temperature reduction during the monitored period of the courtyards ranged between 7 and 10 °C for the Seville case studies, and between 4 and 7 °C for Cordoba. The better performance of the Seville courtyards seems to relate mainly to the higher AR value of these configurations (AR between 2.58 and 7.10) and worse outside climate conditions, which enhances the advantages of courtyards. The results demonstrate that deep courtyards, which have a high AR value, perform better thermally than shallow courtyards.

Courtyards are indicated as a key element in mitigating heat wave effects and improving the passive conditioning of buildings and cities since, for all the case studies analysed, the outdoor peak temperature decreased by a mean value of 7 °C during the week monitored, ranging from 6.8 to 14.8 °C. The data obtained highlight how these urban microclimate configurations can play an important role in diminishing the impact of heat waves on urban overheating and building performance. In

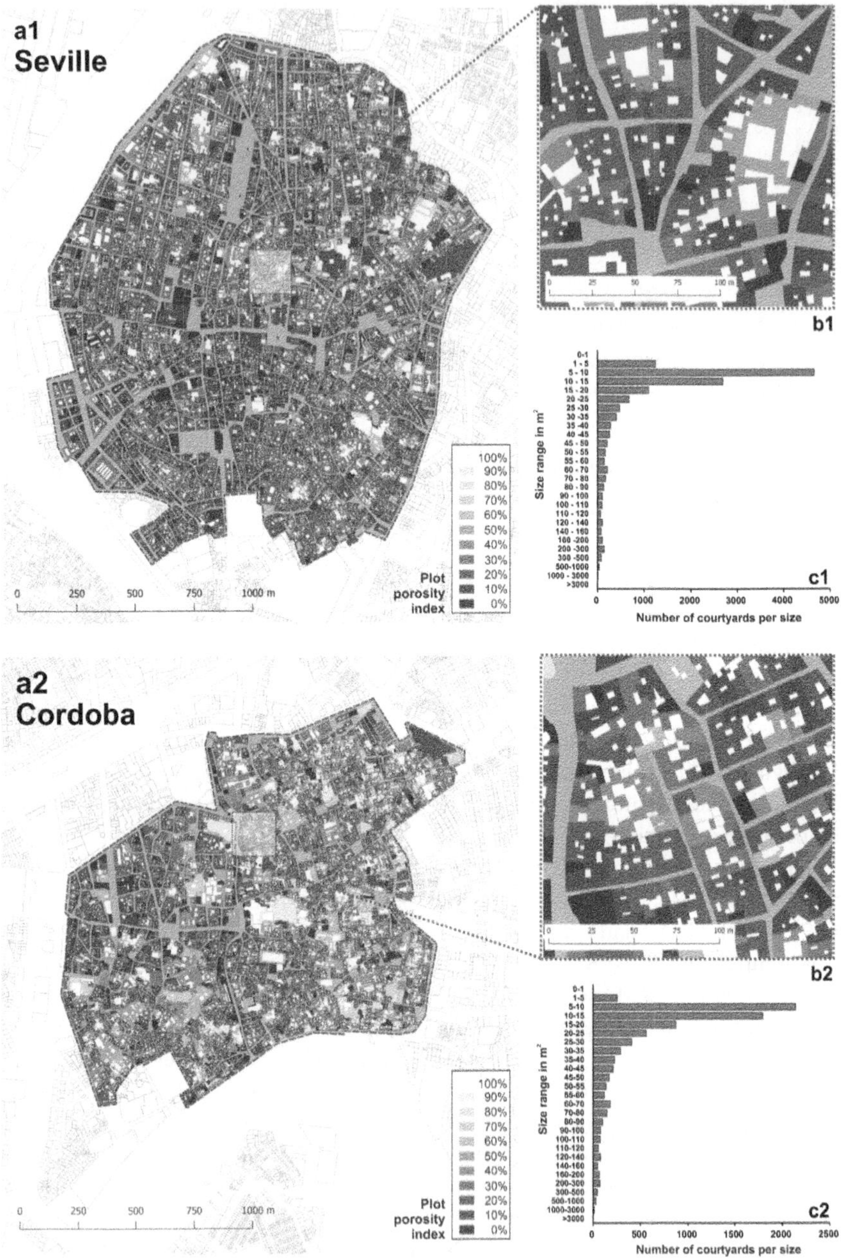

Fig. 4 Distribution and number of courtyards in the city centres of Seville and Cordoba. **a1-2** Illustration of the plot porosity index in the studied area, **b1-2** zoomed picture of each urban context, **c1-2** bar graphs illustrating the number of courtyards according to size

Table 2 Characterisation of the selected courtyards

Case study	Surface (m^2)	Dimensions (m)	Height (m)	AR I	AR II
a. Seville					
CS1A	6.4	3.2 × 2.0	14.2	4.44	7.10
CS2A	11.0	5.5 × 2.0	14.2	2.58	7.10
CS3A	12.3	3.,5 × 3.5	15.9	4.54	4.54
b. Cordoba					
CS1B	14.6	4.3 × 3.4	6.3	1.47	1.85
CS2B	65.5	8.4 × 7.8	6.8	0.81	0.87
CS3B	86.4	9.6 × 9.0	8.1	0.84	0.90

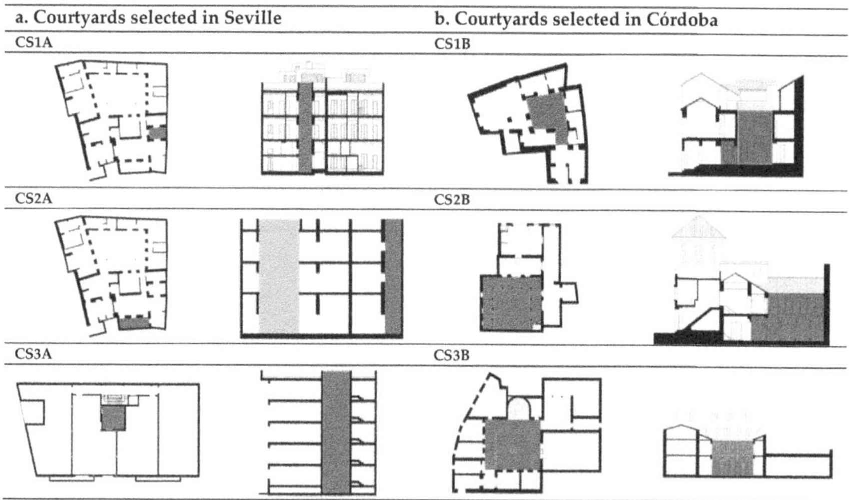

Fig. 5 Overview of the schematic plans and sections of the representative courtyards. **a** Case studies selected in Seville city centre. **b** Courtyards selected in Cordoba city centre

the next section, different solutions to improve the reported benefits are discussed and tested.

3.3 *Enhancing Courtyard Performance to Create Climate-Resilient Cities*

Many factors influence the thermal performance of courtyards, such as geometry and orientation, thermal mass, the albedo of the courtyard surfaces or the presence of vegetation, water and shading devices. This section evaluates the thermal benefits of shading devices to improve courtyard performance. Two case studies in Seville and

Cordoba were tested for one day during which peak temperatures exceeded 40 °C. The schematic plans and sections of the selected case studies are illustrated in Fig. 7. Table 3 summarises the characteristics of the courtyards.

Figure 8 shows the monitored results of both courtyards with, and without, the installation of shading devices.

Without shading devices, the courtyards show a reduction in outside peak temperature (in red) of 7 °C. The same peak temperature mitigation potential seems to relate to similar geometry and surface albedo in both scenarios. The difference between the maximum and minimum daily temperature, or daily thermal oscillation inside the

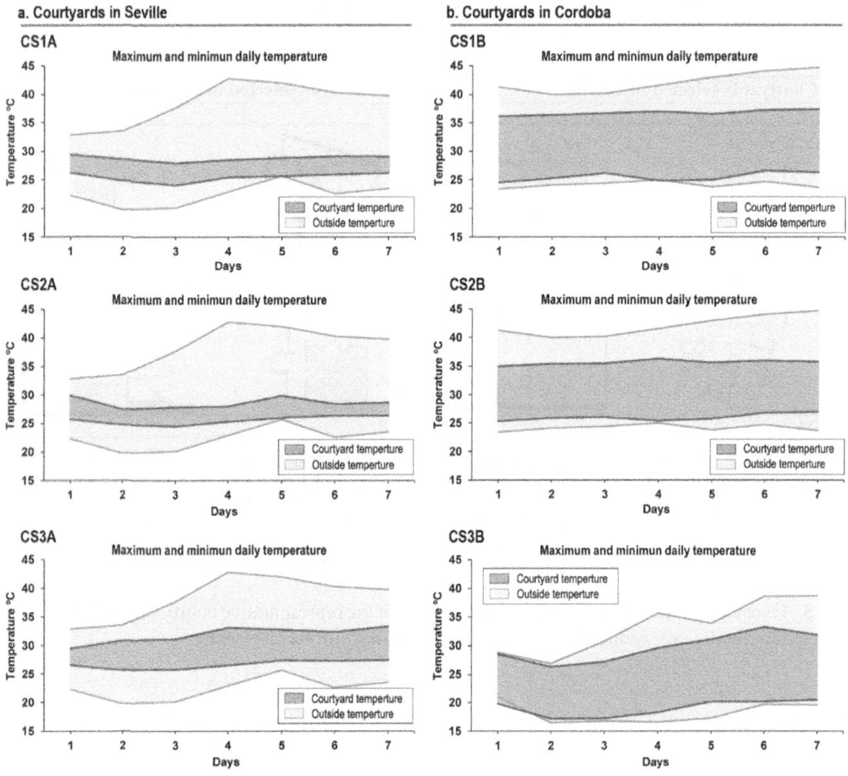

Fig. 6 Maximum and minimum daily temperatures of the courtyards monitored during one summer week, compared to outside urban temperature oscillation. **a** Case studies in Seville city centre. **b** Courtyards in Cordoba city centre

Table 3 Characterisation of the selected courtyards to evaluate the impact of shading devices

Case study	Surface (m^2)	Dimensions (m)	Height (m)	AR I	AR II
a. Seville	35.9	6.9 × 5.2	5	0.72	0.96
b. Cordoba	17.2	4.3 × 4.0	6.3	1.57	1.46

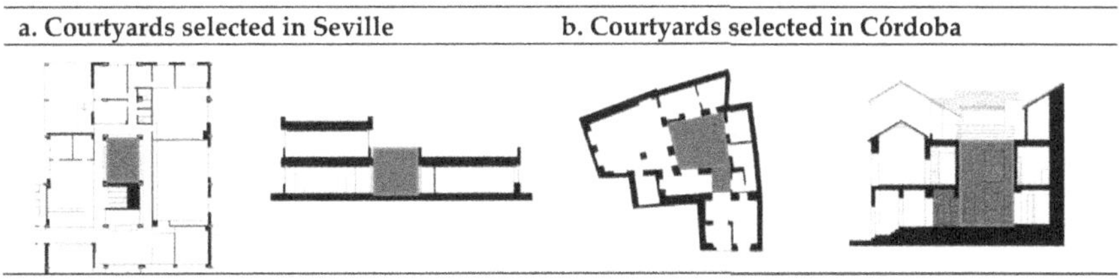

Fig. 7 Overview of the schematic plans and sections of the two case studies to evaluate the impact of shading devices

courtyards (in orange), is 13 and 12 °C, respectively. It can be observed that courtyard b1 performs better throughout the day, which can be explained by its deeper geometry (or higher mean AR value of approximately 1.5) that reduces solar radiation on the courtyard surfaces.

With shading devices, both courtyards show an increased reduction in outside peak temperature by up to 12–13 °C. The thermal oscillation in courtyard a2 is higher than in courtyard b2, which seems to be related to the lower mean AR value of approximately 0.84. However, in both cases the daily thermal oscillation is radically reduced, ranging between 4 °C and 8°, clearly improving the thermal benefits of these microclimates.

The results show the importance of mitigating solar gains inside these inner spaces in order to improve their thermal performance and reduce peak temperatures. It should be emphasized that both shading devices were installed with some separation from the perimeter to enable courtyard ventilation.

3.4 Need for New Approaches for Courtyard Microclimate Simulation

The effectiveness of ENVI-met software to simulate the thermodynamic performance of courtyards was evaluated in two case studies in Seville. The schematic plans and sections of the selected case studies are illustrated in Fig. 9. Table 4 summarises the characteristics of the courtyards.

In both scenarios, two datasets were measured and simulated during a single day, inside the courtyard and in the building surroundings. The results of the monitored data were compared with the simulated data in Fig. 10.

The monitored data in the selected courtyards show the same tempering effect previously reported. Courtyard 1 displays better performance than courtyard 2 since it is deeper and has a higher AR value, of up to 4.1, which prevents solar radiation from reaching the lowest levels. However, courtyard 1 shows a stronger overheating effect during the night due mainly to its depth, since the heat accumulated during the day does not escape easily, and the temperature of the courtyard rises during the night.

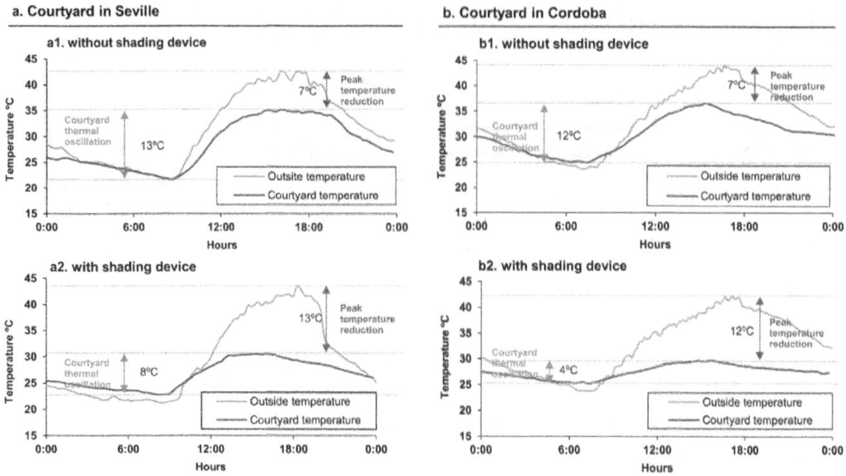

Fig. 8 Evaluation of shading techniques to improve the courtyard microclimate. **a** Courtyard performance with, and without, shading in Seville. **b** Courtyard performance with, and without, shading in Cordoba

Table 4 Characterisation of courtyards to evaluate a CFD urban modelling tool

Case study	Surface (m^2)	Dimensions (m)	Height (m)	AR I	AR II
Courtyard 1	22.9	7.4 × 3.1	12.6	1.70	4.06
Courtyard 2	99.0	7.5 × 13.2	10.7	1.43	0.81

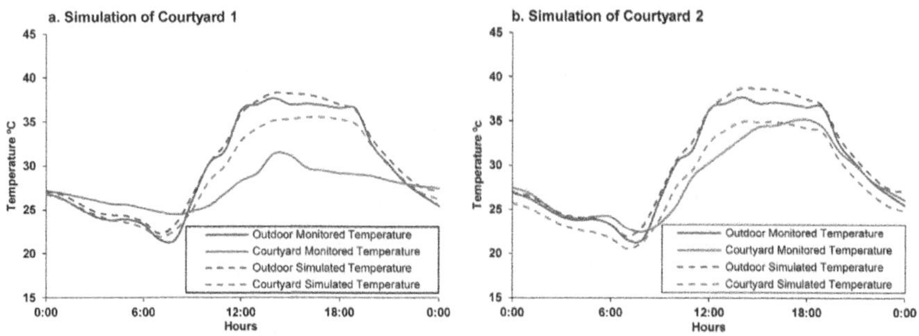

Fig. 9 Overview of the schematic plans and sections of the two case studies to evaluate a CFD urban modelling tool

Comparing the monitored and simulated data, provided in Fig. 10, it can be observed that the measured and simulated outdoor air temperatures have a similar pattern. However, the inner air temperature in courtyards is lower in monitoring than in simulation, above all in the first case, the deeper courtyard. Courtyard 1 shows the maximum divergence between monitored and simulated data, with a difference

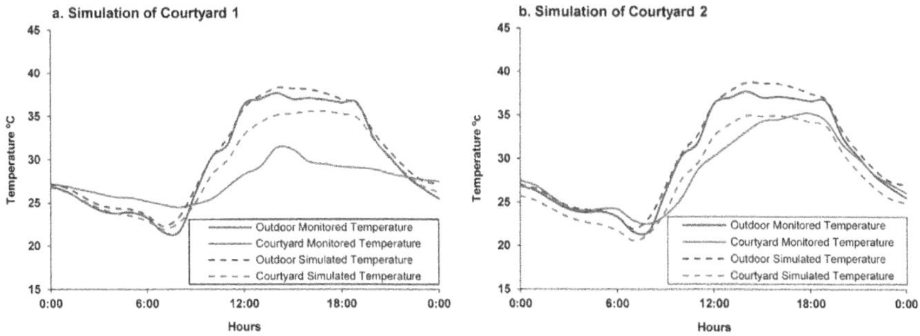

Fig. 10 Evaluation of ENVI-met software to simulate the courtyard microclimate

of up to 6 °C. In this case, ENVI-met was not able to reproduce the daytime courtyard performance. Moreover, in both scenarios, the software failed to consider the overheating effect of courtyards during the night.

The standardised statistical indices of the measured and simulated data using ENVI-met in both cases are reported in Table 5. Statistical indices were calculated for different measured and simulated points outside and inside the building block in order to determine the accuracy and reliability of the software for inner and outer urban microclimates.

The results show a highly accurate simulation of the outdoor temperatures in all models, obtaining a RMSE value of 0.77 and 0.82, and a R2 of 0.99 in both cases. In contrast, the values obtained for the air temperatures in the courtyards, despite displaying a R2 above 0.84, present a RMSE of 3.35 in the worst case. These results highlight the fact that the deeper the courtyard is, the worse the simulated performance obtained. Courtyard microclimates are characterised by small-scale outdoor urban spaces, which represent a limitation for simulation by urban CFD software. Thus, it is necessary to develop additional procedures to accurately evaluate the mitigation potential of peak temperature of these deep, small-scale traditional spaces.

Table 5 Standardised statistical indices of measured and simulated data using ENVI-met

	RMSE	RMSEu	RMSEs	R^2
Courtyard 1				
Outdoor	0.77	1.33	0.62	0.99
Courtyard	3.35	4.50	1.33	0.84
Courtyard 2				
Outdoor	0.82	1.41	0.66	0.99
Courtyard	1.52	2.10	0.88	0.93

3.5 Limitations of the Study

This research has the following limitations. First, the reported results and conclusions on the performance of courtyards are based on the assessment of nine selected representative case studies located in two Mediterranean cities throughout the summer season. Nevertheless, extensive courtyard monitoring of different scale studies and cities in Spain carried out in previous years shows similar results, which are widely detailed in (Rivera Gómez et al. 2019). Further studies with more scenarios and in different climate regions should be carried out to extend the reported conclusions to other climate areas. Second, this study focused on the temperature results at user height. Further research could use measurement points at different heights of the courtyards due to the fact that the microclimate impact of courtyards decreases according to height. Third, the performance of courtyards with and without shading devices was compared on days with peak temperatures higher than 40 °C. However, alternatives were monitored on different days, with different outside hourly temperature profiles, which may have influenced the reported absolute improvement values. And fourth, the simulation of courtyard performance using a CFD urban modelling tool was calibrated through a manual iterative calibration process, which limits optimal model achievement.

4 Conclusions

This paper has evaluated the thermal benefits of inner courtyards to mitigate peak-temperature periods during hot climate events. Courtyard microclimate performance was measured in two historic cities in Spain, Seville and Cordoba, through a top-down approach from city-scale to specific courtyard configuration and simulation in order to highlight the role of these traditional spaces in historic urban contexts, quantify their thermal benefits, identify the optimal courtyard configuration and support decision-making processes towards climate-resilient buildings and cities. Based on the results, it is possible to extract the following conclusions:

At city-scale in the two cities evaluated, Seville and Cordoba (Spain), courtyards are an important feature of both historic urban contexts. Approximately 80% of existing plots have inner courtyards in both cities, with a size ranging from 5 to 15 m^2 in the majority of cases. Moreover, in these cases, the most extended relation between height/width, the aspect ratio (AR), shows a mean value of 3.7 and 3.1 for Seville and Cordoba, respectively.

At courtyard-scale, the monitored results of the six representative courtyards evaluated in both cities highlight the temperature mitigation potential of these inner buffer spaces. Courtyard microclimates can reduce peak temperature from 6.8 °C up to 14.3 °C. The results also show that maximum peak-temperature reduction is obtained during hotter days. Thus, it is demonstrated that this building configuration acts as a powerful climate-resilient strategy to counter the effects of the projected and more frequent heat waves expected in southern and south-eastern Europe due to climate change, especially in the Mediterranean areas. The higher the outdoor temperature, the greater the climate benefit of the courtyard. Furthermore, considering all the courtyards evaluated, the AR value stands out as a representative parameter of the courtyard's thermal tempering potential. Deeper courtyards, with AR higher than 3, provide better thermal regulation than shallow courtyards.

With the aim of improving courtyard performance, this study also demonstrates the high impact of solar radiation on the courtyard microclimate. The use of shading devices boosted peak temperature mitigation during hotter days from 7 to 13 °C. Therefore, courtyard configurations that lower solar incidence will produce a better microclimate performance during hot days.

Regarding courtyard simulation tools, the results obtained confirm that existing urban CFD tools show poor accuracy in simulating the courtyard microclimate. It is demonstrated that the deeper the courtyard, the worse the simulated performance obtained. Courtyards are characterised by small-scale outdoor urban spaces, with specific boundary conditions, which limit simulation by CFD urban modelling software. Thus, additional procedures are needed to accurately evaluate the design of these thermal buffer strategies to mitigate the impact of extreme heat wave events on buildings and cities.

Finally, it can be concluded that the courtyard microclimate is a key urban strategy for mitigating the impact of extreme heat events in cities and buildings, and it could be optimised by a range of strategies to enhance its thermal buffer potential. However, further studies are required to support the efficient implementation of these urban configurations in cities. New research will be carried out to accurately predict courtyard performance and evaluate the impact of these buffer spaces on building performance, with the aim of supporting the decision-making process in the design of these promising spaces towards climate-resilient buildings and cities.

Acknowledgements This work was supported as part of the ROAD Project (RTI2018-093521-B-C33) funded by the Spanish government. The authors gratefully acknowledge the financial support provided by the IUACC and University of Seville, via the internationalisation grant within the VIPPI-US.

References

Acero JA, Herranz-Pascual K (2015) A comparison of thermal comfort conditions in four urban spaces by means of measurements and modelling techniques. Build Environ 93:245–257. https://doi.org/10.1016/j.buildenv.2015.06.028

AEMET (2019) Olas de calor en España desde 1975. Área de Climatología y Aplicaciones Operativas (AEMET)

Akbari H, Cartalis C, Kolokotsa D et al (2016) Local climate change and urban heat island mitigation techniques—the state of the art. J Civ Eng Manag 22:1–16. https://doi.org/10.3846/13923730.2015.1111934

Aldawoud A (2008) Thermal performance of courtyard buildings. Energy Build 40:906–910. https://doi.org/10.1016/j.enbuild.2007.07.007

Almumar MMS (2019) Understanding the environmental performance of the Iraqi traditional courtyard house, Is there an order of façades orientation in randomly oriented land plots? J Build Eng 22:140–146. https://doi.org/10.1016/j.jobe.2018.12.004

Asfour OS (2020) A comparison between the daylighting and energy performance of courtyard and atrium buildings considering the hot climate of Saudi Arabia. J Build Eng 30:101299. https://doi.org/10.1016/j.jobe.2020.101299

Cantón MA, Ganem C, Barea G, Llano JF (2014) Courtyards as a passive strategy in semi dry areas. Assessment of summer energy and thermal conditions in a refurbished school building. Renew Energy 69:437–446. https://doi.org/10.1016/j.renene.2014.03.065

Chi F, Xu L, Peng C (2020) Integration of completely passive cooling and heating systems with daylighting function into courtyard building towards energy saving. Appl Energy 266:114865. https://doi.org/10.1016/j.apenergy.2020.114865

Coakley D, Raftery P, Keane M (2014) A review of methods to match building energy simulation models to measured data. Renew Sustain Energy Rev 37:123–141. https://doi.org/10.1016/j.rser.2014.05.007

Cramer W, Guiot J, Fader M et al (2018) Climate change and interconnected risks to sustainable development in the Mediterranean. Nat Clim Chang 8:972–980. https://doi.org/10.1038/s41558-018-0299-2

Cramer W, Guiot J, Marini K (2019) Risks associated to climate and environmental changes in the Mediterranean region. A preliminary assessment by the MedECC Network Science-policy interface (MedECC)

De Wolff R (2008) Developing an environmental fog potential map using a GIS. In: Technical Report July, KNMI. De Bilt

Dirksen M, Ronda RJ, Theeuwes NE, Pagani GA (2019) Sky view factor calculations and its application in urban heat island studies. Urban Clim 30:100498. https://doi.org/10.1016/j.uclim.2019.100498

EU (2007) Directive 2007/2/EC of The European Parliament and of The Council of 14 March 2007 establishing an Infrastructure for Spatial Information in the European Community (INSPIRE). Available at: http://inspire.ec.europa.eu/legislation-details/directive-20072ec-european-parliament-and-council. Accessed 14 Jun 2017

European Environment Agency (2019) The European environment: State and Outlook 2015. Knowledge for transition to a sustainable Europe

Garshasbi S, Haddad S, Paolini R et al (2020) Urban mitigation and building adaptation to minimize the future cooling energy needs. Sol Energy 204:708–719. https://doi.org/10.1016/j.solener.2020.04.089

Gobierno de España (2017) Servicios INSPIRE de Cartografía Catastral. Dirección General del Catastro. Available at: http://www.catastro.minhap.gob.es/webinspire/index.html. Accessed 14 May 2017

Guedouh MS, Zemmouri N, Hanafi A, Qaoud R (2019) Passive strategy based on courtyard building morphology impact on thermal and luminous environments in hot and arid region. Energy Procedia 157:435–442. https://doi.org/10.1016/j.egypro.2018.11.208

Hassan AM, Lee H, Oh S (2016) Challenges of passive cooling techniques in buildings: A critical review for identifying the resilient technique. J Teknol 78:149–162. https://doi.org/10.11113/jt.v78.5748

IPCC (2018) Global Warming of 1.5 °C. An IPCC Special Report on the impacts of global warming of 1.5°C above pre-industrial levels and related global greenhouse gas emission pathways, in the context of strengthening the global response to the threat of climate change,. IPCC, Geneva, Switzerland

IPCC (2014) Climate Change 2014: Synthesis Report, Contribution of working Groups I, II and III to the Fifth Assessment Report of the Intergovernmental Panel on Climate Change. IPCC, Geneva, Switzerland

IPCC (2007) Climate Change 2007: Synthesis Report, Contribution of Working Groups I, II and III to the Fourth Assessment Report of the Intergovernmental Panel on Climate Change. IPCC, Geneva, Switzerland

Johansson E (2006) Influence of urban geometry on outdoor thermal comfort in a hot dry climate: a study in fez, Morocco. Build Environ 41:1326–1338. https://doi.org/10.1016/j.buildenv.2005.05.022

Junta de Andalucía (2019) Evolución de las principales variables climáticas actualizadas al 4° Informe del IPCC. Junta de Andalucía

Ketterer C, Matzarakis A (2015) Comparison of different methods for the assessment of the urban heat island in Stuttgart, Germany. Int J Biometeorol 59:1299–1309. https://doi.org/10.1007/s00484-014-0940-3

Lopez-Cabeza, Galán-Marín, Rivera-Gómezs (2020) Thermodynamic performance enhancement of courtyards using a shading device. Proceedings 38:17. https://doi.org/10.3390/proceedings2019038017

López-Cabeza VP, Galán-Marín C, Rivera-Gómez C, Roa-Fernández J (2018) Courtyard microclimate ENVI-met outputs deviation from the experimental data. Build Environ 144:129–141. https://doi.org/10.1016/j.buildenv.2018.08.013

Naboni E, Havinga L (2019) Regenerative design in digital practice. a handbook for the built environment. Eurac, Bolzano, IT

Natanian J, Aleksandrowicz O, Auer T (2019) A parametric approach to optimizing urban form, energy balance and environmental quality: the case of Mediterranean districts. Appl Energy 254.https://doi.org/10.1016/j.apenergy.2019.113637

QGIS (2017) Source Geographic Information System QGIS 2.18. Available at: https://www.qgis.org/en/site/. Accessed 14 May 2017

Rivera Gómez C, M. Diz Mellado E, Galán Marín C, López Cabeza V (2019) Tempering potential-based evaluation of the courtyard microclimate as a combined function of aspect ratio and outdoor temperature. Sustain Cities Soc 51:101740. https://doi.org/10.1016/j.scs.2019.101740

Rodríguez-Algeciras J, Tablada A, Chaos-Yeras M, et al (2018) Influence of aspect ratio and orientation on large courtyard thermal conditions in the historical centre of Camagüey-Cuba. Renew Energy 125:840–856. https://doi.org/10.1016/j.renene.2018.01.082

Rojas-Fernández J, Galán-Marín C, Roa-Fernández J, Rivera-Gómez C (2017) Correlations between GIS-based urban building densification analysis and climate guidelines for Mediterranean courtyards. Sustain 9. https://doi.org/10.3390/su9122255

Samuelson HW, Baniassadi A, Gonzalez PI (2020) Beyond energy savings: investigating the co-benefits of heat resilient architecture. Energy. https://doi.org/10.1016/j.energy.2020.117886

Shen P, Braham W, Yi Y (2019) The feasibility and importance of considering climate change impacts in building retrofit analysis. Appl Energy 233–234:254–270. https://doi.org/10.1016/j.apenergy.2018.10.041

United Nations (2015) Sustainable development goal 13. Take urgent action to combat climate change and its impacts (SDG13). Available at: https://sustainabledevelopment.un.org/sdg13. Accessed 7 Sep 2020

WMO and WHO (2015) Heatwaves and health: guidance on warning-system development. WMO and WHO

Yang L, Fu R, He W et al (2020) Adaptive thermal comfort and climate responsive building design strategies in dry–hot and dry–cold areas: case study in Turpan, China. Energy Build 209. https://doi.org/10.1016/j.enbuild.2019.109678

Zamani Z, Heidari S, Hanachi P (2018) Reviewing the thermal and microclimatic function of courtyards. Renew Sustain Energy Rev 93:580–595. https://doi.org/10.1016/j.rser.2018.05.055

Sustainable Land Management for Rural Adaptation in the Mediterranean and Middle Eastern Watersheds

Itxaso Ruiz and María José Sanz

Abstract The Mediterranean Basin and the Middle Eastern Region have witnessed the transformation of their landscapes due to climate change and anthropogenic land-use. Given their geography, land-use changes at the local scale have induced perturbations that, beyond affecting the local-to-regional climate, may have propagated to larger scales. While some of these land-use practices have intensified desertification, others have fostered lasting and sustainable landscapes. Learning how past land-use has affected the regional environment provides insight into shaping holistic measures aimed to combat climate change. With this work, we highlight sustainable land-use practices that assist in mitigating environmental degradation and adapting to current and future climate change in the Mediterranean Basin and the Middle Eastern Region. For that, we have used an integrated assessment of the ecosystem services and the social functions of 141 sustainable land management practices clustered in 23 groups. The assessment indicates that the more successful practices are those that can be mainstreamed on their own, yet their combination facilitates the integrated management of full watersheds. This type of study is relevant in the context of decision-making on land management, as policies at the local and regional levels can benefit from interdisciplinary science and the knowledge of land-users. Moreover, past (unsustainable) environmental impacts resulting from land-use practices are taken into account in the design of future adaptation strategies, thus avoiding greater impacts.

1 Introduction

The Middle East Region and the eastern Mediterranean Basin are home to some of the oldest human settlements in the world (Goring-Morris and Belfer-Cohen 2011).

I. Ruiz (✉) · M. J. Sanz
Basque Centre for Climate Change (BC3), Leioa, Spain
e-mail: itxaso.ruiz@bc3research.org

M. J. Sanz
Ikerbasque – Basque Foundation for Science, 48009 Bilbao, Spain

W. Leal Filho and E. Manolas (eds.), *Climate Change in the Mediterranean and Middle Eastern Region*, Climate Change Management,
https://doi.org/10.1007/978-3-030-78566-6_21

Both anthropogenic land-use and natural climate change fostered the opening and the transformation of their countryside with consequences beyond the biophysical properties of the landscape that affected the regional climate (Ruiz and Sanz 2020). For the past millennia, enlarged societies, commerce, and land-use capacities have further aggravated the environmental challenges of these regions, strongly impacting their water cycles (Pausas and Millán 2019). Thus, nowadays, these regions face water challenges, intensified desertification, biodiversity loss, and acuter climate change (Mirzabaev et al. 2019). We find evidence of intensified climate change in the temperatures of the Mediterranean and Red Sea's surface waters, which have raised ~ 0.4° C/decade for the past recent decades in comparison to the global ~ 0.1 °C/decade (Chaidez et al. 2017; Macias et al. 2013). Additionally, summer precipitation patterns over the past 30–40 years, likewise indicate a downward trend in many locations in the Mediterranean Basin, suggesting that summers are becoming drier (Deitch et al. 2017).

Landscape impacts derived from historical agricultural practices in these regions have been cumulative over time (Roberts et al. 2019) and up to some extent, agriculture has been a key driver of regional landscape transformation (Ruiz et al. 2020). Nonetheless, beyond historical over-exploitation of resources and intensive land-use, many agricultural and land management practices traditionally implemented have also promoted lasting and sustainable landscapes. Examples of those are the millennial multi-functional landscapes of the Mediterranean Basin (Mercuri et al. 2019) and the storage and resource diversification of the Neolithic populations in the Eastern Middle Region (Flohr et al. 2016). In this line, recent studies point at sustainable land management practices as key tools to increase the resilience of socio-ecological systems (e.g. Marques et al. 2016; Sanz et al. 2017), thus placing them at the heart of rural adaptation.

Acknowledging the importance of choosing the best portfolios of land management practices to address climate change and other environmental challenges in the Mediterranean and the Middle East Regions, we review traditional agricultural and land management practices that have historically promoted lasting and sustainable landscapes. These, respond to traditional knowledge and assist in both mitigating environmental degradation and adapting to current and future climate change conditions. Below we also introduce the issue of feedbacks between climate, land and water, that need to be understood and taken into account while considering sustainable land management options with the view to achieve multiple services, such as water regulation, climate adaptation and mitigation.

2 The Mediterranean Basin: Land and Water Feedbacks

The Mediterranean Basin has suffered from the transformation of its landscapes with consequences to the regional climate. Therefore, it is important to understand the relations between the land-use change and the climate in the region and beyond, to inform on land restoration and climate adaptation policies and measures. Let´s take the example of forests. Several publications from different regions conclude that the impact of reforestation and afforestation on the water yield is negative (Calder 2002; Jackson et al. 2005; Malmer et al. 2010) while others argue that planting additional forests can raise downstream water availability and intensify the hydrologic cycle (Millán et al. 2005; Millán 2014; Sheil and Murdiyarso 2009). This last group of studies includes catchment investigations that provide evidence for less local water availability after deforestation, while studies at larger scales suggest increasing levels of forest cover should raise precipitation levels and river runoff. In any case, progressive deforestation and wetland destruction have direct implications for the regionals and global hydrologic cycles (Ellison et al. 2012; Millán 2007). Thus, overall, changing land-use patterns, both in the developing and in the developed world, should be subject to increased scrutiny.

The Mediterranean Basin is a deep sea surrounded by mountains in the subtropical latitudes, making it an ideal test ground for checking land cover and climate feedbacks. It is a small-scale coupled atmosphere–ocean system with larger sensitivity to climate change, which results in amplified climate signals of a more immediate response. Several feedback loops between land-use perturbations in the western Mediterranean Basin and the climatic hydrological system from the local through the regional to the global scales were described by Millán (2007, 2014) (Fig. 1). In Fig. 1, the local loop involves sea breezes over the coastal mountain ranges (Millán et al. 2005). The regional loop influences the evolution of the sea surface temperature in the western Basin during the summer; this warm(er) water then feeds torrential rains in the autumn, and more recently, also in winter and spring, and also in central Europe (Gangoiti et al. 2011; Palau et al. 2017). Finally, the Atlantic-global loop

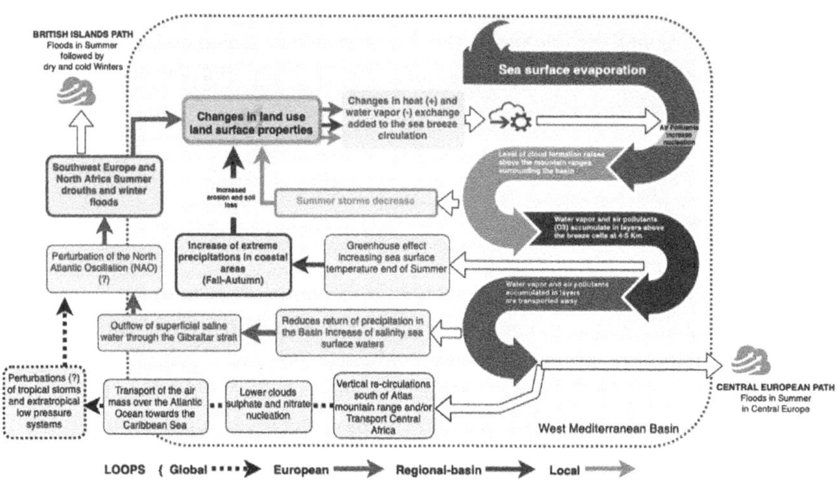

Fig. 1 Feedback loops between land-use perturbations in the western Mediterranean Basin and the climatic-hydrological system from the local through the regional to the global scales. Modified from Millan (2007) and Millan (2014)

has two components which can affect the North Atlantic Oscillation (NAO): the output of saltier water to the Atlantic and possible perturbations to the extra-tropical depressions and hurricanes in the Gulf of Mexico that are generated by changing the characteristics of the Saharan dust transported across the Atlantic (Millán 2007; Prospero and Lamb 2003).

Palau et al. (2017) indicate that the annual evolution and spatial distribution of the Total Precipitable Water (TPW) vapor column over the Mediterranean Sea is seasonally dependent and different on the western and the eastern Basins. The authors also indicate that since there is no correlation either with the evaporation from the sea surface or with the sea surface temperature, solar radiation alone does not explain the annual variability and the spatial heterogeneities of the observed TPW column over the Mediterranean Sea. This fact seems to confirm that the TPW is also driven by mesoscale processes that prevail in the region, for example, the daily cycles of air masses and the poor ventilation conditions that lead accumulation processes of air masses over the Mediterranean Sea and coastal areas (Millán et al. 2000).

Figure 1 shows that due to feedback processes, the local-to-regional perturbations initiated by land-use changes at the local level in the Mediterranean Basin may propagate to the global climate system (Millan 2014). Given the global large expectations on the role of land in mitigating climate change (Roe et al. 2019), particular attention should be made to the atmosphere-land-water interactions of the Mediterranean Basin, as water resources are one of the most threatened ecosystem services in the region. Carbon sequestration, instead, could be seen as a co-benefit of land management practices that focus on the water cycle preservation and recovery, which is of particular importance given the climate change threats to the Mediterranean Basin (IPCC 2014). Cramer et al. (2018) in a recent review of existing research on climate change, found that the recent accelerated climate change has exacerbated existing environmental problems in the Mediterranean Basin, which are caused by the combination of changes in land-use, increasing pollution, and decreasing biodiversity.

3 Methods

Learning from traditional knowledge of land management can help to shape future actions on mitigating environmental degradation and adapting to current and future climate change. To this purpose, we have evaluated 141 Sustainable Land Management practices (SLM) implemented in the rural areas of the Middle East Region and the Mediterranean Basin. The SLM practices have been collected from the World Overview of Conservation Approaches and Technologies network (WOCAT 1992) and are evaluated using the multi-objective assessment (multiple ecosystem services) described in Ruiz et al (2020).

The WOCAT network allows for a direct comparison across the full range of SLM practices and technologies while offers both environmental and social impacts for each SLM practice. However, by solely using this database, our assessment is limited to the availability of SLM actions implemented under the WOCAT network, thus constraining results. Up to 2020, there have been 141 SLM actions implemented within the Mediterranean Basin and Middle East Regions (Fig. 2). These regions include humid-subtropical, Mediterranean, semi-arid, and desert climates, according to the Köppen-Geiger climate classification from observed temperature and precipitation data for the period 1901–1925 (Rubel and Kottek 2010).

We have evaluated which SLM actions might assist best to tackle these region's climate change associated challenges. Deciding which rural practices to pursue requires assessing the benefits and the possible trade-offs both to the environment and society. To do so, we have used the assessment framework developed by Ruiz et al. (2020). that evaluates each practice's potential in assisting five ecosystem services (i.e. Climate Regulation; Soil erosion control; Biodiversity enhancement and pest/disease control; Water regulation; Soil quality enhancement), and four social services (i.e. Economy and production; Management and irrigation; Human well-being; Institutions). Since SLM practices are designed and implemented at the local scale, we have grouped similar ones to facilitated its upscaling, e.g. 'Controlled grazing', 'Rotational grazing', and 'Rangelands resting' have been grouped into 'Eco-graze' (see Box 1). From this process, we have obtained a total of 23 SLM practices out of the 141 initial ones. Following the assessment from Ruiz et al.

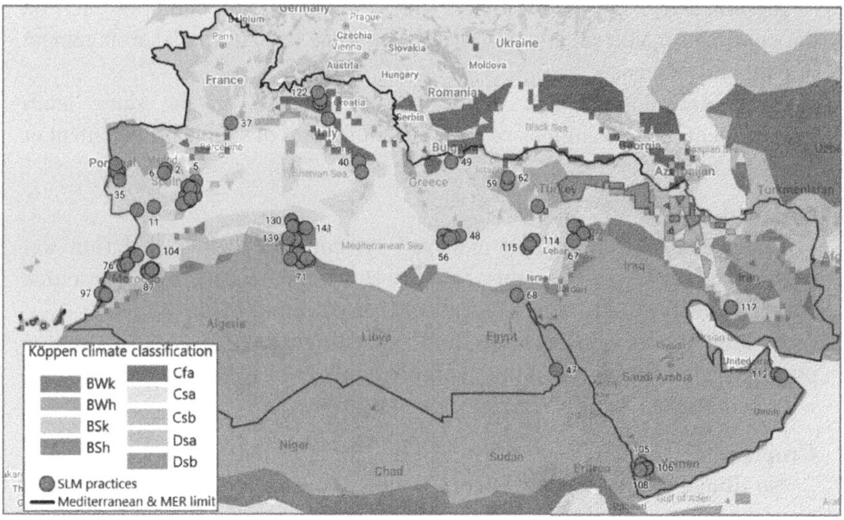

Fig. 2 Geographic location of the 141 SLM practices (blue dots) implemented within the Mediterranean Basin and Middle East Region (black line) under the WOCAT framework. Colours correspond to the Köppen-Geiger climate classification. BWk: Cold desert climate; BWh: Hot desert climate; BSk: Cold semi-arid climate; BSh: Hot semi-arid climate; Cfa: Humid subtropical climate; Csa: Hot-summer Mediterranean climate; Csb: Warm-summer Mediterranean climate; Dsa: Mediterranean-influenced hot-summer humid continental climate; Dsb: Mediterranean-influenced warm-summer humid continental climate. Data source for the Köppen-Geiger climate classification in Rubel and Kottek (2010)

(2020), we have then combined redundant impacts, e.g. 'Invasive alien species' with 'Pest/disease control', and filtered those SLM practices that did not assist with an averaged > 1.25 (from −3 to + 3) to any of the 9 services considered. Finally, we have calculated the mean score of each SLM practice to each function (Table 1).

Box 1 Description of the 23 SLM practices

Forestland

Afforestation: Native trees, shrubs, and grasses planted through participatory action

Reforestation: Establishment of new forest areas in formerly (less than 50 years according to UNFCCC, 2002) deforested lands through sustainable/natural regeneration

Control of wildfires: Includes: (1) prevent forest fire from occurring; (2) extinguish forest fires rapidly while they are still small; (3) use fire only for certain purposes and on a limited scale

Establishment of protected forest areas: Promotion of natural and national parks, protection of forest in reserves, and control of anthropogenic disturbances

Mixedland

Agroforestry: Woody perennials (trees, shrubs, palms, bamboos, etc.) on the same land-management units as crops, in some form of spatial arrangement or temporal sequence

Agro-silvio-pastoralism: Fodder shrubs planted on the same land-management units as animals or crops, in some form of spatial arrangement or temporal sequence

Grazingland

Eco-graze: Rotate grazing managed systems, let them rest during wet seasons and get the right balance between stock numbers and forage resource

Area closure to grazing: Prevent cattle from eating fodder

Fodder plants conservation and reseeding: Techniques such as tree pruning or fodder species pitting and reseeding used to restore grazinglands with an advanced and irreversible state of degradation

Cropland

No tillage and reduced tillage: Growing crops (or pastures) in semi-arid environments with a minimum disturbance of the soil

Use of amendments and fertilizers: Use of compost, straw pen manure, rocks, and/or sand to enhance productivity

Green cover crops: Perennial grasses in orchards and woody crops (vineyards, almond and olive plantations) between rows to permanently cover the soil

Crop rotation / intercropping: Successive cultivation of different crops in a specified order on the same field / Growing two or more crops on the same land simultaneously in a given growing season

Terracing: Construction of either earth-banked terraces by carefully removing a superficial soil layer from one part of a field and concentrating it on the lower end to reduce slope gradient and length; or soil/stone bunds across slopes and following contours

Salinity drop: Growing certain vegetables/cereals, using fertilizers, and/or flooding of the cropping field

Water Bodies

Micro-irrigation systems: Use of systems such as drip irrigation (delivering small amounts of water directly to the plants through pipes), or micro-catchments (water harvesting systems collecting the runoff from hillslopes and rainfall through micro-depressions within a field)

Water harvesting and storage: Water reservoir to allow off-season irrigation

Water harvesting and transport: Water slowdown and/or collection from the runoff of hillslopes and local streams

Several Land-Uses

Vegetation against soil erosion: Plantation of native woody species, strips and/or trees to control wind and water erosion and mitigate landscape degradation

Vegetation against pollution: Introduction of plants to control and/or immobilize pollution

Mulching: In croplands, mulching involves spreading crops waste after harvesting. In forestlands and after forest fires, slash mulch is spread immediately after a wildfire to prevent soil erosion and reduce overland flow

Other Manual and Technological SLM Practices

Variable rate biosolids management: Promote precision farming by optimizing organic amendment inputs in croplands

4 Asessment of Sustainable Land Practices, Ecosystem Services, and Social Functions

Table 1 shows the results of the integrated assessment. All SLM practices score high (i.e. > 1.8 from the −3 to + 3 range) in at least one of the functions, indicating their suitability for a particular purpose, e.g. *Mulching* scores 2.17 in the function 'Soil erosion control' while shows only shy-positive results in the rest of functions (i.e.

Table 1 Results from the integrated assessment of the 23 SLM practices on their ecological and social functions

Sustainable Land Management (SLM) practices	ECOLOGICAL FUNCTIONS: Water regulation	Soil erosion control	Soil quality enhancement	Biodiversity enhancement and Pest/diseas control	SOCIAL FUNCTIONS: Climate regulation	Economy and production	Human well-being	Management and irrigation	Institutions
Afforestation	1.40	1.50	1.67	1.63	1.17	1.17	1.57	0.00	-
Reforestation	**2.29**	**2.09**	1.78	1.14	**2.10**	1.42	1.47	0.67	1.00
Control of wildfires	0.00	0.36	0.09	1.55	1.50	1.21	1.36	0.67	**2.00**
Establishment of protected forest areas	-	1.83	**2.00**	**2.33**	**2.00**	1.29	0.75	-	-
Agroforestry	1.36	**2.13**	1.83	**2.10**	1.80	**2.03**	1.78	-0.92	1.80
Agro-silvio-pastoralism	0.83	1.88	1.29	1.20	1.17	1.61	1.33	0.50	1.00
Eco-graze	-	1.92	**2.00**	1.00	**2.00**	1.50	1.50	-2.00	1.67
Area closure to grazing	**3.00**	1.71	**2.00**	**2.00**	1.75	0.33	1.45	1.67	-
Fodder plants conservation and reseeding	**2.00**	**2.25**	**2.00**	**2.25**	**2.00**	**2.38**	1.83	-	-
No tillage and reduced tillage	1.63	**2.04**	1.43	1.00	1.91	1.24	1.27	0.91	**2.00**
Use of amendments and fertilizers	1.68	1.92	1.92	**2.45**	**2.10**	1.47	1.30	0.15	**3.00**
Green cover crops	1.09	**2.50**	**2.11**	1.71	1.60	0.88	1.67	0.93	1.00
Crop rotation / intercropping	1.67	**3.00**	1.43	1.50	1.83	1.20	1.13	0.00	-
Terracing	1.76	1.96	1.13	1.15	1.50	1.27	1.78	0.08	0.67
Salinity drop	-0.92	1.60	0.00	**2.00**	1.00	1.35	1.10	0.13	1.00
Conservation agriculture	**2.13**	**2.77**	1.00	1.86	**3.00**	**2.08**	**2.33**	**2.00**	**3.00**
Micro-irrigation systems	1.52	**2.27**	1.00	1.50	1.13	1.28	1.22	0.93	-
Water harvesting and storage	1.94	-0.83	**2.50**	-	**2.00**	0.67	1.20	**2.01**	**2.00**
Water harvesting and transport	1.23	1.75	1.10	1.00	1.70	1.16	1.16	0.33	0.00
Vegetation against soil erosion	1.62	**2.05**	1.83	1.54	**2.17**	1.84	**2.11**	**2.00**	1.29
Vegetation against pollution	1.78	1.80	1.67	1.55	**2.00**	1.80	0.22	1.67	1.50
Mulching	1.18	**2.17**	1.75	1.78	1.11	0.63	1.67	1.50	-
Variable rate biosolids management	**2.00**	-	**2.00**	**2.00**	**2.00**	**2.00**	0.00	-	-

The mean scores ranging from -3 (i.e. negative impact) to + 3 (i.e. positive impact) are in coloured cells. The symbol—is used to indicate no data

0.63–1.75). Exceptions to this are the practices of *Afforestation* and *Water harvesting and transport*, which score lower than 1.8 in all functions.

In summary, *Afforestation* practices are in this assessment the plantation of native trees, shrubs, and/or grasses through participatory action in lands of other former use. Thus, this SLM practice implies a change in the use of land, which is not always straightforward to carry out successfully. Several factors such as the potential for natural water recharge, weather variability, adequacy of the soil contents and nutrients, plantation density, etc. need to be taken into account when implementing it (e.g. Maestre and Cortina 2004). *Water harvesting and transport* consist of the slowdown and/or collection of runoff water. This SLM practice is mainly oriented to provide water to croplands and to slow down the water flow from river courses. Because of its specificity and due to its associated labour costs of establishment and maintenance, this practice does not stand out for any of the services, yet it shows a quite high result, i.e. 1.75, to the function 'Soil erosion control'. Eight practices show high scores for the ecosystem services, but low ones for the social functions, whilst the opposite does not occur for any SLM practice. Those practices are: *Reforestation, Establishment of protected areas, Agroforestry, Area closure to grazing, Use of amendments and fertilizers, Green cover crops, Crop rotation/intercropping*, and *Variable rate bio-solids management*. Three other practices show high scores for both ecological services and social functions, indicating that these, are actions that address simultaneously several ecosystem services and they may be mainstreamed in their own across the Mediterranean and Middle East Regions to promote sustainable rural landscapes. The three SLM practices are: *Fodder plants conservation and reseeding, Conservation agriculture*, and *Vegetation against soil erosion.*

If assessed service-wise, different practices emerge as the best choice, e.g. to the function 'Climate regulation' the practice *Area closure to grazing* results as the best one, yet to the function 'Soil erosion control' it is *Crop rotation / intercropping*. However, we aim to discuss those SLM practices that respond well to all ecological and social functions at the time, that is, those SLM practices that offer a holistic answer to the challenges posed by climate change and landscape degradation. To do so, we will discuss all SLM options based on the type of land in where they can be applied, being: forestalnd, mixedland, grazingland, cropland, water bodies, and several land-uses. We have followed the definitions from WOCAT (2002) for each type of land (Box 2).

Box 2 Land types/Classes

Forest / woodlands: land used mainly for wood production, other forest products, recreation, protection;

Mixed: mixture of land use types within the same land unit: agro-forestry (trees and crops), agro-pastoralism (crops and livestock), agro-silvo-pastoralism (crops, trees and livestock), silvo-pastoralism (forest and grazing land);

Grazing land: land used for animal production;

Cropland: land used for cultivation of crops, including fallow (field crops, orchards);

Other: housing, recreation, roads, mining sites, etc. in where we have included 'Water bodies': rivers, lakes, ponds, puddles and artificial reservoirs; and 'Several land-uses': SLM practices that can be applied in more of one land types.

4.1 Forestland

From all SLM practices, four of them are applicable in forest areas, those are: *afforestation; reforestation; control of wildfires*; and *stablishment of protected forest* areas. All four practices aim at either increasing forest cover with native trees species and/or protecting natural and primary forest.

On the one hand side, increasing forest cover can be attained by *reforestation*, which is the sustainable/natural or assisted regeneration of forest areas in formerly deforested lands; and by *afforestation* defined as the plantation of native trees, shrubs, and grasses through participatory action (Box 1). Results in Table 1 show that increasing forest cover strengthens the capacity of vegetation to address climate change mitigation by enhancing carbon sequestration and stock below and above the ground (Jia, et al. 2019). Increasing forest cover also decreases the impacts of extreme weather events by regulating the micro-climate and by offering a natural barrier against the physical forces of such events. Additionally, these practices help rising atmospheric moisture through evapotranspiration. Higher atmospheric moisture, besides decreasing local temperatures in summer, could enhance topographic rainfall (Millán et al. 2005, 2014). These practices also help to preserve soil quality and soil erosion control by rising the water infiltration capacity of the soil through deep rooting, as increased infiltration reduces and slows down surface runoff and transport of sediments (Fernández-Ondoño et al. 2010). Deep rooting, moreover, prevents soil crusting and compacting while promotes soil moisture and improved water quality, assisting to the natural regulation of water resources (Costa et al. 2003). Further off-site benefits from increased water infiltration are downstream flow assurance, groundwater and river pollution prevention, protection from flooding, and reduction of potential damages to infrastructures.

The social benefits of *reforestation* and *afforestation* are lower than the ecological benefits. Reasons for this are the associated reduction of land available for agricultural purposes inducing to economic costs, together with the potential need for economic investment, water, and workload at the first stage of implementation of these practices (e.g. Schwilch et al. 2012 pg. 177–180). These economic disadvantages, however, may be overcome with the integration of communities into forest management tasks. Increasing forest areas, moreover, strengthens the resilience of forest-dependent communities by providing other market products than timber such as honey, cork, and mushrooms, while assisting to human well-being by boosting self-sufficiency and providing communities with cultural and recreational opportunities (Auclair et al. 2011).

On the other hand side, *control of wildfires*, i.e. group of actions that prevent forest fires from occurring, and *establishment of protected forest areas* such as forest reserves and natural and national parks, are SLM options aimed at protecting primary forest. Preserving primary forest is key to biodiversity, as reforested and restored forest areas seek centuries to reach complex ecological systems such as the ones from primary forests (Meli et al. 2017). Moreover, both practices benefit from land-users' traditional knowledge and are supported at the local, national, and international levels by rural development programs and governments (e.g. Natura 2000 Network).

Among the four SLM practices, *reforestation*, as defined here, emerges as the most beneficial to all five ecosystem services, while *control of wildfires* is best for all four social ones (Fig. 3). Accordingly, the set of these two practices appear as a holistic response to best assist all services and functions by providing most benefits and synergies in forests form the Mediterranean and Middle East Regions.

4.2 Mixedland

We have identified two SLM options from mixedlands, i.e. from lands with several types of use in the same area unit. These are: *agroforestry*, a SLM practice that combines crops and fruit and forest trees; and *agro-silvio-pastoralism*, consisting in

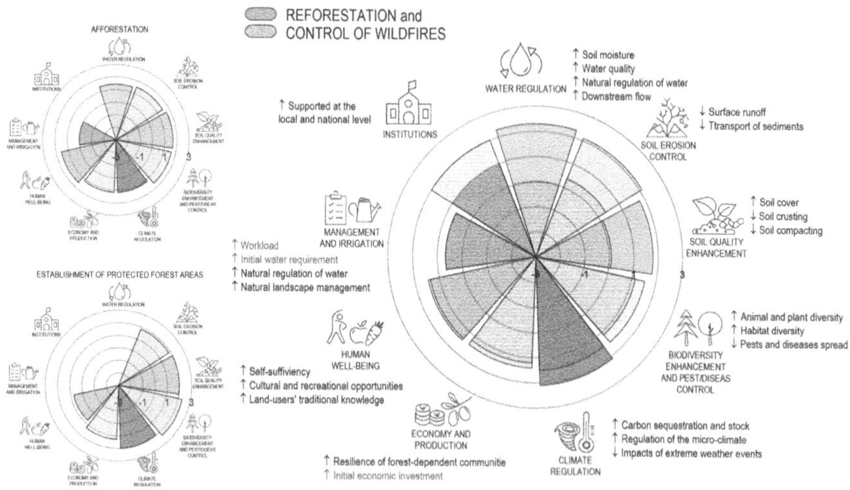

Fig. 3 Descriptive results from the SLM practices 'Reforestation' (solid pattern) and 'Control of wildfires' (stripes) to the nine assessed functions. In small, results from the SLM practices: Afforestation, and Establishment of protected forest areas

the planting of fodder shrubs together with crops and/or trees and animals (Box 1). Both SLM practices are powerful tools preventing soil erosion and improving soil quality due to increased soil cover (Table 1). Soil cover and associated rooting, dampens runoff velocities and sediment transport, especially in slopes and hills, which enhances soil stabilization and crop production (see Forestland). *Agro-silvio-pastoralism systems*, moreover, besides offering fodder to the livestock and shade, naturally introduce organic amendments to the soils, increasing their nutrient content and fertility (FAO 2013). *Agroforestry systems* coupled with gravity irrigation (flooding terraces) help retaining rainwater and conserving groundwater resources while inducing pleasant and better regulated micro-climates. To these, all associated benefits of deep rooting infiltration are likewise acquired with the introduction of trees in the cropland systems, strengthening the natural management of the available water resources and efficiency in its use through access at different depths (Ramachandran Nair 1993). Moreover, the inclusion of trees and shrubs to croplands and grazinglands provides other species with new ecosystems while offers a barrier for pest spreading.

Similar to the case of SLM practices in forestlands, *agroforestry* and *agro-silvio-pastoralism* systems need initial economic investment, water for irrigation, and workload, which can threaten its success in the short-term (Table 1). These practices, however, can provide high benefits to the economy and human well-being in the long term, as they have the capacity to naturally manage the landscape and increase its profitability through output diversification (Schoonhoven and Runhaar 2018). Furthermore, and in spite of its nowadays configuration, the Mediterranean Basin is known for its mosaic-like traditional landscape, i.e. multi-functional with crops in spatial arrangement with forests and settlements. The multi-functionality of the Mediterranean landscape is, therefore, today jeopardized due to crop specialization and intensive land-use (Mercuri et al. 2019). In this way, *agroforestry* and *agro-silvio-pastoralism* systems can be promising practices that help to revert the accumulated effects over the millennia from past land-use change practices and land intensification, a clever way of naturally managing the environment while sustainably exploiting its resources, and a profitable option to develop the rural economy (Ruiz and Sanz 2020). Accordingly, these two practices may represent a more integrated approach to achieving not only healthy but also productive and more diverse ecosystems by integrating the biophysical and socio-cultural values of the region.

Agroforesty and *agro-silvio-pastoralism* systems have been historically more widespread in the tropics than in the temperate zones because of climatic conditions favouring rapid growth (Ramachandran Nair 1993). Nonetheless, in semi-arid regions *silvio-pastoral systems*, windbreaks and shelterbelts, and multipurpose trees on croplands have been largely implemented (Schwilch et al. 2012). Here, both *agroforesty* and *agro-silvio-pastoralism* pratices show high potential to all nine considered services and functions (Fig. 4), a finding that is supported by the literature, which exemplifies how the two SLM options offer a solution to many land-use problems and climate challenges in both the Mediterranean and Middle East Regions (Sanz et al. 2017).

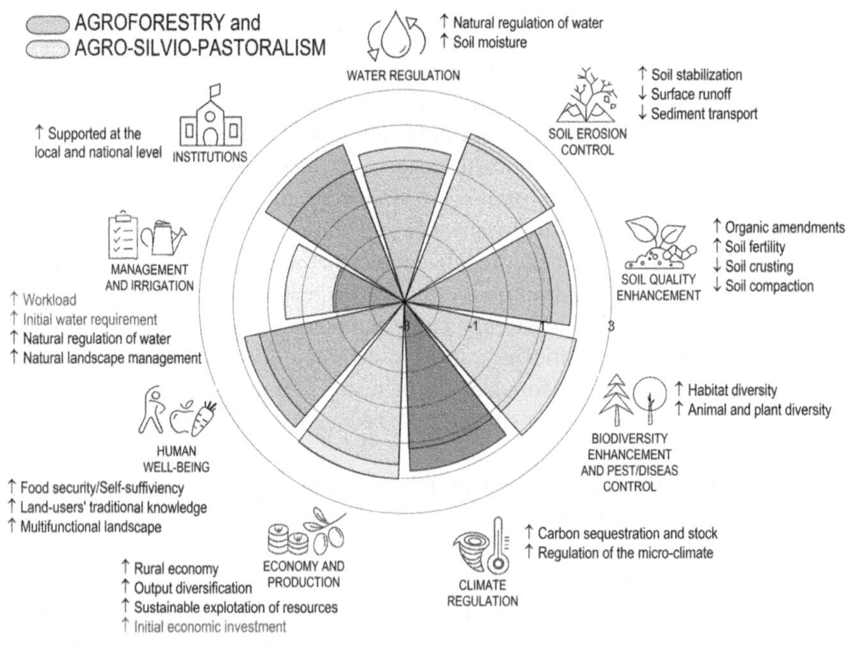

Fig. 4 Descriptive results from the SLM practices 'Agroforestry' (soild pattern) and 'Agro-silvio-pastoralism' (stripes) to the nine assessed functions

4.3 Grazingland

The identified SLM practices of grazinglands are: *eco-graze*, which consist on the resting of pastures in rotation; *fodder plants conservation and reseeding*, consisting on pitting and reseeding vegetation to restore degraded grazinglands; and *area closure to grazing*, which is the inability to access a specific area for grazing (Box 1). The two first SLM options are active approaches to restoration and conservation, while area closure is a passive measure.

On the one hand, *eco-grazing* and *area closure to grazing* are useful measures in cases of moderate soil degradation, when fodder and vegetation still have a spontaneous restoration capacity. Thus, they are applied to prevent excessive stress in the environment (Schwilch et al. 2012). Both practices offer high benefits to soil quality and soil erosion control, promoting soil organic matter and soil cover (Table 1). Soil moisture likewise increases with soil cover, while surface runoff, wind erosion, and soil crusting and sealing are reduced (see Forestland). The two SLM options positively help to regulate water resources by reducing their use and positively impact biodiversity by allowing habitat diversification (Fig. 5a). On the other hand, *area closure to grazing* shows high benefits to climate regulation by allowing a healthy amount of available forage, which, besides storing carbon, decreases the landscape's vulnerability to extreme weather events (Domínguez 2013). Likely, by closing the area to grazing and allowing natural revegetation, anthropogenic activities decease, consequently impacting the local climate in a positive way.

On the social side, both SLM options positively impact human well-being as they reduce social disparities due to their adoptability, and provide SLM and land degradation knowledge to land-users. *Eco-grazing* also increases economic revenues through animal production and fodder production and quality, and it is supported by institutions aiming at combating rural exodus (Schwilch et al. 2012). Moreover, products like the argan oil can be obtained with the collection of argan nuts while the land rests from grazing, otherwise eaten by cattle (SLM practice ID:85). This SLM practice, however, hinders land management, as although it is a traditionally applied practice, it is no longer common and can lead to conflicts among land-users at the beginning of its implementation. Contrastingly, *area closure to grazing* positively impacts land management and workload, as it is a passive measure that does not require maintenance or work, yet it hampers economy and production as the plot's profitability ceases and the initial cost of fencing is usually high (Fig. 5a).

Fodder plants conservation and reseeding, instead, is a SLM practice applied once the state of land degradation is irreversible and native vegetation cannot grow spontaneously (Fig. 5b). In contrast to fodder plants conservation, reseeding is a more laborious measure that may comprise substantial management and workload, and increased water use for irrigation. Although we miss data to contrast these suppositions, implementation experiences show that treating 10–20% of completely denuded plots may be sufficient for a 'starter vegetation' to grow (SLM practice ID: 66). If such challenges appeared, these could be overcome by institutional help through subsidies and support from governments, yet data on how this practice affects

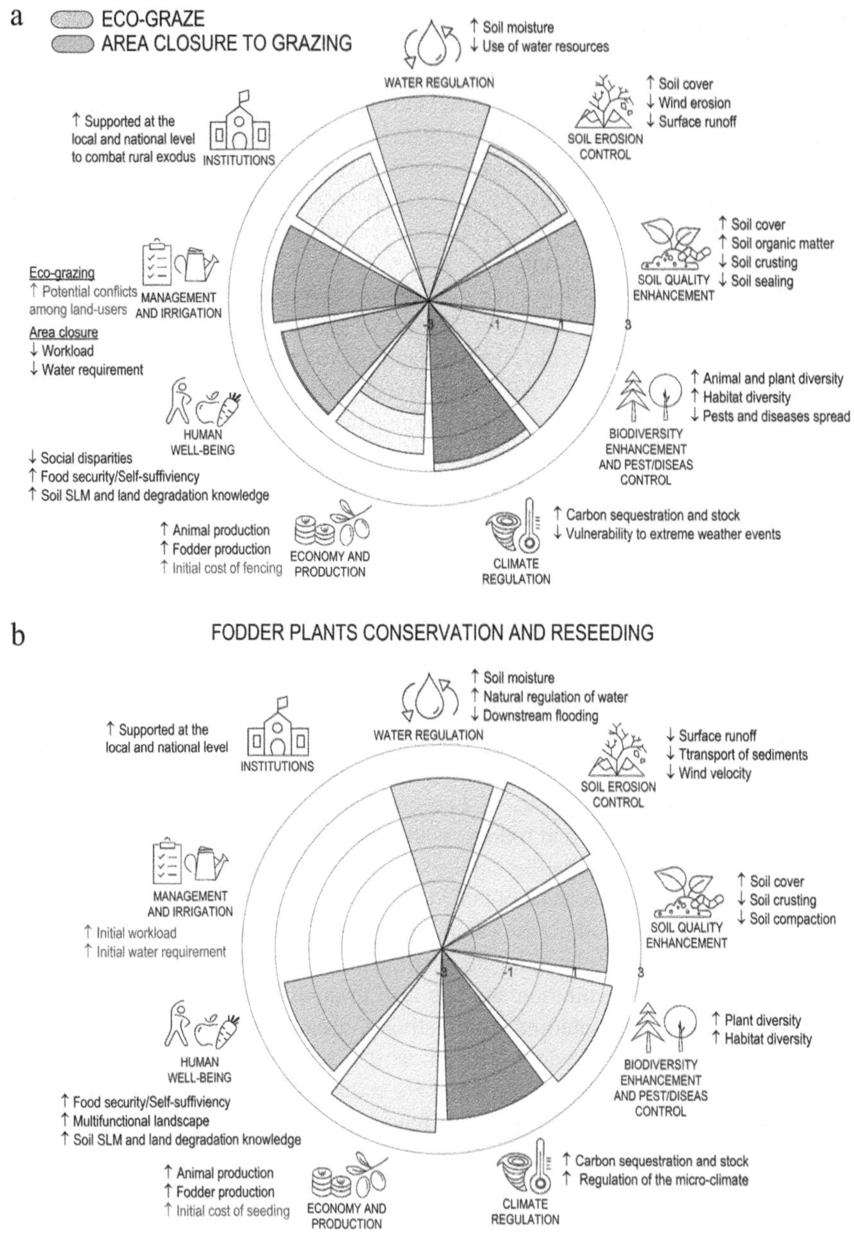

Fig. 5 **a** Descriptive results from the SLM practice 'Area closure to grazing' (soild pattern) and 'Eco-graze' (stripes) to the nine assessed functions. **b** Descriptive results from the SLM practice 'Fodder plants conservation and reseeding' to the nine assessed functions

institutions is also missing. What this SLM option reports is all the high benefits to all the other seven services and functions (Fig. 5b). To start, the five ecosystem services are highly impacted in a positive way due to increase or preserved and ameliorated native vegetation, offering benefits to the health and the structure of the soil, the strengthening of the water cycle, and the regulation of the micro-climate towards milder weather. Furthermore, this SLM option positively impacts farm income with enhanced fodder and animal production and human well-being (Genin et al. 2016).

Grazinglands rank among the most degraded land-use type in the Mediterranean and Middle East Regions, as a result of drought and traditional grazing activities carried out over millennia (Zeder 2008). That is why, beyond the implementation of active approaches to restore grazinglands (i.e. *Fodder plants conservation and reseeding*), preventive passive measures (i.e. *eco-grazing* and *area closure to grazing*) are equally important. Foremost crucial, in any of both cases, is the adjustment of the stocking rate for the sustainable use of grazinglands.

4.4 Cropland

The long agricultural history of the Mediterranean and Middle East Regions is reflected in the numerous SLM solutions applied to croplands. The identified SLM practices of croplands are: *no tillage and reduced tillage*; *use of amendments and fertilizers*; *green cover crops*; *crop rotation/intercropping*; *terracing*; *salinity drop*; *conservation agriculture* (Fig. 6).

Three of these practices, i.e. *use of amendments and fertilizers, green cover crops, and crop rotation/intercropping,* offer several benefits in terms of ecosystem services, while are less strongly impacting the social ones in a positive way (Table 1). The *use of amendments and fertilizers* (compost, straw pen manure, etc.) enhances crop productivity by improving soil structure and fertility while stimulating biological activity. Moreover, with this practice, animal waste is recycled, organic carbon is best preserved, and soils have a higher content of moisture, making them more resilient to weather events (e.g. Herencia et al. 2011). Similarly, *green cover crops* and *crop rotation and intercropping* are maintain the soil covered throughout the year optimizing the use rainfall, regulating the local climate, enhancing carbon sequestration, promoting microbial biomass, and especially, preventing soil erosion and soil loss due to wind and surface runoff (Sastre et al. 2020). However, the three SLM practices imply workload. The use of amendments also implies initial monetary costs as the benefits of applying them may not be immediate, and green cover crops may require a substantial increase in the demand of water for irrigation (Celette et al. 2008).

Three other practices, i.e. *no tillage and reduced tillage*, *terracing*, and *salinity drop*, only show moderately-positive scores to all services and functions (Table 1). *Terracing*, although being a SLM practice with millennial tradition of implementation in the Mediterranean and the Middle East Regions, implies workload and higher expenses on agricultural inputs. In spite of these, *terracing* allows cultivation in step areas and reduces runoff and soil erosion through decreased slope gradient and increased vegetation (Schwilch et al. 2012). *Salinity drop* helps conserving biodiversity and enhancing water and nutrient absorption of plants, yet decreases product diversity and overall water quality and quantity (e.g. Nagaz et al. 2007). *No tillage and reduced tillage* practices maintain the soil partially covered with organic matter throughout the year, keeping soil moisture and reducing workload. Nevertheless, this practice does not disrupt the life cycle of weeds and pests and can hamper crop growth. Moreover, the overall benefits and trade-offs to the carbon budget (i.e. capture of atmospheric carbon vs. release of belowground stored carbon) of this practice are not fully understood (Almagro et al. 2017; Álvaro-Fuentes et al. 2008).

Lastly, *conservation agriculture* shows the highest results (Fig. 6). This system of practices implies the implementation of three SLM actions at the same time: no tillage and reduced tillage, green cover crops, and crop rotation or intercropping. With its implementation, not only the benefits before reported are obtained, but the soil is less disturbed, guaranteeing its structure, chemical fertility, and biodiversity. This practice also eliminates monoculture, favoring multifunctional landscapes, and reduces economic and energetic inputs by saving labor and fuel costs, which reduce greenhouse gas emissions (López-Fando et al. 2007). Per contra, conservation agriculture many tiems makes use of chemical weeding and lacks of training and awareness among land-users in the Mediterranean and the Middle East Regions (Carmona et al. 2015).

4.5 Water Bodies

Three SLM are directly related to water management and water bodies. These are: *micro-irrigation systems*, that include drip irrigation and micro-catchments; *water harvesting and storage*, consisting of reservoirs that allow off-season irrigation; and *water harvesting and transport*, which besides collecting water from the runoff also includes measures to slowdown it down (Box 1).

If well planned and implemented, the three SLM practices help reducing groundwater overexploitation, rainfall water losses, and river pollution while allowing for a more efficient use of the hydrological resources (Table 1). Given the growing desertification on the Mediterranean and the Middle East Regions along with precipitation decreases and storminess enhancement (Mirzabaev et al. 2019; Seneviratne et al. 2012), these benefits by themselves, are key to the sustainable land-use and development of both regions and yet, the three SLM practices score moderately-well overall in the integrated assessment (Fig. 7). Reasons for that may be that on the one hand, and although *micro-irrigation systems* allow for the most efficient use of the water resources, these entail heavy workload to implement and maintain (e.g. drip irrigation pipes), technical knowledge, and local experience. *Micro-irrigation systems* moreover, require cooperation among watershed stakeholders. This, despite challenging, offers the opportunity to establish cooperatives (Mekdaschi Studer and Liniger 2013 pg. 109–140). On the other hand, *water harvesting, storage, and transport* while ensuring downstream flow, increased infiltration, and decreased erosion

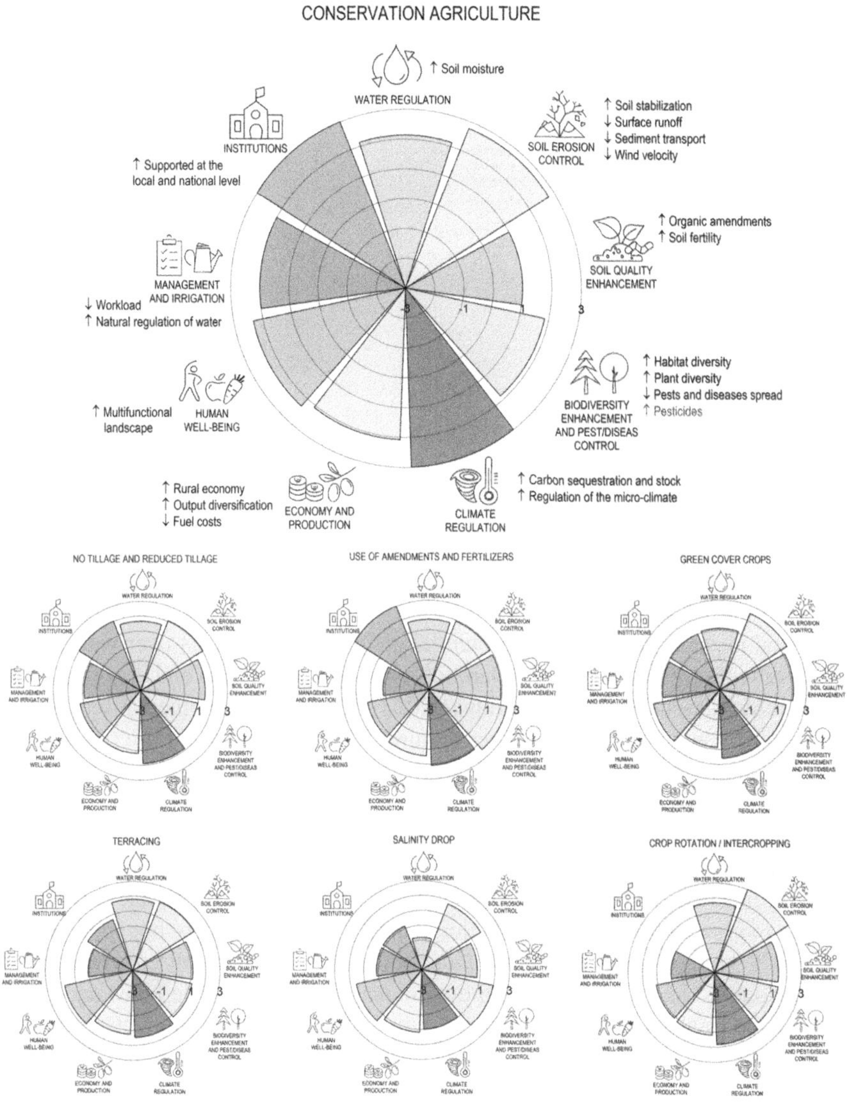

Fig. 6 Descriptive results from the SLM practice 'Conservation agriculture' to the nine assessed functions. In small, results from the SLM practices: No tillage and reduced tillage, Use of amendments and fertilizers, Green cover crops, Terracing, Salinity drop, and Crop rotation / intercropping

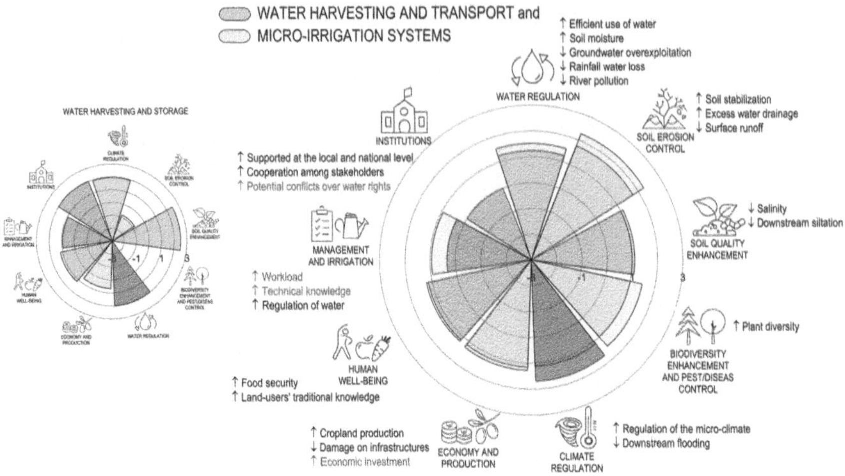

Fig. 7 Descriptive results from the SLM practices 'Water harvesting and transport' (solid pattern) and 'Micro-irrigation systems' (stripes) to the nine assessed functions. In small, results from the SLM practice 'Water harvesting and storage'

and flood hazards, may entangle potential conflicts and disputes over water rights and distribution, especially among land-users of different sectors and along the water stream (Mekdaschi Studer and Liniger 2013). Thus, the implementation of large water storage and transport structures requires clear land and water use rights, coordinated management of the full watershed, and a planned network of water points. Large water infrastructures, such as reservoires, can impact biodiversity and imply population dispacement.

We find that, overall, water regulation is achieved more effectively by implementing practices that promote natural revegetation, such as reforestation and area closure to grazing (Table 1). Nonetheless, these SLM practices are not applicable in croplands while the Mediterranean and Middle East societies are growing both in population and in resource consumption, demanding more arable lands and goods (The World Bank Group 2018). In this context, water management practices applied to croplands and grazinglands are pivotal to achieve long-term sustainable landscapes with productive ecosystems, socio-economic benefits, and increased food security in some areas.

4.6 Other or Several Land Uses

Three SLM practices can be applied in different land-uses, these are: *vegetation against soil erosion*, consisting on planting native woody species, strips and/or trees; *vegetation against pollution*, both in soils and water; and *mulching*, in croplands involving the spread of waste after harvesting and in forestlands the spread of slash mulch after a forest or wildfire (Box 1).

Vegetation against pollution can stabilize soil trace elements and prevent surface water pollution (Fig. 8). It also adds organic matter to the soil, can reduce soil and sediment losses due to erosion, stabilize riverbanks, promote biodiversity, decrease flood damage to crops, and improve landscape quality (Dipartimento Agricoltura e Sviluppo Rurale 2007). However, this practice may decrease crop production and conflict mitigation due to land-use change from cropland to no-profit land (Table 1). Instead, *mulching* improves the biological activity in the soil and helps increasing productivity in croplands while effectively reduces runoff in forestalnds after fire events (Prats et al. 2012). Moreover, mulching and vegetation against soil erosion are SLM practices that promote surface cover, which enhances soil quality and controls soil loss, favouring soil structure and water infiltration. Furthermore, *vegetation against soil erosion* reduces overgrazing, mitigates landscape degradation, and

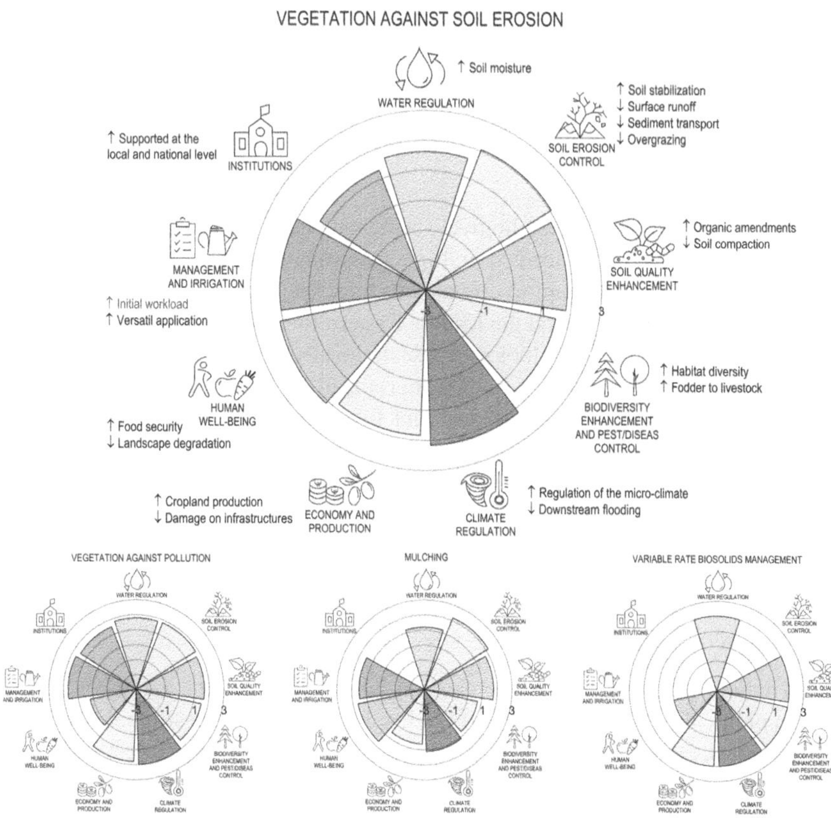

Fig. 8 Descriptive results from the SLM practice 'Vegetation against soil erosion' to the nine assessed functions. In small, results from the SLM practices: Vegetation against pollution, Mulching, and Variable rate biosolids management

prevents flooding (Fig. 8). This is a restoration technology largely used in hillslopes and semi-arid regions where heavy rainfalls cause severe soil loss (Schwilch et al. 2012). It is a cheap SLM practice that can be applied with different available resources (e.g. branches, woody species, wooden fences, trees) and that may promote biodiversity and fodder to livestock. It implies initial workload and investment, which in several cases determines its application. Nevertheless, it usually does not need maintenance, increasing farmer's income in the long-term.

Lastly, *variable rate biosolids management* is a SLM practice that has not been possible to group, as the focus of this practice is not the impact of the application of amendments itself (i.e. use of amendments and fertilizers), but the methodology used for their application. This SLM action consists of the site-specific application of minerals and organic amendments in croplands. It increases soil fertility and soil organic carbon while inhibits pollutant diffusion, thus, positively impacting economy and production (Fig. 8). However, this is a high technological SLM practice that lacks training and awareness among land-users, and that hampers management and irrigation, as variable rates of manure mineralize differently within management techniques, implying higher workload (Moshia et al. 2016).

5 Conclusions

With this work, we aim to qualitatively assessed Sustainable Land Management (SLM) practices that could lead to multiple ecosystem services, in particular with the view of mitigate and adapt to climate changes, while promoting other benefits, in the Mediterranean Basin and the Middle Eastern Region. To do so, we have used an integrated assessment that looks to upscale locally-designed SLM practices for its consideration at the scale of the Mediterranean Basin and the Middle Eastern Region. The assessment takes into account five ecosystem services (i.e. Climate Regulation; Soil erosion control; Biodiversity enhancement and pest/disease control; Water regulation; Soil quality enhancement), and four social functions (i.e. Economy and production; Management and irrigation; Human well-being; Institutions).

The resulting SLM practices can be implemented on the landscapes, for example at the level of a watershed. Our assessment shows that the most promising SLM practices for different locations or land pre-existing uses in a hypothetical watershed in the Mediterranean Basin could be:

(a) Highlands or forestlands: reforestation. This practice offers, among others, the ecological benefits of enhanced biodiversity, carbon sequestration, water-use efficiency, and improved soil structure, while it also boosts forest-dependent communities;
(b) Midlands or grazinglands/mixedlands: agroforestry, vegetation against soil erosion, and fodder plants conservation and reseeding. These SLM actions promote multifunctional landscapes that effectively combat desertification while foster productive ecosystems with socio-economic benefits;

(c) Lowlands or croplands: conservation agriculture. This practice, consisting on three SLM actions (i.e. no tillage and reduced tillage, green cover crops, and crop rotation or intercropping) helps preserving soil quality, enhances the rural economy, and increases food security;
(d) Water bodies: micro-irrigation systems and water harvesting and transport. Above all, water management SLM actions are key to ensure the whole ecological and economic development of a watershed.

The highlighted SLM practices can be implemented on their own or in combination across the Mediterranean and Middle East Regions, and watersheds seem to be geographic units that allow for more integrated approaches to materialize multiple ecosystem services. Also due to the reason that, it is decisive to implement them under an integrated planning and promoting good levels of coordination at the governance diferent levels (local, regional, national). Overall, the main obstacles to SLM implementation showed to be economic costs of implementation and workload, as well as the perception and low adoption by local communities. Thus, the consideration of SLM in the context of waterseeds could allow to address this socio-economic barries, i.e. from its high- to its low-lands, while taking into account the socio-economic benefits of the proposed actions and the knowledge of the local people who will, in the last instance, adopt them.

References

Almagro M, Garcia-Franco N, Martínez-Mena M (2017) The potential of reducing tillage frequency and incorporating plant residues as a strategy for climate change mitigation in semiarid Mediterranean agroecosystems. Agri Ecosyst Environ 246:210–220. https://doi.org/10.1016/j.agee.2017.05.016

Álvaro-Fuentes J, López MV, Cantero-Martinez C, Arrúe JL (2008) Tillage effects on soil organic carbon fractions in Mediterranean Dryland agroecosystems. Soil Sci Soc America J 72(2):541–547. https://doi.org/10.2136/sssaj2007.0164

Auclair L, Baudot P, Genin D, Romagny B, Simenel R (2011) Patrimony for resilience: evidence from the forest agdal in the moroccan high atlas mountains. Ecolo Soc 16(4). https://doi.org/10.5751/ES-04429-160424

Calder IR (2002) Forests and hydrological services: reconciling public and science perceptions. In: Land use and water resources research. University of Newcastle upon Tyne, Centre for Land Use and Water Resources Research, vol 2. pp 1–12. Available at: https://ideas.repec.org/a/ags/luawrr/47860.html

Climate Change (2014) Mitigation of climate change. Contribution of working group III to the fifth assessment report of the intergovernmental panel on climate change [Edenhofer O, Pichs-Madruga R, Sokona Y, Farahani E, Kadner S, Seyboth K, Adler A, Baum I, Brunner S, Eickemeier P, Kriemann B, Savolainen J, Schlömer S, von Stechow C, Zwickel T and Minx JC (eds.)]. Cambridge University Press, Cambridge, United Kingdom and New York, NY, USA

Carmona I, Griffith DM, Soriano MA, Murillo JM, Madejón E, Gómez-Macpherson H (2015) What do farmers mean when they say they practice conservation agriculture? a comprehensive case study from southern Spain. Agr Ecosyst Environ 213:164–177. https://doi.org/10.1016/j.agee.2015.07.028

Celette F, Gaudin R, Gary C (2008) Spatial and temporal changes to the water regime of a Mediterranean vineyard due to the adoption of cover cropping. Eur J Agron 29(4):153–162. https://doi.org/10.1016/j.eja.2008.04.007

Chaidez V, Dreano D, Agusti S, Duarte CM, Hoteit I (2017) Decadal trends in Red Sea maximum surface temperature. Sci Rep 7(1):23955–26900. https://doi.org/10.1038/s41598-017-08146-z

Costa MH, Botta A, Cardille JA (2003) Effects of large-scale changes in land cover on the discharge of the Tocantins River, Southeastern Amazonia. J Hydrol 283(1–4):206–217. https://doi.org/10.1016/S0022-1694(03)00267-1

Cramer W, Guiot J, Fader M, Garrabou J, Gattuso JP, Iglesias A, Xoplaki E et al. (2018) Climate change and interconnected risks to sustainable development in the Mediterranean. Nature Climate Change 8(11):972–980. https://doi.org/10.1038/s41558-018-0299-2

Deitch M, Sapundjieff M, Feirer S (2017) Characterizing precipitation variability and trends in the World's mediterranean-climate areas. Water 9(4):259. https://doi.org/10.3390/w9040259

Dipartimento Agricoltura e Sviluppo Rurale (2007) Programma di sviluppo rurale per il veneto 2007–2013. Available at: https://www.regione.veneto.it/static/www/agricoltura-e-foreste/psr_alleg_02_costimisure131215rev7_ok.pdf

Domínguez P (2013) Culturally mediated provision of ecosystem services: the AGDAL of Yagour, Springer, New York, NY, pp 379–393. https://doi.org/10.1007/978-1-4614-5702-2_16

Ellison D, Futter MN, Bishop K (2012) On the forest cover-water yield debate: From demand-to supply-side thinking. Glob Change Biol 18(3):806–820. https://doi.org/10.1111/j.1365-2486.2011.02589.x

FAO (2013) In: Climate-smart agriculture sourcebook. ERIS–European review of international studies, vol 3. https://doi.org/10.3224/eris.v3i2.14

Fernández-Ondoño E, Serrano LR, Jiménez MN, Navarro FB, Díez M, Martín F, Aguilar J et al. (2010) Afforestation improves soil fertility in South-Eastern Spain. Europ J Forest Res 129(4):707–717. https://doi.org/10.1007/s10342-010-0376-1

Flohr P, Fleitmann D, Matthews R, Matthews W, Black S (2016) Evidence of resilience to past climate change in Southwest Asia: early farming communities and the 9.2 and 8.2 ka events. Quatern Sci Rev 136:23–39. https://doi.org/10.1016/j.quascirev.2015.06.022

Gangoiti G, Gómez-Domenech I, De Cmara ES, Alonso L, Navazo M, Iza J, Millán MM (2011) Origin of the water vapor responsible for the European extreme rainfalls of August 2002: 2. a new methodology to evaluate evaporative moisture sources, applied to the August 11–13 central European rainfall episode. J Geophys Res Atmospheres 116(21). https://doi.org/10.1029/2010JD015538

Genin D, Crochot C, MSou S, Araba A, Alifriqui M (2016) Meadow up a tree: feeding flocks with a native ash tree in the Moroccan mountains. Pastoralism 6(1):11. https://doi.org/10.1186/s13570-016-0058-9

Goring-Morris NA, Belfer-Cohen A (2011) Neolithization processes in the Levant: the outer envelope. Current Anthropol 52(4). https://doi.org/10.1086/658860

Herencia JF, García-Galavís PA, Maqueda C (2011) Long-term effect of organic and mineral fertilization on soil physical properties under greenhouse and outdoor management practices. Pedosphere 21(4):443–453. https://doi.org/10.1016/S1002-0160(11)60146-X

Jackson RB, Jobbágy EG, Avissar R, Roy SB, Barrett DJ, Cook CW, Murray BC et al. (2005) Atmospheric science: trading water for carbon with biological carbon sequestration. Science 310(5756):1944–1947. https://doi.org/10.1126/science.1119282

Jia G, Shevliakova E, Al E (2019) Chapter 2 : land-climate interactions. In : Climate change and land. an IPCC special report on climate change, desertification, land degradation, sustainable land management, food security, and greenhouse gas fluxes in terrestrial ecosystems, pp 1–186

López-Fando C, Dorado J, Pardo MT (2007) Effects of zone-tillage in rotation with no-tillage on soil properties and crop yields in a semi-arid soil from central Spain. Soil and Tillage Res 95(1–2):266–276. https://doi.org/10.1016/j.still.2007.01.005

Macias D, Garcia-Gorriz E, Stips A (2013) Understanding the causes of recent warming of mediterranean waters. how much could be attributed to climate change? PLoS ONE 8(11):e81591. https://doi.org/10.1371/journal.pone.0081591

Maestre FT, Cortina J (2004) Are Pinus halepensis plantations useful as a restoration tool in semiarid Mediterranean areas? For Ecol Manage 198(1–3):303–317. https://doi.org/10.1016/j.foreco.2004.05.040

Malmer A, Murdiyarso D, Sampurno Bruijnzeel LA, Ilstedt U (2010) Carbon sequestration in tropical forests and water: a critical look at the basis for commonly used generalizations. Global Change Biol 16(2):599–604. https://doi.org/10.1111/j.1365-2486.2009.01984.x

Marques MJ, Schwilch G, Lauterburg N, Crittenden S, Tesfai M, Stolte J, Schmiedel U et al. (2016) Multifaceted impacts of sustainable land management in drylands: a review. Sustainability (Switzerland) 8(2). https://doi.org/10.3390/su8020177

Mekdaschi Studer R, Liniger H (2013) In: Water harvesting: guidelines to good practice (centre for). Centre for Development and Environment (CDE). Bern, Rainwater Harvesting Implementation Network (RAIN), Amsterdam; MetaMeta, Wageningen; The International Fund for Agricultural Development (IFAD), Rome

Meli P, Holl KD, Rey Benayas JM, Jones HP, Jones PC, Montoya D, Moreno Mateos D (2017) A global review of past land use, climate, and active versus passive restoration effects on forest recovery. PLOS ONE 12(2):e0171368. https://doi.org/10.1371/journal.pone.0171368

Mercuri AM, Florenzano A, Burjachs F, Giardini M, Kouli K, Masi A, Fyfe R et al. (2019) From influence to impact: the multifunctional land use in Mediterranean prehistory emerging from palynology of archaeological sites (8.0–2.8 ka BP). Holocene 29(5):830–846. https://doi.org/10.1177/0959683619826631

Millán MM, Estrela MJ, Sanz MJ, Mantilla E, Martín M, Pastor F, Versino B et al. (2005) Climatic feedbacks and desertification: The Mediterranean model. J Climate 18(5):684–701. https://doi.org/10.1175/JCLI-3283.1

Millán MM (2007) Climate change impacts on the water cycle resources and quality : research-policy interface. Publications Office of the European Union

Millán MM (2014) Extreme hydrometeorological events and climate change predictions in Europe. J Hydrol 518(PB):206–224. https://doi.org/10.1016/j.jhydrol.2013.12.041

Millán MM, Mantilla E, Salvador R, Carratalá A, Sanz MJ, Alonso L, Navazo M et al. (2000) Ozone cycles in the western Mediterranean basin: interpretation of monitoring data in complex coastal terrain. J Appl Meteorol 39(4):487–508. https://doi.org/10.1175/1520-0450(2000)039<0487:OCITWM>2.0.CO;2

Mirzabaev A, Wu J, Evans J, Garcia-Oliva F, Hussein IAG, Iqbal MM, Weltz M et al. (2019) Climate change and land. Chapter 3: Desertification. IPCC Special Report Global Warming of 1.5 °C. 174. Available at: https://www.ipcc.ch/site/assets/uploads/2019/08/2d.-Chapter-3_FINAL.pdf

Moshia M, Khosla R, Westfall D, Davis J, Reich R (2016) Precision manure management on site-specific management zones: nNitrogen mineralization. J Plant Nutr 39(1):59–70. https://doi.org/10.1080/01904167.2015.1009547

Nagaz K, Masmoudi MM, Mechlia NB (2007) Soil salinity and yield of drip-irrigated potato under different irrigation regimes with saline water in arid conditions of Southern Tunisia. J Agron 6(2):324–330. https://doi.org/10.3923/ja.2007.324.330

Palau JL, Rovira F, Sales MJ (2017) Satellite observations of the seasonal evolution of total precipitable water vapour over the Mediterranean Sea. Adv Meteorol (2017). https://doi.org/10.1155/2017/4790541

Pausas JG, Millán MM (2019) Greening and browning in a climate change hotspot: the Mediterranean basin. Bioscience 69(2):143–151. https://doi.org/10.1093/biosci/biy157

Prats SA, MacDonald LH, Monteiro M, Ferreira AJD, Coelho COA, Keizer JJ (2012) Effectiveness of forest residue mulching in reducing post-fire runoff and erosion in a pine and a eucalypt plantation in north-central Portugal. Geoderma 191:115–124. https://doi.org/10.1016/j.geoderma.2012.02.009

Prospero JM, Lamb PJ (2003) African droughts and dust transport to the Caribbean: climate change implications. Science 302(5647):1024–1027. https://doi.org/10.1126/science.1089915

Ramachandran Nair PK (1993) In: An introduction to agroforestry. Oluwer Academic Publishers, vol 23. https://doi.org/10.1177/003072709402300413

Roberts CN, Woodbridge J, Palmisano A, Bevan A, Fyfe R, Shennan S (2019) Mediterranean landscape change during the Holocene: synthesis comparison and regional trends in population land cover and climate. Holocene 29(5):923–937. https://doi.org/10.1177/0959683619826697

Roe S, Streck C, Obersteiner M, Frank S, Griscom B, Drouet L, Lawrence D et al. (2019, November 1). Contribution of the land sector to a 1.5 °C world. Nature Climate Change. Nature Publishing Group. https://doi.org/10.1038/s41558-019-0591-9

Rubel F, Kottek M (2010) Observed and projected climate shifts 1901–2100 depicted by world maps of the Köppen-Geiger climate classification. 19(2):135–141. https://doi.org/10.1127/0941-2948/2010/0430

Ruiz I, Almagro M, García de Jalón S, Solà MDM, Sanz MJ (2020) Assessment of sustainable land management practices in Mediterranean rural regions. J Environ Manage 276(September). https://doi.org/10.1016/j.jenvman.2020.111293

Ruiz I, Sanz MJ (2020) Effects of historical land-use change in the Mediterranean environment. Sci Total Environ 732:139315. https://doi.org/10.1016/j.scitotenv.2020.139315

Sanz MJ, Vente JD, Chotte J-L, Bernoux M, Kust G, Ruiz I, Akhtar-Schuster M (2017) In: Sustainable Land Management contribution to successful land-based climate change adaptation and mitigation (A Report o)

Sastre B, Álvarez B, Antón O, Pérez MÁ, Marques MJ, Bienes R, García-Díaz A (2020) Groundcovers in olive groves in semiarid climates: are they always beneficial? Water 12(8):2230. https://doi.org/10.3390/w12082230

Schoonhoven Y, Runhaar H (2018) Conditions for the adoption of agro-ecological farming practices: a holistic framework illustrated with the case of almond farming in Andalusia. Int J Agric Sustain 16(6):442–454. https://doi.org/10.1080/14735903.2018.1537664

Schwilch G, Hessel R, Verzandvoort S (Eds) (2012) In: Desire for greener land. options for sustainable land management in Drylands. Bern, Switzerland, and Wageningen, The Netherlands: University of Bern-CDE, Alterra-Wageningen UR, ISRIC-World Soil Information and CTA-Technical Centre for Agricultural and Rural Cooperation

Seneviratne SI, Nicholls N, Easterling D, Goodess CM, Kanae S, Kossin J, Zhang X et al. (2012) Changes in climate extremes and their impacts on the natural physical environment. In: Managing the risks of extreme events and disasters to advance climate change adaptation. A Special Report of Working Groups I and II of the IPCC. A Special Report of Working Groups I and II of the Intergovernmental Panel on Climate Change (IPCC), pp 109–230

Sheil D, Murdiyarso D (2009) How forests attract rain: an examination of a new hypothesis. In: BioScience. Oxford Academic. https://doi.org/10.1525/bio.2009.59.4.12

WOCAT (1992). WOCAT SLM database. Available at: https://qcat.wocat.net/en/wocat/list/?type=technologies. Retrieved September 9 2020

WOCAT (2002) United Nations Convention to Combar Desertification. Available at: https://qcat.wocat.net/en/wocat/. Retrieved September 2, 2020

World Bank (2018) The World Bank Group. Available at: https://data.worldbank.org/indicator/SP.POP.GROW?view=chart (Accessed 6.12.19)

Zeder MA (2008) Domestication and early agriculture in the Mediterranean basin: origins diffusion and impact. Proc Natl Acad Sci USA 105(33):11597–11604. https://doi.org/10.1073/pnas.0801317105

Climate Change Impacts on North Africa: Public Health Perspectives

Rehab A. Rayan, Mohamed Kamal, Christos Tsagkaris, and Loyle Campbell

Abstract Northern Africa is experiencing warmer and longer summers, more frequent and extreme heat waves, and changed precipitations. This is coupled with considerable socio-economic factors, rapid population growth, and increasing demand on hydro-systems that cause additional health risks to North Africans. Climatic changes impact the health of the region's inhabitants. Either directly through severe heat, drought, and storms, or indirectly through atmospheric pollution or alterations in access to quality food and water. The primary impacts to health are caused by societal and ecological aspects, severe climatic conditions, and changes in the patterns in climate-associated diseases. Hence, these limited-resources North African countries are encountering growing risk. Climate change has serious implications to those who are more susceptible to its effects. The elderly—especially the poor and chronically ill—are one such demographic. Therefore, enhancing surveillance and managing systems to mitigate and adapt to climate change is becoming inevitable. This will probably be coupled with the adoption of preventive health action measures that allow for cross-border collaboration among all the region's countries. This chapter introduces the demographics and socioeconomics of countries in North Africa. It begins by presenting tangible issues such as water scarcity, rising sea and CO_2 levels. Following this it discusses how climatic change has impacted public health, focusing specifically on infections and chronic diseases. The chapter concludes by highlighting major adaptation and mitigation strategies emphasizing national and regional policies.

R. A. Rayan (✉)
Department of Epidemiology, High Institute of Public Health, Alexandria University, Alexandria, Egypt
e-mail: rayanr@alexu.edu.eg

M. Kamal
EcoConServ Environmental Solutions (ECS), Cairo, Egypt

C. Tsagkaris
Faculty of Medicine, University of Crete, Heraklion, Greece

L. Campbell
Sciences Po Paris, School of International Affairs, Lewis Katz, USA

W. Leal Filho and E. Manolas (eds.), *Climate Change in the Mediterranean and Middle Eastern Region*, Climate Change Management,
https://doi.org/10.1007/978-3-030-78566-6_22

1 Introduction

1.1 Demographics and Socioeconomics

North Africa is a geographical region that interconnects the Middle East region with Africa. There is not any internationally accepted or agreed definition regarding which countries are covered by this region. The United Nations Statistics Division geographically defines North Africa as Algeria, Egypt, Libya, Morocco, Sudan, Tunisia, and Western Sahara (UN 2011). While this is not the same definition adopted by the African Union or the African Development Bank, this study will use the UN subdivision. Aside from Sudan and Western Sahara, the remaining countries are all adjacent to the Mediterranean Sea, which has historically played a fundamental role in the development of the North African economy (Issawi 2006).

According to the "*World Population Prospects 2019*" published by the UN department of Economic and Social Affairs, North Africa's total population is approximately 246 million people, making it only the third most populous subregion in Africa. The country with the largest population as of 2020 is Egypt at 102 million people. The smallest populations are noted in Morocco and Western Sahara at 6.8 million and 600 thousand respectively (UN DESA 2019). The median age for the region is 25.5 years, with nearly 52% of the population living in urban areas.

Regarding size, three of the continent's five largest countries by area are in North Africa. Prior to South Sudan's independence in 2011, Sudan was the largest country in Africa. Now, the largest is Algeria (2,381,741 km^2) followed by Sudan at third (1,861,484 km^2) and Libya at fourth (1,759,540 km^2). Most of this area is covered by the Sahara Desert, which is the largest scorching desert on the planet (Pausata et al. 2020). Otherwise, making most of the area in every country unusable. In Egypt, only 7.7% of land is inhabited, with most of the population densely settled along the Nile and the coastal areas. The same scenario exists in other North African countries, as most of the populations live on the coast of the Mediterranean.

The average GDP per capita in the region is \$3602, with Libya having the highest at \$7683.8 and Sudan recording the lowest at \$441.506 (World Bank 2019a). The total nominal GDP of Northern Africa is 721 billion USD, which makes up approximately 30% of Africa's total GDP (African Development Bank 2020). The national economies of the region have faced a diverse set of obstacles, from political instability to security challenges. As of 2020, Algeria and Sudan have dropped their economic classification from upper-middle income to lower-middle income and lower-middle income to low income respectively (Serajuddin and Hamadeh 2020).

The rate of youth unemployment for the region is 33%—a figure that is inflated by Libya's youth unemployment rate of 51% (ILO 2020). There is no consistent trend across the region, with countries such as Egypt and Morocco improving during the last five years, while Algeria and Tunisia have declined dramatically (Hameda 2019). During these economic challenges, the region's population growth has not slowed down with the average fertility rate of 2.9 children per woman. This population growth is paired with North Africans living longer. In 2019, the average life expectancy in the region was recorded to be 73 years, which has grown dramatically since the 1990s (World Bank 2019b).

1.2 Water Scarcity

Both ends of the climate change spectrum, droughts, and flooding, will negatively affect agriculture in North Africa. It is predicted that climate change will cause a significant decrease in agricultural land in North Africa, which will be accompanied by a decline in crop growth seasons. Otherwise, meaning that there will be less anticipated precipitation and longer periods of drought (Radhouane 2013; Schilling et al. 2012). The impact of water scarcity, especially for irrigation, will severely damage the economic prospects of North Africa as the region struggles to grapple with food and water security challenges (African Development Bank 2020; Schilling et al. 2012). As precipitation decreases and temperatures increase, countries within the area will struggle to find crops able to grow in these conditions and could eventually face a food crisis if a strategy is not in place. This can cause direct implications on public health in the region, as reliable access to food and water plays a major role in mitigating many public health problems (Rom and Pinkerton 2014).

In Egypt, there are predictions of a large decrease in the Nile River's flow because of climate change and the expected Grand Renaissance Dam (GRED) (Abdelhaleem and Helal 2015; Schilling et al. 2012). The reduction in flow is also anticipated to affect Sudan during times of drought. However, Sudan will gain benefits as lower water levels will lower the chance of extreme flooding during the wet seasons. As for Egypt, the country has historically depended on the Nile River as its primary source of livelihood. The river has been the engine for all agricultural activity and provided the population a safe source of drinking water. With the expected decrease in flow, the Egyptian agricultural sector faces a significant threat to water security, which may force the country to explore alternative sources for safe drinking water (Droogers et al. 2012). In Morocco, agricultural activity has been dependent on seasonal precipitation and groundwater irrigation sources (Schilling et al. 2012). However, this is also threatened. Groundwater stocks are projected to decline with a reduction in precipitation, which farmers in Morocco are already experiencing. As irrigation water becomes less available in an already water-scarce region, living conditions will deteriorate, and public health problems will be magnified. With less affordable food and water available, Chronic health conditions will worsen, and acute health conditions will become more apparent as populations in the region struggle to find food and water.

1.3 Rise in Seawater Levels and Increased CO_2 Emissions

As mentioned earlier, most North Africans live near coastal areas (mainly the Mediterranean Sea and the Red Sea). Both shores are subjected to a rise in sea level and are projected to experience gradual but consistent coastal erosion (McGranahan et al. 2007; Radhouane 2013). This will cause the direct loss of land for many major cities in the region, as cities such as Alexandria, Algiers, Casablanca, Tunis, and Tripoli are all within 200 m from the shoreline. Aside from the land loss, a rise in sea levels will negatively affect the water quality in the area and cause serious losses in the freshwater sources (groundwater) for those coastal cities. This can cause serious public health problems, with diseases such as Cholera becoming more prevalent (Rom and Pinkerton 2014).

Climate change will also affect the overall air quality, as more CO_2 is released into the atmosphere making the air, we breathe less much polluted (Rom and Pinkerton 2014). In North Africa, rapid urbanization has caused the air quality in some major capital cities to deteriorate massively over the years. However, as the climate becomes less stable and more emissions are released into the atmosphere, North Africans will become exposed to more particulate matter air pollution (Farzanegan and Markwardt 2018).

2 Aims and Objectives

This chapter aims to discuss some climatic change impacts on public health, focusing specifically on infections and chronic diseases. It highlights major adaptation and mitigation strategies emphasizing national and regional policies.

3 Methods

A review of the publications was undertaken to determine climate change effects on health in North Africa, which guided planning this paper's detailed structure. We focused on outcomes from research on public health and climate-associated diseases in North Africa. The paper emphasizes that climate change and ecology in Northern Africa comprises a key characteristic in the emerging diseases and hence, public health. The structure of this paper offers a perspective on the mechanisms of climate changes influencing climate-associated diseases. It presents some policy challenges encountered by Northern Africa countries in adopting national and regional mitigating solutions to prevent and control such diseases. It delivers recommendations on adaptation and mitigation strategies to manage climate changes effects on such diseases in North Africa.

4 Results

A new challenge on the horizon for North African countries is the impact of climate change. As sea levels rise and overall temperature increases, more existential and resource scarcity challenges will be magnified as North African countries attempt to adapt. Climate change in North Africa will have a diverse range of impacts, ranging from scarcity of drinking water to increased heat waves and unstable weather (Chiarelli et al. 2015; Radhouane 2013). Since the region is considered one of the most water-scarce areas on the planet (Droogers et al. 2012), a future with less water will further deteriorate the fragile economies in the region that are only growing in population.

5 The Effects of Climate Change on Health

The impact of climate change on the social and environmental determinants of health has become increasingly relevant over the last few decades. The numbers speak for climate change, urging for a push to move the world away from its consequences. By 2017, the emission of greenhouse gases caused a 1.0 °C global temperature increases from pre-industrial levels, which inadvertently caused sea levels to rise by about 20 cm since 1880 (WHO 2018a). These ecological changes have caused serious harm to humanity. The coronavirus pandemic can be considered one example of this. As many scholars have identified that environmental deregulation pays back in human health. In 2020, while fighting a once-in-a-century pandemic, several scholars pointed out that environmental deregulation pays back in human health. Between 2030 and 2050, climate change is anticipated to result in approximately 250,000 additional deaths per year. The etiology of these deaths will vary from malnutrition, malaria, diarrhea, and heat stress to cancer. The triggering environmental factors include but are not limited to air quality, extreme weather, and vector-borne diseases (Manisalidis et al. 2020; WHO 2018b).

Climate change affects the quality of air both indoors and outdoors. Higher and unstable temperatures can reduce air quality, which can compromise the quality of life for patients with respiratory and cardiovascular diseases (US EPA 2014). It has been shown that in this frame rising carbon dioxide levels and warmer temperatures can increase airborne allergens, such as ragweed pollen. The quantity of ozone in the air is also an important component of unhealthy air quality (Brey et al. 2018). Ground-level ozone is a harmful air pollutant that can be found in city smog and exposure can increase the risk of premature death and respiratory diseases (Zhang et al. 2019). Extreme weather is probably the most striking feature of climate change's impact on health. Over the last few decades, the increasing frequency and severity of extreme precipitation, flooding, droughts, storms, and other extreme weather events, has created a new set of public health risks. Groups that are vulnerable to this include young children, older adults, people with disabilities or medical conditions, and the poor (Curtis et al. 2017).

5.1 Infectious Diseases

While hepatitis C (HCV) has been a major public health problem for some countries in North Africa, the prevalence differs (Daw et al. 2016). Prior to a national intervention in 2017, Egypt had the highest rate of HCV cases in the world, with a survey from 2008 showing 14.7% of the population being infected (Elgharably et al. 2016; Esmat 2013). Where the climate gets warmer and more unpredictable, even more people will be exposed to HCV because there is already a high rate of infection in the region (Rom and Pinkerton 2014; World Health Organization 2017). The higher rate of infection will also bring problems as colder temperatures can amplify the symptoms of patients diagnosed with HCV.

As North African winters have increased precipitation and summer has warmer temperatures, there will be an increased risk of exposure to vector-borne diseases (Radhouane 2013; Rom and Pinkerton 2014). There is a direct correlation between harmful vector's actual potential to the total number of health risks. Some climatic changes will bring an ideal environment for harmful vectors that carry diseases and increase the transmission of infectious disease.

Vector-borne diseases are transmitted by disease vectors, such as mosquitoes, ticks, and fleas. Infectious pathogens, such as viruses, bacteria, and protozoa are carried by these organisms from animals to humans. The COVID-19 pandemic has been considered a vector-borne disease, with either bats or pangolins transmitting the SARS-CoV-2 virus to humans. The geographic spread, the survival, and the population dynamics of vectors are affected by changes in temperature, precipitation, and extreme events (Maragakis 2020). Some well-known examples of vector-borne diseases include waterborne pathogens (bacteria, viruses, and parasites such as *Cryptosporidium* and *Giardia*); toxins produced by harmful algal and cyanobacterial products in the water; and chemicals that end up in water from human activities. A change in water temperature patterns results in an elongated presence of pathogens and toxins in water and seafood throughout the year. Regions not touched by these diseases in the past may be affected as well (Campbell-Lendrum et al. 2015; WHO 2020). This applies particularly to Vibrio bacteria and algal toxins (Lipp et al. 2002).

Like other aspects, nutrition is also affected by climate change. Elevated air temperature has been associated with food poisoning caused by *Salmonella*, given that bacterial growth is faster in warm environments. Depending on the severity of the disease and healthcare accessibility, the clinical spectrum of these diseases ranges from gastrointestinal distress to death. Conventional food control practices are challenged by climate change and food safety protocols need to be continuously updated (Kaput et al. 2015; Sabaté et al. 2016). Finally, climate change has been linked to an increased risk of exposure to chemical contaminants in food. Elevated sea surface temperatures can increase mercury concentrations in seafood, while extreme weather events affecting the oceans, seas and lakes are likely to spread contaminants from the local ecosystems' food chain to predators and eventually humans. Other studies have shown altered arsenic availability, uptake, and allocation in rice under elevated temperature (US EPA 2014).

5.2 Chronic Diseases

The impact of climate change will directly affect public health in North Africa. Considering that the region's population is growing and aging, a range of commonly found chronic conditions could be amplified by the negative repercussions of climate change (Rom and Pinkerton 2014). The current prevalence for diabetes in the Middle East and North Africa region is 9.2% for the adult population, which is amongst the highest in the world. There are many reasons this occurred, including a significant reduction in physical activity and other major lifestyle changes that accompanied urbanization of the region (Majeed et al. 2014). Diabetics are particularly vulnerable to climate change because they can have an increased sensitivity to heat stress. This will become a common challenge that public health systems will have to consider, particularly if they experience extreme heat waves. Beyond this, climate change will push the healthcare sector to shift medical supplies to more acute public health issues, which could eventually cause a shortage of medical supplies for diabetics (Zilbermint 2020).

Like diabetes, people with heart disease will have negative consequences of climate change because their medications make them more susceptible to heat stress (Traina et al. 2017). The rapid urbanization over the past 40 years has led to major lifestyle and dietary changes, leading to heart disease becoming an increasingly pervasive cause of death as North Africa's population begins to age (Rom and Pinkerton 2014).

In a similar vein, it is also very difficult for estimations to calculate deaths inflicted by disruptions to transportation systems, healthcare structures, pharmacies, day clinics, and hospitals. Interrupted communication networks are expected to become an even more serious concern as we move towards digital health, smart and Internet of Things–based hospitals. In a similar note, power outages following natural phenomena have been linked to carbon monoxide poisoning from improper use of portable electric generators and gastrointestinal distress because of poorly conserved food (Conti et al. 2005; Curtis et al. 2017). Extreme heat results in heatstroke and dehydration, with potential cardiovascular, respiratory, and cerebrovascular implications. Populations used to lower temperatures such as those in northern latitudes are less prepared to cope with excessive temperatures (Keatinge and Donaldson 2004). On top of these, the stress of experiencing and surviving extreme weather phenomena can trigger or worsen mental health conditions such as depression and post-traumatic stress disorder (PTSD) (WHO 2018a).

The impact of modifiable and non—modifiable factors on health is a fundamental concept of public health. While the physical environment is usually classified as a non—modifiable factor, climate change can also be considered a modifiable one to the extent to which it relates to human activity. The severity of these health risks will depend on public health and safety systems' ability to address or prepare accordingly. At the same time, social and individual behaviors combined with eco-friendly policymaking can counteract the worsening impacts of climate change on health (Curtis et al. 2017; Manisalidis et al. 2020).

6 Discussions

Reduction of the negative consequences caused by climate change on the public health of North Africans will require a mixed approach of adaptation and mitigation (Ludwig et al. 2014). This will require each country to develop an integrated strategy to tackle pressing issues such as water availability and sanitation services while building national resilience to the uncertainty of climate change. Public Health Management will need to integrate the impacts of climate change in every decision-making process (Zilbermint 2020), as public health systems will be seriously affected by climate change (Rom and Pinkerton 2014).

While there are many methods to mitigate specific elements contributing to climate change and increased health risks, it is crucial to have clear national and local strategies to tackle the impacts of climate change beyond just public health (Campbell et al. 2016; Fawzy et al. 2020). Strategies developed are split into methods of adaptation and methods of mitigation. Methods of adaptation are primarily focused on building resilience to the changes that occur from climate change. In contrast, methods of mitigation focus on planning and designing interventions that lower greenhouse gas emissions by enforcing policies with clear restrictions and science-based emission goals (IPCC 2015).

6.1 Redesigning National and Regional Policies

In 2017, the Joint Monitoring Program (JMP) for Water Supply and Sanitation published a progress report highlighting the global trends for the Water Sanitation and Health (WASH) sector. This report illustrated the total percentage of the population with access to safe drinking water and improved sanitation services. In this category, North African countries perform well except for Sudan where only 60% of the population has access to clean drinking water and only 37% of the population has access to sanitation services (WHO and UNICEF 2017). If things continue to deteriorate within the country WASH services will only be further depleted for Sudan and North African countries.

To address this issue Sudan would need to design a national policy to achieve higher levels of safe drinking water and accessible sanitation services. For this policy to achieve its true potential, it must integrate the projected or expected impacts of climate change on the country (Puppim de Oliveira 2009). It must also aim to implement small adaptation and mitigation initiatives to ensure that the policy solutions suggested are aiding the country in tackling WASH-related problems while building resilience to climate changes predicted challenges. In countries such as Egypt and Morocco, substantial resources are required to reshape and redesign the national water policies. Efforts to do so must address the respective reductions in water flow for the Nile River and groundwater reserves (Abd Ellah 2020; Schilling et al. 2012). Not only are the drinking water sources for both countries under threat, but they are consistently being contaminated and polluted (Hamed et al. 2018). This has put water resources in North African countries under threat from multiple angles, with little guidance provided to tackle these issues.

There have already been several national initiatives to provide policies or strategies that tackle the problems related to the projected water scarcity in Egypt. These include the 2017 "*Water for the Future*" national water resource plan and the 2030 strategic vision for "*treated wastewater reuse in Egypt*". Both acknowledge the impacts of climate change, aside from other external factors, on water scarcity.

However, there is no clear picture of how these initiatives fit on the larger scale. For this, Egypt must design a comprehensive national water policy covering all the different water usage and how they must adapt to climate change over the coming years (Zekri 2020). This must include programs for the safe reuse of wastewater in agriculture, methods to use alternative sources of drinking water during droughts and initiatives to increase the efficiency of water usage throughout the country. The policymakers must ensure that the water-energy-food nexus is sufficiently covered within this policy and that it truly provides decision-makers with sustainable alternatives and guidelines on how to pursue those alternatives. The policy must also state clear standards and goals for public health because water availability plays a significant role in reducing disease (Rom and Pinkerton 2014).

The ClimaSouth involves some North African nations, such as Morocco, Algeria, Tunisia, Libya, and Egypt. Such a program focuses on recovery; however, public health is not an interest in this framework. Other regional collaborations including the North Africa region are the Mediterranean Action Plan (MAP). The MAP is an ongoing collaboration of 22 bodies to the Barcelona Convention and the Mediterranean Experts on Environmental and Climate Change (MedECC). This partnership accesses a global network of over 400 professionals who coordinate with state institutions, like the Climate Experts Group of the Union for the Mediterranean and the Barcelona Convention.

6.2 *Methods of Adaptation and Mitigation*

Some notable methods of adaptation have been recently applied in Africa. This can be seen in the development of national disaster risk management departments and public health management divisions (IPCC 2015). However, adaptation initiatives are usually isolated and lacking an integrated approach or direction. This is where regional cooperation can play a crucial role in adaptation because regional countries will usually share similar challenges.

North African countries must implement mitigation strategies that reduce the consumption of hydrocarbons, as they are primary contributors to pollution and greenhouse gas emissions (U. N. Environment 2020). This can include replacing polluting energy sources with renewable energy technologies such as hydro, wind, solar or biogas by implementing incentives for the energy producers at a national level (Radhouane 2013). This would also mitigate the negative impacts polluting energy sources have on public health by improving the overall air quality (Rom and Pinkerton 2014). It would also improve the energy security of remote rural areas in the North African region, which are common and usually lack a consistent energy supply. Any strategy to replace conventional energy sources with non-conventional sources must also consider emissions from the industrial sector, as these emissions can be even more detrimental to air quality than domestic energy consumption (WMO 2019). This must also include efforts of reducing methane emissions (CH_4) that are either vented or leaked into the atmosphere as this gas's greenhouse effect is 28 times stronger than CO_2.

One crucial sector that plays a major element in the region's public health and food security is agriculture. Agricultural activity is currently under threat in North Africa, as water scarcity is becoming apparent and more farmland either lacks sustained access to irrigation or is being transformed into residential land (World Bank 2016). For this specific challenge, there are three key approaches that North African countries must adopt. First, the agricultural sectors must build resilience regarding natural disasters and their impact on crops and yield. As heavy precipitation becomes more regular during the wet season and more extreme weather, farmers must adapt and have a resilience plan to ensure that they do not lose their crops to the harsh conditions (Schilling et al. 2012). This could include simple approaches such as building protections for the crops and soil, from the external environment. Other more advanced adaptation measures include growing crops that would not be as affected by an overflow of water or that have stronger roots ensuring they can survive extreme weather.

The second approach revolves around using treated wastewater for agricultural irrigation, to lower the impact on the already depleting groundwater and surface water reservoirs. In the case of Algeria, where there is a notable depletion of groundwater sources (Samia et al. 2013), the usage of treated wastewater could mitigate the impact on groundwater reservoirs. All the while providing the farming industry with a reliable source of irrigation, that is usually rich with nutrients (Benzater et al. 2019). This can play an even larger role in building resilience for isolated communities and rural areas by creating a circular model for the consumption of treated wastewater. All North African countries must exploit the potential within wastewater reuse for irrigation, especially countries like Egypt, where the predicted impact of surface water reduction can be catastrophic (Abdelhaleem and Helal 2015). Not only would providing wastewater treatment infrastructure mitigate the impacts of water scarcity, but it would also provide a crucial public health and environmental protection service by which less untreated wastewater is disposed into the environment and our natural waterways (Hamdard 2012). If implemented within an overarching program, it could also enhance the access to improved sanitation services in rural areas in North Africa, where there is usually less service coverage than urban areas.

The third approach is to introduce more efficient irrigation methods that conserve water while also ensuring that crops get sufficient irrigation (Sakellariou-Makrantonaki and Papanikolaou 2008). This is a mixed-methods approach, by which it is a measure of adaptation and mitigation simultaneously. Rather than using the common flood irrigation method in areas where surface water is available in North Africa, such as Egypt and Sudan, farmers could be trained on utilizing drip irrigation systems that consume much less water (El-Kilani and Sugita 2017). Farmers could be trained to utilize digital applications and technology to monitor the crops and know how to maximize their efficiency regarding water usage to reduce their overall water consumption (Sakellariou-Makrantonaki and Papanikolaou 2008).

Countries must also adapt to potential droughts and famine, where there could be a serious shortage in the food supply chain, by having a national plan that can be triggered once the dangers are anticipated (IPCC 2015). Extreme weather conditions can further expose people with chronic conditions to more risk, and North African countries must ensure that national public health systems are knowledgeable of the dangers of climate change on patients. Furthermore, efforts must be made to create R&D divisions focused on mitigating the impacts of climate change on people with chronic diseases, especially those in areas with low health service coverage (Rom and Pinkerton 2014). Mitigation measures would include alternative medication plans that expose patients to fewer risks. The country must also have national awareness programs of the public health implications of climate change, to further educate the population (Radhouane 2013). This must be contextualized since certain rural areas across North Africa would not comprehend the scope of the problem or relate to it unless it is brought to their attention in a relatable approach.

7 Limitations

Central authorities in North Africa lack awareness about the link between health and climate, which stems from the broader absence of work connecting the impact of climate change on health. If the implications of climate change are not incorporated into the sectoral analysis for healthcare, then climate-related public health risks may be neglected. To address this, government agencies across the region should be provided access to the relevant literature and use it to develop evidence-based policies. Research that might be included could be technical reports from the UN and policy briefs or seminars from international conferences. The multimedia can play an important role in bringing such documents to the attention of relevant stakeholders. Especially in the case of identifying escalating concerns to public health. Accordingly, it is imperative to provide budgets for preventative measures to counteract climate-related health hazards. Yet, public funding is likely to be insufficient to cover the costs of such measures. In this case variable funding sources are a viable solution to fill any gaps in the evaluation, management, and prevention of climate-related health hazards. However, further research monitoring the status and trends of climate-related health hazards classified by causes of death in North Africa is required.

8 Conclusions and Future Perspectives

The North Africa region is struggling to grapple with internal and cross-national disputes, which have restricted regional and international collaborations. Climate change affects the North African population's health directly via extreme temperatures, drought, or storms, or indirectly via altering in water and food availability or quality, air pollution and other triggers. These direct and indirect aspects can induce the spread of climate-associated diseases, which are further aggravated by ecological and social circumstances. The poorer countries of North Africa are at the greatest risk. Climate change can modify the dissemination of primary pollutants and increase the production of secondary pollutants, like near-surface ozone. At the same time, it impacts the most vulnerable the hardest. Individuals with preexisting conditions, the elderly and the poor are the most susceptible to extreme heat. For such categories, improved surveillance and control systems are required. The effects of climate change are beyond the borders of North Africa. Higher temperatures and droughts are increasingly global, which challenges nations that lack the sufficient political resources to collaborate. However, environmental and health bodies, such as the MAP, are seeking to address this by prioritizing regional collaboration. Regarding the climatic projections and the susceptibility of North Africa nations, climate change adaptation and mitigation should be implemented especially for countries with no action plans for prevention. Respecting health, among the highly successful adaptation measures, could be decreasing air pollution or offering shading solutions. North African nations should collaborate on cross-borders levels for adapting several health hazards such as vector-borne diseases and droughts.

References

Abd Ellah RG (2020) Water resources in Egypt and their challenges, Lake Nasser case study. Egypt J Aquatic Res 46(1):1–12. https://doi.org/10.1016/j.ejar.2020.03.001

Abdelhaleem F, Helal E (2015) Impacts of grand ethiopian renaissance dam on different water usages in upper Egypt. British J Appl Sci Technol 8(5):461–483. Available at: https://www.academia.edu/12587251/Impacts_of_Grand_Ethiopian_Renaissance_Dam_on_Different_Water_Usages_in_Upper_Egypt

African Development Bank (2020) African economic outlook 2020: developing Africa's workforce for the future. Available at: https://africa.sis.gov.eg/english/library/reports/african-economic-outlook-2020-developing-africa-s-workforce-for-the-future/

Benzater B, Hachemaoui A, Elouissi A, Benaricha B (2019) The reuse of wastewater in the context of climate change: case of mascara wilaya (Algeria). In: Chaminé HI, Barbieri M, Kisi O, Chen M, Merkel BJ (eds) Advances in sustainable and environmental hydrology, hydrogeology, hydrochemistry and water resources. Springer International Publishing, pp 153–156. https://doi.org/10.1007/978-3-030-01572-5_38

Brey SJ, Barnes EA, Pierce JR, Wiedinmyer C, Fischer EV (2018) Environmental conditions, ignition type, and air quality impacts of wildfires in the southeastern and western United States. Earth's Fut 6(10):1442–1456. https://doi.org/10.1029/2018EF000972

Campbell-Lendrum D, Manga L, Bagayoko M, Sommerfeld J (2015) Climate change and vector-borne diseases: what are the implications for public health research and policy? Philos Trans Royal Soc B: Biolog Sci 370(1665):20130552. https://doi.org/10.1098/rstb.2013.0552

Campbell BM, Vermeulen SJ, Aggarwal PK, Corner-Dolloff C, Girvetz E, Loboguerrero AM, Ramirez-Villegas J, Rosenstock T, Sebastian L, Thornton PK, Wollenberg E (2016) Reducing risks to food security from climate change. Glob Food Sec 11:34–43. https://doi.org/10.1016/j.gfs.2016.06.002

Chiarelli DD, Davis KF, Rulli MC, D'Odorico P (2015) Global change in Africa: climate change and large scale land acquisition. Rivista Internazionale Di Scienze Sociali, 131(4):463–472. Available at: https://ideas.repec.org/a/vep/journl/y2015v131i4p363-472.html

Conti S, Meli P, Minelli G, Solimini R, Toccaceli V, Vichi M, Beltrano C, Perini L (2005) Epidemiologic study of mortality during the summer 2003 heat wave in Italy. Environ Res 98(3):390–399. https://doi.org/10.1016/j.envres.2004.10.009

Curtis S, Fair A, Wistow J, Val DV, Oven K (2017) Impact of extreme weather events and climate change for health and social care systems. Environ Health 16(1):128. https://doi.org/10.1186/s12940-017-0324-3

Daw MA, El-Bouzedi A, Ahmed MO, Dau AA, Agnan MM (2016) Hepatitis C virus in North Africa: an emerging threat [Review Article]. The Scientific World Journal; Hindawi. https://doi.org/10.1155/2016/7370524

Droogers P, Immerzeel WW, Terink W, Hoogeveen J, Bierkens MFP, van Beek LPH, Debele B (2012) Water resources trends in middle East and North Africa towards 2050. Hydrol Earth Syst Sci 16(9):3101–3114. https://doi.org/10.5194/hess-16-3101-2012

Elgharably A, Gomaa AI, Crossey MM, Norsworthy PJ, Waked I, Taylor-Robinson SD (2016) Hepatitis C in Egypt—past, present, and future. Int J General Med 10:1–6. https://doi.org/10.2147/IJGM.S119301

El-Kilani RMM, Sugita M (2017) Irrigation methods and water requirements in the Nile Delta. In: Satoh M, Aboulroos S (eds) Irrigated agriculture in Egypt: past, present and future. Springer International Publishing, pp 125–151 https://doi.org/10.1007/978-3-319-30216-4_6

Esmat G (2013) Hepatitis C in the eastern mediterranean region. Eastern Mediterranean Health Journal = La Revue De Sante De La Mediterranee Orientale = Al-Majallah Al-Sihhiyah Li-Sharq Al-Mutawassit, 19(7), 587–588

Farzanegan MR, Markwardt G (2018) Development and pollution in the Middle East and North Africa: democracy matters. J Policy Model 40(2):350–374. https://doi.org/10.1016/j.jpolmod.2018.01.010

Fawzy S, Osman AI, Doran J, Rooney DW (2020) Strategies for mitigation of climate change: a review. Environ Chem Lett 18(6):2069–2094. https://doi.org/10.1007/s10311-020-01059-w

Hamdard M (2012) Freshwater swaps: potential for wastewater reuse—Alexandria. LAP LAMBERT Academic Publishing, Egypt

Hameda E (2019, December 16) Youth unemployment in North Africa: a short overview—IFAIR. Available at: https://ifair.eu/2019/12/16/youth-unemployment-in-north-africa/

Hamed Y, Hadji R, Redhaounia B, Zighmi K, Bâali F, El Gayar A (2018) Climate impact on surface and groundwater in North Africa: a global synthesis of findings and recommendations. Euro-Mediterranean J Environ Integrat 3(1):25. https://doi.org/10.1007/s41207-018-0067-8

ILO (2020) Unemployment, youth total (% of total labor force ages 15–24) (modeled ILO estimate)|Data. Available at: https://data.worldbank.org/indicator/SL.UEM.1524.ZS

IPCC (2015) AR5 synthesis report: climate change 2014. Available at: https://www.ipcc.ch/report/ar5/syr/
Issawi C (2006) An economic history of the middle East and North Africa (1st edition). Routledge
Kaput J, Kussmann M, Radonjic M, Virgili F, Perozzi G (2015) Human nutrition, environment, and health. Genes and Nutrit 10(5). https://doi.org/10.1007/s12263-015-0489-8
Keatinge WR, Donaldson GC (2004) The impact of global warming on health and mortality. South Med J 97(11):1093–1099. https://doi.org/10.1097/01.SMJ.0000144635.07975.66
Lipp EK, Huq A, Colwell RR (2002) Effects of global climate on infectious disease: the cholera model. Clin Microbiol Rev 15(4):757–770. https://doi.org/10.1128/CMR.15.4.757-770.2002
Ludwig F, van Slobbe E, Cofino W (2014) Climate change adaptation and integrated water resource management in the water sector. J Hydrol 518:235–242. https://doi.org/10.1016/j.jhydrol.2013.08.010
Majeed A, El-Sayed AA, Khoja T, Alshamsan R, Millett C, Rawaf S (2014) Diabetes in the middle-east and North Africa: an update. Diabetes Res Clin Pract 103(2):218–222. https://doi.org/10.1016/j.diabres.2013.11.008
Manisalidis I, Stavropoulou E, Stavropoulos A, Bezirtzoglou E (2020) Environmental and health impacts of air pollution: a review. Front Public Health, 8.https://doi.org/10.3389/fpubh.2020.00014
Maragakis LL (2020) Coronavirus, social and physical distancing and self-quarantine. Available at: https://www.hopkinsmedicine.org/health/conditions-and-diseases/coronavirus/coronavirus-social-distancing-and-self-quarantine
McGranahan G, Balk D, Anderson B (2007) The rising tide: assessing the risks of climate change and human settlements in low elevation coastal zones. Environ Urban 19(1):17–37. https://doi.org/10.1177/0956247807076960
Pausata FSR, Gaetani M, Messori G, Berg A, Maia de Souza D, Sage RF, deMenocal PB (2020) The greening of the sahara: past changes and future implications. One Earth 2(3):235–250. https://doi.org/10.1016/j.oneear.2020.03.002
Puppim de Oliveira JA (2009) The implementation of climate change related policies at the subnational level: an analysis of three countries. Habitat Int 33(3):253–259. https://doi.org/10.1016/j.habitatint.2008.10.006
Radhouane L (2013) Climate change impacts on North African countries and on some Tunisian economic sectors. J Agric Environ Int Dev (JAEID) 107(1):101–113. https://doi.org/10.12895/jaeid.20131.123
Rom WN, Pinkerton KE (2014). Introduction: consequences of global warming to the Public's Health. In Pinkerton KE, Rom WN (eds) Global climate change and public health. Springer, pp 1–20. https://doi.org/10.1007/978-1-4614-8417-2_1
Sabaté J, Harwatt H, Soret S (2016) Environmental nutrition: a new frontier for public health. Am J Public Health 106(5):815–821. https://doi.org/10.2105/AJPH.2016.303046
Sakellariou-Makrantonaki M, Papanikolaou C (2008) Water saving by using modern irrigation methods. /paper/Water-saving-by-using-modern-irrigation-methods.-Sakellariou-Makrantonaki-Papanikolaou/9e87b9db7c5a4b5b0e58356aad8d13c1d53bffa3
Samia TB, Houria KB, Nacer K, Marc C (2013) Assessment and management of water resources in Northeastern Algeria: case of watersheds Kebir West Safsaf and Guebli rivers Skikda. Appl Water Sci 3(2):351–357. https://doi.org/10.1007/s13201-013-0085-2
Schilling J, Freier KP, Hertig E, Scheffran J (2012) Climate change, vulnerability and adaptation in North Africa with focus on Morocco. Agr Ecosyst Environ 156:12–26. https://doi.org/10.1016/j.agee.2012.04.021
Serajuddin U, Hamadeh N (2020) New World Bank country classifications by income level: 2020–2021. Available at: https://blogs.worldbank.org/opendata/new-world-bank-country-classifications-income-level-2020-2021
Traina MI, Almahmeed W, Edris A, Murat Tuzcu E (2017) Coronary heart disease in the Middle East and North Africa: current status and future goals. Curr Atheroscler Rep 19(5):24. https://doi.org/10.1007/s11883-017-0659-9

UN (2011) Standard Country or Area Codes for Statistical Use (M49)—Global Inventory of Statistical Standards. Available at: https://unstats.un.org/unsd/iiss/Standard-Country-or-Area-Codes-for-Statistical-Use-M49.ashx

UN DESA (2019) World population prospects 2019: Highlights. Available at: https://read.un-ilibrary.org/population-and-demography/world-population-prospects-2019-highlights_13bf5476-en

U. N. Environment (2020) United in science report: a multi-organization high-level compilation of the latest climate science information. Available at: http://www.unenvironment.org/resources/report/united-science-report-climate-change-has-not-stopped-covid-19

US EPA O (2014) Volatile organic compounds' impact on indoor air quality [Overviews and Factsheets]. Available at: https://www.epa.gov/indoor-air-quality-iaq/volatile-organic-compounds-impact-indoor-air-quality

WHO (2018a) Climate change and health. Available at: https://www.who.int/news-room/fact-sheets/detail/climate-change-and-health

WHO (2018b) Heat and Health. Available at: https://www.who.int/news-room/fact-sheets/detail/climate-change-heat-and-health

WHO (2020) Vector-borne diseases. Available at: https://www.who.int/news-room/fact-sheets/detail/vector-borne-diseases

WHO & UNICEF (2017) Progress on drinking water, sanitation and hygiene: 2017 update and SDG baselines. World Health Organization. Available at: https://apps.who.int/iris/handle/10665/258617

WMO (2019) 2019 concludes a decade of exceptional global heat and high-impact weather. Available at: https://public.wmo.int/en/media/press-release/2019-concludes-decade-of-exceptional-global-heat-and-high-impact-weather

World Bank (2016) Agriculture in Africa: telling facts from Myths [Text/HTML]. Available at: https://www.worldbank.org/en/programs/africa-myths-and-facts

World Bank (2019a) GDP (current US$)—Middle East & North Africa | Data. Available at: https://data.worldbank.org/indicator/NY.GDP.MKTP.CD?locations=ZQ

World Bank (2019b) Life expectancy at birth, total (years). Available at: https://data.worldbank.org/indicator/SP.DYN.LE00.IN

World Health Organization (2017) Global hepatitis report 2017. World Health Organization. Available at: https://apps.who.int/iris/handle/10665/255016

Zekri S (ed) (2020) Water policies in MENA countries. Springer International Publishing. Available at: https://doi.org/10.1007/978-3-030-29274-4

Zhang J, Wei Y, Fang Z (2019) Ozone pollution: a major health hazard worldwide. Front Immunol 10. https://doi.org/10.3389/fimmu.2019.02518

Zilbermint M (2020) Diabetes and climate change. J Commun Hosp Int Med Perspect 10(5):409–412. https://doi.org/10.1080/20009666.2020.1791027

Exploring the Beliefs, Concerns and Understandings About Climate Change of Greek University Students from the Social Sciences and Humanities

Maria Daskolia

Abstract Higher education institutions have a critical role to play in preparing young generations to become better aware and more responsible citizens to cope with climate change. Universities must align their curricula to help students develop the necessary understandings and skills. Yet, to be able to organize this, it is necessary to know, among other things, what students believe, feel and understand about climate change and how informed they are about it. However, most of the research conducted so far has focused mainly on primary and secondary school students' beliefs, attitudes and understandings, while the few studies addressing university students are limited to science students or students enrolled in science courses. The study reported here was conducted within a Greek University. It surveyed 250 undergraduate students from the Social Sciences and Humanities. Students were asked to fill out a questionnaire which aimed to assess their beliefs and concerns about climate change, their self-reported degree of knowledge and their understandings about how the climate system works, factors that contribute to global warming and actions to reduce it. Data analysis showed that most students believe that climate change is real and largely human-induced, and express their concern about it. Yet, there are gaps, uncertainty and misconceptions in the students' understanding of climate science and about the impact of different factors and measures. Most students agree that both schools and Universities must teach youths about the causes, consequences and solutions to climate change, and express their need to learn more about the issue. The study's findings lay out the path for possible integration of climate change education in the Social Sciences and Humanities academic programs.

M. Daskolia (✉)
Environmental Education Lab, Department of Educational Studies, School of Philosophy, National and Kapodistrian University of Athens, University Campus, 15722 Zografou, Greece
e-mail: mdaskol@eds.uoa.gr

W. Leal Filho and E. Manolas (eds.), *Climate Change in the Mediterranean and Middle Eastern Region*, Climate Change Management,
https://doi.org/10.1007/978-3-030-78566-6_23

1 Introduction

Climate change is one of the direst, most pressing and multifaceted challenges of our times. All countries and societies across the globe are being affected by the manifold harmful impacts of global warning on human-nature systems (Hansen and Stone 2016). The Mediterranean Basin and all countries pertaining to it are among the regions that are prone to be affected most by the impacts of actual and future climate changes (Ozturk et al. 2015; Cramer et al. 2018). To address the risks and consequences of climate change and promote sustainable development, informed climate action at a local, national and international scale is needed. Climate Change Education (CCE) is at the core of this action aiming to enable future generations to make responsible decisions and adopt sustainable lifestyles in order to confront the challenges raised by climate change (UNESCO 2017). To this end, '*changing mindsets, not the climate*' is the communication message eloquently put forward by UNESCO to briefly delineate the role education is envisaged to play in empowering people, youths in particular, to get involved in meaningful action against climate change.

Fostering new competences and strengthening the voices of the so-called "*climate change generation*" (Feldman et al. 2010) has never been more important. The Paris Agreement in Article 12 stresses the need for fostering "*climate change education, training, public awareness, public participation and public access to information*" across all sectors (UNFCCC 2015). It also calls for the development of new academic programs to ensure that young generations will develop a comprehensive understanding of the interrelated challenges put forth by climate change and the actions necessary to be activated (UNESCO 2017). While education at all levels, from preschool to higher education, is acknowledged as a key dimension of international climate action (UNFCCC 2015), Universities have a critical role to play in supporting students to acquire the necessary knowledge, attitudes and skills (Leal Filho 2010; Wachholz et al. 2014; Leal Filho et al. 2018; Leal Filho and Hemstock 2019). Apart from leading multi-disciplinary research, contributing to the debate or setting the climate change policy agenda, Universities are teaching institutions that address large numbers of (young) people. They are, therefore, in a key position to raise awareness and educate their students on a lifelong understanding of climate change issues.

University students are a key target audience for CCE because of their dual identity of being young adults and future professionals at the same time. As 'emergent adults', Ojala (2020) argues that they belong to a 'transitional' age group, which is more likely to develop new knowledge and to re-examine previously acquired norms and habits. Such a learning potential is coupled by the fact that they are trained to become professionals who will need to base their future work-related decisions on an informed climate change perspective. For all these reasons, higher education curricula need to provide university students with a range of CCE opportunities (Molthan-Hill et al. 2019), taking into consideration both the particularities of the disciplinary field and the professional qualifications set by different academic programs and the students' current levels of awareness, concerns and understandings about the climate change issue. Some important first steps have been taken towards this direction but there are still many more to work through (Leal Filho et al. 2018; Molthan-Hill et al. 2019).

Understanding how young people approach climate change can inform the design of CCE courses and programs. Research so far has been mainly focused on investigating the general public's perception of climate change (see among others, Weber and Stern 2011; Smith and Joffe 2013; Whitmarsh and Capstick 2018; Ballew et al. 2019) and on informing communication and policy actions. A considerable bulk of research has also been directed towards documenting and exploring primary and secondary school students' misconceptions about climate change (i.e., Boyes et al. 1993; Skamp et al. 2009; Shepardson et al. 2012; Stevenson et al. 2016; Hermans and Korhonen 2017; Jackson and Pang 2017). Comparatively fewer studies have been focused on investigating university students' beliefs, concerns and understandings of climate change.

Earlier studies in the field are mostly surveys of US American samples of graduate students with a science background or enrolled in sciences courses (Wachholz et al. 2014). They intended to measure the students' depth and accuracy of scientific knowledge about the greenhouse effect, global warming or climate change (these terms were used interchangeably), sometimes in conjunction with other environmental issues, mainly the ozone hole. Most studies report a wide range of misunderstandings about climate change among geology students (Wilson and Henson 1993) or students enrolled in an environmental science course (Morgan and Moran 1995). Misconceptions concerning the impact of the ozone hole on climate change were also reported among US students with a chemistry major (Kerr and Walz 2007) or students enrolled in a meteorology course (Cordero et al. 2008). In one study with college students majoring in elementary education (Khalid 2003), measurement of knowledge around the greenhouse effect, atmospheric ozone, and acid rain also brought up similar misconceptions and misunderstandings. All these studies reveal a rather high degree of confusion and uncertainty, mainly among science students, over several dimensions of the climate change issue, despite the fact that students identify climate change as an important environmental problem and express a high level of concern (Cordero et al. 2008).

A more recent US study, although quite influential on an international level, is the one by Wachholz et al. (2014). They surveyed 375 US college students representing a cross-section of disciplines, ranging from the humanities, social sciences, sciences, business, and other applied professions. The study attempted to go beyond previous research in the field by mostly focusing on measuring higher education students' scientific understanding of climate change and their views on and willingness to combat it. With the use of a questionnaire, the study explored the students' knowledge, attitudes, and understandings about the causes and possible consequences of climate change; their intentions and commitment to reduce their personal greenhouse gas emissions; and their degree of satisfaction with the CCE provision at university level and their suggestions for improving it. The study concluded that the majority of students believed climate change is occurring and that it is basically a human-induced environmental issue. Approximately two-thirds of the respondents reported being concerned about climate change, even though their worries were not reflected in some kind of action to mitigate it. Although most of them perceived their knowledge about climate change to be sufficient, they did not answer some basic questions about it correctly. Almost half of the students believed that there was a high degree of disagreement among scientists about whether climate change is real, and many of them viewed it as a slow and distant problem, unrelated to their current well-being. Finally, a high percentage of students reported being dissatisfied with the CCE opportunities provided by their University.

In a Mediterranean context, there are a few national and cross-national examples of research focusing on university students' beliefs, concerns and understandings of climate change. Among the first studies and the only one conducted so far with a Greek university student population is that by Manolas et al. (2010). The sample consisted of 245 forestry students enrolled in one University in Northern Greece. A questionnaire was designed and administered to the students during the examination period, asking them, among other things, about their views on the seriousness of the climate change issue, the factors contributing to the greenhouse effect, the causes of global temperature increase, and the efforts to solve the problem. The study's findings indicated that the students recognized the climate change problem as important. Among the factors contributing the most to the greenhouse effect, sources that produce greenhouse gases and human actions and policies that intensify the problem were identified. Global temperature increase was perceived by the students to be caused mainly by industrial activity, CFCs, deforestation, energy plants and transportation. Among the factors rated as least important were central heating installations, the increase in global population, household waste dumps, and agricultural and cattle-raising activities. Respondents expressed the view that global temperature rise caused by the greenhouse effect is under the individual's control. They expected that adverse consequences will be particularly obvious in the future and that their generation will be negatively affected by them. Despite their pessimistic forecasts, one in three students deemed that all present efforts are useful. Even though they declared willingness to support all measures for reducing carbon emissions, the majority of them was rather reluctant or negative to bear the brunt of such measures.

Most of the studies conducted either with a Spanish or a Portuguese university student population are cross-national, involving at least one more, Latin American or other European, university student sample. In one of the first studies in such context, Bråten et al. (2009) explored and compared the personal epistemologies with respect to climate change across 225 Norwegian and 217 Spanish undergraduates attending studies in Psychology and Education with the use of a 49-item questionnaire. They also examined the relationships between epistemic beliefs about climate change and the students' gender, knowledge and interest in the topic. The study indicated the cultural embeddedness of epistemic beliefs about climate change, and pointed to some factors restricting or enhancing such beliefs. More particularly, simplicity beliefs were found to be more evident among Norwegian students, while certainty beliefs among Spanish students. However, the study concluded that further research was necessary to explore the cultural variables at the root of observed differences and to discern whether such differences are related to a different emphasis of public discourse in Northern Europe in terms of the climate change's negative consequences, compared to a focus on the certainty aspect of knowledge about climate change in Southern Europe.

In another study that was carried out with a cross-national sample of 484 young university students, students from Spain and Italy (together with students from Brazil, China and the United States) were asked to examine how they perceive media communication in relation to the scientific consensus over climate change (Díaz Estévez et al. 2014). Data were collected through a survey questionnaire that was designed and distributed to the students during class hours. The study suggested that despite the fact that climate change seems to be part of a global culture and related to the values defining this generation's interests, respondents showed they had a rather superficial knowledge of climate change causes. They also had a deficient knowledge regarding scientific consensus over climate change. The majority recognized there is human intervention in climate change, while more than half considered that climate change is also caused by natural factors. When asked about the most reliable media to expand their knowledge about climate change, they named films (both fiction and documentaries), although social media is where they usually search to get information about the topic, followed by online press and television. Finally, respondents expressed the view that an interdisciplinary approach is essential to tackle problems related to climate change.

A more recent study conducted on a Spanish students sample is the one by Arto-Blanco et al. (2017). 219 Mexican and 284 Spanish undergraduate students from various faculties of two national Universities participated in the study. The researchers used both qualitative (drawings and text) and quantitative (questionnaire) measures and analysis to explore the students' social representations of human responsibility regarding climate change. The study's results indicated that although most participants identified human agency as the cause of climate change and recognized that everyday human activities aggravate the problem, there were gaps in the students' understandings about it. Greenhouse emissions were primarily associated with industry and the use of automobiles, while domestic consumption (in energy or food) was not among the factors identified. General pollution or specific forms of it, along with other environmental degradation processes, such as deforestation, were acknowledged to contribute to the problem. Spanish students referred to the use of vehicles followed by recycling and waste management as the individual activities mostly contributing to climate change, even though they could not specify how. They also emphasized economic growth and industrialization as the main culprits of climate change compared to Mexicans who identified exclusively individual activities. Spanish students tended to acknowledge different levels of citizen responsibility compared to Mexicans. The study concluded that 'cultural variables' seem to have a more decisive role in the students' conceptualization of climate change, over-exceeding the one 'academic field' or 'year of study' have on climate literacy patterns.

Another recent study conducted exclusively on a Portuguese university sample is the one by Santos et al. (2016). The study addressed the views and attitudes about climate change among 124 postgraduate students attending three master programs in environmental sciences in 3 Universities respectively. A self-completion questionnaire consisting of 30 closed-ended and two open questions adapted from Wachholz et al. (2014) and Manolas et al. (2010) was administered to the students. The study's findings indicated that nearly all respondents believed that climate change is real and largely induced by humans. They also expressed their concern about climate change and the view that the scientific community is in agreement. However, less than half of them believed that their Master academic training had focused enough on the topic, and expressed the need for more information. The majority reported they had a moderate technical knowledge about the issue. The study also pointed to some misconceptions the students held about the causes and consequences of climate change, such as about the effect of the ozone hole on it.

The same 124 Portuguese students were also part of the cross-national sample in the study reported by Morgado et al. (2017). 40 Mexican and 40 Mozambican postgraduates completed the same questionnaire in a study aiming to investigate the students' perceptions, motivations, attitudes and knowledge about climate change in three cultural contexts. The study's findings indicated that most respondents across all three nationalities believed that climate change is happening, although Mexican students were significantly less certain compared to the Portuguese and the Mozambicans. Mozambican students were significantly much more interested in climate change. They all believed that biotic communities were being impacted by climate change in all parts of the world and that their country was already experiencing these impacts. All were, nevertheless, quite optimistic that humanity can mitigate climate change effects. They also believed that their training was quite focused on climate change issues. However, the students were not so much knowledgeable about how their actions could influence climate change and expressed the need for more information about it.

In our literature review three more studies were identified focusing on university students' beliefs and understandings about climate change originating from other Mediterranean countries, and more specifically, Turkey, Croatia and Egypt. In their research, Ateş et al. (2017) investigated Turkish undergraduate students' beliefs about the occurrence, causes and consequences of global climate change, and how and to what extent future time perspective along with perceived knowledge and environmental attitudes could be predictive factors. A large sample of 1,580 students distributed across 5 Faculties in one state Turkish university participated in the study and were asked to complete an adapted 14 closed-ended questionnaire. The majority of the students agreed that global warming is occurring, it is caused mainly by human activities and would bring about negative consequences, with female students holding stronger beliefs over these issues. However, the authors underline the fact that smaller percentages of students were uncertain or did not believe that any of these three dimensions was true. The study concluded that the majority of the students felt they had moderate to limited knowledge about the causes and consequences of global climate change.

The study by Nefat and Benazić (2019) produced similar results to the ones in previous studies. Their survey was conducted on a sample of 253 graduate and undergraduate students in the Faculty of Economics and Tourism of one Croatian University. The data were collected with the use of a questionnaire that was adapted from more instruments developed in previous studies. Descriptive data analysis demonstrated that the large majority of Croatian students agree that climate change is actually occurring, it is human-induced and there is consensus among scientists about it. The researchers commented that the relatively higher percentages derived from their study compared to previous research are possibly due to the growing awareness about climate change worldwide. However, although more than half of the students consider climate change a serious threat to humankind, themselves and their families, they feel moderately informed and they would be interested in learning more about it. They stated that climate change is not part of their academic curricula, even though most of them had dealt with the topic in school.

Finally, the only study carried out in a non-European Mediterranean context is the one reported by Ibrahim et al. (2018). A large number (1,300) of students from four Faculties of an Egyptian University took part in a survey, which aimed to investigate their knowledge and attitudes about global warming. The data were collected with the use of a questionnaire which was based on instruments used in previous studies to assess school students' knowledge and attitudes about global warming. The study results indicated that the majority of Egyptian held positive attitudes toward the issue, however their level of awareness differed depending on their discipline: a higher percentage of students with a background from the Sciences and Agriculture were better aware compared to students pursuing studies in Commerce or Law. In terms of their knowledge about what causes global warming, most students identified car emissions, industrial pollution, and the burning of domestic and agricultural waste as the main factors contributing to it.

The study reported here is part of a greater research project initiated by the Environmental Education Lab research center of the National and Kapodistrian University of Athens (NKUA), the oldest and largest institution of Higher Education in Greece and among the largest and most well-known universities in Europe and the Mediterranean region. Comprising 32 Faculties and their respective Departments, and functioning within 8 Schools, the NKUA offers a wide range of undergraduate courses and almost 100 postgraduate programs. Based on a survey conducted during 2015–2016 to identify the degree to which Greek Universities had incorporated climate change issues into their curricula (Skanavis et al. 2017), the NKUA was found to have a rather limited outreach. Although these data may have been altered considerably in the meantime, it is worth investigating both the baseline and the potential for promoting CCE across the NKUA programs of study. The research project that is being conducted by the Environmental Education Lab aims to explore the climate change literacy profiles and types of learning engagement of the NKUA students from a wide range of disciplinary fields and programs of academic studies and to assess their levels of awareness, concern, understanding and learning disposition in relation to climate change. It also aims to determine differences in the knowledge, understanding and learning of climate change among students from different Faculties and based on various socio-demographic and behavioral factors.

2 Method

The survey was conducted between November 2019 and January 2020 with a sample of first-year graduate students from the Faculty of Social Sciences and Humanities (namely the School of Philosophy), and more particularly students enrolled in the Departments of Psychology and Philosophy. Convenience sampling was used and 250 students were contacted in the context of their participation in introductory courses, after personal communication with the Lecturer(s) responsible for the course(s) and permission granted. The students were approached in class, before the start of the lecture. The topic and purpose of the study were explained to them and they were asked to provide their free and informed consent prior to completing the questionnaire. Students were assured that their responses would remain anonymous and that their overall performance in the questionnaire would not affect their academic grades.

Data collection was based on a questionnaire composed of 26 closed-ended and 2 open questions. For close-ended questions a 4-point Likert scale or a single-choice answer format from a list was used. Seven more questions were included to collect basic socio-demographic information (gender, age, university department, year/semester of academic studies, place of origin, father's and mother's education level). Twenty questions were closely modeled on the survey instrument used by Leiserowitz et al. (2011), with a nationally representative sample of 517 American teens (aged 13–17) and 1,513 American adults, to measure and compare their knowledge, concern, and beliefs about climate change. The questionnaire was developed and used in several national and international opinion poll surveys within the context of the Yale Project on Climate Change Communication (https://climatecommunication.yale.edu/). An appendix with the correct answers to the questions, all supported by well-established scientific evidence and references to peer-reviewed scientific sources, is also provided in the same report. All questions selected from the Yale Project's survey instrument were translated into Greek by the author and accuracy was checked by a second English-speaking researcher.

The questionnaire was pre-tested with 23 graduate students from the Department of Educational Studies (NKUA) to identify any comprehension and language issues and to improve its face and content validity. Questionnaire completion took on average 30 min. Data collected were analyzed from a total of 238 questionnaires (Table 1): 161 Psychology students (67.6%) and 77 Philosophy students (32.4%). 12 questionnaires were removed from analysis since they were completed by students from other academic degrees. Respondents were all undergraduates between 18 and 23 year old (95%, n = 225), with the majority of them being first-year (89.9%, n = 214), aged 18 to 19 years old (86.5%, n = 205). Unsurprisingly enough, the majority

Table 1 The study's sample profile

		Frequency	Percent
Gender	Males	32	13.4
	Females	206	86.6
Department	Psychology	161	67.6
	Philosophy	77	32.4
Study year	1st	214	89.9
	2nd, 3rd, 4th	24	10.1
Age	18	182	76.5
	19	23	9.7
	20	11	4.6
	>20	22	9.0
	Total	238	100.0

of the respondents were females (86.6%, n = 206) with only a 13.4% males (n = 32), following a similar trend in the Greek student population enrolled in these degrees.

In the following section we present the study's results related to questions surveying the students' beliefs and concerns about climate change, their self-reported knowledge and understandings about how the climate system works, what factors contribute to the global temperature rise (causes), and potential actions that would help mitigate global warming (solutions). Data analysis reported in this paper includes descriptive statistics. Differences in the respondents' answers were found to be non-significant for 'gender' and 'disciplinary field/ Department' variables.

3 Results and Discussion

3.1 Beliefs About Climate Change

The absolute majority of our study's sample believe that climate change is happening (97.9%, n = 232) and an equally high percentage confirms that this is beyond any doubt (89.1%, n = 212) (Fig. 1). This is in accordance with previous studies, such as the one by Wachholz et al. (2014), reporting that a high percentage of American college students (75%, n = 255) believe they are experiencing climate change. A similar certainty about climate change is expressed among almost all postgraduate Portuguese (98%) and Mozambican students (100%) and to a slightly lesser extent among Mexican students (82%) as reported in the studies by Santos et al. (2016) and Morgado et al. (2017). An equally high percentage of Croatian students from the Faculty of Economics and Tourism expressed their certainty about climate change (95%) in the study by Nefat and Benazić (2019). Turkish undergraduate students

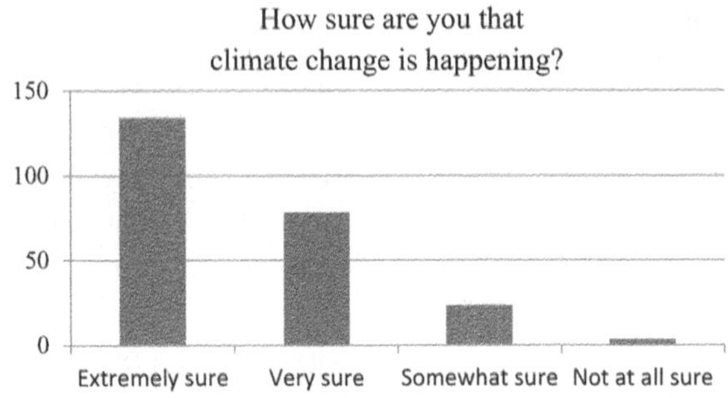

Fig. 1 Degree of students' expressed certainty that climate change is real

also confirm that the phenomenon of global warming is real (87.7%) in the study by Ateş et al. (2017).

More than half of the Greek respondents identified climate change as basically a human-induced environmental problem (66%, n = 157). This percentage was somewhat higher in the study by Wachholz et al. (2014) addressing American students (78%, n = 273); and in the study by Nefat and Benazić (2019) investigating Croatian graduate and undergraduate students (79%). Santos et al. (2016) reported an even higher percentage of Portuguese postgraduate students (96%) who identified climate change as human-induced, and a similar trend was revealed in the results of the study by Arto-Blanco et al. (2017), with most participants considering human agency as the main cause of climate change. As opposed to these results, the study by Díaz Estévez et al. (2014) indicated that only 29.96% among university students of different nationalities (including Spanish and Italian students) recognized an anthropogenic origin in climate change, compared to a much higher percentage (67.98%) who viewed it as the result of a combined nature and human agency. In our study, the respective percentage of Greek students who perceived climate change as being due to both human activities and natural changes was considerably smaller (31.5%, n = 75).

Contrary to what other studies report, Greek students in our study believe that there is a high degree of consensus in the scientific community that climate change is real (64.3%, n = 153), with only a smaller percentage saying that they perceive a lot of disagreement among scientists (14.3%, n = 34) (Fig. 2). This finding offers a more optimistic perspective compared to what, almost one decade ago, studies such as the one by Wachholz et al. (2014) reported about American students. The fact that they framed the very existence of climate change as a matter of controversy among scientists was argued to be the result of a deliberate attempt from the media and specific interest groups to stir up a sense of doubt and denial about the climate issue in public opinion.

Similar trends are shown in previous research, as for example the one by Nefat and Benazić (2019), reporting an even higher percentage of Croatian students (81%) who view that the scientific community is in agreement about climate change. Likewise, the majority of Portuguese students in the study by Santos et al. (2016) expressed their certainty that there is consensus among scientists about climate change (75%).

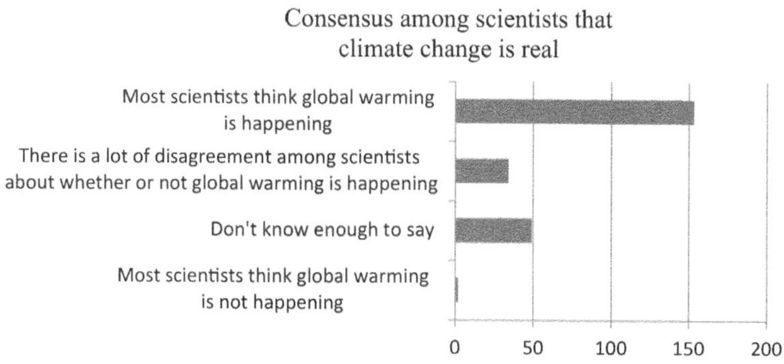

Fig. 2 Students' beliefs about scientific consensus on climate change

3.2 *Concern About Climate Change*

Expressed concern about climate change was quite high in our study's sample (Fig. 3). When asked how worried they felt about climate change, four out of five Greek students (81.1%, n = 193) replied they were either very worried (22%, n = 54) or somewhat worried (58.4%, n = 139). Similar tendencies in the students' concern, although with slightly lower percentages, were identified in previous studies with American students (Wachholz et al. 2014) or students from other Mediterranean countries (i.e., Cordero et al. 2008; Santos et al. 2016; Morgado et al. 2017; Nefat and Benazić 2019). Nevertheless, almost 20% of the Greek students say they are not particularly worried or not worried at all about climate change, which is quite an alarming figure.

To further explore the students' concern about climate change, two new questions were posed towards the end of the survey, asking them how they rate the climate issue in terms of seriousness; and how much they had thought and worried about climate change until they completed this questionnaire. In relation to how they rate climate change, the vast majority (94.9%, n = 222) identify it as a highly serious (42.4%, n = 101) or very serious problem (50.8%, n = 121), confirming their previously

Fig. 3 Students' level of expressed concern about climate change

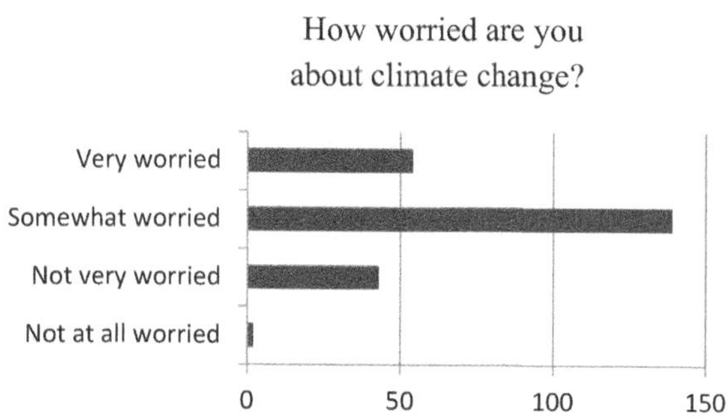

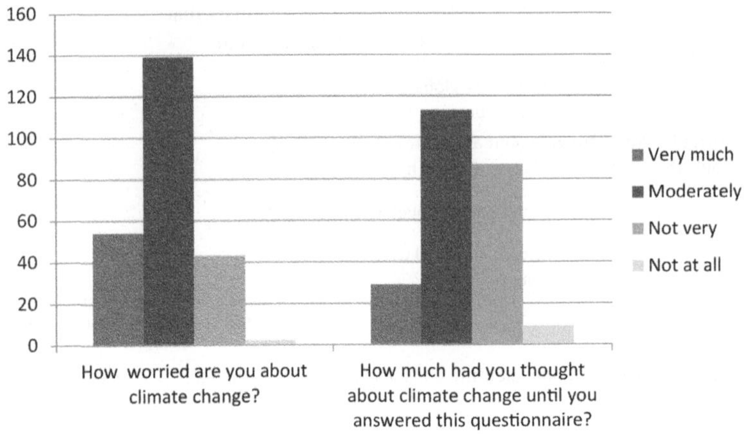

Fig. 4 Students' level of expressed concern about climate change compared to their level of self-reported engagement with the issue prior to questionnaire completion

expressed concern about climate change. In the study by Manolas et al. (2010) study Greek forestry students, when asked to rate the seriousness of the problem also gave a high score. A similar high rate was assigned by the majority of Croatian students in terms of how serious they view climate change as a threat to humankind compared to being a threat to themselves or their families (Nefat and Benazić 2019), and in other studies (Morgado et al. 2017; Wachholz et al. 2014). Nevertheless, when asked again, much fewer students in our study admitted they had thought a lot (12.2%, n = 29) or even moderately (47.5%, n = 113) about climate change until their participation in the study (Fig. 4).

3.3 *Self-reported Knowledge About Climate Change*

Students were also asked how knowledgeable they consider themselves to be about various aspects of the climate change issue (Fig. 5). Higher percentages responded they felt confident enough about their knowledge level regarding the various impacts of global warning (81.9%, n = 184). However, comparatively fewer students answered the same about their understanding of how the earth climate system works (60.3%, n = 143) or about the causes (67.2. n = 158%) and potential solutions to reduce the global temperature rise (59.7%, n = 142). A similar finding is described in the studies by Santos et al. (2016) and Morgado et al. (2017), in which Portuguese, Mozambican, Mexican samples of students reported a moderate or minimal knowledge of how to deal with climate change. Most Turkish undergraduate students also reported a moderate (49%) to limited (25%) knowledge about the causes and consequences of global climate change (Ateş et al. 2017). Based on results of their cross-national study, Díaz Estévez et al. (2014) argue that university students seem to

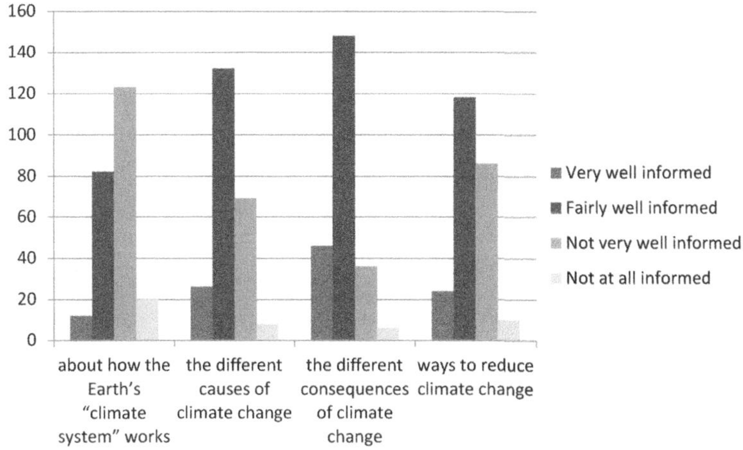

Fig. 5 Level of self-reported knowledge about dimensions of climate change

have a rather superficial knowledge about the causes of climate change. Further to this, Bråten et al. (2009) argued about the cultural embeddedness of climate change knowledge, pointing to the fact that students' beliefs seem to be shaped by the degree, content and emphasis given by public discourse on the topic in different parts of Europe.

The vast majority of our study's sample expressed their need to get more information about the climate change issue (94.1%, n = 222). Similar needs were articulated by students in other Mediterranean countries (i.e., Santos et al. 2016; Morgado et al. 2017; Nefat and Benazić 2019) as well as by American students (Wachholz et al. 2014). Finally, although most Greek students agree that both schools and Universities have the responsibility to teach youths about climate change, the majority of students ascribe such a role more to schools (91.6%, n = 217) compared to higher education institutions (63.1%, n = 149).

3.4 Understanding the Science of Climate Change

3.4.1 Weather Versus Climate

The majority of students correctly understand that weather often changes from year to year (71.7%, n = 170) and that 'climate' means the average weather conditions in a region (75.6%, n = 180) (Fig. 6). However, almost half the respondents (42.9%, n = 102) incorrectly believe that the climate also changes from year to year. In similar percentages students correctly reject the idea that 'climate' and 'weather' mean the same (50.8%, n = 121) and correctly understand with a varying degree of confidence that 'weather' does not mean the average climate conditions in a

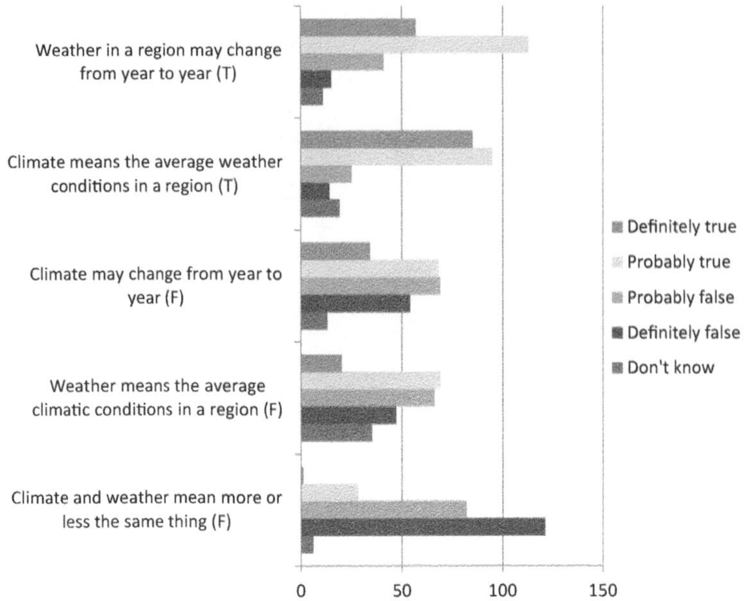

Fig. 6 Students' level of understanding climate Versus weather

region (47.6%, n = 113), even though 14.8% of them (n = 35) say they do not have sufficient knowledge to identify the correct answer. Research evidence indicates that distinguishing between weather and climate is a challenge and people often encounter difficulties in realizing where the differences between the two lie (Lombardi and Sinatra 2012).

3.5 Understanding of Climate Science Facts

Regarding the students' understanding of the flow of heat across the planet, only few among them seem to accurately know that ocean currents carry heat from the equator toward the north and south poles (31.1%, n = 74). The majority of students (60.9%, n = 145) cannot discern whether this is correct or not. Much fewer students correctly understand that the atmosphere does not carry heat from the poles (which are cold) towards the equator (which is warm) (18.9%, n = 45), with a high percentage admitting that they don't know the answer (63.4%, n = 151). In the study by Leiserowitz et al. (2011), Americans provide correct answers to these two questions in slightly higher percentages (54% and 33% respectively).

The majority of our study's respondents (77.2%, n = 183) correctly understand that the Earth's climate has not remained the same through the millions of years as outlined by Lamb (1977) among other scholars. They also correctly believe that climate change has played an important role in the advance or collapse of some past human civilizations (78.5%, n = 186), as many scientists have pointed out, among whom Ellenblum (2012), whose research on how climatic disturbances affected the prosperity or decline of old civilizations in the Mediterranean Basin shed new light on our historic knowledge of the region.

A smaller percentage of Greek students answers accurately that rising levels of carbon dioxide in the atmosphere have caused global temperatures to increase in the past (56.5%, n = 134), although 31.2% (n = 74) say they don't know if this is true or not. 77.2% of the students correctly reject the statement that current global temperature conditions are warmer than in previous periods in the past, while 66.3% rightly affirm that the 2009–2019 decade was the warmest since 1850, although 29.7% say they don't know. Finally, a high percentage of students incorrectly believe that in the past, the Earth's climate always shifted gradually between warm and cold periods (65.5%, n = 154), while 20% (n = 47) says they don't know. The graph in Fig. 7 summarizes the students' responses to each of the above mentioned climate science facts.

Previous research also indicates that even University students (as well as school students) tend to hold conceptions that are not in agreement with scientifically accepted conceptions about how the climate system works (Versprille et al. 2017;

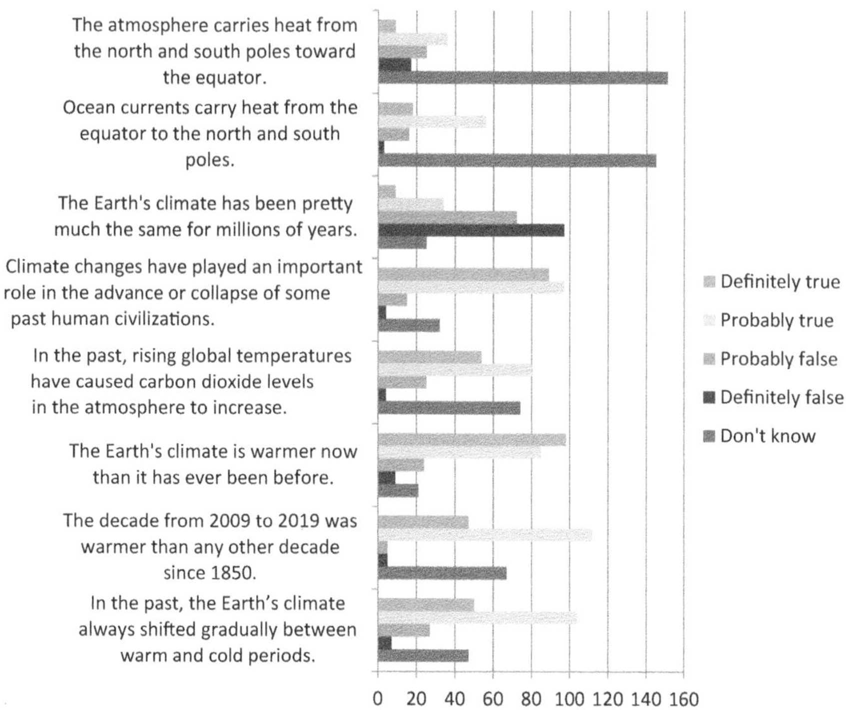

Fig. 7 Students' level of understanding about various 'climate science facts'

Shepardson et al. 2012). As these studies suggest, this may lead to a poor understanding of the phenomena and to a faulty perception of the mechanisms involved. However, the argument put forth by this study points to a different direction, by saying that any of these indications of what students understand about climate change need to be treated cautiously and in conjunction with how research on learning frames them. One such approach, building on Bowen and Rodger (2008), argues that even when the students' knowledge is aligned with what climate scientists claim, students may still lack the necessary "interpretive framework for making sense of the knowledge held" (p. 97).

3.6 Beliefs About Factors Contributing to Global Warming

Respondents in our study were then provided with a list of factors and asked to what degree each of them affects the average global temperature of the Earth. The vast majority (89.8%, n = 212) correctly answered that greenhouse gases have a major (47.5%, n = 112) or some impact (42.4%, n = 100) on it (Fig. 8). Only a very small percentage of students (5.4%, n = 13) reported that greenhouse gases affect the Earth's temperature a little or not at all. Other studies with university students in other Mediterranean countries produced similarly high percentages by identifying greenhouse gases, especially car emissions, as the main cause of global temperature rises (Ibrahim et al. 2018). In the study by Arto-Blanco et al. (2017) 42.8% of Spanish students related greenhouse emissions primarily to the use of automobiles although they tended to attribute a higher degree of causality to industrial pollution (50.7%). Greek forestry students also singled out greenhouse gases as the main factor contributing to global warming, distinguishing between emissions resulting from industrial activity (mean 8.60) and transportation (mean 6.38) (Manolas et al. 2010).

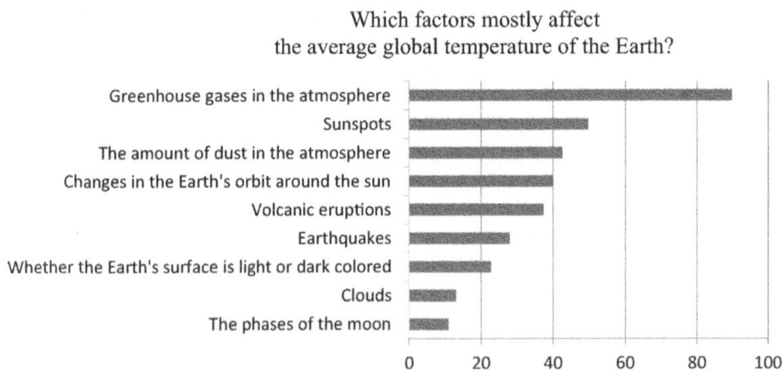

Fig. 8 Students' beliefs about the factors that mostly affect the average global temperature of the Earth

Regarding other factors with a contribution to global warming, Greek students' beliefs were mixed to a greater or lesser extent. For example, those who correctly understand that sunspots affect the Earth's temperature are higher (49.8%) compared to those who wrongly believe they do not (13.3%). Students' responses are divided between those who correctly believe that changes in the Earth's orbit around the sun affect the global Earth temperature (39.5%) and those who think this is false (33.5%). A slightly higher percentage of students correctly understand that the amount of dust in the atmosphere has some impact on the temperature of the Earth (42.6%) compared to those who wrongly believe there is little or no impact at all (35.3%).

On the other hand, significantly more are those who falsely think that clouds do not have some kind of contribution to global warming (57.3%) compared to those who correctly believe they do (13.1%). Similar tendencies, although with a less sharp gap, are indicated between those who wrongly believe that volcanic eruptions do not affect the Earth's temperature (46%) compared to those who correctly believe they do (37.4%); and between those who wrongly say that light or dark colored Earth surface areas have no impact on global temperature (31.9%) compared to those who correctly identify there is (22.8%). Those who correctly see no connection between earthquakes and the rise of global Earth temperature are almost half the sample (46.6%) compared to a much smaller percentage who incorrectly believes there is some connection (28%). Likewise, more students correctly reject that the phases of the moon have some kind of an impact on the global temperature (38.4%) compared to those who falsely believe they do (11%).

However, the study indicates that there are many knowledge gaps in the students about the possible contributors to the Earth's global temperature. Apart from identifying the role of the greenhouse gases, comparatively high numbers of students admit they don't know enough to say whether the lightness/darkness of the Earth's surface areas (45.1%) or sunspot phenomena (36.2%) have any kind of contribution; whether atmospheric clouds can trap the heat beneath them (29.3%); what the actual responsibility share for any changes in the Earth's orbit around the sun is (27%); whether earthquakes (25.3%) or the amount of dust in the atmosphere (21.5%) have any impact; and what the role of volcanic eruptions (16.2%) in affecting global temperature is.

Misconceptions about what causes global temperature rise have also been reported in previous studies conducted with university students samples (i.e., Cordero et al. 2008; Kerr and Walz 2007; Khalid 2003; Morgan and Moran 1995; Wilson and Henson 1993; Wachholz et al. 2014). Although most Greek students in our study seem to recognize that greenhouse gases have a major contribution to global warming compared to what, for example, the study by Wilson and Henson (1993) reported with a US sample more than two decades ago, several false conceptions or uncertainties still exist. The fact that respondents are less knowledgeable about the natural causes of climate change is supported by Lorenzoni et al. (2007), arguing that most people tend to believe that climate change is mostly caused by anthropogenic and natural causes although they do not understand the details. Moreover, findings based on the study by Morgado et al. (2017) indicate that students are more able to understand causes stemming from individual behaviors instead of natural causes. Nevertheless, it is worth keeping in mind what Díaz Estévez et al. (2014) turned our attention to, which is that university students seem to have a superficial knowledge of the causes of climate change, despite the fact that climate change is part of a global culture and intrinsically related to the values that define this generation's interests.

3.7 Beliefs About Actions to Mitigate Global Warming

The majority of our study's respondents (92.3%, n = 216) correctly understand that if the switch from fossil fuels to renewable energy sources were applied as a universal measure, there would be a considerable reduction in the global temperature of the Earth. When asked for a second time to select only one option from a list that would do the most to reduce global warming, again switching from fossil fuels to renewable energy sources was by far the one selected by most students (58.1%, n = 137).

Planting trees is another action a high percentage of students (88%, n = 206) correctly consider that has a positive impact on mitigating climate change and the same applies to measures attempting to diminish tropical deforestation (87.6%, n = 205) or aim to switch from gasoline to electric cars (88.5%), n = 207). Driving less is also identified by a high percentage of students as a solution (71.5%, n = 168), while 63.2% believe that increasing public transportation would reduce global warming (n = 148). Half of the participants believe that switching from regular incandescent to compact fluorescent bulbs (52.3%, n = 123) has some kind of contribution to reducing global temperature. Fewer think that insulating buildings may have a positive impact (30.3%, n = 71) compared to those who do not believe this would be of some use (38.9%, n = 91). A similar tendency is observed in students correctly saying that switching from fossil fuels to nuclear power would be a solution (29.9%, n = 70) compared to those who disagree with this (33.8%, n = 79).

On the other hand, a high percentage of students incorrectly believe that reducing toxic waste would have a positive impact on climate change (89.7%, n = 210). This misconception is also reflected in the percentage of students choosing this as the best option from a list of factors reducing global warming (14.4%, n = 34), although with a great distance from those who select switching from fossil fuels to renewable energy sources (58.1%, n = 137) (Fig. 9). Another misconception is identified when more than half of the students incorrectly believe that banning aerosol spray cans would reduce global warming (61.5%, n = 144). More studies have identified that young people often link several environmental problems, pollution in particular, with anthropogenic climate change (Andersson and Wallin 2000; Pekel and Özay 2005).

Respondents are divided among those who do believe that placing a large tax on all fossil fuels would lower global warming (50.9%, n = 119) and those who do not (37.2%, n = 87) or do not know (12%, n = 28). A high enough percentage (76.1%, n = 178) disagrees that the option of having at most 2 children per family would be a solution. A similar trend in the Greek students' perception is indicated in relation to the option of immediately stopping the consumption of red meat, with more than

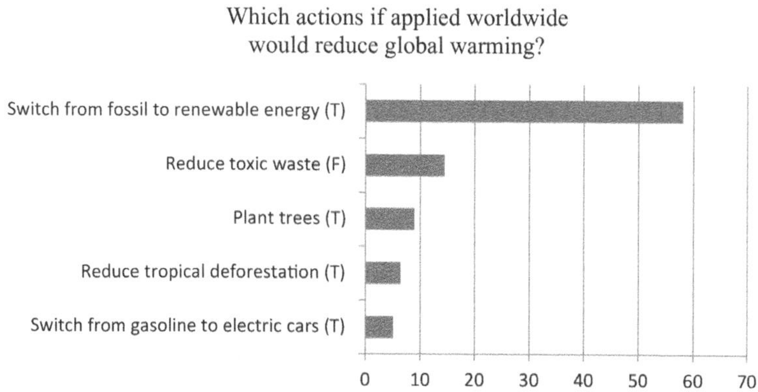

Fig. 9 The top 5 actions students believe would reduce global warming if applied worldwide

half of the respondents rejecting it (60.4%, n = 142) and a quite steeper percentage admitting they don't know about it (22.1%, n = 52).

Previous studies have reported similar tendencies in the university students' perception of the factors contributing to climate change which are at the same time part of the solutions to it. For example, in the study by Manolas et al. (2010), Greek forestry students rated higher human actions and policies aiming to regulate the production of greenhouse gases at the level of industrial activity, deforestation, energy production and transportation. Respectively, they rated lower the possibility of taking measures such as central heating installations, global population rise, household waste dumps and agricultural activity.

In another study (Santos et al. 2016), Portuguese students expressed their willingness to support specific measures to reduce greenhouse gases (57%), including paying more for fuel and electric energy, although 31% were not sure about it and 13% would not support such actions. Among the proposed mitigation measures would be reducing the use of cars and fossil fuels and sharing the car with friends and colleagues (24%), increasing recycling and the reutilization of materials, such as the use of plastic bags (18%), as well as cutting down energy (18%) and water consumption (10%). Other measures proposed by smaller percentages of students were raising the environmental awareness of people close to them (8%), lessening meat consumption (7%), promoting local products (4%), and engaging in nature conservation activities and the planting of trees (3%).

4 Conclusions

The study reported here was carried out with a sample of first-year undergraduate students enrolled in two academic programs of the Faculty of Social Sciences and Humanities at the National and Kapodistrian University of Athens (NKUA). Being part of a wider research project aiming to explore the climate change literacy and learning profiles of the NKUA students across a range of disciplinary fields, this study offers a glimpse of how Greek University students with a non-science disciplinary background perceive, understand and feel about climate change. The aim for conducting this survey was twofold: to contribute with new empirical data to a scant but continuously growing research area on university students' beliefs, concerns and understandings of climate change, both at a national and international level, with a special focus on the Mediterranean region; and to inform future decision-making about whether and how to integrate new CCE courses into the curricula of particular academic programs of study.

Data were collected with the use of a questionnaire, which was closely modeled on the survey instrument developed by Leiserowitz et al. (2011) from the Yale Project on Climate Change Communication team. Data analysis focused on exploring the students' beliefs and concerns about climate change, and demonstrated their self-reported knowledge and understandings about some basic climate science concepts and facts, and the causes and actions against global warming. The main study's results are summarized below:

- Almost all students believe that climate change is occurring and that it is a fact beyond any doubt. More than half of the respondents identify climate change as basically a human-induced problem and think that there is consensus among scientists on this. A large majority perceives climate change as an alarming problem with most students feeling worried about it to a moderate or high degree.
- Even though most students declare themselves confident regarding their knowledge about climate change, many of them admit they have a less accurate understanding of how the Earth's climate system works or the causes and potential solutions to climate change. At the same time, the vast majority expresses their need to get more information about the causes, impacts and possible solutions to climate change, while they also recognize that this role pertains primarily to schools and next to Universities.
- In terms of their understanding of some basic climate science concepts and facts, most students are successful in distinguishing between climate and weather although some still encounter difficulties. The same applies to perceiving that climate conditions had been changing along human history although not in similar patterns as currently; and that climate has been a decisive factor in shaping culture and history. Their accuracy in understanding is even less acute in relation to other climate science claims, such as the heat flow channels across the planet.
- In terms of the factors causing global temperature to rise, a large majority of students correctly identify greenhouse gases. Many, though, hold misconceptions or feel uncertain to a greater or lesser degree about the role of other factors, especially nature-related ones, such as sunspots, the Earth's orbit around the sun, the amount of dust in the atmosphere, colored Earth surface areas, and volcanic eruptions. There are also students who mistakenly view others factors, such as earthquakes or the phases of the moon, as impacting global temperature although they don't.

- The large majority of students recognize that switching from fossil fuels to renewable energy sources would be a solution to the problem. Planting trees, reducing tropical deforestation, switching from gasoline to electric cars, driving less and increasing public transportation are among the actions identified by students as having a positive impact on mitigating global warming. However, many students fail to perceive that actions like switching to compact fluorescent bulbs, domestic insulation or nuclear power would contribute too. Misconceptions regarding the role of measures such as reducing toxic waste or banning aerosol spray cans show that they still hold strong among this sample. Finally, students feel ambivalent about whether the adoption of fiscal measures, birth control, or changes in dietary habits could help mitigate climate change.

Extended scientific evidence asserts that anthropogenic climate change is occurring and that it represents one of the greatest challenges of our times (Hansen and Stone 2016). This study's findings are in accordance with previous studies (Feldman et al. 2010; Morgado et al. 2017; Santos et al. 2016; Wachholz et al. 2014), showing that Greek students are along the same line of thought: they do perceive climate change as a real and largely human-induced problem. They also feel concerned about climate change and acknowledge it as a major current issue. Moreover, contrary to previous studies (Wachholz et al. 2014; Morgado et al. 2017), Greek respondents do not endorse inconsistent views about lack of consensus among the scientific community on climate change, which is another promising fact. However, since high levels of awareness and concern do not always correspond to high levels of knowledge, Greek students admit they lack important knowledge about various aspects of the climate change issue and express their need to develop a better understanding of them. The students' openness and willingness to learn more about climate change combined with their high levels of awareness and concern about the issue confirms that CCE has a place in the curricula of at least these two degrees (Psychology and Philosophy) and possibly in more academic areas from the Social Sciences and Humanities.

Knowledge gaps are identified by the students themselves on fundamental climate science information that would allow them to understand how the climate system works, and on more socially-driven inputs and perspectives, such as climate change causes and solutions. Our study also confirms the existence of a certain degree of confusion, uncertainty or lack of knowledge among students, such as in relation to 'weather versus climate' (Lombardi and Sinatra 2012) or specific climate science facts and claims (Versprille et al. 2017; Shepardson et al. 2012). Apart from that almost all students identify greenhouse gases as the main cause of global warming, they seem bewildered to a greater or lesser extent about the role of other factors, with many of the respondents holding inaccurate conceptions about them. The same applies to how they approach actions of mitigation, with, again, the majority of them feeling more confident about options leading to a more fossil fuel-free and energy-saving society. However, they stand indecisive in front of more options or tend to reject some of them, probably out of the nuanced frames they have developed about other environmental issues (such as nuclear power or toxic pollution) or based on their negatively biased perceptions about certain individual or social activities (such as driving less, excluding red meat from diet, or promoting birth control).

However, what is important to keep in mind is that many of these misconceptions and biases stem from the inherent complexity and the accompanying uncertainty climate change induces as a 'wicked problem' (Incropera 2016). Kollmus and Agyeman (2002) are quite explicit arguing that climate action, as any other environmental actions, is grounded on a combination of fundamental scientific knowledge, pro-environmental attitudes and emotional involvement about the issue, all shaped by the individuals' values and norms and influenced by a range of socio-psychological parameters and external contextual factors, intervening to affect a climate-specific consciousness and decision-making. Moreover, scant understanding and skepticism about climate change are identified among the main barriers impeding the students' engagement with climate action (Lorenzoni et al. 2007). An action competence approach (Jensen and Schnack 1997) that focuses on facilitating students to overcome such barriers through critical reflection and by integrating climate action in their everyday and local activities and in their future professional practices, through learner-centered, action-oriented, and transformative pedagogies, can be an appropriate frame for CCE in Higher Education institutions. Such considerations have to be taken into account when thinking about how research findings can inform the design of CCE opportunities at this level.

A final note on the limitations this study raises is shared here. One such limitation concerns the tool employed for collecting data on Greek university students' beliefs, concerns and understandings of climate change. The choice of using a questionnaire closely modeled on the one developed by the Yale Project on Climate Change Communication was deliberate, based on the criteria that (a) it was singled out as a reliable survey instrument, which also provided access to the correct answers, all supported by well-established scientific evidence; and (b) it allowed for a large-scale assessment among hundreds of respondents and the possibility to compare results with those of other studies following a similar methodology. However, one consequence of this choice is that the students' beliefs, concerns and understandings are restricted by the content of this questionnaire, with possibly other relevant aspects and nuances of the topic, which might have been captured through a different methodology, to be left out. It is therefore argued that future research should be based on a multi-method research design, combining a large-scale, quantitative approach with more, small-scale case studies of a qualitative perspective.

A second limitation of the study is that the results are not generalizable to the overall student population either on a national or an institutional level. It is thus relevant that future research addresses student samples from other Faculties, degrees and programs of academic study at the NKUA as well as other Greek HE institutions. A comparative study among students of various profiles would also be meaningful to determine statistically significant differences. 'Disciplinary field of study', 'age', 'year *or* level of study' and 'gender' are some profiling variables that need to be more thoroughly investigated as to whether they can explain statistical variance in Greek students' beliefs, concerns and understandings of climate change. Although such statistical analyses were beyond the scope of the study reported here, which was mainly of a descriptive nature, it is deemed important to take more than one of these variables into account in the selection and composition of the sample in future research and in order to avoid various biases, such as the gender bias related to certain degrees that concentrate higher numbers of either male or female students. Finally, a more culturally-oriented and sensitive research design would allow us to explore more thoroughly and explain climate change beliefs and understandings within the NKUA students population or across different university samples, both on a national and cross-national level, including the various Mediterranean countries.

Addressing climate change at the Universities is an imperative for CCE (Leal Filho et al. 2018; Leal Filho and Hemstock 2019), not only to raise awareness among students but also to promote ways to build new and sound interdisciplinary knowledge, to critically refute an incomplete and faulty one, and to empower students to act on it. This applies to all countries of the Mediterranean Basin sharing a region that is particularly prone to be affected by the impacts of actual and future climate changes (Ozturk et al. 2015; Cramer et al. 2018), including Greece. Addressing this goal needs to be based on a research-based comprehension of how students across a range of disciplinary fields perceive and understand climate change; and use this knowledge as a starting point to further explore the barriers impeding learning about climate change as well as the educational strategies to overcome them (Azeiteiro and Leal Filho 2017). Our study points to that direction, by bringing to the fore how Greek students from the Social Sciences and Humanities view and understand climate change. Although not generalisable to the overall population, the study's results provide useful insights about the necessity, readiness and possible routes for incorporating CCE into the academic curricula of these disciplinary fields in Greece as well as opening up the discussion about similar endeavors in other Mediterranean countries.

References

Andersson B, Wallin A (2000) Students' understanding of the greenhouse effect, the societal consequences of reducing CO_2 emissions and the problem of ozone layer depletion. J Res Sci Teach: Offic J Nat Assoc Res Sci Teach 37(10):1096–1111

Arto-Blanco M, Meira-Cartea PÁ, Gutiérrez-Pérez J (2017) Climate literacy among university students in Mexico and Spain: influence of scientific and popular culture in the representation of the causes of climate change. Int J Glob Warm 12(3–4):448–467

Ateş D, Teksöz G, Ertepınar H (2017) Exploring the role of future perspective in predicting Turkish university students' beliefs about global climate change. Discourse Commun Sustain Educ 8(1):32–52

Azeiteiro UM, Leal Filho W (eds) (2017) Climate literacy and innovations in climate change education. Int J Glob Warming. Inderscience Publishers. IF 1.043 (2015/16) Q2

Ballew MT, Leiserowitz A, Roser-Renouf C, Rosenthal SA, Kotcher JE, Marlon JR, Lyon E, Goldberg MH, Maibach EW (2019) Climate change in the American mind: Data, tools, and trends. Environ: Sci Pol Sustain Dev 61(3):4–18

Bowen GM, Rodger V (2008) Debating global warming in media discussion forums: strategies enacted by persistent deniers and implications for schooling. Can J Environ Educ 13(1):89–106

Boyes E, Chuckran D, Stanisstreet M (1993) How do high school students perceive global climatic change: what are its manifestations? What are its origins? What corrective action can be taken? J Sci Educ Technol 2(4):541–557
Bråten I, Gil L, Strømsø HI, Vidal-Abarca E (2009) Personal epistemology across cultures: exploring Norwegian and Spanish university students' epistemic beliefs about climate change. Soc Psychol Educ 12(4):529–560
Cordero EC, Todd AM, Abellera D (2008) Climate change education and the ecological footprint. Bull Am Meteor Soc 89(6):865–872
Corner A, Roberts O, Chiari S, Völler S, Mayrhuber ES, Mandl S, Monson K (2015) How do young people engage with climate change? The role of knowledge, values, message framing, and trusted communicators. Wiley Interdisc Rev: Clim Change 6(5):523–534
Cramer W, Guiot J, Marini K (2018) Risks associated to climate and environmental changes in the Mediterranean Region. MedECC Report
Dawson V (2015) Western Australian high school students' understandings about the socioscientific issue of climate change. Int J Sci Educ 37(7):1024–1043
Díaz Estévez M, de Frutos García RA, Peña Moya J (2014) Communication of the scientific consensus on climate change to the citizenship: knowledge and perception of young university students from five countries regarding the media treatment of global warming. Interactions: Stud Commun Cult 5(1):51–70
Ellenblum R (2012) The collapse of the eastern mediterranean: climate change and the decline of the east, 950–1072. Cambridge University Press, Cambridge
Feldman L, Nisbet MC, Leiserowitz A, Maibach E (2010) The climate change generation. Survey analysis of the perceptions and beliefs of young Americans. Yale Project on Climate Change & George Mason University Center for Climate Change Communication
Hansen G, Stone D (2016) Assessing the observed impact of anthropogenic climate change. Nat Clim Chang 6(5):532–537
Hermans M, Korhonen J (2017) Ninth graders and climate change: attitudes towards consequences, views on mitigation, and predictors of willingness to act. Int Res Geograph Environ Educ 26(3):223–239
Ibrahim AA, Fahmy HD, Mahmoud SR (2018) Knowledge and attitude regarding global warming phenomenon among Assiut University Students. Assiut Sci Nurs J 6(14):1–11
Incropera FP (2016) Climate change: a wicked problem: complexity and uncertainty at the intersection of science, economics, politics, and human behavior. Cambridge University Press, Cambridge
Jackson L, Pang MF (2017) Secondary school students' views of climate change in Hong Kong. Int Res Geograph Environ Educ 26(3):180–192
Jensen BB, Schnack K (1997) The action competence approach in environmental education. Environ Educ Res 3(2):163–178
Kerr SC, Walz KA (2007) "Holes" in student understanding: addressing prevalent misconceptions regarding atmospheric environmental chemistry. J Chem Educ 84(10):1693–1696
Khalid T (2003) Pre-service high school teachers' perceptions of three environmental phenomena. Environ Educ Res 9(1):35–50
Kollmuss A, Agyeman J (2002) Mind the gap: why do people act environmentally and what are the barriers to pro-environmental behavior? Environ Educ Res 8(3):239–260
Lamb HH (1977) Climate. Present, past and future - Volume. 2: Climatic history and the future. London, Methuen
Leal Filho W (ed) (2010) Universities and climate change: introducing climate change to University Programmes. Springer, Berlin
Leal Filho W, Hemstock SL (2019) Climate change education: an overview of international trends and the need for action. In: Leal FW, Hemstock S (eds) Climate change and the role of education. Springer, Cham, pp 1–17

Leal Filho W, Morgan EA, Godoy ES, Azeiteiro UM, Bacelar-Nicolau P, Ávila LV, Mac-Lean C, Hugé J (2018) Implementing climate change research at Universities: barriers, potential and actions. J Cleaner Prod 170:269–277

Leiserowitz A, Smith N, Marlon JR (2011) American teens' knowledge of climate change. Yale University. Yale Project on Climate Change Communication: New Haven, CT

Lombardi D, Sinatra GM (2012) College students' perceptions about the plausibility of human-induced climate change. Res Sci Educ 42(2):201–217

Lorenzoni I, Nicholson-Cole S, Whitmarsh L (2007) Barriers perceived to engaging with climate change among the UK public and their policy implications. Glob Environ Chang 17(3–4):445–459

Manolas EI, Tampakis SA, Karanikola PP (2010) Climate change: the views of forestry students in a Greek university. Int J Environ Stud 67(4):599–609

Molthan-Hill P, Worsfold N, Nagy GJ, Leal Filho W, Mifsud M (2019) Climate change education for universities: a conceptual framework from an international study. J Clean Prod 226:1092–1101

Morgado F, Bacelar-Nicolau P, von Osten JR, Santos P, Bacelar-Nicolau L, Farooq H, Alves F, Soares AM, Azeiteiro UM (2017) Assessing university student perceptions and comprehension of climate change (Portugal, Mexico and Mozambique). Int J Clim Change Strateg Manage 9(3):316–336

Morgan MD, Moran JM (1995) Understanding the greenhouse effect and the ozone shield: an index of scientific literacy among university students. Bull Am Meteor Soc 76(7):1185–1190

Nefat A, Benazić D (2019) The student perceptions, concern and knowledge with regard to climate changes in Croatia. Ekonomski Pregled 70(5):723–742

Ojala M (2020) To trust or not to trust? Young people's trust in climate change science and implications for climate change engagement. Children's Geographies, pp 1–7

Ozturk T, Ceber ZP, Türkeş M, Kurnaz ML (2015) Projections of climate change in the Mediterranean Basin by using downscaled global climate model outputs. Int J Climatol 35(14):4276–4292

Pekel FO, Özay E (2005) Turkish high school students' perceptions of ozone layer depletion. Appl Environ Educ Commun 4(2):115–123

Santos PT, Bacelar-Nicolau P, Pardal MA, Bacelar-Nicolau L, Azeiteiro UM (2016) Assessing student perceptions and comprehension of climate change in Portuguese higher education institutions. In: Filho WL, Adamson K, Dunk R, Illingworth S, Azeiteiro UM, Alves F (eds) Implementing climate change adaptation in cities and communities integrating strategies and educational approaches. Springer, Basel, pp 221–236

Shepardson DP, Niyogi D, Roychoudhury A, Hirsch A (2012) Conceptualizing climate change in the context of a climate system: implications for climate and environmental education. Environ Educ Res 18(3):323–352

Skamp K, Boyes E, Stannistreet M (2009) Global warming responses at the primary secondary interface. 1. Students' beliefs and willingness to act. Aust J Environ Educ 25:15–30

Skanavis C, Kounani A, Ntountounakis I (2017) Greek universities addressing the issue of climate change. In: Leal Filho W (ed) Climate change research at universities: addressing the mitigation and adaptation challenges. Springer, Cham, pp 333–344

Smith N, Joffe H (2013) How the public engages with global warming: a social representations approach. Public Underst Sci 22(1):16–32

Stevenson KT, Peterson MN, Bondell HD (2016) The influence of personal beliefs, friends, and family in building climate change concern among adolescents. Environ Educ Res 25(6):832–845

UNESCO (2017) Changing minds, not the climate: the role of education. UNESCO, Paris

UNFCCC (2015) Adoption of the Paris agreement. 21st conference of the Parties. UN, Paris

Versprille A, Zabih A, Holme TA, McKenzie L, Mahaffy P, Martin B, Towns M (2017) Assessing student knowledge of chemistry and climate science concepts associated with climate change: Resources to inform teaching and learning. J Chem Educ 94(4):407–417

Wachholz S, Artz N, Chene D (2014) Warming to the idea: university students' knowledge and attitudes about climate change. Int J Sustain High Educ 15:128–141

Weber EU, Stern PC (2011) Public understanding of climate change in the United States. Am Psychol 66(4):315

Whitmarsh L, Capstick S (2018) Perceptions of climate change. In: Clayton S, Manning C (eds) Psychology and climate change. Academic Press, London, pp 13–33
Wilson K, Henson B (1993) Learning about global warming: a study of students and journalists. Learning about science easily and readily series. Boulder CO, National Centre for Atmospheric Research

Hydrological Impacts of Projected Climate Change on Northern Tunisian Headwater Catchments—An Ensemble Approach Addressing Uncertainties

Hamouda Dakhlaoui, Kirsti Hakala, and Jan Seibert

Abstract The present study addresses the hydrological impact of climate change, and associated uncertainties, on Northern Tunisian headwater catchments, which serve as the main source of surface water for Tunisia. This study is based on a wide range of hydrological projections resulting from the factorial combination of modelling chain components comprising: eleven high-resolution EURO-CORDEX regional climate models (forced by general circulation models; GCM-RCMs), three rainfall-runoff models (RRM) (GR4J, HBV and IHACRES), two time periods (mid-term: 2040–2070 and long-term: 2070–2100) and two climate scenarios (RCP 4.5 and RCP 8.5). A quantile mapping bias correction method was applied to correct the climate simulations. Projections of discharge were analysed according to several metrics commonly used by water managers. We applied the analysis of variance (ANOVA) technique to quantify the contribution of the uncertainties related to different components in the modelling chain to the overall uncertainty. The results show a marked decrease of precipitation and an increase of temperature especially for the most pessimistic climate scenario (RCP 8.5) towards the end of the twenty-first century, reaching −26% and + 4 °C, respectively. The modelling results showed that the local hydrology of the study catchments will be substantially influenced by climate change. Overall, by the end of the century, discharge and soil moisture are expected to decrease

H. Dakhlaoui (✉)
LMHE, Ecole Nationale d'Ingénieurs de Tunis, Université Tunis El Manar, Tunis, Tunisia
e-mail: hammouda.dakhlaoui@laposte.net

Ecole Nationale d'Architecture Et d'Urbanisme, Université de Carthage, Sidi Bou Said, Tunisia

K. Hakala
Department of Infrastructure Engineering, University of Melbourne, Victoria, Australia
e-mail: kirsti.hakala@unimelb.edu.au

J. Seibert
Department of Geography, University of Zurich, Zürich, Switzerland
e-mail: jan.seibert@geo.uzh.ch

Department of Aquatic Sciences and Assessment, Swedish University of Agricultural Sciences, Uppsala, Sweden

W. Leal Filho and E. Manolas (eds.), *Climate Change in the Mediterranean and Middle Eastern Region*, Climate Change Management,
https://doi.org/10.1007/978-3-030-78566-6_24

by up to 58 and 30%, respectively. We find a significant similarity between the different RRMs in terms of hydrological projections despite their different structures. ANOVA partitioning among the sources of uncertainty shows that the uncertainty induced by the GCM-RCMs is the dominant source of uncertainty and is anticorrelated with the area of the catchment. This demonstrates the challenge of applying GCM-RCM output at fine spatial scales, especially for catchments smaller than their grid. In addition, uncertainty stemming from hydrological models was found to be greater during the dry season.

1 Introduction

1.1 Assessing Impacts of Climate Change

Climate change has already affected water availability worldwide, (e.g. IPCC 2014; Bijl et al. 2018; Qin et al. 2019). Climate change poses a serious concern for agricultural production, challenges socio-economic systems, and threatens environmental sustainability (Gosling and Arnell 2016; Flörke et al. 2018). In this context, adaptation to climate change is critical, which calls for a detailed understanding of the impacts of climate change on water resources and a robust consideration of the uncertainties involved in climate change impact analyses. Since climate change impacts can vary significantly across different regions, it is important to conduct impact studies at the regional and local scale to develop useful adaptation and mitigation strategies.

To conduct climate change impact studies, researchers often use the combination of multiple models in so-called modelling chains. This allows for the transfer of large-scale (e.g. synoptic) information to the local scale (e.g. catchment). The end goal of a typical modelling chain is to produce projections of discharge under the influence of climate change. For catchment-scale impact studies, a typical modelling chain includes future climate scenarios, general circulation models (GCMs), downscaling techniques, post-processing methods, and hydrological models (Hakala et al. 2019). GCMs are capable of simulating ocean and atmospheric circulation, the global carbon cycle, dynamic vegetation, and atmospheric chemistry (IPCC 2014). A variety of potential climate scenarios can be used to force GCMs, such as greenhouse gas (GHG) emission scenarios based on specific socio-economic assumptions or so-called representative concentration pathway scenarios based on forcing projections that could, in theory, include more than one socio-economic scenario (IPCC, 2014). Due to the low spatial resolution of GCMs (horizontal grid spacing about 100–300 km), climate simulations from GCM are usually downscaled to a finer resolution using statistical techniques or dynamical techniques—the latter utilises regional climate models (RCMs). The downscaled data may still show a significant bias compared to the observed data. Various post-processing methods (often called "bias correction") have been developed to reduce biases in the simulated time series (Teutschbein and

Seibert 2012). Hydrological models, which have been calibrated using observed meteorological and hydrological data, are finally used to predict future discharge series based on the bias-corrected climate simulations. For a more detailed introduction to modelling chains, we refer to Hakala et al. (2019). The concrete modelling chain used in this study is described in Sect. 3.

Discharge projections stemming from modelling chains are subject to uncertainties, which need to be considered when developing and implementing adaptation strategies. Uncertainties arise from the different components of the modelling chain such as the GHG emission scenarios, climate models, downscaling methods, post-processing methods, rainfall-runoff models (RRMs), and natural variability (Addor et al. 2014). The discharge projection uncertainty is generally assessed by using an ensemble approach, in which the components of the modelling chain are varied to characterize the resulting impact on the discharge projections. The spread of the resulting projections is interpreted as uncertainty (Hakala et al. 2019). To quantify the uncertainty contribution of each modelling chain component to the total uncertainty of the discharge projections, an analysis of variance (ANOVA) (Hawkins and Sutton 2009) can be used (e.g. Addor et al. 2014; Vetter et al. 2017).

1.2 Hydroclimatology of Tunisia

Tunisia's climate is characterized by a substantial interannual and seasonal variability of precipitation. Annual precipitation presents a north–south gradient. It can exceed 1500 mm/year in the extreme North and less than 100 mm/year in the South. Potential evapotranspiration (PET) varies from 1300 mm/year in the North to more than 2500 mm/year in the South (Mansour and Hachicha 2014). Water resources are constrained by the irregularity in rainfall between months and years (Fader et al. 2020). Due to the Mediterranean climate, precipitation mainly contributes to river discharge during autumn, winter and spring. Within Tunisia, rivers generally have an ephemeral or intermittent behaviour (Tramblay et al. 2021). Further, water resources in Tunisia are unequally distributed, and their spatial distribution often does not match human and environmental demand. Indeed, 83% of Tunisia's surface water resources are located in the northern region, while the demand is mainly located along the eastern coast (Ben Fraj et al. 2019). To ensure water supply despite these circumstances, many dams and water transfer installations have been constructed during recent decades. Water transfers from Northern Tunisia to water demand centres located along the eastern coastline satisfy around 60% of the national potable water demand and 15% of irrigation demand (SECADENORD, 2020). Irrigation by surface water mainly occurs in Northern Tunisia and the coastal plains to produce wheat, barley, tomatoes, vegetables, grapes and oranges (Ben Abdallah 2007). Furthermore, some important ecological systems are dependent upon surface water supply. For example, the wetlands of Lake Ichkeul and Sebkhet Soliman (Northern Tunisia) are supplied by freshwater from rivers. Lake Ichkeul and its marshes, are some of the most important wetlands of the western Mediterranean basin and are an essential stopover point for birds migrating between Eurasia and Sub-Saharan Africa (Hamdi et al. 2012). Sebkhet Soliman is one of the few places in the Cap Bon peninsula where water is retained during the dry season, and it is a great refuge for waterfowl and home to breeding populations (RSIS 2020).

1.3 Expected Climatological Changes in Tunisia

The Mediterranean region has been identified as a susceptible region to climate change, especially Tunisia (Cramer et al. 2018; Fader et al. 2020), which has already undergone marked climate shifts during the last century (Tramblay et al. 2016; Dakhlaoui et al. 2017). Several recent studies have shown important changes to Tunisia's future climate, including an expected decrease in total precipitation and an increase in mean annual temperature and evapotranspiration (Milano et al. 2013; IPCC 2014; Moussa et al. 2018; Dakhlaoui et al. 2019b). Hachani et al. (2017) utilized several GCMs from CMIP5 and compared model projections with the reference period (1996–2005). They found that projected temperatures increased by 1 and 5 °C, projected precipitation decreased by 5.4 and 20%, and projected reference evapotranspiration (ETo) increased by 3% and 9% until 2030 and 2090, respectively. Allani et al. (2019) found an average increase in annual ETo of between 4.7 and 9.3% and a decrease in annual rainfall of between 4.7 and 17.3%, in the Nebhana dam system (Central Tunisia), using future climate change scenarios obtained from five GCMs of CMIP5 under two GHG emission scenarios—Representative Concentration Pathways (RCP 4.5 and 8.5) for the periods 2021–2040, 2041–2060 and 2061–2080, compared to the reference period 1985–2004. Haj-Amor et al. (2020) analysed projections from one GCM. They found a significant trend of increasing annual maximum and minimum temperatures, relative humidity, wind speed, and solar radiation, and a significant decreasing trend of future rainfall in a coastal oasis situated in southern Tunisia (Metouia).

1.4 Climate Change Impacts on Hydrology in Tunisia

The projected decrease of precipitation, temperature and PET shown by the existing literature could result in critical water stress in the future (Fader et al. 2020). Sellami et al. (2015 and 2016) applied the SWAT model to a small catchment located in eastern Tunisia (Chiba), to provide hydrological projections. They forced the hydrological model with climate projections based on the A1B emission scenario (Deidda et al. 2013). Daily discharge projections of the Chiba catchment were found to decrease in response to drier and warmer climatic conditions in the 2050s. Allani et al. (2019) proposed an evaluation of the impacts of climate change on water supply and demand of the Nebhana dam system (Central Tunisia), using the GR2M hydrological model, forced by RCP 4.5 and 8.5 emission scenarios for the periods 2021–2040, 2041–2060 and 2061–2080. Results showed a decrease in inflow between 15 and 37% compared to the reference period, 1985–2004. They found that the supply does not meet the system demand. Dakhlaoui et al. (2019b) proposed an evaluation of the impact of the selection of calibration period on hydrological projection in Northern Tunisia. They used a delta-change monthly correction to perturb the observed climate series according to climate simulations under two emission scenarios (RCP 4.5 and RCP 8.5) for one medium-term horizon (2040–2070) and one long-term horizon (2070–2100). Climate projections were obtained from eight regional climate models forced by general circulation models (GCM-RCMs) from EURO-CORDEX. They found that when RRMs were calibrated using the entire historical period (versus using sub-periods similar to that of future climate conditions), they subsequently underestimated the impact of climate change on mean annual runoff by 5−20%.

The literature review presented above shows that studies that investigated the hydroclimatic impacts of climate change in Tunisia have proposed a limited assessment of the uncertainty stemming from the different components of the modelling chain. Uncertainties associated with hydroclimatic projections are generally evaluated using an ensemble approach, i.e., by varying the different components of the modelling chain (Hakala et al. 2019). However, the impact studies for Tunisia have used either one single catchment for analysis and one single hydrological model (e.g. Sellami et al. 2015; 2016; Allani et al. 2019; Hachani et al. 2017), one climate model (e.g. Nasr et al. 2008; Adham et al. 2019; Haj-Amor et al. 2020), one radiative hypothesis (Sellami et al. 2015, 2016), one future period (Sellami et al. 2015 and 2016), or have used the previous generation of climate models (e.g. Nasr et al. 2008; Sellami et al. 2015, 2016). To our knowledge, no study has evaluated the contribution of the uncertainties stemming from the different components of the modelling chain to the overall uncertainty imbedded within the discharge projections for this region of the world. Instead, the uncertainty from a given element of the modelling chain has been evaluated separately (e.g. Allani et al. 2019; Dakhlaoui et al. 2019b). Understanding uncertainties related to climate change projections is especially necessary for Northern Tunisia, due to its strategic role as a water supplier for the rest of the country, and the sensitivity of its ecological system to the water supply.

This study's overall goal was to provide a robust estimate of climate change impacts on water resources in Northern Tunisia by considering a wide variety of uncertainty sources. For this purpose we: (1) compiled an ensemble of precipitation and temperature simulations for different climate models and GHG emission scenarios, (2) simulated discharge using these ensemble time series as input to force three RRMs and (3) analysed the spread and mean behaviour of these hydrological simulations. As part of 3 we also performed an ANOVA to attribute the uncertainties to the different steps in the model chain.

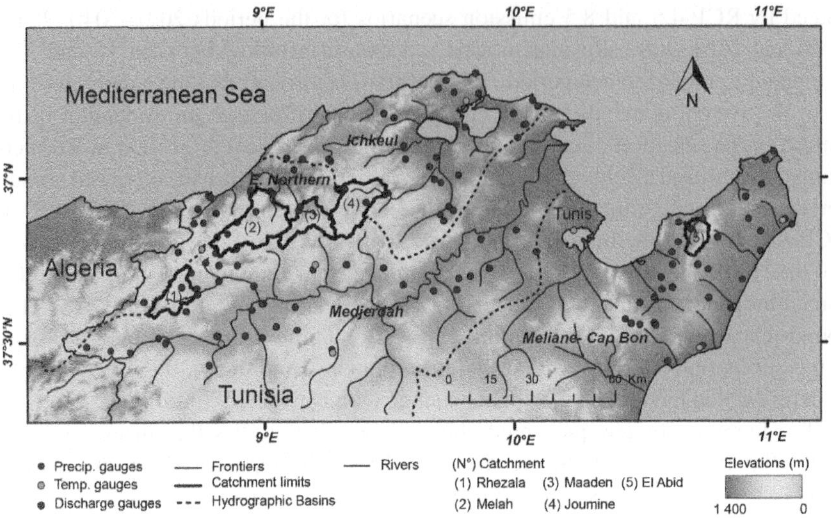

Fig. 1 Locations of study basins are indicated by a bold black line. The locations of the precipitation (blue circle), temperature (red circle) and discharge gauges (purple circle) are also shown

2 Data

2.1 *Study Areas*

The study area is located in Northern Tunisia, which contains the main basins of Tunisia: Extreme North, Ichkeul, High Medjerdah and Cap Bon, that produce the main surface water in Tunisia, and provide the highest water quality. From the 59 gauge stations situated in Northern Tunisia, we selected five representative catchments with negligible influences of major hydraulic installations, such as dams and water transfers, and with good observation data availability (daily gaps of less than 30% of the total reference period of 1970–2000). The five catchments (Fig. 1) are: (1) Rhezala at Fernana station, (2) Melah at Ouchtata station; (3) Maaden at BouBrima station; (4) Joumine at Jebel Antra station, and (5) El Abid basin at pont route station. The catchment sizes ranged from 81 to 315 km^2. The study catchments were characterized by a semi-arid to humid Mediterranean climate with a warm season (Dakhlaoui et al. 2019a). The hydro-climatic characteristics are reported in Table 1.

2.2 *In-Situ Hydro-Climatic Data*

Rainfall data were provided from 123 rain-gauges situated in the study area (see Fig. 1), having daily gaps of less than 30% of the total data period of 1970–2000, and

Table 1 The main hydro-climatic characteristics of the study catchments, averaged over the period 1970–2000

Catchment	Gauge	DGRE Code	Area (Km^2)	Average elevation (m)	Precipitation (mm)	Temperature (°C)	PET (mm)	Runoff (mm)
Rhezala	Fernana	5303510	138	470	857	16.3	1020	101
Melah	Ouchtata	3700250	315	423	928	16.5	1023	179
Maaden	Bou Brima	3700350	145	337	783	16.6	1029	135
Joumine	Jebel Antra	3600110	234	375	742	16.3	1016	116
El Abid	Pont-Route	3600110	81	202	530	17.6	1061	40

providing a good stationary network for the spatial interpolation. For the temperature forcing data, eight meteorological stations with monthly mean series of air temperature were used. Climate forcing data was obtained by a spatial interpolation on a 2 × 2 km grid with the inverse distance weighting technique. Precipitation data were interpolated according to the inverse distance weighting technique proposed by Valéry et al. (2010). The temperature was interpolated using the inverse distance weighting technique proposed by Ruelland et al. (2014), with a lapse rate of −0.65 °C/100 m. PET was estimated using the Oudin formula (Oudin et al. 2005).

2.3 *High-Resolution Climate Simulations*

Climate model data used in this study include simulations and projections of daily mean temperature and precipitation from 11 pairs of GCM-RCMs (general circulation model forcing regional climate model) from the EURO-CORDEX project (Table2), with a 0.11° resolution (~12.5 × 12.5 km). The GCM-RCM historical simulation spans the period 1970–2005, and the future period corresponds to 2006–2100. Two RCPs scenarios were considered for the future time period, RCP 4.5 and RCP 8.5, which are an optimistic and pessimist scenario, respectively.

3 Methods

The modelling chain is composed of two emission scenarios (RCP 4.5 and RCP 8.5), two future periods (2040–2070 and 2070–2100), 11 regional climate models (RCMs) from the European domain (EURO-CORDEX), forced by five general circulation models (GCMs), five catchments located in Northern Tunisia, a bias correction method (Quantile Mapping), and three bucket-type rainfall-runoff models (GR4J: Perrin et al. 2003; HBV: Bergström and Lindström 2015; and IHACRES: Jakeman et al. 1990; Table 3) (see Fig. 2).

Table 2 Overview of the 11 EURO-CORDEX simulations used in this study

N°	General circulation model	Regional climate model
1	CNRM-CERFACS-CNRM-CM5	CLMcom-CCLM4-8–17
2	CNRM-CERFACS-CNRM-CM5	SMHI-RCA4
3	ICHEC-EC-EARTH	CLMcom-CCLM4-8–17
4	ICHEC-EC-EARTH	SMHI-RCA4
5	ICHEC-EC-EARTH	KNMI-RACMO22E
6	ICHEC-EC-EARTH	DMI-HIRHAM5
7	IPSL-IPSL-CM5A-MR	IPSL-INERIS-WRF331F
8	IPSL-IPSL-CM5A-MR	SMHI-RCA4
9	MOHC-HadGEM2-ES	CLMcom-CCLM4-8–17
10	MPI-M-MPI-ESM-LR	CLMcom-CCLM4-8–17
11	MPI-M-MPI-ESM-LR	SMHI-RCA4

Table 3 Overview of the characteristics of the three tested models

	GR4J	HBV	IHACRES
Number of parameters	4	8	5
Production module	Precipitation interception by PET, Non-linear soil moisture accounting store, Actual evapotranspiration parabolic function of soil moisture,	Evapotranspiration extracted from stored soil moisture The level of soil moisture accounting store determines the quantity of precipitation intercepted by the soil Actual evapotranspiration piecewise linear function of soil moisture	Precipitation interception by PET, Non-linear soil moisture accounting store, Actual evapotranspiration exponential function of soil moisture
Water exchange	Yes Inter-catchment groundwater flow	No	No
Runoff generation and routing module	Two unit hydrographs, non-linear routing	Two reservoirs, the upper is non-linear and the lower is linear (subsurface runoff) One unit hydrograph	One reservoir One unit hydrograph
Reference	Perrin et al. (2003)	Bergström and Lindström (2015)	Jakeman et al. (1990)

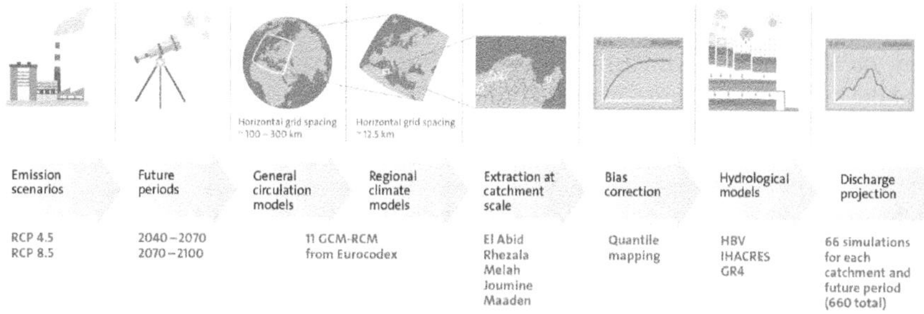

Fig. 2 Modelling chain to simulate discharge as used in this study. Projections of soil moisture were based on the same chain. Each box represents a component of the modelling chain, with different alternatives used for various components (modified from Hakala et al. 2019)

3.1 *Hydrological Models*

Three bucket-type, rainfall-runoff models, running at the daily time step were used in this study: GR4J, HBV and IHACRES. They were selected because they are parsimonious models and all were calibrated based on historical precipitation, PET and runoff data, but differ in how they conceptualise hydrological processes and in their complexity (4–8 free parameters). This makes accounting for model structure uncertainty possible.

These models have different production functions, which allows for an inter-comparison of climate impacts on surface water resources. In GR4J, the production routine accounts for precipitation and potential evapotranspiration and calculates the effective precipitation that contributes to flow and feed the production reservoir. The soil moisture is accounted for by a non-linear storage. Actual evapotranspiration (AET) is accounted for according to a parabolic function of soil moisture. Only one free parameter is used for the production routine. In HBV, the fraction of effective precipitation is calculated according to a parabolic shape function depending on soil moisture. AET equals PET when the soil storage is filled to a certain fraction and decreases linearly for smaller soil moisture values. In IHACRES the input rainfall is partitioned explicitly into effective precipitation, PET, and changes in catchment moisture. The AET is calculated according to an exponential function of soil moisture, which is accounted for itself in a non-linear storage. In IHACRES two free parameters are dedicated to the production module.

The recent successful applications of the three models in other catchments in Tunisia (Bargaoui et al. 2008; Dakhlaoui et al. 2009, 2012; Abbaris et al. 2014) were the basis for their inclusion within this study. Further description of the hydrological models can be found in Dakhlaoui et al. (2017, 2020). The model parameters were calibrated using the Shuffle Complex Evolution (SCE) algorithm (Duan et al. 1992) with the Kling-Gupta Efficiency (KGE, Gupta et al. 2009) as a goodness-of-fit measure. The hydrological models were run at a daily time step. Still, for calibration and validation, the performance was evaluated using a 10-day time step (i.e., 36 values per year) since the focus of the study is on water resources and not on day-to-day variability. Nash–Sutcliffe Efficiency (*NSE*) criteria (Nash and Sutcliffe 1970) and the cumulated volume error (*VE*) were also used to evaluate RRM performance.

3.2 Bias Correction of GCM-RCM Climate Variables

Quantile mapping (QM) was applied as a bias correction method to correct the daily precipitation and temperature of the GCM-RCMs. Previous studies have shown QM to outperform other bias correction methods (Teutschbein and Seibert 2012, 2013; Chen et al. 2013). The aim of QM is to correct the distribution of the climate model data so that it matches the distribution of the observational data. It consists of estimating quantiles for both an observation and modelled climate variable under a control period. A transfer function is then created by interpolation between corresponding quantile values, which is applied to the projected climate variable. The number of quantiles is a free parameter. Here the empirical cumulative distribution functions (CDFs) of observed and modelled climate variables were estimated using empirical percentiles. Values in between the percentiles were approximated using linear interpolation. In cases where new GCM-RCM values (such as those from the projected period) were larger than the control values used to estimate the empirical CDF, a linear regression fit was used to extrapolate beyond the range of observations. The transfer function of the QM was inferred using a 30-year control period (1970–2000), which is the recommended time-length for climate applications (WMO 2011).

3.3 Analysis of Variance of Hydrological Projection (ANOVA)

We considered the variance of the hydrological projections (change in mean annual discharge) as an estimate of their uncertainty, and we used an analysis of variance (ANOVA, Hawkins and Sutton 2009) technique to quantify the contribution of the uncertainties related to different aspects in the modelling chain to the overall uncertainty. Three main sources of uncertainty were compared in our study: the hydrological models (GR4J, HBV and IHACRES), the emission scenarios (RCP 4.5 and RCP 8.5), the GCM-RCMs (11 models), as well as the sum of the significant interaction between factors and residual error. The fraction of explained variance for each factor was computed by dividing the variance of each factor by the total variance. The analysis was performed for each catchment and each future period, and for both annual and seasonal change in discharge.

The combination of the two emission scenarios, eleven climate models, and three hydrological models in a factorial way, led to a total of 66 model chains applied to five catchments and two future periods. In total, 660 hydrological projections were produced and analysed. The factorial design of this modelling chain allows for a separation of the components of the model chain from each other and the subsequent analysis of how each element contributes to the total variance and quantification of the interactions between these components.

4 Results and discussion

4.1 Calibration of RRM for the Reference Period

The RRMs were calibrated over the period 1970–2000 by SCE-UA according to the *KGE* criteria (Table 4). The *KGE* values range from 0.76 to 0.91, *VE* was close to 0%, and *NSE* was between 0.76 and 0.81, for the five study catchments. The performance of all three models was generally similar between models. However, the HBV model had the highest performance according to *VE* and *KGE*, and the GR4J model had the highest performance according to NSE. Series of observed and simulated discharge covering the whole reference period (not shown here) show that seasonal patterns of runoff are all well captured by all RRMs for all the catchments.

4.2 Climate Projection

The climate projections over two future 30-year time periods: mid-term (2040–2070) and long-term (2070–2100), and two RCP scenarios (4.5 and 8.5) were considered. In the first two rows of Fig. 3 we present the projected seasonal precipitation and temperature as the median of the 11 GCM-RCMs, for the four combinations of future periods and RCPs (shown together with the historical period; black line), over the study catchments. The relative changes in precipitation between the historical period and the projection periods are presented in Fig. 4.

Results show that the climate change signal is very different from one GCM-RCM to another. However, all GCM-RCMs depict a warmer future climate, and almost all climate models project dryer conditions for the future. For precipitation, for the mid-term period, the RCP 4.5 scenario results in a decrease of about −8% in total precipitation and −15% for the RCP 8.5 scenario. For the long-term period, a decrease of about −16% of total precipitation is expected for RCP 4.5 and −26% for RCP 8.5. These changes mainly occur during the wet season (November to April). For temperature, for the mid-term period, an increase of + 0.44 to + 2.3 °C is projected with the scenario RCP 4.5 and + 1.2 to + 2.6 °C with scenario RCP 8.5. For temperature, over the long-term period, an increase of + 1.2 to + 2.3 °C is projected with the scenario RCP 4.5, and + 2.1 to + 4 °C with the scenario RCP 8.5. Temperature changes are likely to have little impact on discharge since AET is mainly

Table 4 Performance of rainfall-runoff models calibrated over the reference period (1970–2000) using the observed climatic data, for the five study catchments

Criteria	*KGE*			*NSE*			*VE*		
RRM / Catchment	GR4J	HBV	IHACRES	GR4J	HBV	IHACRES	GR4J (%)	HBV (%)	IHACRES (%)
El Abid	0.78	0.89	0.76	0.81	0.78	0.74	6.7	−0.1	8.1
Rhezala	0.91	0.88	0.86	0.82	0.77	0.72	−1.9	2.0	2.0
Melah	0.83	0.85	0.83	0.70	0.70	0.71	3.1	0.2	4.6
Joumine	0.88	0.89	0.88	0.79	0.78	0.77	2.0	−0.3	−0.8
Maaden	0.88	0.88	0.87	0.76	0.76	0.74	−0.3	−0.2	0.9

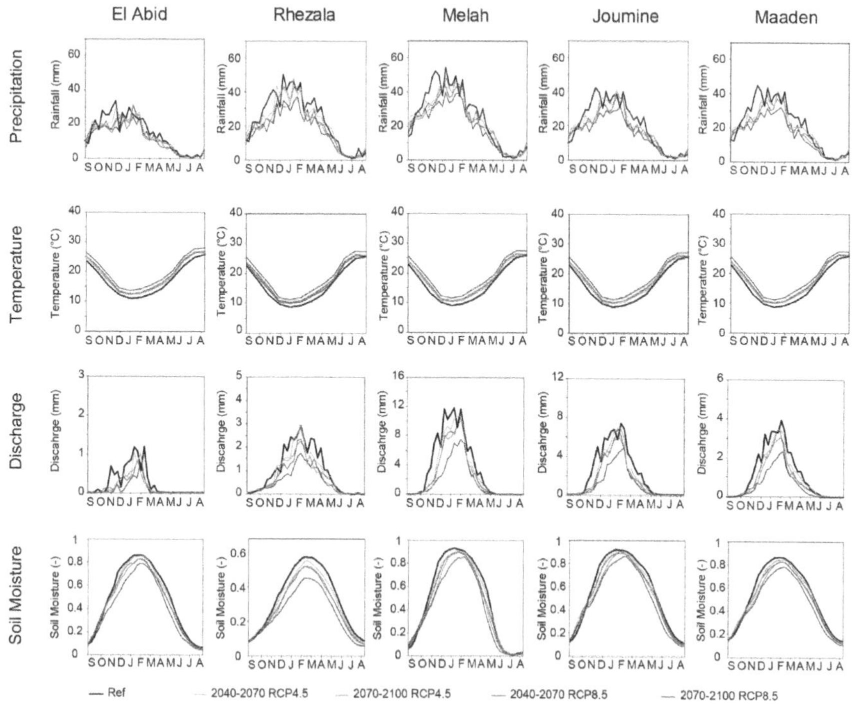

Fig. 3 Projected precipitation, temperature, discharge (HBV model) and Soil Moisture (HBV model) for the five study catchments, under RCP 4.5 and RCP 8.5, over the two future periods, 2040–2070 and 2070–2100 (shown as red lines), compared to the historical period (shown as a black line; Ref)

controlled by the available soil moisture. In contrast, the decrease in precipitation during winter months may have a critical impact on water resources.

The results of future climate change over Northern Tunisia confirm those of previous studies based on an earlier generation of climate models, which have indicated a likely increase of temperature and a decrease of precipitation under the influence of climate change (e.g. Deidda et al. 2013; Bargaoui et al. 2014; Moussa et al. 2018). However, the high resolution of the recent climate models used in this study allowed a better spatial representation of the projected climatic change. The resolution of EURO-CORDEX climatic models (12.5 km ·12.5 km = 156.25 km^2) is more adapted to the size of the study catchment, ranging from 81 to 315 km^2, than the earlier generation of RCMs, such as those from the ENSEMBLES project (Deidda et al. 2013), which have a resolution of 24 km ·24 km = 576 km^2 over the study region.

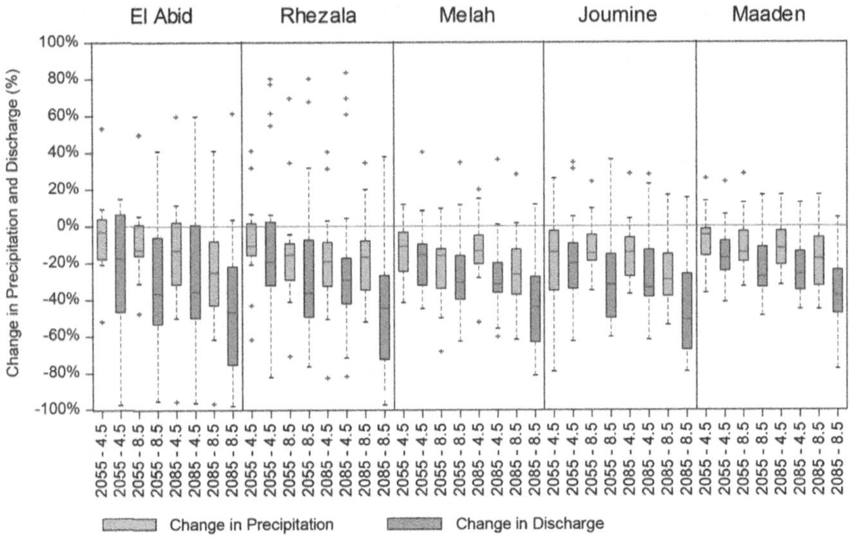

Fig. 4 The boxplots summarise the change in precipitation and discharge (for GR4J, HBV and IHACRES) for the studied catchments; between the reference periods 1970–2000 and the projected periods: 2040–2070 (2055) and 2070–2100 (2085); for RCP 4.5 and RCP 8.5. In total, there are 11 values per boxplot for precipitation and 33 values for discharge

4.3 Hydrological Projections in Future Hydroclimatology

In the third row of Fig. 3, projected discharge from the HBV model is shown as the median of the 11 GCM-RCMs, together with the historical period (black line) for the study catchments. For the mid-term period (2040–2070), the predicted changes in precipitation and temperature show a decrease in runoff between 10 and 30% under scenario RCP 4.5 and a 25–38% decrease in runoff under scenario RCP 8.5. For the long-term period (2070–2100) the volume changes found were a 20–37% decrease in runoff under scenario RCP 4.5 and a 41–58% decrease in runoff under scenario RCP 8.5 (Fig. 4). We found generally similar behaviour between the different RRMs. The hydrological response of Northern Tunisian catchments has presented a high sensitivity to any change in precipitation. The assessment of the elasticity of discharge to precipitation results in a hydrological multiplication factor of around two, which means that any future change rate in precipitation would cause a change of twice this rate in discharge. However, this result needs to be interpreted with care since the hydrological modelling was performed using a daily time step. Northern Tunisia is well known for its torrential pluvial events. Any future change in the precipitation intensity could have a significant impact on the hydrological response and affect this multiplication factor. For example, an increase in precipitation intensity could lead to more runoff, which could compensate for the decrease in total precipitation.

In the last row of Fig. 3, we present the projections of soil moisture as the median of 11 GCM-RCMs from the HBV model for the study catchments. We consider here the state variable representing the filling rate of the soil box (for more details about the used version of HBV model, see Dakhlaoui et al. 2017). For the mid-term period (2040–2070), predictions show a decrease in mean annual soil moisture of about 3–12% under scenario RCP 4.5 and 7–15% under scenario RCP 8.5. For the long-term period (2070–2100) a decrease in mean annual soil moisture was found to be between 7 to 17% under scenario RCP 4.5 and 14–30% under scenario RCP 8.5. This decrease is more apparent during spring and winter. However, we notice a slight increase in projected soil moisture during autumn, which seems to be the result of the increase of projected precipitation during this season. We found a generally similar soil moisture behaviour amongst the different RRMs (not shown here).

Although RRMs are capable of simulating realistic discharge, their soil moisture routines are highly simplistic. We recognise that the state variables representing the soil moisture in the RRM used herein could not be easily interpreted as soil moisture, because of the conceptual nature of these models. However, the change in these state variables could give an idea about the change in soil moisture in the study catchments. The similarity between the different RRMs in representing soil moisture projections, despite their different structure, could help to increase confidence in the projected change in soil moisture provided by these RRMs. Informing the state variables representing soil moisture during calibration, by using soil moisture observations from field measurements or satellite information, could help to improve the representation of soil moisture in these models (Shahrban et al. 2018).

The significant future decrease in discharge in Northern Tunisia could cause important future stress for the whole country, which is reliant upon this region to provide almost all surface water resources. The future decrease in precipitation and soil moisture in Northern Tunisia could affect agricultural production in this region, where more than 90% of the land area is cultivated under rain-fed conditions, mainly olives and cereals (Mansour and Hachicha 2014). The decrease in soil moisture could be more harmful to cereal production because it is more important during spring and winter, corresponding to the growing phase (Mougou et al. 2011).

4.4 *Analysis of Variance of Hydrological Projection (ANOVA)*

The decomposition of the projection variance by ANOVA partitioning among the three sources of uncertainty (Fig. 5) shows that the uncertainty induced by the GCM-RCMs is the dominant source of uncertainty. We can see that uncertainty stemming from the GCM-RCMs is negatively correlated to the catchment area. El Abid and Rhezala catchments having the smallest area (81 and 138 km^2, respectively), show the highest contribution of GCM-RCMs to total uncertainty, compared to the other catchments. However, the Melah catchment, which has the largest area (315 km^2), also has the lowest contribution of GCM-RCMs to the total uncertainty. The anti-correlation of the uncertainty of GCM-RCMs to the area of the catchments found in

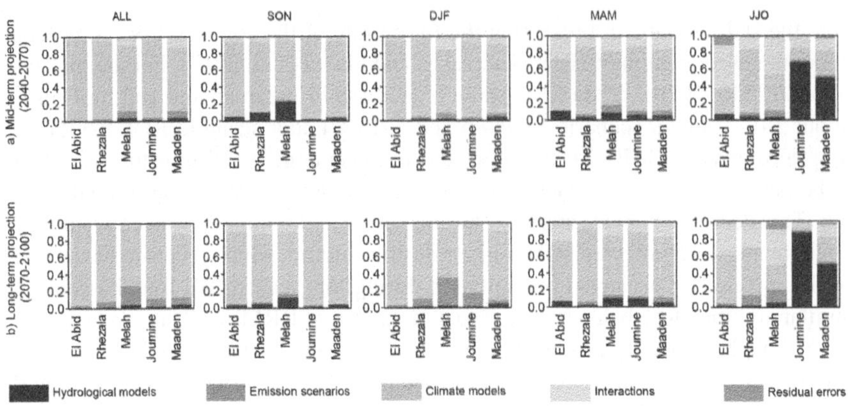

Fig. 5 Decomposition of the projection variance. ANOVA partitioning among the three sources of uncertainty, the significant interactions, and the residual errors for discharge change are shown for the two 30-year future periods (2010–2070 and 2070–2100). Results are shown according to the following seasons: annual (ALL), autumn (SON), winter (DJF), spring (MAM), and summer (JJA)

this study, could be explained by the inability of the resolution of the climate models used to correctly reproduce the spatial variability of the climate variable (especially precipitation). For example, El Abid catchment, which showed the highest proportional uncertainty stemming from GCM-RCMs, has an area smaller than one entire EURO-CORDEX grid cell. This makes the use of GCM-RCM output challenging since they were not designed to represent features at the spatial scale smaller than their grid (Hakala et al. 2018). This infers that RCMs with higher resolution would improve the reliability of impact studies of climate change at the catchment scale.

RCPs were found to be a relatively greater source of uncertainty over time (i.e. contributing greater uncertainty by the end of the century). Hydrological models were found to be a greater source of uncertainty during the dry season than in the wet season. This could be explained by the difficulty of RRMs to reconstruct discharge under dry season conditions compared to wet season (e.g. Choi and Beven 2007; Zhang et al. 2011) and by the low transferability of RRM parameters toward drier conditions, shown by previous studies (Coron et al. 2012; Dakhlaoui et al. 2017, 2019a).

5 Conclusion

The aim of this study was to evaluate the hydrological impact of climate change in Northern Tunisia and to assess the uncertainty stemming from the different components of the modelling chain. Our results show a significant future decrease in discharge and soil moisture that could reach up to 58 and 30%, respectively. These results indicate that water stress in Tunisia is expected to become more dire and complicated, which should motivate the creation of adaption strategies to climate change to ensure that water demand can be met, and to manage the competition between ecological water requirements and human demand.

The decomposition of the projection variance of change in discharge by ANOVA partitioning among the three sources of uncertainty (RCPs, GCM-RCM and RRM) shows that the uncertainty introduced by the GCM-RCMs is the dominant source of uncertainty and is anticorrelated to the catchment area. This indicates that RCMs with higher resolution could improve the conclusions drawn from hydrological climate change impact studies at the catchment scale. Also, RRMs can contribute a significant proportion of uncertainty, especially for the dry season.

A univariate bias correction method was used in this study, which implies that climate variables were bias-corrected separately, without taking into consideration intervariable dependence between these variables. More advanced bias correction techniques that tackle this limitation, such as the multivariate bias correction (Vrac 2018; Cannon 2018), could be explored in future studies. State variables representing the soil moisture in the conceptual RRM could not be easily interpreted as soil moisture. Future studies should focus on constraining these state variables during calibration by soil moisture observations from field measurements or remote sensing data. Future studies should also focus on hydrological modelling at a shorter time increment to consider the impact of change in the magnitude of precipitation intensity on discharge.

Acknowledgements This work was partly performed during a postdoctoral research visit of the first author at the University of Zurich, Switzerland, which was supported by a Swiss Government Excellence Scholarship. The authors are also grateful to INM (Institut National de la Météorologie) and to DGRE (Direction Générale des Ressources en Eau) in Tunisia for providing the necessary hydro-climatic data for this study. We also recognize the European Coordinated Regional Downscaling Experiment (EURO-CORDEX) for the climate modelled data. We thank Dr. Tracy Ewen for reviewing and correcting the language.

References

Abbaris A, Dakhlaoui H, Thiria S, Baragaoui Z (2014) Variational data assimilation with the YAO platform for hydrological forecasting evolving water resources systems. In: Understanding predicting and managing water-society interactions, IAHS Publ. vol 364. pp 1–6

Addor N, Rössler O, Köplin N, Weingartner R, Seibert J (2014) Robust changes and sources of uncertainty in the projected hydrological regimes of Swiss catchments. Water Resour Res 50(10):7541–7562. https://doi.org/10.1002/2014WR015549

Adham A, Wesseling JG, Abed R, Riksen M, Ouessar M, Ritsema CJ (2019) Assessing the impact of climate change on rainwater harvesting in the Oum Zessar watershed in Southeastern Tunisia. Agric Water Manag 221(2019):131–140

Allani M, Mezzi R, Zouabi A, Béji R, Joumade-Mansouri F, Hamza ME, Sahli A (2019) Impact of future climate change on water supply and irrigation demand in a small mediterranean catchment. case study: Nebhana dam system, Tunisia. J Water Climate Change. https://doi.org/10.2166/wcc.2019.131

Bargaoui Z, Dakhlaoui H, Houcine A (2008) Modélisation pluie-débit et classification hydroclimatique. Rev Sci Eau 21:233–245

Bargaoui Z, Tramblay Y, Lawin EA, Servat E (2014) Seasonal precipitation variability in regional climate simulations over northern basins of Tunisia. Int J Climatol. https://doi.org/10.1002/joc3683

Ben Abdallah S (2007) The water resources and water management regimes in Tunisia. agricultural water management: proceedings of a workshop in Tunisia (Series: Strengthening Science-Based Decision Making in Developing Countries). In: Holliday L (ed) Proceedings of a workshop in Tunisia. The National Academies Press, Washington, D.C., pp 81–87

Ben Fraj W, Elloumi M, Molle F (2019) The politics of interbasin transfers: socio–environmental impacts and actor strategies in Tunisia. Nat Resour Forum 43:17–30. https://doi.org/10.1111/1477-8947.12165

Bergström S, Lindström G (2015) Interpretation of runoff processes in hydrological modelling: experience from the HBV approach. Hydrol Processes 29:3535–3545. https://doi.org/10.1002/hyp.10510

Bijl DL et al (2018) A global analysis of future water deficit based on different allocation mechanisms. Water Resour Res 54(8):5803–5824. https://doi.org/10.1029/2017wr021688

Cannon AJ (2018) Multivariate quantile mapping bias correction: an N-dimensional probability density function transform for climate model simulations of multiple variables. Clim. Dynam. 50(31–49):2018a. https://doi.org/10.1007/s00382-017-3580-6

Chen J, Brissette FP, Chaumont D, Braun M (2013) Performance and uncertainty evaluation of empirical downscaling methods in quantifying the climate change impacts on hydrology over two North American river basins. J Hydrol 479(200–214):2013. https://doi.org/10.1016/j.jhydrol.2012.11.062

Choi HT, Beven KJ (2007) Multi-period and multi-criteria model conditioning to reduce prediction uncertainty in distributed rainfall-runoff modelling within GLUE framework. J Hydrol 332:316–336

Coron L, Andréassian V, Perrin C, Lerat J, Vaze J, Bourqui M, Hendrickx F (2012) Crash testing hydrological models in contrasted climate conditions: an experiment on 216 Australian catchments. Water Resour Res 48:W05552. https://doi.org/10.1029/2011WR011721

Cramer W, Guiot J, Fader M, Garrabou J, Gattuso JP, Iglesias A, Lange MA, Lionello P, Llasat MC, Paz S, Peñuelas J, Snoussi M, Toreti A, Tsimplis MN, Xoplaki E (2018) Climate change and interconnected risks to sustainable development in the Mediterranean. Nat Clim Chang 8:972–980. https://doi.org/10.1038/s41558-018-0299-2

Dakhlaoui H, Bargaoui Z, Bárdossy A (2009) Comparaison de trois méthodes d'usage de la technique des voisins les plus proches en vue d'amélioration de la performance de l'algorithme SCE-UA appliqué pour le calage du modèle pluie-débit HBV. In: Hydroinformatics in hydrology, hydrogeology and water resources, IAHS Publ. 331, pp 139–153

Dakhlaoui H, Bargaoui Z, Bárdossy A (2012) Toward a more efficient Calibration schema for HBV rainfall-runoff model. J Hydrol 444–445:161–179. https://doi.org/10.1016/j.jhydrol.2012.04.015

Dakhlaoui H, Ruelland D, Tramblay Y, Bargaoui Z (2017) Evaluating robustness of conceptual rainfall-runoff models under climate variability in Northern Tunisia. J Hydrol 550(2017):201–217. https://doi.org/10.1016/j.jhydrol.2017.04.032

Dakhlaoui H, Ruelland D, Tramblay Y (2019) A bootstrap-based differential split-sample test to assess the transferability of conceptual rainfall-runoff models under past and future climate variability. J Hydrol. https://doi.org/10.1016/j.jhydrol.2019.05.056

Dakhlaoui H, Seibert J, Hakala K (2019b) Hydrological impacts of climate change in Northern Tunisia. advances in sustainable and environmental hydrology, hydrogeology, hydrochemistry and water resources. In: Chaminé HI, Barbieri M, Kisi O, Chen M, Merkel BJ (eds) Proceedings of the 1st springer conference of the arabian journal of geosciences (CAJG-1), Tunisia 2018. Series: advances in science, technology and innovation. https://doi.org/10.1007/978-3-030-01572-5_71

Dakhlaoui H, Seibert J, Hakala K (2020) Sensitivity of discharge projections to potential evapotranspiration estimation in northern Tunisia. Reg Env Change. https://doi.org/10.1007/s10113-020-01596-8

Deidda R, Marrocu M, Caroletti G, Pusceddu G, Langousis A, Lucarini V, Puliga M, Speranza A (2013) Regional climate models' performance in representing precipitation and temperature over selected Mediterranean areas. Hydrol Earth Syst Sci 17:5041–5059

Duan QY, Sorooshian S, Gupta V (1992) Effective and efficient global optimization for conceptual rainfall-runoff models. Water Resour Res 28:1015–1031

Fader M, Giupponi C, Burak S, Dakhlaoui H, Koutroulis A, Lange MA, Llasat MC, Pulido-Velazquez D, Sanz-Cobeña A (2020) Water. In: Climate and environmental change in the Mediterranean Basin—current situation and risks for the future. First Mediterranean Assessment Report [Cramer W, Guiot J, Marini K (eds)] Union for the Mediterranean, Plan Bleu, UNEP/MAP, Marseille, France, pp 57, in press

Flörke M, Schneider C, McDonald RI (2018) Water competition between cities and agriculture driven by climate change and urban growth. Nature Sustain 1(1):51–58. https://doi.org/10.1038/s41893-017-260006-8

Gosling SN, Arnell NW (2016) A global assessment of the impact of climate change on water scarcity. Clim Change 134(3):371–385. https://doi.org/10.1007/s10584-013-0853-x

Gupta HV, Kling H, Yilmaz KK, Martinez GF (2009) Decomposition of the mean squared error and NSE performance criteria Implications for improving hydrological modelling. J Hydrol 377:80–91

Hachani A, Ouessar M, Zerrim A (2017) Chapter 1 : a study of water stress on olive growing under the effect of climate change in South East of Tunisia. In: Ouessar M et al. (eds) Water and Land Security in Drylands. https://doi.org/10.1007/978-3-319-54021-4_1

Haj-Amor Z, Acharjee TD, Dhaouadi L, Bouri S (2020) Impacts of climate change on irrigation water requirement of date palms under future salinity trend in coastal aquifer of Tunisian oasis. Agric Water Manag. https://doi.org/10.1016/j.agwat.2019.105843

Hakala K, Addor N, Seibert J (2018) Hydrological modeling to evaluate climate model simulations and their bias correction. J Hydrometeor. https://doi.org/10.1175/JHM-D-17-0189.1,inpress.10.1002/joc.1779

Hakala K, Addor N, Teutschbein C, Vis M, Dakhlaoui H, Seibert J (2019) Hydrological climate change impact modeling. In: Maurice P (ed) Encyclopedia of water: science technology and society. https://doi.org/10.1002/9781119300762.wsts0062

Hamdi N, Touihri M, Charfi F (2012) Diagnostic Ecologique du parc National Ichkeul (Tunisie) après la construction des barrages; Cas des oiseaux d'eaux. Rev. Écol. (Terre Vie), vol 67

Hawkins E, Sutton R (2009) The potential to narrow uncertainty in regional climate predictions. Bull Am Meteorol Soc 90(8):1095–1107

IPCC—Intergovernmental panel on climate change (2014). Climate Change 2014: Synthesis Report. Contribution of Working Groups I, II and III to the Fifth Assessment Report of the Intergovernmental Panel on Climate Change [Core Writing Team, Pachauri RK, Meyer LA (eds)]. IPCC, Geneva, Switzerland, pp 151

Jakeman AI, Littlewood IG, Withehead PG (1990) Computation of the instantaneous unit hydrograph and identifiable component flows with application to two small upland catchments. J Hydrol 117:275–300

Jones RN (2000) Managing uncertainty in climate change projections–issues for impact assessment. Climatic Change 45(3–4):403–419

Mansour M, Hachicha M (2014) Chapter 21—the vulnerability of tunisian agriculture to climate change. In: Ahmad Pz, Saiema Rasool S (eds) Emerging technologies and management of crop stress tolerance, Academic Press, pp 485–500. https://doi.org/10.1016/B978-0-12-800875-1.00021-1

Milano M, Ruelland D, Fernandez S, Dezetter A, Fabre J, Servat E, Fritsch J-M, Ardoin-Bardin S, Thivet G (2013) Current state of Mediterranean water resources and future trends under climatic and anthropogenic changes. Hydrol Sci J 58:498–518

Mougou R, Mansour M, Iglesias A, Zitouni-Chebbi R, Battaglini A (2011) Climate change and agricultural vulnerability: a case study of rain-fed wheat in Kairouan, Central Tunisia. Reg Environ Change 11:137–142. https://doi.org/10.1007/s10113-010-0179-4

Moussa S, Sellami H, Mlayh A (2018) Climate change impact projections at the catchment scale in Tunisia using the multi-model ensemble mean approach. Arab J Geosci 11:181. https://doi.org/10.1007/s12517-018-3532-4

Nash JE, Sutcliffe JV (1970) River flow forecasting through conceptual models—part I: a discussion of principles. J Hydrol 10:282–290

Nasr Z, Almohammed H, Gafrej Lahache R, Maag C, King L (2008) Drought modelling under climate change in Tunisia during the 2020 and 2050 periods. In: López-Francos A (ed) Drought management: scientific and technological innovations. Zaragoza : CIHEAM, pp 365–37 0 (Options Méditerranéennes : Série A. Sémin aires Méditerranéens; n . 80)

Oudin L, Hervieu F, Michel C, Perrin C, Andréassian V, Anctil F, Loumagne C (2005) Which potential evapotranspiration input for a lumped rainfall-runoff model? part 2: towards a simple and efficient potential evapotranspiration model for rainfall-runoff modelling. J Hydrol 303:290–306

Perrin C, Michel C, Andréassian V (2003) Improvement of a parsimonious model for streamflow simulation. J Hydrol 279:275–289

Qin Y et al. (2019) Flexibility and intensity of global water use. Nature Sustain 2(6):515–523, 8. https://doi.org/10.1038/s41893-019-0294-2

RSIS–Ramsar Sites Information Service (2020) Sebkhet Soliman. Available at: https://rsis.ramsar.org/ris/1713?language=en [consulted in 29 November 2020]

Ruelland D, Dezetter A, Hublart P (2014) Sensitivity analysis of hydrological modelling to climate forcing in a semi-arid mountainous catchment. Hydrol Changing World: Environ Human Dimensions IAHS Publ 363:145–150

Secadenord (2020) La Société d'Exploitation du Canal et Adductions des Eaux du Nord. Available at: https://www.secadenord.com.tn/english/home.html [consulted in 03 December 2020]

Sellami H, Benabdallah S, La Jeunesse I, Vanclooster M (2015) Quantifying hydrological responses of small Mediterranean catchments under climate change projections. Sci Total Environ 543:924–936

Sellami H, Benabdallah S, La Jeunesse I, Vanclooster M (2016) Climate models and hydrological parameter uncertainties in climate change impacts on monthly runoff and daily flow duration curve of a Mediterranean catchment. Hydrol Sci J 1415–1429. https://doi.org/10.1080/02626667.2015.1040801

Shahrban M, Walker JP, Wang QJ, Robertson DE (2018) On the importance of soil moisture in calibration of rainfall–runoff models: two case studies. Hydrol Sci J 63(9):1292–1312. https://doi.org/10.1080/02626667.2018.1487560626667.2015.1100302

Teutschbein C, Seibert J (2012) Bias correction of regional climate model simulations for hydrological climate-change impact studies: review and evaluation of different methods. J Hydrol 456–457:11–29. https://doi.org/10.1016/j.jhydrol.2012.05.052

Teutschbein C, Seibert J (2013) Is bias correction of regional climate model (RCM) simulations possible for non-stationary conditions? Hydrol Earth Syst Sci 17(5061–5077):2013. https://doi.org/10.5194/hess-17-5061-2013

Tramblay Y, Ruelland D, Hanich L, Dakhlaoui H (2016) Hydrological impacts of climate change in North African countries. Sub-chapter 2.3.1. The Mediterranean Region under Climate Change. A scientific Update, IRD Éditions, pp 736

Tramblay Y, Rutkowska A, Sauquet E, Sefton C, Laaha G, Osuch M, Albuquerque T, Alves MH, Banasik K, Beaufort A, Brocca L, Camici S, Csabai Z, Dakhlaoui H, DeGirolamo AM, Dörflinger G, Gallart F, Gauster T, Hanich L, Kohnová S, Mediero L, Plamen N, Parry S, Quintana-Seguí P, Tzoraki O, Datry T (2021) Trends in flow intermittence for European rivers. Hydrol Sci J. https://doi.org/10.1080/02626667.2020.1849708

Valéry A, Andréassian V, Perrin C (2010) Regionalization of precipitation and air temperature over high-altitude catchments: learning from outliers. Hydrol Sci J 55:928–940

Vetter T, Reinhardt J, Flörke M et al. (2017) Evaluation of sources of uncertainty in projected hydrological changes under climate change in 12 large-scale river basins. Climatic Change 141:419–433. https://doi.org/10.1007/s10584-016-1794-y

Vrac M (2018) Multivariate bias adjustment of high-dimensional climate simulations: the rank resampling for distributions and dependences (R2D2) bias correction. Hydrol Earth Syst Sci 22(3175–3196):2018. https://doi.org/10.5194/hess-22-3175-2018

World Meteorological Organization (2011) Guide to climatological practices (WMO-No. 100), Geneva

Zhang H, Huang GH, Wang D, Zhang X (2011) Multi-period calibration of a semi-distributed hydrological model based on hydroclimatic clustering. Adv Water Res 34:1292–1303

Addressing Climate Change in the Mediterranean Basin: Some Thoughts for the Future

Walter Leal Filho and Evangelos Manolas

Abstract This short paper provides some perspectives on climate change adaptation in the Mediterranean basin, paving the way for further initiatives.

1 Introduction

The Mediterranean Sea is in between three continents: Europe, Asia and Africa. In the West connects with the Atlantic Ocean via the Strait of Gibraltar, in the east with the Red Sea, via the Suez Canal. The size of the Mediterranean Sea is 2.966.000 Km^2. Its maximum length, from Gibraltar to the coastline of Syria is 3.860 km, while its maximum width is 1.800 km. Throughout history, a great civilization flourished in the Mediterranean Basin. The cultural richness of the Mediterranean basin is exceptional.

There are important differences among Mediterranean countries. Northern Mediterranean counties are much more developed than southern Mediterranean and Middle East countries. These differences are made worse by rapid population growth in the north Africa and Middle East, which has increased from 105 million in 1960 to 444 million in 2017.

W. Leal Filho (✉)
International Climate Change Information and Research Programme, Research and Transfer Centre "Sustainable Development and Climate Change Management", Hamburg University of Applied Sciences, Hamburg, Germany
e-mail: walter.leal2@haw-hamburg.de

Department of Natural Sciences, Manchester Metropolitan University, Chester Street, Manchester M1 5GD, UK

E. Manolas
Department of Forestry and Management of the Environment and Natural Resources, School of Agricultural and Forestry Sciences, Democritus University of Thrace, 193 Pantazidou Street, 68200 Orestiada, Greece
e-mail: emanolas@fmenr.duth.gr

W. Leal Filho and E. Manolas (eds.), *Climate Change in the Mediterranean and Middle Eastern Region*, Climate Change Management,
https://doi.org/10.1007/978-3-030-78566-6_25

In this complex situation, several new challenges from climate change arise, including warming, droughts, extreme weather events, sea-level rise and ocean acidification. These challenges are related with other environmental issues like pollution and urban growth (Risks Associated to Climate and Environmental Changes in the Mediterranean Region 2019).

2 Climate Change and the Mediterranean Basin

The Mediterranean Region is warming 20% faster than the global average (To Vima team 2019). Although global temperature has increased by 1.1 °C, in the Mediterranean it has increased by 1.4 °C. Scientists predict increasing risks in the future. In the next decades we expect increase of heat waves, increase of desertification, reduction of rainfall, significant increase of the intensity and the frequency of extreme weather events, while about 100 million citizens will be affected by the availability of water resources (To Vima team 2019).

The Mediterranean has been impacted by climate change in many different ways. More specifically, the Mediterranean basin has been described as a hotspot for climate change and current predictions indicate an increased warming and drying of the area (Fraga et al. 2021).

Many scientists believe that the increase of temperature in the Mediterranean will always be higher than the average temperature of the planet. Water will become precious in the decades ahead. As a result of climate change big parts of Southern Europe and Northern Africa will face serious water shortages in the future.

Many countries may not know the threats they will face. Although some developed countries in the area take serious action, other countries do much less. Countries in the South and Eastern Mediterranean are less equipped in order to sufficiently address the threats of climate change (Kyriatsoulis 2019).

Mediterranean countries account for around 6% of the world greenhouse emissions. However, the Mediterranean basin offers significant possibilities for the production of energy from renewable sources (UNEP n.d.).

Woetzel et al. (2020) provide a useful analysis of climate change with regard to heat, drought, water stress, wildfires, death rate and health issues. It is worth summarizing their effort: past European heat waves have resulted in deaths. In the EU in total, up to 35,000 deaths were attributed to the heat wave of August 2003 (Bhattacharya 2003). Without appropriate mitigation measures, EU researchers predict that heat related mortality could increase twofold by 2080 (Ciscar et al. 2018). Other items of relevance are as follows.

Heat. With warm, dry summers and mild winters, the Mediterranean climate has long favored tourism and the cultivation of valuable crops, in particular wine grapes. However, rising temperatures may disrupt established industries and norms. The temperature in the Mediterranean basin has climbed 1.4°, and the rise is predicted to continue to exceed global rates (Cramer et al. 2018). The number of days with a maximum temperature above 37° will affect the entire Mediterranean region, in particular northern Africa, the Middle East, southern Spain, and Turkey (Mediterranean-1) (EURO-CORDEX 2014).

Drought. In Italy, Portugal, Spain, and parts of Greece and Turkey, rainfall from April to September is predicted to decrease by as much as 10% by 2030 and 20% by 2050. (EURO CORDEX 2014). By 2050, drought conditions could be prevalent for at least six months a year in these countries (Mediterranean-2) (Palmer Drought Severity Index). These changing temperature and rainfall patterns will change the Mediterranean climate. Scientists have projected Madrid's climate in 2050 will be more like Marrakech's in Morocco today, while in France Marseille's climate will resemble that of Algier in Algeria (Bastin et al. 2019).

Water stress. In the Mediterranean many basins could see a decline of approximately 10% in water supplies by 2030 and of up to 25% by 2050 (Mediterranean-3). Water stress already seriously impacts most countries in the Mediterranean. The countries most affected are Morocco and Libya (Woetzel et al. 2020).

Due to climate conditions, an increased demand for water resources will increase by 4 to 18% until the end of the century, while the increase of the population may lead to escalation of these numbers. To such escalation may also contribute factors such as tourist development, new industries and extension of urban zones (Cramer et al. 2018).

In the decades ahead, more than 250 million people will be deprived of the necessary amounts of water they need. In addition, 10 of the 20 cities in the world which are expected to suffer most are located in the Mediterranean (UNEP n.d.)

Wildfires. Temperature and drought conditions in Southern Europe are factors which favor severe wildfires such as the ones that killed 99 people in July 2018 in the popular holiday resort of Mati, in Greece (The Conversation 2018). In 2017, fire destroyed more than 500,000 hectares of forest in Portugal, over 5% of the country's area. The largest fire, in and around the town of Pedrógão Grande, killed 66 people (Turco et al. 2019).

Now, new research in Nature Communications suggests that the summer fire season in Mediterranean Europe is going to get worse. Under the warmest climatic predictions, the area that is currently burned annually would double. One might go as far as to say that 40% more area would be burnt even if the Paris Climate Agreement is respected and warming stays below 1.5 °C. (The Conversation 2018).

Death rate and health issues. The annual death rate which is due to the increase of the temperature in the European side of the Mediterranean will increase by 1.8 and 2.6% by mid-twenty-first century while at the end of the century its increase will fluctuate between 3 and 7%. There is increased danger that transmissive diseases will increase by 2050. For example, by 2050, high danger areas with regard to the western Nile virus will extend further and so will the seasons the virus will be transmitted (MedECC 2020a).

Flooding. In terms of excessive flooding, future adaptation trends include prioritizing low-risk action to mitigate the effects of flooding (Benito et al. 2020). This includes more nature-based solutions in urban areas to prevent flood effects. These solutions have been used to increased water filtration, surface water retention and evaporation which decreases flow connectivity over the area and reduces the flooding hazards caused by climate change (Ferreira et al. 2020).

With regard to flood fatalities between 1980 and 2018 the MEditerranean Flood Fatality Database (MEFF DB) reports 1809 fatalities (Vinet et al. 2019).

The majority of fatalities (1242, i.e., 69%) occurred in Turkey (Biricik 1997). After Turkey, Southern France is the second region contributing to the human toll, with 275 fatalities, followed by Greece and Catalonia (Vinet et al. 2019).

Agriculture, pre-dominantly olive farming, is adversely affected due to a mixture of over-cultivation and climate effects. Under future climatic conditions agriculture and olive farming are expected to suffer severe adverse effects in many aspects from water relations to quality attributes (Fraga et al. 2021).

The wine industry is also under pressure. The wine industry is important to many economies, and countries such as France, Italy, and Spain produce about half of the world's wine and 74.1% of the wine made in the European Union (European Commission 2017).

The production from traditional winemaking areas could suffer as the Mediterranean climate changes, since grapevines are very sensitive to fluctuations in temperature and rainfall. Also, the quality of both grapes and wine is dependent on the weather (Jones et al. 2005). Higher temperatures most often reduce grape quality by increasing sugar levels and decreasing acidity. Researchers have also predicted a wide range of possible impacts of climate change on grape yields. Some scientists think that the Mediterranean area suitable for viticulture could fall by up to 70% but others do not see negative impacts at all (Hannah et al. 2013).

On the other hand, as the temperature in the region increases, it is also possible that specific grape varieties will no longer grow where they do now, while, at the same time, there will be possibilities to plant new varieties. (Wolkovich et al. 2018). Areas in Italy, Portugal, and Spain which grow grapes could experience great reduction in production or even collapse (Fraga et al. 2016).

Population pressure in coastal areas continues to increase. In the last 50 years (1965–2015) built areas in a distance of 1 km from the sea doubled. As a result, biodiversity, coastal ecosystems and, in particular, the protection services provided by nature against floods have been downgraded. According to the Barcelona Protocol on the Management of Coastal Areas countries need to completely prohibit building activities for a distance of at least 100 m from the sea. In Greece, 7–9% of the coastline has been built, in Turkey 10–12%, in Italy and Spain 13–17%, in Israel 18–26% while the situation is worse in Lebanon where the percentage is 27–38%.

3 Conclusions and Recommendations

In Mediterranean riparian areas, new adaptation methods are being employed. In order to reduce the effects of climate change, riparian areas are being restored using nature-based solutions. Biobased solutions include soil and water engineering. Furthermore, sustainable management plans are being incorporated into restoration and maintenance. This includes the development of new databases to store information about hydrology and ecology in the area. This may aid in development of novel adaptation methods specific to the area (Zaimes 2020).

As here outlined, the effects of climate change are seen directly through heat, drought, storms and floods while indirect effects are seen through changes in food and water availability, air pollution and other climatic stressors. Such phenomena impact the health of people in the region with vulnerable population experiencing more severe effects. Future adaptation plans include more control and surveillance plans. Furthermore, more Health Action Plans need to be implemented with more cross-border collaboration (Linares et al. 2020).

With regard to olive farming, sustainable crop management was applied which resulted in increased crop production and further decreased the carbon-environmental footprint of production. Such methods have been described as essential for climate change adaptation and can be integrated into the national agricultural and environmental policy. Furthermore, a circular sustainable economy can be established (Michalopoulos et al. 2020).

Short term adaptation plans included managing irrigation and protection against extreme weather. It has been suggested that the uses of kaolin clay may aid in reducing stress against extreme climate events. Long term plans include relocation of crops to more suitable areas. Such methods can be translated to other agricultural sectors (Fraga et al. 2021).

In more recent instances innovative adaptation methods to agricultural problems are being developed. This includes inducing epigenetic changes in the plants that targets specific traits in the plants. This helps to build resilience to extreme weather events. This has been carried out to produce drought resistant crops however, more genes are being assessed for future adaptation. Such methods are more advanced than conventional adaptation methods (Martinelli et al. 2020). Finally, we need climate action which will make the region more resilient to challenges. We need actions which will include renewable sources of energy, integrated management of water resources, sustainable agriculture, fishing, alternative tourism (MedEEC 2020b).

References

Bastin JF, Clark E, Elliott T, Hart S, van den Hoogen J et al (2019) Understanding climate change from a global analysis of city analogues. PLOS ONE 14(10):e0224120. https://doi.org/10.1371/journal.pone.0224120

Benito G, Sanchez-Moya Y, Medialdea A, Barriendos M, Calle M, Rico M, Sopeña A, Machado MJ (2020) Extreme floods in small mediterranean catchments: Long-term response to climate variability and change. Water 12(4):1008. https://doi.org/10.3390/w12041008

Bhattacharya S (2003) The 2003 Europe heat wave caused 35,000 deaths, New Scientist, October 10

Biricik AS (1997) Flood disasters in Senirkent (13 July 1995–18,19 July 1996). Marmara Geogr J 1:9–30

Ciscar JC, Feyen L, Ibarreta D, Soria A (coordination) (2018) Climate impacts in Europe. Final report of the JRC PESETA III project. European Commission. file:///C:/Users/Owner/AppData/Local/Temp/kjna29427enn_1.pdf

Cramer W, Guiot J, Fader M et al (2018) Climate change and interconnected risks to sustainable development in the Mediterranean. Nature Clim Change 8:972–980. https://doi.org/10.1038/s41558-018-0299-2

EURO-CORDEX (2014) Regional climate model (RCM) ensemble. Available at: https://euro-cordex.net/060378/index.php.en

European Commission (2017) Vineyards in the EU—Statistics. Available at: https://ec.europa.eu/eurostat/statistics-explained/pdfscache/58459.pdf

Ferreira C, Mourato S, Kašanin-Grubin M, Ferreira AJD, Destouni G, Kalantari Z (2020) Effectiveness of nature-based solutions in mitigating flood hazard in a Mediterranean Peri-Urban Catchment. Water 12(10):2893. https://doi.org/10.3390/w12102893

Fraga H, Cortazar-Atauri IG, Malheiro AC, Santos JA (2016) Modelling climate change impacts on viticultural yield, phenology and stress conditions in Europe. Glob Change Biol 22(11). https://doi.org/10.1111/gcb.13382

Fraga H, Moriondo M, Leolini L, Santos JA (2021) Mediterranean olive orchards under climate change: a review of future impacts and adaptation strategies. Agronomy 11(1):56. https://doi.org/10.3390/agronomy11010056

Hannah L, Roehrdanz PR, Ikegami M, Shepard AV et al (2013) Climate change, wine, and conservation. Proc Nat Acad Sci 110(17)

Jones GV, White MA, Cooper OR, Storchmann K (2005) Climate change and global wine quality. Clim Change 73:319–343. Available at: https://link.springer.com/content/pdf/. https://doi.org/10.1007/s10584-005-4704-2.pdf

Kyriatsoulis A (2019) Climate change and its effects on the Mediterranean Basin and Greece. Proto Thema. 13 November. Available at: https://www.protothema.gr/world/afieromata/to-idaniko-heimoniatiko-spiti/article/923217/i-klimatiki-allagi-kai-oi-sunepeies-tis-pagosmios-kai-eidikotera-sti-mesogeio-tin-ellada-kai-tin-rodo/

Linares C, Díaz J, Negev M, Martínez GS, Debono, R, Paz S (2020) Impacts of climate change on the public health of the Mediterranean Basin population-current situation, projections, preparedness and adaptation. Environ Res 182:109107. https://doi.org/10.1016/j.envres.2019.109107

Martinelli F, Ollero FJ, Giovino A, Perrone A, Bekki A, Sikora S, El Nabbout R, Bouhadida M, Yucel D, Bazzicalupo M, Mengoni A (2020) Proposed research for innovative solutions for chickpeas and beans in a climate change scenario: the Mediterranean Basin. Sustainability 12(4):1315. https://doi.org/10.3390/su12041315

MedECC (2020a) Climate and environmental change in the mediterranean basin—current situation and risks for the future. In: Cramer W, Guiot J, Marini K (eds) First mediterranean assessment report. Union for the Mediterranean, Plan Bleu, UNEP/MAP, Marseille, France. Available at: https://www.medecc.org/first-mediterranean-assessment-report-mar1/

MedEEC (2020b) First mediterranean assessment report, summary for policy makers 2020. Available at: https://www.medecc.org/wp-content/uploads/2020/11/MedECC_MAR1_SPM_ENG.pdf-

Michalopoulos G, Kasapi K, Koubouris G, Psarras G, Arampatzis G, Hatzigiannakis E, Malliaraki S (2020) Adaptation of Mediterranean olive groves to climate change through sustainable cultivation practices. Climate 8(4):54. https://doi.org/10.3390/cli8040054

Risks Associated to Climate and Environmental Changes in the Mediterranean Region (2019) A preliminary assessment by the MedECC Network Science-policy interface. Available at: https://ufmsecretariat.org/wp-content/uploads/2019/10/MedECC-Booklet_EN_WEB.pdf

The Conversation (2018) Wildfires in Mediterranean Europe will increase by 40% at 1.5°C warming, say scientists. 2 October. Available at: https://theconversation.com/wildfires-in-mediterranean-europe-will-increase-by-40-at-1-5-c-warming-say-scientists-104270

To Vima team (2019) SOS for climate change: sea-level rise in the mediterranean might reach two meters. To Vima. 4 November. Available at: https://www.tovima.gr/2019/11/04/science/sos-gia-tin-klimatiki-allagi-anodos-tis-stathmis-tis-mesogeiou-eos-2-metra/
Turco M, Jerez S, Augusto S (2019) Climate drivers of the 2017 devastating fires in Portugal. Sci Rep 9. (Article number: 13886). Available at: https://www.nature.com/articles/s41598-019-50281-2.pdf
United Nations Environment Programme (n.d). Sea-level rise in the Mediterranean Region. Available at: https://www.un.org/depts/los//consultative_process/contribution21/22UNEPMAP.pdf
van Leeuwen C et al (2013) Why climate change will not dramatically decrease viticultural suitability in main wine-producing areas by 2050. Proc Nat Acad Sci 110(33). PNAS April 23. 110(17):6907–6912. https://doi.org/10.1073/pnas.1210127110
Vinet F, Bigot V, Petrucci O, Papagiannaki K, Llasat MC et al (2019) Mapping flood-related mortality in the Mediterranean basin: results from the MEFF v2.0 DB. Water MDPI 11(10). (Art. no 2196). Available at: https://www.mdpi.com/2073-4441/11/10/2196
Woetzel J, Pinner D, Samandari H et al (2020) A Mediterranean basin without a Mediterranean Climate? McKinsey Global Institute. Available at: https://www.mckinsey.com/~/media/mckinsey/business%20functions/sustainability/our%20insights/a%20mediterranean%20basin%20without%20a%20mediterranean%20climate/a-mediterranean-basin-without-a-mediterranean-climate.pdf
Wolkovich EM, Cortazar-Atawri IG, Morales CI (2018) From Pinot to Xinomavro in the world's future wine-growing regions. Nat Clim Change 8(1). https://doi.org/10.1038/s41558-017-0016-6
Zaimes G (2020) Mediterranean riparian areas–climate change implications and recommendations. J Environ Biol 41(5):957–965. https://doi.org/10.22438/jeb//41/5/MRN-1454

The manufacturer's authorised representative in the EU is Springer Nature Customer Service Centre GmbH, Europaplatz 3, 69115 Heidelberg, Germany. If you have any concerns regarding our products, please contact ProductSafety@springernature.com

Printed and bound by CPI Group (UK) Ltd, Croydon, CR0 4YY
30/09/2026
02225290-0018